Carbon Nanomaterials in Clean Energy Hydrogen Systems - II

NATO Science for Peace and Security Series

This Series presents the results of scientific meetings supported under the NATO Programme: Science for Peace and Security (SPS).

The NATO SPS Programme supports meetings in the following Key Priority areas: (1) Defence Against Terrorism; (2) Countering other Threats to Security and (3) NATO, Partner and Mediterranean Dialogue Country Priorities. The types of meeting supported are generally "Advanced Study Institutes" and "Advanced Research Workshops". The NATO SPS Series collects together the results of these meetings. The meetings are co-organized by scientists from NATO countries and scientists from NATO's "Partner" or "Mediterranean Dialogue" countries. The observations and recommendations made at the meetings, as well as the contents of the volumes in the Series, reflect those of participants and contributors only; they should not necessarily be regarded as reflecting NATO views or policy.

Advanced Study Institutes (ASI) are high-level tutorial courses intended to convey the latest developments in a subject to an advanced-level audience

Advanced Research Workshops (ARW) are expert meetings where an intense but informal exchange of views at the frontiers of a subject aims at identifying directions for future action

Following a transformation of the programme in 2006 the Series has been re-named and re-organised. Recent volumes on topics not related to security, which result from meetings supported under the programme earlier, may be found in the NATO Science Series.

The Series is published by IOS Press, Amsterdam, and Springer, Dordrecht, in conjunction with the NATO Emerging Security Challenges Division.

Sub-Series

A.	Chemistry and Biology	Springer
B.	Physics and Biophysics	Springer
C.	Environmental Security	Springer
D.	Information and Communication Security	IOS Press
E.	Human and Societal Dynamics	IOS Press

http://www.nato.int/science
http://www.springer.com
http://www.iospress.nl

Series C: Environmental Security

Carbon Nanomaterials in Clean Energy Hydrogen Systems - II

edited by

Svetlana Yu. Zaginaichenko
Institute of Hydrogen and Solar Energy,
Kiev, Ukraine

Dmitry V. Schur
Institute for Problems of Materials Science of NAS,
Kiev, Ukraine

Valeriy V. Skorokhod
Institute for Problems of Materials Science of NAS,
Kiev, Ukraine

Ayfer Veziroglu
International Association for Hydrogen Energy,
University of Miami, Miami, USA

and

Beycan İbrahimoğlu
Gazi University, Department of Mechanical Engineering,
Ankara, Turkey

Springer

Published in Cooperation with NATO Emerging Security Challenges Division

Proceedings of the NATO Advanced Research Workshop on
Carbon Nanomaterials in Clean-Energy Hydrogen Systems
Yalta, Crimea, Ukraine
June 24–30, 2010

ISBN 978-94-007-0901-0 (PB)
ISBN 978-94-007-0898-3 (HB)
ISBN 978-94-007-0899-0 (e-book)
DOI 10.1007/978-94-007-0899-0

Published by Springer,
P.O. Box 17, 3300 AA Dordrecht, The Netherlands.

www.springer.com

Printed on acid-free paper

Preface

The 2010 ARW "Carbon Nanomaterials in Clean-Energy Hydrogen Systems" (CNCEHS'2010) was held in June 24–30, 2010 in the remarkable city of Yalta (Crimea, Ukraine) known for its heroic and unusual history. In the tradition of the earlier conferences, CNCEHS'2010 meeting served as a multidisciplinary conference for the presentation and discussion of the most recent research on transition to hydrogen-based energy systems, technologies for hydrogen production, storage, utilization, carbon nanomaterials processing and chemical behavior, energy and environmental problems. The aim of CNCEHS'2010 was to provide the wide overview of the latest scientific results on basic research and technological applications of hydrogen interactions with carbon materials. As well-known, energy shortage and environmental pollution is the major problem in the current word, energy-saving characteristics of hydrogen occupies an important position in the development of new energy, because the energy is a basic and necessary component in the development of the communities. The active delegates from different universities, research/academic organizations and governmental agencies could meet, discuss and present the most recent advances in hydrogen concepts, processes and systems, to evaluate current progress and to exchange academic information, to identify research needs and future developments in this important area. This ARW should help further the progress of hydrogen-based sciences and promote the use of hydrogen and carbon nanomaterials as essential to make the Hydrogen Economy a reality.

According to the resolution of the International Advisory and Organizing Committee, it has been established the awards "Oscars of Hydrogen Energy" for scientists who have made major contributions to the promotion of hydrogen energy worldwide. The world-famous scientist Prof. T. Nejat Veziroglu, who is the author of a great number of books and papers, the President of the International Association for Hydrogen Energy (IAHE), the Founding Director of Clean Energy Research Institute of the University of Miami, USA, and the Founding Editor-in-Chief of the International Journal of Hydrogen Energy was the first recipient of the Award "Oscar of Hydrogen Energy". Beycan Ibrahimoglu, the Director of

Dr. Dmitry V. Schur, chair of Organizing Committee of CNCENS'2010 and Prof. Beycan İbrahimoğlu, Director of Plasmochemical Center in Ankara, Turkey, presented the first award "Oscar of Hydrogen Energy" to Prof. T. Nejat Veziroglu, the President of the International Association for Hydrogen Energy

Plasmochemical Center in Ankara, Turkey, was a producer of the "Oscar of Hydrogen Energy" and will be the provider of such awards in the future.

Moreover according to the resolution of the International Advisory and Organizing Committee, a number of outstanding scientists who have considerable achievements in the field of carbon nanostructures and hydrogen energy have been rewarded with diplomas of NATO ARW "Carbon Nanomaterials in Clean-Energy Hydrogen Systems". Among them were such world-wide well-known scientists as Prof. Nikolay Kartel (Institute of Surface Chemistry of National Academy of Sciences of Ukraine, Kiev, Ukraine), Prof. Nikola Koprinarov (Central Laboratory for Solar Energy and New Energy of Bulgarian Academy of Sciences, Sofia, Bulgaria), Prof. Marcin Leonowicz (Warsaw University of Technology, Poland), Prof. Vladimir I. Shapovalov and Prof. Raouf O. Loutfy (Materials & Electrochemical Research Corporation, Tucson, USA), Prof. Sergey S. Ivanchev (St.-Petersburg branch of Institute of Catalysis named after G.K. Boreskov of the Siberian Branch of the Russian Academy of Science, St.-Petersburg, Russia), Dr. Yuriy M. Shulga (Institute of Problems of Chemical Physics of Russian Academy of Sciences, Chernogolovka, Russia), Zhanna Mileeva (The University of Salford, United Kingdom), Nariman F. Javadov (Experimental-Industrial Plant of Institute for Petroleum Chemical Processes (IPCP) of the Azerbaijan National Academy of Sciences (NAS), Baku, Azerbaijan) and others.

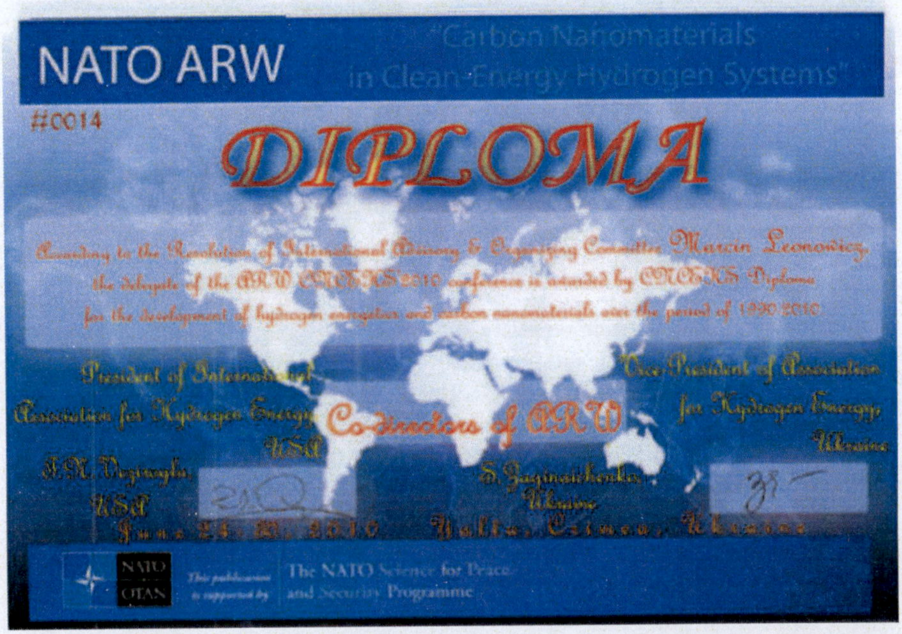

Members of International Advisory & Organizing Committee Dr. Yuriy M. Shulga, Prof. Svetlana Yu. Zaginaichenko, Prof. T. Nejat Veziroglu and Dr. Dmitry V. Schur, presented the NATO ARW diploma to Prof. Sergey S. Ivanchev, director of St.-Petersburg branch of Institute of Catalysis named after G.K. Boreskov of the Siberian Branch of the Russian Academy of Sciences, St.-Petersburg, Russia.

Thus, this event has attracted international researchers as well as younger scientists to strengthen connections among scientists in the area of hydrogenated carbon nanomaterials research. The CNCEHS'2010 is a solid example of how NATO is promoting cooperation among scientists from developed countries of Europe, America and NATO Partner countries. Since fossil fuel resources are limited and harmful to the environment, more attentions have been paid on renewable energy recently. Therefore, conversion and utilization of renewable energy, such as hydrogen energy, became one of the most important topics of this conference since 1993 (1993, 1995, 1997, 1999, 2001, 2003, 2005, 2007). One goal in organizing this conference was to prepare a discussion ambient for the related people and scientists working on renewable energy especially on hydrogen energy. We are planning to repeat the conference after 2 years. We hope these ARW and the one that are being planned for future will help the development of the technology on renewable and hydrogen energy.

The Proceedings of CNCEHS'2010 include some of the selected works presenting in the course of oral presentations, at the poster sessions, and during the round-table discussions at the conference in Yalta and representing the wide spectra of hydrogen energy and carbon nanomaterials related themes, but also latest progress in many key areas.

Finally, the organization of this conference was successful thanks to the support by the Scientific and Environmental Affairs Division of NATO as the NATO Science for Peace and Security Programme of the NATO Science Programme. Their contribution is gratefully acknowledged and the Organizing Committee and all ARW participants want to overflow with effusive thanks to NATO for the financial support of the CNCEHS'2010 ARW, and to **Mr. Jean Fournet**, Assistant Secretary General, Chairman of NATO Science Committee, and **Dr. Deniz Beten**, Section Head NATO Science for Peace Security, for the displayed mutual understanding and the comprehension of significance of problems under discussions at the CNCEHS'2010 ARW.

Svetlana Yu. Zaginaichenko
Dmitry V. Schur
Beycan İbrahimoğlu
Valeriy V. Skorokhod
Ayfer Veziroglu

Contents

Memorable Dates

85th Anniversary of
T. Nejat Veziroglu
Honorary Chairman of ICHMS
President of International Association
for Hydrogen Energy

Dr. Veziroglu, a native of Turkey, graduated from the City and Guilds College, the Imperial College of Science and Technology, University of London, with degrees in Mechanical Engineering (A.C.G.I., B.Sc.), Advanced Studies in Engineering (D.I.C.) and Heat Transfer (Ph.D.).

After serving in some Turkish government agencies as a Technical Consultant and Deputy Director of Steel Silos, and then heading a private company, in 1962 he joined the University of Miami Engineering Faculty, and served as the Director of Graduate Studies, Mechanical Engineering (initiating the first Ph.D. Program in the College of Engineering), Chairman of the Department of Mechanical Engineering, and the Associate Dean for Research. During May 2004 – May 2007, Dr. Veziroglu established the United Nation Industrial Development Organization International Centre for Hydrogen Energy Technologies – as its Founding Director – in Istanbul, Turkey. Presently, he is the Director of the Clean Energy Research Institute.

Dr. Veziroglu teaches Heat Transfer, Mass Transfer, Nuclear Engineering, Solar Energy and Hydrogen Energy. His research interests are instabilities in Boiling Water Reactors, Interstitial Heat Transfer, Renewable Energy Sources and Hydrogen Energy System. He has published some 350 scientific reports and papers, edited 200 volumes of proceedings, and is the Editor-in-Chief of the monthly scientific journals International Journal of Hydrogen Energy. He has been an invited lecturer and/or consultant on energy research and education to many countries, including Argentina, Australia, Bahrain, Brazil, Canada, China, Columbia, Egypt, England, France, Germany, India, Italy, Japan, Kuwait, Malaysia, Nepal, Pakistan, the Philippines, Russia, Saudi Arabia, Switzerland, Turkey, Ukraine and Venezuela, and to several universities and research organizations in the United States.

Dr. Veziroglu organized the first major conference on Hydrogen Energy: The Hydrogen Economy Miami Energy (THEME) Conference, organized several conferences and symposia on Alternative Energy Sources, Environment, Miami Beach, March 1974, and proposed the Hydrogen Energy System. Subsequently, he Hydrogen Energy, Heat and Mass Transfer, and Remote Sensing.

Dr. Veziroglu has membership in some twenty scientific organizations, has been elected to the Grade of Fellow in the British Institution of Mechanical Engineers, the American Society of Mechanical Engineers and the American Association for the Advancement of Science, and is the Founding President of the International Association for Hydrogen Energy.

Dr. Veziroglu has been the recipient of several international awards, including Turkish Presidential Science Award, 1975, Honorary Professorship, Xian Jiaotong University, Xian, China, 1981, I. V. Kurchatov Medal, Kurchatov Institute of Atomic Energy, Moscow, U.S.S.R, 1982, Energy for Mankind Award, 1986, Twenty-Five Years' Service Award, American Nuclear Society, 1987, Turkish Superior Service to Mankind Award, 1991, Honorary Doctorate, Anadolu University, Eskisehir, Turkey, 1998, Honorary Member, Argentinean Academy of Sciences, 2000, and Honorary Doctorate, Donetsk State Technical University, Donetsk, Ukraine, 2001. In 2000, he was nominated for the Nobel Prize in Economics for both envisioning the Hydrogen Economy, and striving towards its realization.

75th Anniversary of Academician V.V. Skorokhod

This year the 75th birthday of Valeriy Vladimirovich Skorokhod was in July 28, 2009; he is an Academician of National Academy of Sciences of Ukraine, a director of the Institute for Problems of Materials Science of NASU, co-chairman of International Organizing Committee of ICHMS'2009. Following his light hand, our conference received the name "Hydrogen Materials Science and Chemistry of Metal Hydrides".

The activity of V.V. Skorokhod is closely associated with the formation of I.N. Frantsevich Institute for Problems of Materials Science, one of leading centres of USSR and in the whole world.

Valeriy Vladimirovich Skorokhod was born on July 28, 1934 in Nikopol, an industrial town of Dnepropetrovsk region on the south of Ukraine in the family of teachers. On graduating from the Kiev Polytechnical Institute, Metallurgical department, V.V. Skorokhod got down to work in 1956 year in the IPMS NASU (at that time institute of metal ceramics and special alloys of NAS of Ukraine) in the department headed by outstanding scientist in the field of powder metallurgy academician I.M. Fedorchenko.

The fifties in Ukraine were the years of intensive development of scientific principles of powder metallurgy. V.V. Skorokhod actively joined the scientific work and investigation of important production operations of powder metallurgy. The research pursuance in the area of powder materials science permitted V.V. Skorokhod to state the main principles of control of sintering bodies structure that received the name "structural engineering". This formed the basis for development of new type of composite materials with the fine-crystal structure.

In 1985 V.V. Skorokhod was elected the Corresponding Member of the NAS of Ukraine in speciality "Materials Science" and in 1990 – the Full Member of the NAS of Ukraine in speciality "Materials Science and Powder Metallurgy".

Scientist V.V. Skorokhod is one of the gifted natures, organizers and research leaders of our time and was the founder of a scientific school famous on an international scale. He gave lectures in Kiev Polytechnical Institute and in IPMS NASU. 12 doctors and 30 candidates of sciences are among his scientific followers. He is the principal editor of the journal "Powder Metallurgy", member of editorial board of International journal "Science of Sintering", honorary member of Serbian

Academy of Sciences and Art, honorary member of Polish ceramic society, laureate of Ya. I. Frenkel International prize winner and Nikola Tesla medal, he was also decorated with the Yaroslav Mudriy prince order of IV degree.

The untriviality of his decisions, great intuition, retentive memory and surprising erudition in the most different areas of knowledges together with high human qualities win the respect of all colleagues. V.V. Skorokhod is an author of more than 500 scientific papers including 8 monographs.

At present time the elaboration of new technologies in the field of energetics is intensified in our institute in order to neutralize the acuteness of problem of energetic crisis that arises periodically and shall appear evidently later on. Many analysts are sure that the world will walk away the petro-mania and find the equivalent for it over the next decades. The alternative sources of energy extend the possibilities of production of optional energetic capacities and by doing so they reduce the dependence on the present deficiency of petroleum.

In recent 10 years under the supervision of V.V. Skorokhod the works on hydrogen materials science, hydrogen power engineering and investigation of various nanostructures are initiated and receive all kinds of support in the Institute for Problems of Materials Science. The edition of journal "Nanostructural materials science" was organized and he is the principal editor of this journal.

The Organizing Committee thanks V.V. Skorokhod officially for the warm support of these two important scientific lines, that shall form the basis of economy in the future, greets him on his remarkable anniversary and wishes him new successes, good health, the creative and physical longevity.

70th Birthday of
Professor D. K. Ross

Professor Keith Ross was born in Ballymoney, Co Antrim in Northern Ireland on 16th July 1939, just before the outbreak of the Second World War. His early interest in science was stimulated by the fact that his father, then the headmaster of the local grammar school (Dalriada), was a physical chemist who had gained his PhD at University College London, where he built an early Infra Red spectrometer and who had subsequently worked for ICI on the development of commercial sulphuric acid production prior to turning to school teaching.

Having attended Dalriada, Portadown College and Campbell College in Belfast, he gained an open exhibition to Pembroke College, Cambridge where he read Natural Sciences. On graduating, he did an MSc in Reactor Physics and Technology in Birmingham University and then went to work for the UK Atomic Energy Authority at Winfrith Heath in Dorset. Here he worked on experimental sub-critical nuclear assemblies, developing neutron chopper techniques for measuring the internal neutron spectra. After 3 years at Winfrith, he was invited back to Birmingham by Professor John Walker to apply his knowledge of neutron physics in the newly emerging field of neutron scattering. The original objective of this work was to measure the inelastic scattering cross sections of potential reactor moderators – information required to model the neutronics of power reactors. This programme involved the construction of a neutron chopper facility at the Herald Reactor at AWRE (now AWE), Aldermaston. This was 5 Mwt light water/enriched uranium reactor equipped with an H2/D2 cold source. With his early research students, he was able to build his own novel neutron scattering instruments. Starting from a beryllium filter, cold neutron chopper, time-of-flight spectrometer, they subsequently built a fast neutron chopper/beryllium detector spectrometer, a rotating crystal spectrometer and double graphite monochromator chopper spectrometer. The first experiments were on high pressure water at temperatures up to 250oC and on graphite up to 2000oC. Once the data requirements for reactor moderators had been satisfied, he was able to apply the neutron scattering technique to a variety of condensed matter problems. His early students at AWRE included Colin Carlile, subsequently Director of the Institute Laue Langevin, Grenoble and Ian Anderson, currently Director for Neutron Sciences at Oak Ridge National Laboratory in the USA who also used their experience in designing and building neutron scattering instruments on Herald in their subsequent careers. He obtained his PhD in 1975 on some of this work. In 1973, the UK joined the ILL and from then until the operation of ISIS at the Rutherford Appleton Laboratory in 1985, his research was centred on the ILL. Thereafter, he has used both these world-leading sources as appropriate for different experiments. In the area of hydrogen in metals has since this time D.K. Ross had a continuing collaboration with Prof. Rex Harris in Birmingham. He obtained a DSc in 1985 and was appointed to a Readership in Birmingham in 1989. In 1991, he moved to the University of Salford to take up the established chair in Physics. Here he was at different times the Head of the Physics Department and Director of the Institute for Materials Research and is currently the Director for the Centre in Functional Materials. He has currently authored about 180 refereed journal articles with an h-index of 27 and is currently leading four significant research projects.

Because of hydrogen's anomalously large neutron cross section, it was obvious to apply neutrons to the study of its behaviour in solids. An early interest was in the use of quasi-elastic neutron scattering to investigate the diffusion of hydrogen in palladium, analysing the observed broadening using the Chudley-Elliott model. The extension of this method to cover different situations became a theme of his career. Having demonstrated that in the α-phase of palladium, hydrogen diffuses as a lattice gas by jumping between nearest neighbour octahedral sites (Carlile and Ross 1974), he demonstrated theoretically how correlation effects influence the shape of the quasi-elastic scattering, both in the incoherent and coherent scattering cases (Ross and Wilson 1977). He also applied the technique to the case of water diffusing between alumino-silicate layers in clays, here developing theories to describe quasi-elastic scattering from atoms diffusing between fixed boundaries (Hall and Ross, 1978,1981) and these theories have since been widely used to analyse this type of experiment. Another extension of the theory dealt with coherent quasi-elastic diffusion where the diffusing atoms interact with each other, culminating in the publication of a density response function treatment with Sinha (1988). With Cook and others (1990), he used the technique of neutron spin analysis to separate the coherent and incoherent quasi-elastic scattering from Nb-D, demonstrating directly the process of "critical slowing down". One interesting aspect of an interacting lattice gas is how it behaves on cooling down. With Bond (1982), he developed the use of Monte Carlo simulations to investigate the ordered structures that appear in f.c.c. lattice gases on cooling. This approach explains the well known 50K anomaly in the Pd-H(D) system where a superlattice with the symmetry I41/amd is formed showing a superlattice reflection at (1,1/2,0) (Anderson, Ross, Carlile, 1976,1978). The ordering process is second order and occurs at a rate conveniently studied with neutrons. With McKergow and others, D.K. Ross observed the transition between short and long range order in this system and formulated the process by which the tetragonal distortion of the ordered compound generates tensile stresses that in turn limit its growth. A similar situation applies in the α-phase of a number of rare earth hydrides particularly yttrium which show complex short range ordering (with McKergow, Anderson and others 1987). The Monte Carlo approach was also applied to Fick's Law Diffusion and coherent quasi-elastic scattering (with Faux, 1987 and Bull, 2001) and to the similar problem in NMR (with Faux and Scholl, 1986).

Inelastic neutron scattering is also well suited to the study of hydrogen in metals because in many systems, the hydrogen vibrates as in a simple harmonic oscillator. The resulting peak in the inelastic scattering was measured in many systems. With Oates and others (1979), he showed that the peak energy for tetrahedral site occupation (fluorite structure) varies as R-3/2 where R is the hydrogen-metal distance and, with Fernandez and others (1999), he showed how the vibration frequencies observed in Laves Phase hydrides could be related to the metallic constituents. More recently the techniques of density functional theory have advanced to the point where rather precise predictions of the proton quantum states, and hence of the inelastic scattering are possible from first principles. With Totolici, Kemali and Morrison (2000), he demonstrated that the measured scattering from a single crystal of PdH very closely matched the calculations up to the third excited state with no arbitrary parameters. With Li, inelastic neutron scattering has also been applied to the study of the vibrations of various ice systems, where the complication exists due to the large number of different phases and the fact that in the proton disordered phases, the proton arrangement is described by the Ice Rules. A simple rationalisation of this complexity in terms of weak and strong force constants (Nature 1993) continues to be a source of controversy. With Li and Benham (1989, 1994) he also adapted the Small Angle Neutron Scattering/contrast matching technique to study how the pores in Vycor lose a liquid phase as the external partial pressure is reduced, a method that has proved to be a powerful way of understanding the interconnection of pores in solids.

In recent years his main interest has been in the use of neutron scattering to study the behaviour of hydrogen in potential hydrogen storage systems. In a series of papers with Georgiev and others, he observed the scattering from para-hydrogen adsorbed on surfaces. Here the scattering is dominated by the transition between the molecular rotational states $l = 0$ (para) to $l = 1$ (ortho) at 14.7 meV. When the molecules are trapped in a potential well, the substates of the $l = 1$ become split and the nature of the splitting provides direct information on the trapping potential. The technique has been used on a number of potential molecular hydrogen stores (nanotubes, activated carbon and zeolites) and has indicated the possible ways of increasing the interaction potential, crucial to producing a viable hydrogen store.

Professor Keith Ross has played a significant role in the development of the subjects he is interested in, giving a considerable number of invited lectures. He is currently Secretary of the International Steering Committee of the International Symposia on Hydrogen-Metal systems.

Zh.A. Mileeva

Svante Arrhenius, Chemist and Nobel Laureate – 150th Anniversary

The mankind had celebrated 150-th anniversary of the famous Swedish chemist Svante Arrhenius on February, 19 this year.

Svante Arrhenius received the 1903 Nobel Prize in Chemistry for his discovery on how chemical compounds can carry electric current (the theory of electrolytical dissociation).

Svante Arrhenius was one of the most important scientists of his time. His electrolytical dissociation theory came to completely change chemists' conception of acids, bases and salts. Electrolytical dissociation means that compounds fall apart into electrically charged ions, for example, when ordinary rock salt is dissolved in water into natrium and chloral ions. Thus, solutions can work as electrolytes and be carriers of electric current.

Thanks to Svante Arrhenius' theory, a number of mysterious chemical and physical phenomena could be explained and be described in a simpler and more homogeneous way than previously. Even though his theory has been modified in the twentieth century, it still remains a major discovery within chemistry. Yet, Svante Arrhenius' most important contribution might be the so-called Arrhenius equation, which formulates the connection between how quickly there is a reaction and the energy that must be supplied for it to occur. This connection is of fundamental importance for the understanding of how chemical reactions really occur. Svante Arrhenius was also one of the very first to make the connection between the amount of carbon dioxide in the atmosphere and global temperature – what we today call the greenhouse effect.

Svante Arrhenius began his studies in Uppsala. The theories he developed in his doctoral dissertation were first treated with such scepticism that he passed with the lowest possible grade. After some time, his theories were revaluated, however, and he was employed as a lecturer at Stockholms University College in 1891. He soon became Professor of Physics here and was also Vice-Chancellor of Stockholm University College for 7 years. Today, his name lives on in the Arrhenius laboratory, which houses many of the scientific departments at Stockholm University.

A. Pomytkin, Ph.D., assoc.prof.

Who Was Pierre Auger?

The Auger Observatory experiment was named after Pierre Victor Auger (1899 - 1993), who can be considered the discoverer of giant airshowers generated by the interaction of very high-energy cosmic rays with the earth's atmosphere.

Most of his professional life was devoted to the following fields of experimental physics:

- Atomic physics (photoelectric effect);
- Nuclear physics (slow neutrons);
- Cosmic ray physics (atmospheric air-showers).

During the Second World War, he joined the Free French Forces, and participated in the creation of a French-British-Canadian group on atomic energy research, becoming the head of this department in Montreal.

After the war, he became Director of the Department of Sciences for UNESCO. He strongly campaigned for the creation of international research organizations.

More information on the life and work of Pierre Auger can be found at the CNRS website.

Photos courtesy of Mariette Auger. Top: Jungfrau, 1935. Bottom: New York, 1960.

V.M. Adeev

In Memory of Yu. F. Shmalko (14.04.1948–12.10.2008)

In April, 2008, Yuri Fedorovich celebrated his 60th jubilee. Full of life and creative plans, Professor Yu. F. Smalko worked on a monograph, was in charge of Faculty of Physics and Energy and Department of unconventional energy technologies and ecology in Karazin Kharkov National University he created and also of laboratory of hydrogen energy technologies in A. Podgorny Institute for Mechanical Engineering Problems of National Academy of Sciences of Ukraine. Sudden death ends his life abruptly.

After graduation from post-graduate studies in Low Temperature Institute of National Academy of Sciences of Ukraine (Kharkov) 1976 in specialty 'experimental physics' Yuri Fedorovich Shmalko worked in A. Podgorny Institute for Mechanical Engineering Problems of National Academy of Sciences of Ukraine. He was one of initiators and organizers of the first in Ukraine Faculty of Physics and Energy subordinated to Ministry of Ukraine for Education and Science and to National Academy of Sciences of Ukraine. In 2003, he was elected as professor and head of department in Kharkov National University; over a period of years he worked as a Dean of Faculty of Physics and Energy, lectured on new trends in the field of physical and technical problems, guided workshops. Faculty of Physics and Energy is a new science and educational structure intended to integrate academic science and high school science in order to rise the level of high education, to draw young scientists and training highly qualified professionals for to-day and future energetic industry.

Yu. F. Shmalko was a recognized specialist in the field of hydrogen energy, worked as a visit-professor at University of Illinois, Chicago, USA; Technische Universität Hamburg-Harburg, Germany; Jilin University (Changchun, China). He was a full Member of International Association for Hydrogen Energy and a foreign Member of National association of the USA in hydrogen energy; he was elected as a vice-president of Ukrainian Association in hydrogen energy. Yuri Fedorovich was a member of editorial boards of several international scientific journals and a member of standing Organizing Committees of international science conferences in the field of hydrogen energy, in particular, ICHMS Organizing Committee. He authored 190 scientific works including several monographs.

Professional interests of Yu. F. Shmalko covered a wide range of problems associated with physical and physical and chemical properties of 'hydrogen isotopes - hydride-forming material' systems, and also development of metallic-hydride installations for energy conversion. He was the first to found and study experimentally the effect of thermodesorptional activation of hydrogen isotopes by metallic hydrides and work out a concept for multifunctional metallic-hydride devices. Yu. F. Shmalko paid a lot of attention to numerical simulation of phase equilibriums in 'hydrogen isotopes - metallic hydride' systems; his last monograph was published based on the results of this study.

Optimism, ability to overcome difficulties in solving scientific problems, ability to get on well with people, kindliness and courtesy towards people, and high self-discipline distinguished Yu. F. Shmalko.

We will always hold the memory of Yuri Fedorovich in our hearts.

V.S. Marinin
K.R. Umerenkova

In Memory of N.S. Astratov (10.02.1944–19.08.2008)

Nicolay S. Astratov – outstanding speaker and tutor, scientist, public figure, active and task-oriented man.

During the last 10 years Nicolay S. Astratov was a member of ICHMS Organizing Committee, executing a lot of work for the conference organization and its conducting.

Nicolay S. Astratov had been graduated from the Chemical Technology Faculty of the National Technical University of Ukraine “KPI” (department of cellulose-paper fabrication) in 1967. He began his professional activity at the Balahna’s cellulose-paper plant at the position of technologist (1967-1971). In 1971 he became post-graduate student of the Ukrainian Scientific-Researching Institute of Paper. Later Nicolay S. Astratov was a chief of scientific group at the same institute. From 1985 he began to teach students at the department of organic chemistry and technology of cellulose-paper fabrication of Kiev Polytechnic Institute. In 1986 he was assigned by scientific degree – Ph.D., and in 1990 – scientific title of associated professor. He became a member of the Ukrainian Technological Academy in 2003.

Nicolay S. Astratov is author of more than one hundred scientific papers, 3 monographs and series of patents and author’s certificates.

The last 10 years of his life Nicolay S. Astratov devoted to new carbon modifications studying. Because of the cellulose-paper fabrication problems good knowing he with the group of his students began scientific investigations of carbon nano-structures (fullerenes and nano-tubes) addition into various types of paper. More than two tens of fullerene-containing and more than one ten carbon nano-tubes containing types of paper and carton were designed and fabricated. Patents were received on a part of technologies which were developed.

Nicolay S. Astratov was a talented tutor and scientist, had authority and respect among colleagues and his students. Bright memory about Nicolay Sergeevich will be kept in hearts of everyone who knew and loved him.

Schur D.V.
Zaginaichenko S. Yu.
Pomytkin A.P.
Movchanyuk O.M.

In Memory of Yurii Andreevich Ossipyan (1931–2008)

Outstanding scientist and organiser of science, Yury Andreevich Ossipyan, passed away on September 10, 2008, at the age of 78. His principal scopes covered condensed matter physics and physics of strength and plasticity of solids. He was the founder of the Institute of Solid State Physics RAS in Chernogolovka, a member of the Presidium of the Russian Academy of Sciences and member of the Presidium of the RAS Research Centre in Chernogolovka. The death of Yury Andreevich is an irreparable loss to the scientific community of Russia and the world.

Yu. A. Ossipyan was born in Moscow on February 15, 1931. In 1955, he graduated from Moscow Institute for Steel and Alloys, having majored in metallurgical engineering. He began his research career at the Institute of Metal Science and Physics of Metals, affiliated with the Central Research Institute of Ferrous Metallurgy, while at the same time having his theoretical studies at the Faculty for Mechanics and Mathematics of Lomonosov Moscow State University.

The focus of his whole life was organisation, growth and development of the Institute of Solid State Physics RAS (ISSP), established by Academician G.V. Kurdyumov in 1963. From 1963 to 1973, Yury Andreevich was a deputy of Georgy Vyacheslavovich for research at ISSP, and was ISSP director from 1973 until 2002. In 2002, Academician Yu.A. Ossipyan became science supervisor of the Institute of Solid State Physics RAS.

Yury Andreevich had more than 200 papers published, dedicated to the theory of phase transitions, physics of materials strength, physics of electric and magnetic phenomena, physics of semiconductors, optics of dielectrics and semiconductors, and other fields of solid state physics. In the 1960s, Yu.A. Ossipyan commenced his pioneering experimental studies regarding interaction between electrons and extended defects in crystals. During that period, he discovered an unexpected and fascinating phenomenon referred to as the photoplastic effect in modern literature. Together with his students, Yu.A. Ossipyan discovered the electroplastic effect and the occurrence of electric charge on dislocations in semiconducting II-VI compounds; the existence of clusters of "dangling" valent bonds in dislocation cores in silicon, electron spin resonance and spin-dependent recombination on dislocations. Elegant experiments on high-frequency conduction led them to the discovery of quasi-one-dimensional electron bands bound up with dislocations, and combined electron resonance on dislocations in silicon. The pioneering experiments on the dangling-bond electron paramagnetic resonance have by now engendered a powerful tool for semiconductor diagnostics – the electron paramagnetic resonance spectroscopy of defects in semiconductors.

The contribution of Yu.A. Ossipyan and his research school to the physics of dislocations in semiconductor crystals won international recognition. He was elected a foreign member of national academies of Bulgaria, Hungary, Poland and Czechoslovakia. He was also a member of the National Academy of Engineering of the USA and the International Academy of Astronautics. For many years, Yury Andreevich successfully headed the International Union of Pure and Applied Physics (IUPAP). In 1972, he was elected Corresponding Member of the USSR Academy of Sciences and in 1981 – Full Member of the Academy. His works in the physics of dislocations brought him in 1984 one of the most prestigious awards of the USSR Academy of Sciences – Lebedev Gold Medal. In 2005, Yu. A. Ossipyan was awarded the highest distinction of the Russian Academy of Sciences – Lomonosov Grand Gold Medal.

Yury Andreevich Ossipyan was an excellent teacher and tutor for young scientists. He fully understood the importance of attracting and training new research personnel, and devoted a great deal of his energy, time and attention to this activity. He established an ISSP-based Chair of Solid State Physics of Moscow Institute for Physics and Technology (MIPT). For many years, he lectured to students and postgraduates at MIPT and remained Head of the Chair until his very last days. Dozens of students, who later grew into outstanding Russian scientists, prepared their PhD and Doctoral theses under his supervision. It was again him who initiated organization of a Faculty of Physical Chemistry of Moscow State University in Chernogolovka.

From 1985 to 2008, Yury Andreevich was Editor-in-Chief of "Kvant" (Quantum), a popular scholarly journal for children. He also headed the editorial board of the no less famous series of "Bibliotechka Kvant" (Quantum Little Library). The role of such periodicals in attracting young people to scientific careers cannot be overestimated.

Yury Andreevich always offered his best human qualities – friendliness, intelligence, wisdom, delicate sense of humour, sympathy for human suffering, thoughtful and responsive attitude towards people surrounding him, with deep insight into real merits of a person regardless of his rank. He chose to work with intelligent, honest and talented people, had creative imagination and could well formulate goals and plans for the future. He was never afraid of shouldering responsibility and knew how to achieve the goals he set for himself. The uniquely friendly and welcoming atmosphere surrounding him was always supportive of creativity and success of talented young researchers. The brilliant talent of Yu.A. Ossipyan as a science administrator and the system of recruiting and training research staff he had developed ensured ISSP's growing into one of the largest and most efficient academic institutions of physics research in Russia.

V.V. Kveder

On 150 Anniversary of M. Planck Birthday

"Discovery of quantum physical properties is my generaldiscovery"

M. Planck

In April 23, 2008 all scientific community celebrated 150 anniversary of Max Planck birthday, the outstanding German scientist, physicist-theoretician, the founder of quantum theory, a laureate of Nobel Prize in Physics of 1918.

In 1900, he introduced the concept of quantum of action (Planck's constant) and, based on the quantum ideas, deduced the radiation law. The works on thermodynamics, relativity theory and natural philosophy belong to M. Planck.

In one of the latest days of the passing century, 14 December 1900, the scientist gave a lecture "On the theory of radiant energy distribution for a normal spectrum" at the session of German Physical Society.

Energy emission and absorption by atoms and molecules occurs not continuously, as believed, but discretely (only in quantized form) by certain "portions" or quanta. That was a distillation of Planck's idea. It undermined the grounds of classic physic dogmata.

In that way quantum physics appeared.

Subsequently, Einstein characterized that event: " It is the Planck radiation law that provided the first exact definition of absolute values for atoms... It showed convincingly that besides atomistic structure of matter, there existed in a way atomistic structure of energy controlled by a universal constant... This discovery became a basis for all researchers in physics in XX century. Establishment of an actual theory of molecules and atoms and energetic processes controlling their transformations is impossible without this discovery. It destroyed the frame of classical mechanics and electrodynamics and posed the problem of finding a new cognitive basis for the whole physics".

At first, the quantum theory was accepted frostily. Some considered it unrealistic the others – not necessary. It was Einstein who equipped with this theory for the first time. In 1905, he introduced the concept of light quanta, photons and afterwards elaborated the quantum theory of heat capacity of solids.

Planck returned the favor. He belonged to those who approved and understood the relativity theory (apropos, he proposed the term "relativity theory" in 1906).

Reserved, dryish and somewhat stand-offish, Planck was a romantic in science and a loving husband and a careful farther in his family. He had two daughters and two sons. The elder son was killed at Verdun in 1916, the younger was shoot in 1945 for participation in the conspiracy against Hitler.

M. Planck died on 4 October 1947, half a year to his ninetieth birthday and three years to the fiftieth anniversary of the quantum theory".
[http://science.ng.ru/safe/2000-06-21/7_plank.html].

M. Planck Society for assistance sciences with its headquarter in Berlin has supported fundamental research in natural and humanitarian fields for 60 years. The Society, an independent, non-government organization in German, has 80 own institutes, scientific and research laboratories and working groups

which results supplement the work of colleges and other scientific and research institutions and are accessible for society.

At present, about 12600 collaborators and 11300 invited scientists, post-graduates and students work in the M. Planck Society under guidance of Peter Gruss, the scientists in the field of developmental biology. Budget for 2007 was 1433 milliards euro; it is mostly financed from public finance.
[http://www.germania-online.ru/index.php/2008-04-14-18-15-54/45].

In this way German nation memorialized one of the outstanding physicist.

On 115 Anniversary of P.L. Kapitsa Birthday

If theory coincides with experiment,
it is not a discovery, but closure.

P.L. Kapitsa

9 July 2009 was 115 anniversary of Pyotr Leonidovich Kapitsa birthday, the founder of native experimental physics and the forefather of a scientific dynasty, who made important discoveries in several different areas.

Pyotr Leonidovich is an outstanding physicist, academician (1939), a member of Presidium of the Soviet Academy of Sciences (since 1957), Hero of Socialist Work two times (1945, 1974), a laureate of Nobel Prize in Physics (1978), a laureate of Stalinist prize two times (1941, 1943), a holder of M.V. Lomonosov big gold medal (1959).

P.L. Kapitsa was born on 9 July 1894 in the city of Kronshtadt. His farther was a building engineer of Kronshtadt fortress and died in 1921, his mother was engaged in pedagogical and literature activities in folklore. In 1905 Kapitsa entered Kronshtadt gymnasium and was expelled for poor progress in 1906, and moved to a real college that he left in 1912. In the same year he entered the electrotechnical faculty in the Petrograd Polytechnical Institute and graduated in 1918, afterward he defended a graduate work on a physical theme and was left at the institute as a lecturer of physics and mechanics.

He was engaged in scientific work in Polytechnical Institute at a physical department from 1918 to 1921. In 1921, he and academicians of Academy of Sciences, A.N. Krilov and A.F. Ioffe, took part in a foreign mission to purchase instrumentation.

In 1921, Kapitsa was admitted to the Cavendish laboratory as a researcher and worked there from 1921 to 1934. In 1923, he defended a doctoral thesis "Transmission of α-beams through material medium and methods of producing strong magnetic fields" and received a PhD degree at Cambridge university.

In 1923, he received a Maxwell prize at the university. In 1925, he was elected as a member of English Royal Society, and later, in the same year – a Corresponding Member of Academy of Sciences of USSR.

He was a deputy director of the Cavendish laboratory since 1924.

He was a Professor of RS and the first director of the Mond laboratory at Cambridge university since 1930.

P.L. Kapitsa was a laureate of Liege university medal in 1934.

He was the director of Institute for Physical Problems of AS of USSR since 1935.

P.L. Kapitsa wrote this autobiography for the institute personnel department in 1950. Short lines with dates and positions hide the life like a scientific detective or a fascinating novel.

P.L. Kapitsa was awarded to many rewards and honorary titles: he was a honorary doctor at eleven universities in four continents, a member of many scientific societies and academies. Not stopping for any minute, "tireless academician" Peter Leonidovich Kapitsa tried to do as much as possible in the latest years of his life. He died on 8 April 1984 in Moscow not living three months till his 90 years. [http://www.tvkultura.ru/news.html?id=6002&cid=54].

In 80 years, the problems in which P.L.Kapitsa was engaged and the technical solutions which he proposed still remain topical and essential. In the framework of our conference subjects, many researchers use low temperatures, liquid helium and plasma conditions to synthesize different nanostructures.

At present, the environmental problems to which P.L. Kapitsa paid great attention during his latest years become the problem number one for the whole mankind. There exists no problem that attracts greater attention of scientific community than the environmental problem. Preserving environment, we retain ourselves. It is incorrect to follow the slogan of some modern "businessmen": "Viva pragmatism, away with romantics!" The main idea is to try to harmonize economical and ecological expediency.

The great scientist's astuteness appeared in his prediction of ecological catastrophes. 10 years before Chernobyl accident, P.L. Kapitsa warned about the necessity of taking into account a human element in using mathematical methods of calculating probability of accidents at atomic power plants.

The great scientist's creative gust and deeds remain in descendant memory forever.

On Centenary of L.D. Landau Birthday

Considering conciseness of our life we cannot afford luxury to go into the questions not promising new results

L.D. Landau

Lev Davidovich Landau, a Soviet physicist, an academician of Academy of Sciences of USSR (elected in 1946), a laureate of Nobel, Leninist and three Stalinist prizes, a Hero of Socialist Work, an academician of Academies of Sciences in Denmark, Netherlands, American Academy of Sciences and Arts (USA), French Physical Society, London Physical Society and London Royal Society, was born in Baku, on January 22, 1908.

Scientific life of Lev Davidovich Landau, an undoubted and acknowledged leader of Soviet theoretical physics, began very early and his scientific achievements were so significant and practically unattainable for the most of scientists.

The results achieved by L.D.Landau in Kharkov became classic, many of them bear his name.

A number of results that can be literally referred to L.D. Landau's fundamental contribution to theoretical physics for that period are: the theory of second-order phase transitions ("Landau theory of second-order phase transformations"), the kinetic equation for systems of particles with Coulomb interaction, the theory of an intermediate state for superconductors, the dispersion theory for magnetic permeability of ferromagnetics, where Landau and Lifshitz deduced a known motion equation for magnetization ("Landau-Lifshitz equation"), the theory of sound dispersion and absorption, the theory of monomolecular reactions, the theory of metals at ultralow temperatures, the theory of light diffusion, the statistical theory of atomic nuclei. Moreover, interpreting L.V. Shubnikov experiments, L.D. Landau expressed an idea of existing antiferromagnetics, and predicted a possibility of electron autolocalization in crystals. Therefore an unprecedented degree of L.D. Landau's scientific activity in UPTI becomes understood. Respective publications made his name world-known, and Kharkov – one of the leading centers of theoretical physics not only in USSR, but also in Europe. Many of famous physicists, N. Bohr, P. Dirac, V.O. Fok, Ya. I. Frenkel, I.E. Tamm, G.A. Gamov, V. Weisskopf, F. Houtermans, G. Placzek, R. Peierls and others arrived and worked in UPTI.

L.D. Landau is thought to be one of the creators of modern physics and, undoubtedly, one of the most outstanding figures in the XX century science. Scientific findings achieved by L.D.Landau are still valuable, and books of "Course" in many volumes remain popular and are used extensively.

Chapter 1
Problems of Development of Hydrogen Power Engineering

L.F. Kozin, S.V. Volkov, S.G. Goncharenko, and B.I. Daniltsev

Abstract The kinetics and mechanism of the interaction of aluminium and magnesium of the ternary system Al-Mg-Bi with water have been studied by high-temperature volumetry. The kinetic parameters of the reaction (rate constants, activation energies and degrees of transformation) have been calculated, and a mechanism of corrosion dissolution of bismuth-activated aluminium and magnesium in water with hydrogen evolution at a high rate is proposed. Kinetic parameters in the case of interaction between bismuth- and magnesium-activated aluminium and water, involving transformation of aluminium and magnesium into boehmite (AlOOH) and magnesium into hydroxide ($Mg(OH)_2$) with hydrogen evolution at a rate of 3,196–4,033 *l*/(m^2·min) in the temperature range in question have been determined. Micro- and nanostructured particles of Al-Mg-Bi alloys have been detected with the aid of a JSM 6490 LV electron microscope (Japan).

Keywords Kinetics · Interaction · Bismuth · Magnesium · Aluminium

1.1 Introduction

The deficiency of organic fuels, which has been rapidly growing in the last two decades, in combination with politicized global environmental problems caused a huge interest in hydrogen power engineering, which is based on the efficient production of a universal eco-friendly energy carrier, hydrogen. Hydrogen is a very abundant element on the Earth. The mass of hydrogen reserves is $18.367 \cdot 10^{18}$ t. At the average hydrogen content of the Earth crust of 1.4 g/kg, its main sources on the Earth are water (hydrosphere mass $1.664 \cdot 10^{18}$ t), coal, oil, natural gas and biomass

L.F. Kozin (✉), S.V. Volkov, S.G. Goncharenko, and B.I. Daniltsev
V.I. Vernadskii Institute of General and Inorganic Chemistry of NAS of Ukraine,
Palladin av. 32/34, 03680 Kyiv-142, Ukraine
e-mail: kozin@ionc.kiev.ua

S.Yu. Zaginaichenko et al. (eds.), *Carbon Nanomaterials in Clean Energy Hydrogen Systems - II*, NATO Science for Peace and Security Series C: Environmental Security 2,
DOI 10.1007/978-94-007-0899-0_1,

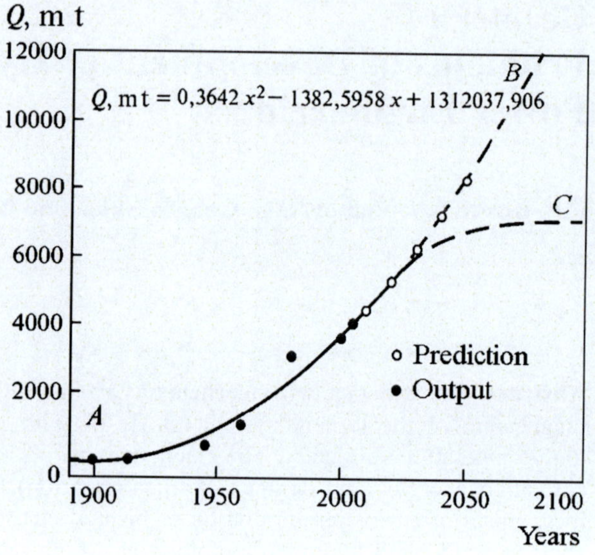

Fig. 1.1 Time dependence of oil output in the world in 1900–2005 and predicted data for the period 2010–2050: *AB*, oil output under "free" market relations without restrictions on production; *AC*, in the case of introduction of plan targets for oil output by the UN Power Committee

(mass of biosphere $1.148 \cdot 10^{13}$ t). At the present time, the technology for large-scale hydrogen production using hydrocarbons (natural gas, oil), the reserves of which on the Earth are finite, has become most commonly used [1, 2]. The increase in oil output (Q, million t) in the world during 1900–2005 is shown in Fig. 1.1. It turned out that the relation Q-τ (years) corresponds to the curve described by the equation:

$$Q(\mathrm{mt}) = 0,3642\tau^2 - 1382,5958\tau + 1312037,906. \tag{1.1}$$

Graphical integration of Eq. 1.1 in the Q-τ coordinates showed that 216.60 billion t of oil has been produced in the world during 100 years. The proved oil reserves in the world are (billion t): 210 (1999) [3], 206.1 (2005) [4] and 140.134–164.500 [5]. It follows that as much oil has been spent during 105 years as its proved reserves in deposits at the beginning of the third millennium.

The reserves of liquid and gaseous hydrocarbon fuels on the Earth are very limited, though they remain so far not explored to the full in many countries, including Ukraine and Russia. The predicted future discoveries of oil deposits (with 0% probability) amount to 110 billion t, so that the overall oil reserves with proved and predicted ones will be 316–320 billion t.

Equation (1.1) can be used as a prediction equation for the calculation of volumes of oil output in the world in the future. It will be 4,423, 5,276, 6,200, 7,197 and 8,267 million t/year in the world in 2010, 2020, 2030, 2040 and 2050 respectively. It follows that theoretically oil output must double by 2050 (8,267 million t) as against 2005 (4026.4 million t).

The question: how many years will oil really last in the world with its proved reserves is important. Integration of Eq. 1.1 between 2005 and 2050 shows that from 2006 to 2010, 21,319 million t of oil will be produced. From 2011 to 2050, 48,495 million t (2011–2020), 57,380 million t (2021–2030), 66,985 million t (2031–2040), 77,320 million t (2041–2050) will be produced in each decade. From 2006 to 2050, 271 billion and 107 million t of oil in all will be produced. This amount of oil is larger by 23.2% than the proved oil reserves in the world.

As was pointed out above, however, the overall oil reserves with due regard for probable reserves of 110 billion t are 316–320 billion t; therefore, it can be stated with certainty that the oil reserves in the world will last till 2050. Pessimistic results were obtained in Refs [2, 6, 7] in the prediction of exhaustion of oil; according to the prediction [8], all oil will be produced in the world as early as 2020.

The very rapid rise of the AB curve in Fig. 1.1, the curve of oil output in 2030–2050, is noteworthy; therefore, it seems to us that it is necessary to make use of the positive experience of the IAEA of controlling the production and consumption of the world's "black gold", oil. Restrictions on oil production as definite quotas must be imposed in 2020–2030.

When planned restrictions on oil production are imposed, e.g. by the UN Power Committee or World Committee on Power Resources, the plot of oil output against time will turn into a "saturation" plateau and will become AC curve (Fig. 1.1). In case of planned restriction on oil production, oil reserves in the world will last 100 years and more.

Figure 1.2 presents data on the cost of a barrel of oil in the period of 160 years. As can be seen, beginning from 1900, a slow reduction in price per barrel of oil was observed, which reached in 1974 10–12 dollars per barrel. Then an energy crisis

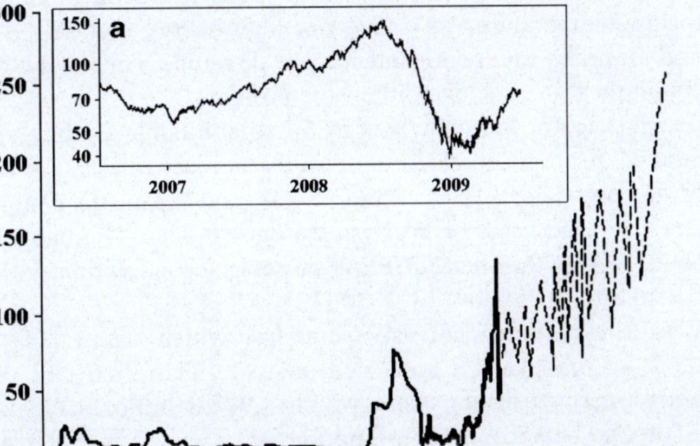

Fig. 1.2 Cost of a barrel of oil in the period 1890–2009 and prediction of prices for the period 2010–2050. The insert **(a)** shows the variation of the cost of a barrel of oil in the period from July 2006 to May 2009

began, and the price increased from 10 dollars per barrel to $75 in December 1980. Then a sheer fall in the price of oil started again, which reached a minimum of about 20 dollars per barrel in September 1980. The minimum cost of a barrel of oil persisted during the 1990s. Then a rise in the price of oil began, which continued during 2001–2006 and reached 145 dollars per barrel in July 2009. The high price of oil persisted for about a month. Then a sheer fall in the price of oil started again, which continued, as is seen from the insert in Fig. 1.2, during 6 months till December 2009. In the period from December 2009 to March 2010, the prices of oil stopped at the level of $42–45 per barrel. In July 2010, a new price of $70–75 per barrel was fixed. It is predicted that in 2011 the price of oil will be 92 dollars per barrel [5–7].

The development of the industry and economy of one or another country depends directly firstly on provision with energy carriers and secondly on the development level of power engineering. Figure 1.3 shows a scheme of the balance of power consumption in such developed countries as USA (a), Germany (b), Russia (c) and Ukraine (d) for comparison. To construct Fig. 1.3, the data obtained in Refs [9–11] were used. The total power consumption in the USA in 2004 was $112{,}990.5 \cdot 10^{12}$ J. The share of oil in 2004 was 39%, of coal 23%, of natural gas 23%, of nuclear power 8%. The share of renewable energy sources (RES) is 8%, including (%): water power 4.0, biomass 3.44, geothermal power 0.4, solar power 0.08, wind power <0.04. The quantity of electrical power produced in the NSA in 2004 was 4,904.9 billion kWh. The share of coal in electrical power production was 4.7% [10, 11]. It is believed that the annual increase in electrical power consumption in the USA will be 1.75–1.80%, and this will reach 5,207 billion kWh by 2025.

In the USA, Germany, Russia, Ukraine, Spain, Japan, France, etc., intensive research and development are being carried out on the conversion of solar energy to electricity (photoelectric systems (PES), solar batteries, solar modules and solar cells) and to low-temperature (90–150°C) or high-energy (>1,200°C) heat: solar collectors and furnaces are being intensively developed in the USA and other countries though they are a science-intensive problem.

At the present time, the total power of PESs, solar modules, cells and batteries in the world balance of produced electrical power has reached 1,630 MW. The share of the USA (the companies BPSolar, Shell Solar) and Japan (the companies Sharp and Kuosecha) is ⅔, and that of Western Europe (France, Germany, Spain, etc.) is ⅓. It is believed that the share of PESs in power consumption will increase to 15–20% by 2015 and to 25–30% by 2030 [12].

Germany is a very power-intensive country, which has 1.42% population, occupies 0.26% of the planet's area, consumes 1.193 billion t of conventional fuel of primary energy carriers, from which 3,977.8 billion kWh is produced. The scheme of the balance of consumption of primary energy carriers, from which the power-generating plants of the country produce 3,977.8 billion kWh of electrical power, is shown in Fig. 1.3b. The main primary energy carriers are organic ones: oil (36%), natural gas (22%), black coal (14%), brown coal (11%), and the share of nuclear power is 13%, which corresponds to 5.171×10^{11} kWh of electrical power and heat.

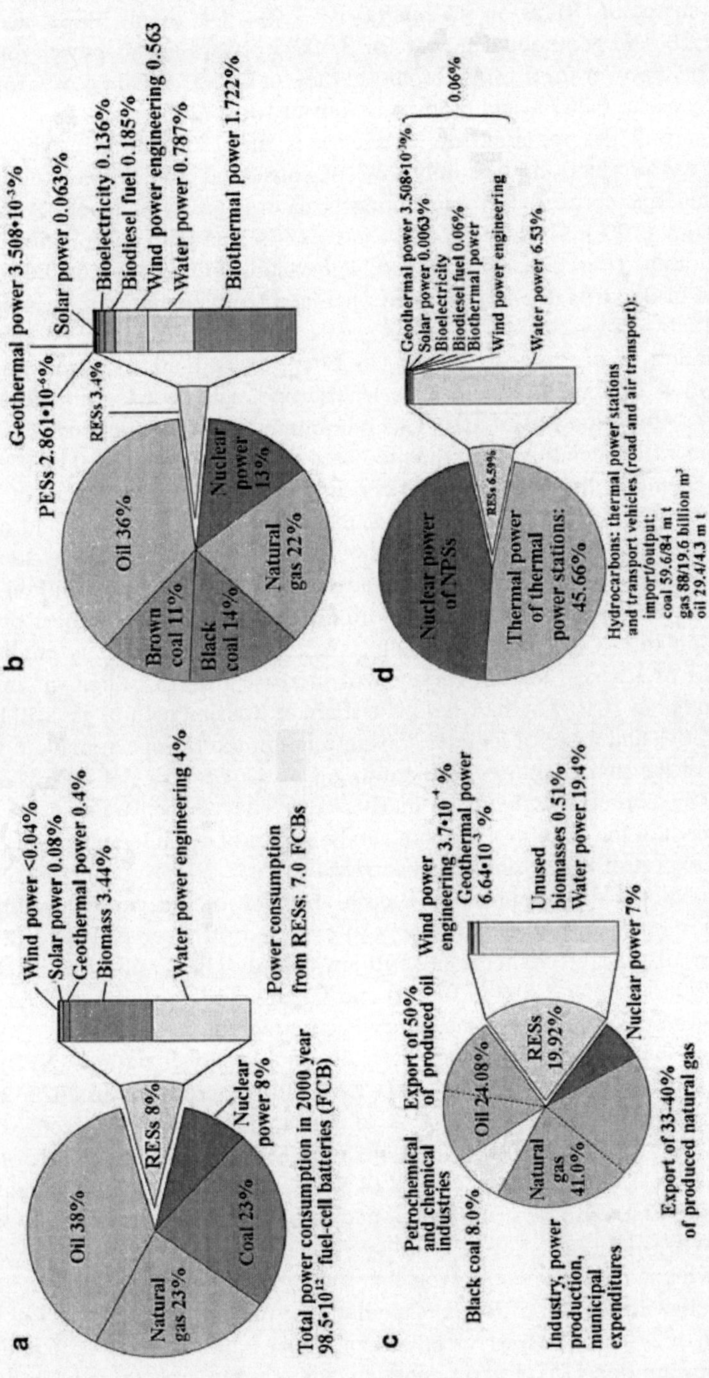

Fig. 1.3 Scheme of the balance of power consumption in such developed countries as USA (**a**), Germany (**b**), Russia (**c**) and Ukraine (**d**) for comparison

The share of RESs in Germany is 3.4%. Of them, PESs account for 2.861×10^{-6}%, geothermal power for 3.508×10^{-3}%, solar power for 0.063%, bioelectrical power for 0.136%, biodiesel fuel for 0.185%, wind power for 0.536%, water power for 0.787% and biothermal power for 1.722%.

The natural gas requirements of Germany are 135 billion m^3/year, whereas national gas production gives only 22% of consumed gas. To meet the needs for organic energy carriers, Germany imports natural gas from Great Britain (7%), Netherlands (19%), Norway (25%), Russia (34%) and 14 million t of oil. Since supplies of gas from Great Britain and Netherlands will soon terminate because of depletion of deposits, the import of Russian and Norwegian gas may increase by a factor of 2.0–2.5 by 2015.

According to the forecast of the US Energy Information Administration, the consumption of natural gas in Western Europe will be 645 billion m^3 and in Germany 149 billion m^3 in 2010. Gas requirements will continuously grow in the future and will exceed by 2030 the predicted consumption in 2010 by 72%.

The scheme of the balance of the consumption of power resources in Russia is shown in Fig. 1.3c. It can be seen that the volume of natural gas and oil output reached 41% and 24.08%, respectively, of the total amount of energy carriers. The export of hydrocarbons in 2001 was: natural gas, 33–40% (166.5 billion m^3); oil, 45% (144 million t); oil products, 62 million t. The main consumers of Russian gas (billion m^3): Germany, 20.4; Italy, 13.5; Ukraine, 18.7. It is predicted that the export of Russian gas to Turkey will increase from 3 billion m^3 in 2001 to 16 billion m^3 in 2010. On the whole, the share of Russian natural gas will be a long time 25–30% and more of its total consumption in the European market. Russia is also one of the main exporters of natural gas to the countries of the Asian-Pacific region. The natural gas industry in Russia is one of the strategic ones in the development of the country's economy and a source of considerable money receipts from the export of this hydrocarbon carrier.

The predicted oil resources of Russia are 13% of the world resources [13]. Russia is inferior only to Saudi Arabia in explored oil reserves [14]. The biggest consumers of oil are: Germany (14 million t), Italy (11.5 million t), Netherlands (11.2 million t), Poland (9.9 million t) and Ukraine (12–15 million t). Russia also exports oil to the countries of the Asian-Pacific region.

The heat and electricity production by the nuclear power stations (NPS) reaches 152 TWh. It is predicted that 206–400 TWh will be produced in 2020–2050 [2]. RESs account in Russia for 19.92% of power. The share of electrical power produced by the water power stations (WPS) is 18.8–19.4% (165.4–169.94 billion kWh), using biomass 0.51% (4.5 billion kWh), by geothermal power stations $6.64 \cdot 10^{-3}$% (58.2 m kWh) and wind power stations 2.17×10^{-4}% (1.9 m kWh) [2].

It is evident from the data given that about 1/5 of electrical power in Russia is produced with the aid of RESs. Calculations showed that RESs in Russia have a potential of 270 million tons of conventional fuel, which exceeds 25% of domestic power consumption [15]. Russia conducts actively research to develop hydrogen power engineering.

The scheme of the balance of the consumption of power resources in Ukraine is shown in Fig. 1.3d. The consumption of primary energy carriers in Ukraine in 2005 was 161.5 million t of conventional fuel. The share of natural gas, coal and its processing products (oil and oil products) was 88.5 million t of conventional fuel (44.6%); nuclear fuel, 52.6 million t of conventional fuel (26.6%); RESs, 20.4 million t of conventional fuel (10.3%) respectively. It is seen from the scheme that the main primary energy carriers are organic ones: coal (52.6 million t), natural gas (88 billion m^3), oil (20.4 million t) and other fuel types (4.7 million t). Ukraine possesses deposits of organic energy carriers with considerable reserves of black and brown coal and shales, as well as large oil and natural gas reserves, which have not been exploited as yet [2]. The development of RESs and hydrogen power engineering in Ukraine is considered in detail in [2].

1.2 Hydrogen Power Engineering

It is clear from the foregoing that the organic resources of our planet are limited in time and finite. Therefore, efforts must be made to intensively develop hydrogen power engineering, ecologically clean thermonuclear power engineering, as well as power sources that seem to be incredible at present [2]. Hydrogen power engineering is in principle versatile and is power engineering of the future since its functioning is not limited by raw material resources. Hydrogen production in the industrially developed countries (IDC) in 1990, 1995, 2000 and 2005 was 1.0×10^{12}, 1.6×10^{12}, 3.63×10^{12} and 8.82×10^{12} m^3.

Hydrogen production in the USA in this period was 39–45% of the volume of production in the IDCs. It is believed that in the IDCs 14.1×10^{12} and 43.7×10^{12} m^3 and in the USA 4.4×10^{12} and 15.8×10^{12} m^3 will be produced by 2010 and 2020. These volumes of hydrogen production seem to be considerable; however, they account only for 25% of the annual natural gas output. An important disadvantage is that the main source of raw material for hydrogen production is natural gas (85%). The cost of natural gas and oil increased by a factor of about 4–5 by 2010. The cost of electrical power produced by NPSs, thermal power stations burning coal and WPSs, however, has not practically changed, therefore the price of hydrogen produced, e.g. by electrolysis, remained the same and approached the cost of "hydrocarbon" hydrogen. This indicates economic stability of hydrogen power engineering.

The concept of working medium of energy-storing substances (ESS) which function in a closed cycle is of special interest for the development of hydrogen power engineering. This article considers the kinetics and mechanism of hydrogen extraction from water by means of ESSs based on aluminium and magnesium, which are activated to impart to them a high reactivity against water by adding various metals-activators, alloy constituents which change the structure of alloys and allow one to achieve high hydrogen evolution rates, which reach 20–30 m^3 of hydrogen/($m^2 \cdot$min). Many problems of hydrogen power engineering as well as modern power engineering and ecology have been analyzed in monographs by the authors [1, 2].

The additives that form with the activated metal stratifying systems, limited solid solutions or eutectics are the most efficient for imparting high reactivity against water to ESSs [1, 2]. The metals-additives are divided into two types according to their physicochemical action on the reactivity of aluminium against water [1, 2, 16, 17]: chemical type (gallium, alkali metals, alkaline-earth and rare-earth metals) and structural type (indium, thallium, tin, lead, cadmium, zinc, mercury, ferrosilicon, etc.). Up to now, such metals-activators as gallium, indium, thallium [18–22], which have a high market value as they are rare disseminated elements and are used in electronics [23], were used for aluminium activation.

To produce hydrogen from water, powdered aluminium and high temperatures (≥1,000°C) [24], as well as pressed mixtures of powdered aluminium and anhydrous caustic soda are used, which react vigorously when in contact with water at moderate temperatures (≤250–340°C) [25, 26].

Investigations carried out by us [27–29] showed that the rate of corrosion dissolution of ESS alloys in water depends on the temperature conditions of their fabrication, the structure of the phase diagrams of binary (Al-M_i), ternary (Al-M_i-M_j) (where M_i , M_j = metals-activators) and other more complex systems, as well as on temperature drop on the crystallization (supercooling) of alloys. The reactivity of aluminium activated with additions of metals against water is the higher, the finer-crystalline structure is formed during the preparation of ESSs. It has been found that at high alloy cooling rates in the case of casting into molds, which reach 500–800 deg/s, a fine-crystalline structure of aluminium alloys and other metal alloys is formed up to obtaining nanostructured ESS alloys [27–30].

In view of the developing world energy crisis and rise in prices for organic heat carriers, as well as of the ever-growing awareness in the world that the reserves of organic heat carriers are finite, technologies for the production of hydrogen, which is an energy carrier of the future, began to be intensively developed in the USA [31], Germany [32], Russia [33–35] and other countries. We have found, e.g., that the dissolution of activated aluminium, containing additions of metals-activators and crystallized under nonequilibrium conditions, in water with hydrogen evolution is caused by the intercrystalline corrosion, which intensively develops in water, of aluminium-bismuth, aluminium-gallium, aluminium-gallium-indium and other alloys inhomogeneous throughout their volume, whose unit cell surface boundaries are rich in metals-activators. This distribution of metals-activators is due to the values of their grain boundary distribution coefficients of crystals, which results in the enrichment of the crystal boundaries of the ESS base metal (aluminium, magnesium, boron, etc.) with bismuth, gallium and other metals [27–30]. The difference in the concentration of the electropositive metals-activators in the cells of the crystal structure of activated aluminium reaches large values, which results in the formation of microgalvanic cells, which determine the emf value of the corrosion process. In this case, aluminium, magnesium and other electronegative elements act as the anode in the microgalvanic cells formed, and the interfaces between crystallite grains rich in electropositive metals-activators act as the cathode [1, 2, 27–30].

Due to the said above, aluminium-based ESSs show a high reactivity against water: they evolve gaseous hydrogen at a predetermined pressure at a high rate, interact with oxygen, especially in humid air, to form hydroxides and oxides, and with other reactants [1, 2, 16, 17, 27–30]. Energy-storing substances can find application in various fields of science and technology, e.g. in the manufacture of metal-containing high-energy rocket fuel [1, 2, 17] and protectors for the corrosion protection of metallic structures in sea water [36–39], for the removal of dissolved oxygen from liquid steels [40], for the manufacture of solid electrolytes based on aluminium oxides for high-temperature electrochemical power sources [41, 42], etc.

We have shown in [27–30] that bismuth is a more efficient metal-activator than gallium for imparting a high reactivity against water to aluminium [27]. The addition of bismuth to aluminium-magnesium alloys imparts a high reactivity not only to aluminium, but also to magnesium.

1.3 Kinetics of ESS Dissolution in Water

Bismuth-activated aluminium-magnesium alloys were fabricated in argon atmosphere in a special setup consisting of a vertical crucible furnace, a double-walled alundum crucible, the walls being thermally insulated from each other with basalt wool, an alundum blade mixer, a quartz cover for the alundum crucible with special holes for argon supply and the placement of the alundum mixer. Since aluminium and bismuth form an immiscible system Al-Bi up to 1,050°C, and magnesium and bismuth form intermetallics Mg_3Bi and Mg_3Bi_2, highly reactive aluminium- and magnesium-base alloys with metal-activator bismuth were fabricated under hydrodynamic mixing. To fabricate Al-Mg-Bi alloys, high-purity metals were used. Kinetic curves of the evolution of hydrogen, which is formed by interaction in time (p-τ curves) between activated aluminium-magnesium alloys and water, were recorded in a high-pressure reactor. The experimental procedure is described in detail in [27–30].

Kinetic curves of the corrosion dissolution of aluminium and magnesium (1.5–5.0 wt.%), activated with 3.0 wt.% bismuth, in water with hydrogen evolution at 225–325°C are shown in Fig. 1.4. As is seen from Fig. 1.4, the kinetic p-τ curves at the temperatures in question are pronounced sigmoids for topochemical reactions proceeding in water with hydrogen formation and growth of nuclei of the solid reaction product phase, boehmite (AlOOH), on the dissolution of activated aluminium and magnesium hydroxide formation on the dissolution of magnesium in the ternary system Al-Mg-Bi [43]. In the binary system Al-Mg, two intermetallics are formed: Al_3Mg_2, incongruently melting at 451°C, and $Al_{12}Mg_{17}$, congruently melting at 462°C [44]. In the system Mg-Bi, an intermetallic Mg_3Bi_2 is formed, which melts congruently at 821°C [44]. Proceeding from the principles developed in [45, 46], it follows that in metal systems with congruently melting compounds, slightly dissociated intermetallics are formed, which are stable in the liquid phase of melts above the liquidus curve. In our case, the intermetallic Mg_3Bi_2 is such a compound.

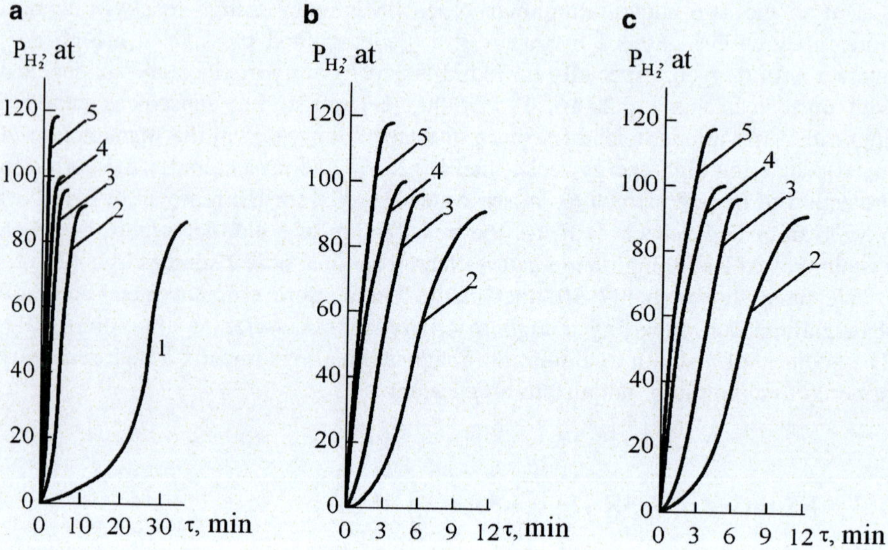

Fig. 1.4 Kinetic curves of hydrogen extraction from water with aluminium activated with 3.0 wt.% bismuth and 1.5 wt.% magnesium (**a**), 3.0 wt.% bismuth and 3.0 wt.% magnesium (**b**), 3.0 wt.% bismuth and 5.0 wt.% magnesium (**c**) at the temperatures (°C): (*1*) 225, (*2*) 250, (*3*) 275, (*4*) 300, (*5*) 325

As follows from the data presented in [44], in the system Mg-Bi, a very low activity of both magnesium ($a_{Mg} = 1.1 \times 10^{-5}$ at 0.05 mole fraction of Mg) and bismuth ($a_{Bi} = 2.1 \times 10^{-4}$ at 0.05 mole fraction of Bi and 700°C) is observed in the case of the compositions used by us.

As an analysis of the obtained results has shown, the interaction between activated aluminium and water with hydrogen evolution proceeds by the following exothermic reactions:

$$2Al^* + H_2O \Rightarrow AlOH + AlH, \tag{1.2}$$

$$AlH + H_2O \Rightarrow AlOH + H_2, \tag{1.3}$$

$$2AlOH + 2H_2O \Rightarrow 2AlOOH + 2H_2 \tag{1.4}$$

with the overall reaction:

$$2Al^* + 4H_2O \Rightarrow 2AlOOH + 3H_2 + Q, \tag{1.5}$$

where Q is the thermal effect of the reaction of corrosion dissolution of activated aluminium (Al^*) in water to form boehmite. The exothermic effect of reaction (1.5) is caused by the fact that the formation enthalpy of AlOOH ($\Delta H^\circ_{form} = -985$kJ/mol)

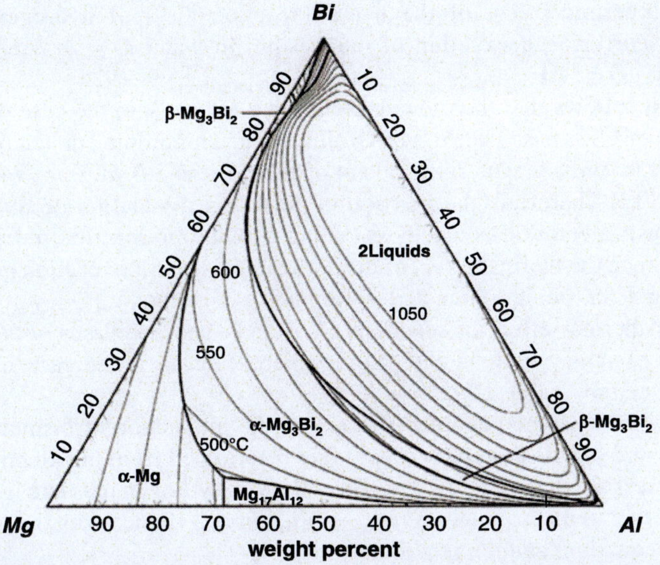

Fig. 1.5 Phase diagram of the ternary system Al-Mg-Bi

is much lower than ΔH°_{form} of water (–285.83 kJ/mol), and hence heat liberation corresponding to Q = 826.68 kJ is observed during the proceeding of reaction (1.5).

Magnesium, which actively interacts with bismuth in the binary system Mg-Bi and in the ternary system Al-Mg-Bi, also shows a high reactivity against water. The structure of the three-component system Al-Mg-Bi is shown in Fig. 1.5. Magnesium interacts with water at the temperatures in question by the following consecutive-reaction mechanism:

$$2Mg + H_2O \Rightarrow MgH_{ads} + MgOH, \tag{1.6}$$

$$MgH_{ads} + H_2O \Rightarrow MgOH + H_2, \tag{1.7}$$

$$2MgOH + 2H_2O \Rightarrow 2Mg(OH)_2 + 2H_2 \tag{1.8}$$

with the overall reaction:

$$2Mg + H_2O \Rightarrow 2Mg(OH)_2 + 2H_2. \tag{1.9}$$

As follows from the data presented in [47], at moderate temperatures (25–150°C), magnesium can interact with water and with hydrogen in the cathodic process with the formation of MgH_2 hydride. Therefore the mechanism of hydrogen extraction from water with aluminium- and magnesium-based ESSs with the participation of both magnesium dihydride MgH_2 and monohydride MgH in chemical corrosion dissolution reactions is possible.

The exothermic effect of the overall reaction (1.9) of hydrogen evolution during the corrosion dissolution of magnesium in water Q = $-\Delta H^{\circ}_{form}$, hence we obtain: $Q = (\Delta H^{\circ}_{Mg(OH_2) formation} - 2\,\Delta H^{\circ}_{H_2O\ formation}) = -925 - (2 \cdot -285.34) =$ 354.32kJ. It follows that the overall exothermic effect Q_i in the case of corrosion dissolution of ESSs based on Al-Mg-Bi alloys at an aluminium content of 95.5, 94.0 and 92.0 wt.% and a magnesium content of 1.5, 3.0 and 5.0 wt.% is 794.79, 787.71 and 778.26 kJ. Therefore, the reaction of corrosion dissolution of ESSs in water with hydrogen evolution is a self-sustaining exothermic reaction proceeding at a high rate. Since the thermal effect of the corrosion dissolution reaction of the metals constituting ESS (magnesium and aluminium) is different ($Q_{Mg(OH)_2} \ll Q_{Al_2O_3}$), the overall thermal effect of reactions (1.5) and (1.9) decreases with increasing proportion of magnesium in the alloy, which leads to a decrease in hydrogen evolution rate (see Table 1.1).

The rates of hydrogen evolution (V_{H_2}) on the interaction of bismuth-activated aluminium and magnesium with water were determined from the kinetic curves of hydrogen pressure shown in Fig. 1.4. The values of hydrogen pressure in the reactor were converted to molar volumes ($P_{H_2} \rightarrow V_{H_2}$) in accordance with the Mendeleev-Clapeyron equation, and the relation:

$$V_{H_2} = \Delta V_0/(S \cdot \Delta\tau) \tag{1.10}$$

was then employed for calculations, where ΔV_0 is the volume, reduced to normal conditions, of hydrogen (L) evolved during a definite reaction time $\Delta\tau$(min), and S is the surface area of the sample under investigation (m^2).

Table 1.1 Temperature dependence of the maximum hydrogen evolution rates $\upsilon^{H_2}_{max}$ (L/m^2·min), the time of reaching them, τ_{max}(min), effective rate constants $k^e i$ (min^{-1}), induction period τ_i (min). Values of activation energy E_a(kJ/mol) and the heat of corrosion dissolution of Al* in water

Kinetic parameters	Temperature (°C) 225	250	275	300	325	Q (kJ)	E_a (kJ/mol)
Al 95.5% – Bi 3.0% – Mg 1.5%							
$\upsilon^{H_2}_{max}$, L/m^2·min.	561	1,170	1,674	2,735	4,033	794.79	
τ_{max}, min.	27	4.5	1.8	MIBR[1*]	MIBR		
k^e_i, min^{-1}.	0.153	0.197	0.247	0.281	0.341		19.5
τ_i, min.	24.7	3.3	1.0	ABS[2*]	ABS		
Al 94.0% – Bi 3.0% – Mg 3.0%							
$\upsilon^{H_2}_{max}$, L/m^2·min.	AI[3*]	903	1,244	2,597	3,215	787.71	
τ_{max}, min.	AI	5.7	2.5	MIBR	MIBR		
k^e_i, min^{-1}.	AI	0.189	0.228	0.262	0.295		15.4
τ_i, min.	AI	3.9	1.4	ABS	ABS		
Al 92.0% – Bi 3.0% – Mg 5.0%							
$\upsilon^{H_2}_{max}$, L/m^2·min.	AI	834	1,089	2,346	3,196	778.26	
τ_{max}, min.	AI	6.7	3.2	MIBR	MIBR		
k^e_i, min^{-1}.	AI	0.179	0.202	0.242	0.271		14.8
τ_i, min.	AI	4.7	1.9	ABS	ABS		

MIBR[1*] manifests itself at the beginning of reaction; *ABS*[2*] is absent; *AI*[3*] alloy is inactive

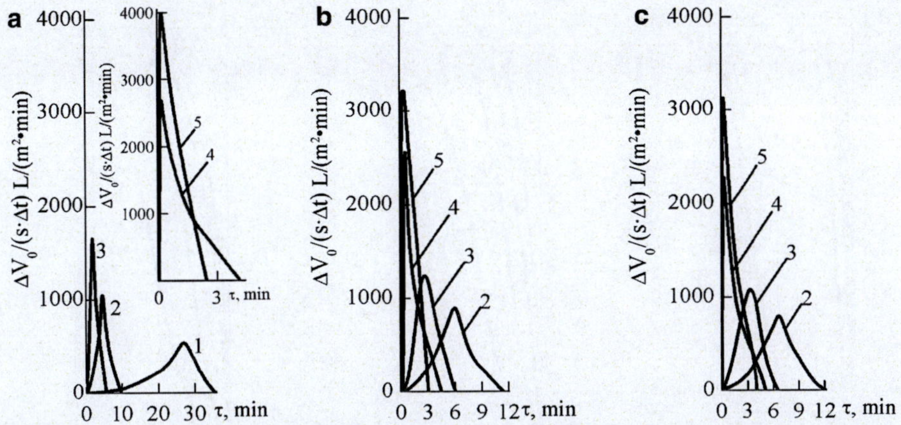

Fig. 1.6 Time dependence of the rate of hydrogen evolution on the interaction with water of aluminium activated with 3.0 wt.% bismuth and 1.5 wt.% magnesium (**a**), 3.0 wt.% bismuth and 3.0 wt.% magnesium (**b**), 3 wt.% bismuth and 5.0 wt.% magnesium (**c**) at the temperatures (°C): (*1*) 225, (*2*) 250, (*3*) 275, (*4*) 300, (*5*) 325

The kinetic $\Delta V_0/(S\cdot\Delta\tau) - \tau$ curves obtained are shown in Fig. 1.6. As can be seen, these kinetic curves pass through a maximum. The determining factor affecting the trend of the kinetic curves of the interaction of activated aluminium and magnesium with water is temperature. For instance, at low temperatures, a reaction of uniform dissolution of cell faces of microcrystals and nanocrystals of the aluminium surface with the formation of a boehmite film and of the magnesium surface with the formation of magnesium hydroxide $Mg(OH)_2$ hinder the access of water and withdrawal of the gaseous reaction product (hydrogen) from the reaction surface, which manifests itself as induction period and by slow hydrogen evolution rates: curves 1–3 in Fig. 1.4 and curves 1–3 in Fig. 1.6. The maximum rate of hydrogen evolution on the interaction of aluminium and magnesium (1.5 wt.%), activated with 3.0 wt.% bismuth, with water at 225, 250 and 275°C is 561, 1,170 and 1,674 L/($m^2\cdot$min) and reaches these values within 27, 4.0 and 1.8 min respectively. These values and hydrogen evolution rate values for other ESS compositions are listed in Table 1. To determine the induction period (τ_i), the degree of transformation of activated aluminium (α_{Al*}) in the reaction of hydrogen evolution from water was calculated. The degree of transformation was determined from the ratio of hydrogen actually evolved at the current moment to the final value. The induction period of the reaction was determined by plotting curves on the $\alpha_{Al*}-\tau$ coordinates and drawing an inflectional tangent to a sigmoid curve. The plots of α_i against τ obtained are shown in Fig. 1.7. It can be seen that the degree of transformation α_i is a function of temperature. The degree of transformation reaches unity the faster, the higher the temperature. For instance, the induction period in the case of interaction of aluminium activated with 3.0 wt.% bismuth and 1.5 wt.% magnesium with water at 225, 250 and 275°C is 24.7, 3.3 and 1.0 min (see Table 1.1).

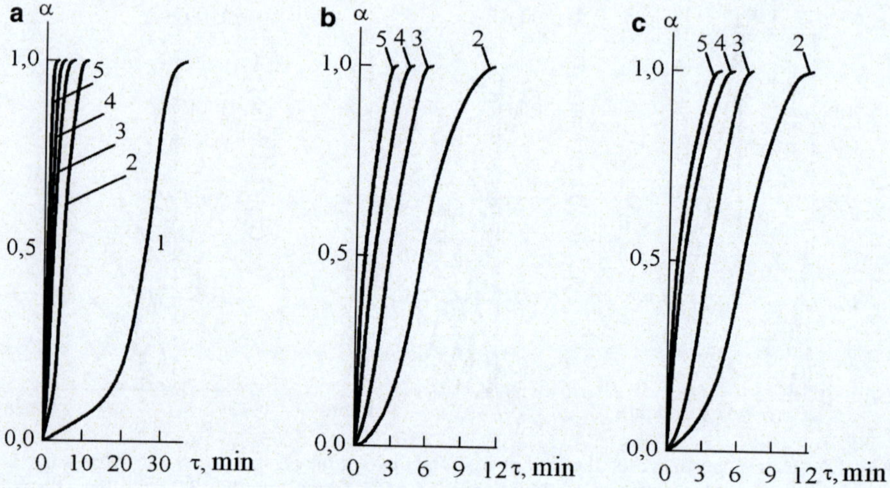

Fig. 1.7 Time dependence of the degree of transformation (α_i) on the interaction with water of aluminium activated with 3.0 wt.% bismuth and 1.5 wt.% magnesium (**a**), 3.0 wt.% bismuth and 3.0 wt.% magnesium (**b**), 3.0 wt.% bismuth and 5.0 wt.% magnesium (**c**) at the temperatures (°C): (*1*) 225, (*2*) 250, (*3*) 275, (*4*) 300, (*5*) 325

At high temperatures, the dissolution reaction of activated aluminium and magnesium proceeds at the grain boundary of nano- and microcrystallites enriched with additions of metal-activator bismuth, whose concentration exceeds the volume concentration due to a low coefficient of bismuth distribution (K_{Al-Bi}) in aluminium. Therefore, when the temperature is increased, an abrupt increase in hydrogen evolution rate is observed. In this case, owing to vigorous hydrogen evolution at 300–325°C, boehmite and magnesium hydroxide films, which hinder the access of the reactant water (curves 4 and 5 in Figs. 1.6 and 1.7) to the reaction surface of Al-Mg-Bi alloys, have no time to form on the surface of these alloys (they are blown away by hydrogen flow). The maximum rate of hydrogen evolution on interaction between aluminium with 1.5% wt% magnesium, activated with 3.0 wt. % bismuth, and water at 300 and 325°C is 2,735 and 4,033 L/(m^2·min). The hydrogen evolution rate is a maximum at the beginning of interaction between alloy and water (see Table 1.1), and at 300–325°C the induction period disappears (see Fig. 1.7 and Table 1.1). The reaction rate constants of Al* dissolution in water ($k^e i$) were calculated from a known first-order equation for heterogeneous reaction:

$$\ln[1/(1-\alpha)] = k_i^e \cdots \tau - C, \tag{1.11}$$

where $k^e i$ is the effective rate constant of the heterogeneous reaction of dissolution of activated Al* and Mg in water, τ is time (min), C is an integration constant.

Figure 1.8 presents experimental data on the interaction of aluminium and magnesium (1.5–5.0 wt.%), activated with 3.0% bismuth, with water in the ln $1/(1-\alpha) - \tau$ coordinates. At 300–325°C, the ln $1/(1-\alpha)-\tau$ curves are linear

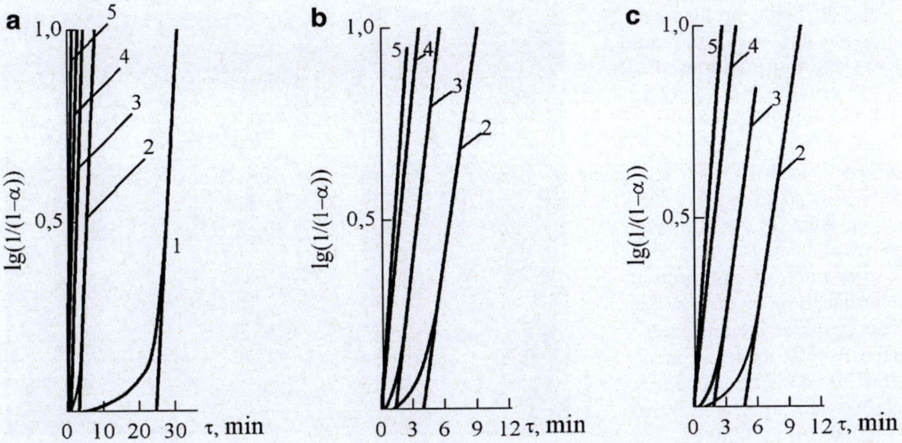

Fig. 1.8 Dependence of ln[1/(1 – α)] on the time of interaction with water of aluminium activated with 3.0 wt.% bismuth and 1.5 wt.% magnesium (**a**), 3.0 wt.% bismuth and 3.0 wt.% magnesium (**b**), 3.0 wt.% bismuth and 5.0 wt.% magnesium (**c**) at the temperatures (°C): (*1*) 225, (*2*) 250, (*3*) 275, (*4*) 300, (*5*) 325

(curves 4, 5 in Fig. 1.8) and at 225–275°C in the region ln 1/(1 – α) ≤ 0.10 (which corresponds to α ≤ 0.20), the kinetic curves are nonlinear (curves 1–3 in Fig. 1.8). The nonlinear portion of the curve, which is observed at α ≤ 0.20, is due to the induction period of the reaction. The linear portions of ln 1/(1 – α)– τ curves were used to calculate the effective reaction rate constants of interaction between Al-Mg-Bi alloys and water with hydrogen evolution. Rate constant values of the interaction of bismuth-activated aluminium and magnesium (with the formation of the ternary system Al-Mg-Bi) are listed in Table 1.1.

The temperature dependence of rate constants was used to calculate the effective energies of the interaction of activated aluminium and magnesium with water from the equation:

$$E = tg\alpha \cdot R \cdot 2.303 \cdot 4.184 \cdot 10^3, \quad (1.12)$$

where R is gas constant (1.987 kcal/mol), and *tg* α is the angular coefficient of experimental curves in the ln k_i – 1/T coordinates.

The E_a values obtained are listed in Table 1.1. The activation energy of the reaction of hydrogen extraction from water with aluminium and magnesium (3 wt.%) activated with 3 wt.% bismuth E_a = 14.8–19.5 kJ/mol, which indicates diffusion rate control.

Investigations showed that aluminium alloys with nanostructured surface and bulk nanostructure of the ESS alloy formed show the highest activity against water and, e.g., reactions (1.5) and (1.9). Figure 1.9 shows the microstructures of a fresh fracture of ESS alloys based on bismuth-activated aluminium and magnesium, which have been photographed on a JSM 6,490 LV scanning electron microscope (Japan) in the regime of making images from a secondary electron detector (SEI).

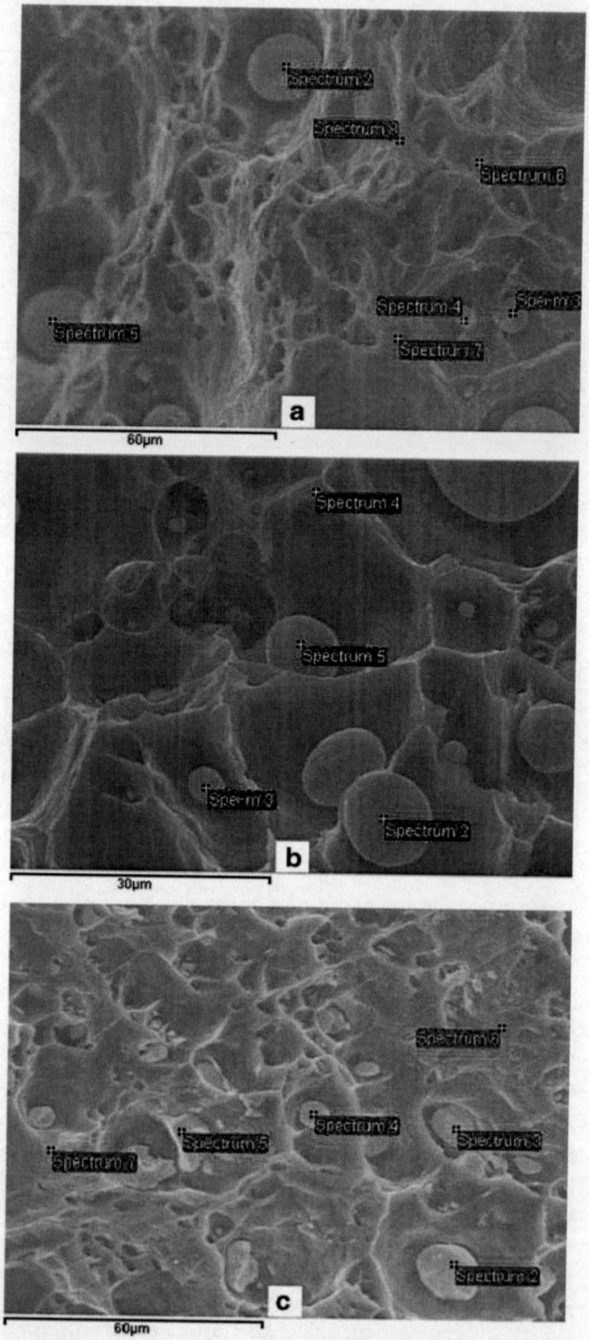

Fig. 1.9 Points on the fresh-fracture surface of aluminium activated with additions of 3.0 wt.% bismuth and 1.5 wt.% magnesium (**a**), 3.0 wt.% bismuth and 3.0 wt.% magnesium (**b**), 3.0 wt.% bismuth and 5.0 wt.% magnesium (**c**), for which an elemental analysis has been carried out by X-ray spectrum microanalysis by means of an energy dispersive analyzer (INCA 450), mounted on a JSM 6,490 LV microscope, at the magnification: ×2,000 (**a**), ×3,000 (**b**), ×2,000 (**c**)

Table 1.2 Elemental composition obtained by X-ray spectrum microanalysis by means of an energy dispersive analyzer (INCA 450), mounted on a JSM 6,490 LV microscope, for points on the surface of a fresh fracture of binary Al-Mg alloy activated by additions of 3 wt.% bismuth and containing 1.5–3.0 and 5.0 wt.% magnesium

Spectrum	Al	Mg	Bi	Total
95.5 wt.% Al–3.0 wt.% Bi–1.5 wt.%Mg; the points are shown in Fig. 1.9a				
Spectrum 2	95.19	1.25	3.56	100.00
Spectrum 3	96.93	0.82	2.25	100.00
Spectrum 5	91.30	1.68	7.02	100.00
Spectrum 6	98.99	0.57	0.44	100.00
Spectrum 7	98.64	0.65	0.71	100.00
Overall spectrum	86.21	3.36	10.43	100.00
94.0 wt.% Al – 3.0 wt.% Bi – 3.0 wt.%Mg; the points are shown in Fig. 1.9b				
Spectrum 2	26.61	7.47	65.92	100.00
Spectrum 3	86.73	1.89	11.38	100.00
Spectrum 4	96.33	0.81	2.86	100.00
Spectrum 5	46.88	3.93	49.19	100.00
Overall spectrum	83.09	4.80	12.11	100.00
92.0 wt.% Al – 3.0 wt.% Bi – 5.0 wt.%Mg; the points are shown in Fig. 1.9c				
Spectrum 2	79.38	5.62	15.00	100.00
Spectrum 3	82.59	6.03	11.38	100.00
Spectrum 4	89.86	4.23	5.91	100.00
Spectrum 6	95.48	4.29	0.23	100.00
Spectrum 7	96.77	1.64	1.59	100.00
Overall spectrum	84.29	6.24	9.47	100.00

As is seen from Fig. 1.9, the fractures of alloys show clearly globules of size 2–12 μm and the ribbed surface of Al-Mg-Bi alloy. Table 1.2 presents the results of an elemental analysis of starting ternary alloys and fresh fractures of samples of Al-Mg-Bi alloy of predetermined composition, obtained by X-ray spectrum microanalysis by means of an energy dispersive analyzer (INCA 450), mounted on the JSM 6,490 LV microscope. Comparison of the data given in Table 1.2 with the analysis points shown in Fig. 1.9 shows that almost spherical globules are rich in bismuth, and that the alloy surface is poor both in magnesium and in bismuth.

It was of interest to study the distribution of the ternary Al-Mg-Bi alloy constituents by X-ray mapping, which allows one to obtain maps of distribution of chemical elements on the surface of the sample and to bring them in coincidence with its electron microscope image. Figure 1.10 shows the microstructure of a fresh fracture of a ternary alloy of the composition: 94.0 wt.% Al, 3.0 wt.% Bi, 3.0 wt.% Mg with clearly discernible spherical globules (Fig. 1.10a) and the microstructures of the surface of alloys, which were obtained using characteristic radiation of aluminium (Fig. 1.10b), magnesium (Fig. 1.10c) and bismuth (Fig. 1.10c). In Fig. 1.10b, a nanostructured region of the alloy is clearly discernible, which consists of white nanoparticles and clusters of both free aluminium and the intermetallics Al_3Mg_2 and $Al_{12}Mg_{17}$ [44]. In Fig. 1.10c, white microdots are discernible, which are magnesium nanoparticles in the intermetallic Mg_3Bi_2 (6–8 nm) and nanoclusters, consisting of about 5–12 nanoparticles comprising bismuth and magnesium,

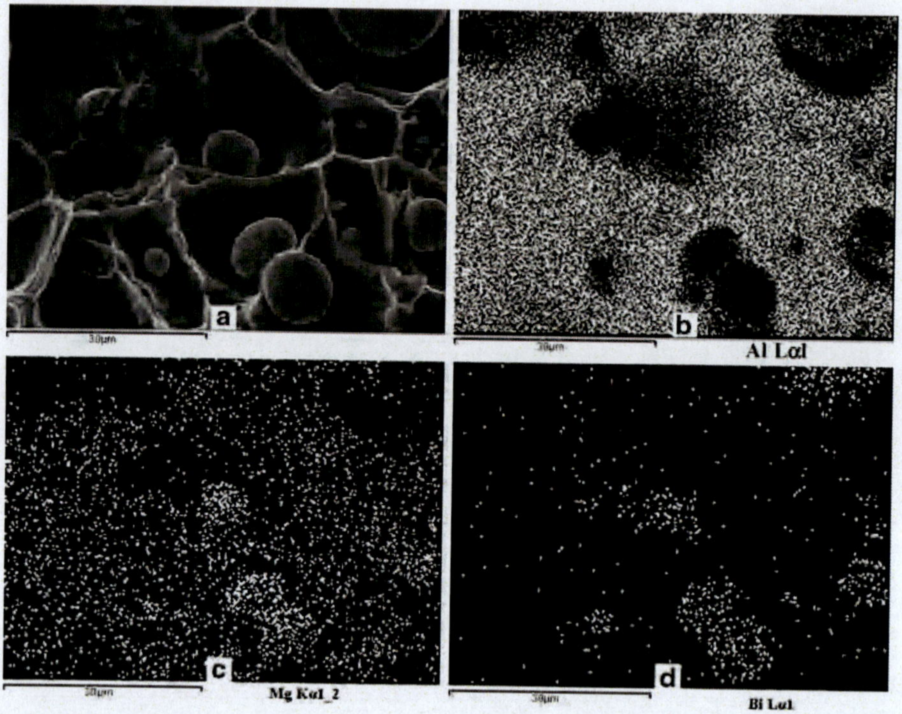

Fig. 1.10 Microstructure of a fresh fracture of aluminium, activated with additions of 3.0 wt.% bismuth and 3.0 wt.% magnesium, in secondary electrons (SEI) (**a**), characteristic radiation of aluminium (**b**), characteristic radiation of magnesium (**c**), characteristic radiation of bismuth (**d**): at ×1500 magnification

as well as magnesium bound to aluminium in intermetallics of the above compositions Al_3Mg_2 and $Al_{12}Mg_{17}$. Naturally, free magnesium nanoparticles and clusters are also present in the alloy under investigation. In Fig. 1.10d, white microdots are discernible, which are bismuth nanoparticles in nanoclusters (5–8 nm), consisting of 2–4 bismuth nanoparticles, and in Mg_3Bi_2 nanoparticles (4–6 nm). The high crystallization rate of Al-Mg-Bi alloys ensures formation of their bulk nanostructure and their high reactivity against water, as well as a high rate of corrosion dissolution of aluminium and magnesium in water.

It should be noted that the heterogeneous distribution of aluminium, magnesium and bismuth in globules, nanoclusters and nanostructured particles results in the formation of a huge number of nanosized galvanic cells (GC). These GCs have a high emf value. This is due to the fact that according to the data presented in [48], aluminium has an electronegative potential of $E^0_{Al^{3+}/Al^0} = -1.662V$; magnesium has a more electronegative potential of $E^0_{Mg^{2+}/Mg^0} = -2.363V$, bismuth has an electropositive potential of $E^0_{Bi^{3+}/Bi^0} = 0.200V$, and, as a consequence, the emf of

GCs formed in the Al-Bi and Mg-Bi alloys under investigation has high values. For instance, the emf of nanosized Al-Bi cell ΔE = 0.200 – (–1.662) = 1.862 V, and the emf of Mg-Bi ΔE = 0.200 – (–2.363) = 2.563 V. This and other above-mentioned factors account for the high efficiency of bismuth as activator of aluminium and magnesium in the reaction of hydrogen evolution from water in comparison with other metals [1, 2]. As a consequence of the formation of nano-sized GCs with high emf, the corrosion dissolution of Al-Mg-Bi alloy in water with hydrogen evolution proceeds at a high rate. The rate of hydrogen evolution from water is 3,293–4,033 L of $H_2/(m^2 \cdot min)$.

1.4 Conclusions

The kinetics and mechanism of the interaction with water of the ternary alloyAl-Mg-Bi, in which bismuth acted as an activator, which imparts a high reactivity against water to aluminium and magnesium, have been studied by high-temperature volumetry at high pressures and temperatures of 225–325°C. The kinetic parameters of the reaction of hydrogen evolution from water: rate constants, activation energies and degrees of aluminium and magnesium transformation on interaction with water by a corrosion mechanism and of the dissolution of the ternary alloy Al-Mg-Bi in water with hydrogen evolution have been calculated.

A mechanism of the corrosion dissolution of activated aluminium and magnesium in water with the formation of boehmite (AlOOH) and magnesium hydroxide ($Mg(OH)_2$) and with hydrogen evolution from water at high rates, which reach 3,293–4,033 L of $H_2/(m^2 \cdot min)$, is proposed. Nanostructured particles of the Al-Mg-Bi alloy constituents have been detected with the aid of a JSM 6,490 LV electron microscope (Japan), and the peculiarities of their chemical composition have been established.

References

1. Kozin LF, Volkov SV (2004) Hydrogen power engineering and ecology. Naukova Dumka, Kyiv, p 331 (in Russian)
2. Kozin LF, Volkov SV (2006) Modern power engineering and ecology: problems and prospects. Naukova Dumka, Kyiv, p 773 (in Russian)
3. Krylov OV (1997) Russ khim zhurn 46 (3):124–136 (in Russian)
4. Makarov AA, Fortov V. Ye.; Vesnik Rossiiskoi Akademii Nauk, 74 (3):195–208 (in Russian)
5. The statistical values (2002) Erdol-Erdgas-Kohle p 32
6. Devins D (1985) Energy. M.: Energoatomizdat 360p. (in Russian)
7. Selivanov NV (2000) Probl Nauki 7:44–51
8. Dovhyi S (1999) Visn NAN Ukrayiny 11:6–9

9. Silvestrov LK, Korenev VM (2006) Energiya; ekonomika, tekhnika. Ekologiya 9:9–12
10. Artemchuk IO, Baranovskyi MI, Bilyk SF (1997) Oil and gas of Ukraine. Naukova Dumka, Kyiv, p 383 (in Ukrainian)
11. Tikhomolova KP (1989) Electroosmosis. Leningrad: Khimiya, 248p. (in Russian)
12. Naumov AV, Zadde VV (2006) Energiya: ekonomika; tekhnika. Ekologiya 6:25–33
13. Nekrasov AS, Sinyak YuV (2006) Energiya: ekonomika; tekhnika. Ekologiya 6:2–11
14. Nekrasov AS, Sinyak YuV (2006) Energiya: ekonomika; tekhnika. Ekologiya 2:2–12
15. Bezrukikh PP (2006) Energiya: ekonomika, tekhnika. Ekologiya 6:25–33
16. Varshavskii IL (1980) Energy-storing substances and their use. Naukova Dumka, Kyiv, p 240 (in Russian)
17. Kozin LF, Volkov SV (2009) Abstracts of the XIth international conference "Hydrogen materials science and chemistry of hydrocarbon materials. ICMS." Yalta-Crimea-Ukraine, Kyiv: AHEV, pp 926–929 (in Russian) 25–31 Aug 2009
18. Parmuzina AV, Kravchenko OV (2008) Activation of aluminium metal to evolve hydrogen from water. Int J Hydrogen Energ 33(12):3073–3076
19. Kravchenko OV, Semenenko KN, Bulyhev BM, Kalmykov KB (2005) Activation of aluminum metal and its reaction with water. J Alloy Comp 397(1–2):58–62
20. Sokolskii DV, Kozin LF, Barmin VP et al (1976) USSR inventor's certificate 535364, IPC C 22C 21/00. Aluminium-base alloy for hydrogen production. Bull Invent 42
21. Sokolskii DV, Kozin LF, Barmin VP et al (1978) USSR inventor's certificate 618920, IPC[4] C 01 B 1/07. Method for hydrogen production. Bull Invent 29:195
22. Kozin LF, Sakharenko VA, Troshenkin BA (1983) USSR inventor's certificate 1108773, IPC C 22C 21/00. Aluminium-base alloy for hydrogen production. Bull Invent 30:184
23. Gulyaev BB (1980) Physicochemical fundamentals of the synthesis of alloys. L: Leningrad University Publishers, Leningrad, p 190 (in Russian)
24. Sheidlin AYe, Zhuk AZ (2006) Concept of alumohydrogen power engineering. Rus Khim Zh, 50(6):105–108 (in Russian)
25. Pin Kerton FE, Meyer MS, Meifner GP (2008) Hydrogen-generating mixed material. General Motors Corp. US Patent 7341703, IPC C 01 B 21/092, C 01 B 3/04. NPC 423/413. Published 11 Jun 2008
26. Azatyan VV, Kozlyakov VV (2003) Tyazh. Mashinostr 9:14–26 (in Russian)
27. Volkov SV, Kozin L.Kh, Honcharenko SH, Daniltsev BI (2008) Method for hydrogen production. Ukrainian Patent no 35192, IPC C 01 B 3/012. Published 10 Sept 2008 (in Ukrainian)
28. Kozin LF, Volkov SV, Goncharenko SG et al (2009) Ukr Khim Zhurn 75 (11):3–9 (in Russian)
29. Kozin LF, Volkov SV, Goncharenko SG, Daniltsev BI (2008) Abstract, special-purpose comprehensive program of scientific research of the Ukrainian NAS "Fundamental problems of hydrogen power engineering". Scientific report session. Kyiv, p 28 (in Russian)
30. Kozin LF, Sakharenko VA (1990) Zhurn Prikl Khimii 63(3):542 (in Russian)
31. Schulz R, Huot J, Liang G, Boily S (2003) Method for the production of gaseous hydrogen by chemical reactions of metals or metal hydrides. HYDRO- QUEBEC. US Patent 6572836, IPC[7] C 01 B 3/02, C 01 B 3/04. NPC 423/648. Published 3 June 2003
32. München DE (2003) Patent DE10258072 / Art des Empfangs des Wasserstoffs mit Anwemdung des amorphen Siliziums Wacker-Chemie GmbH, 81737; Dow Corning Corp., Midland, Mich, US. 11 Dec 2003
33. Avakov VB, Zinin VI, Ivanitskii BA et al (2003) Method for hydrogen storage and production by aluminum hydrolysis for self-contained power plants with electrochemical generators. Russian Patent 2260880, (in Russian)
34. Glukhikh IN (2004) Hydrogen generator, space-rocket corporation "Energiya". Russian patent 2266157, IPC[7] B 01J 7/02, C 01 B 3/08. Filed 24 Feb 2004, published 20 Dec 2005 (in Russian)
35. Tereshchuk VS (2004) Hydrogen generator. Russian Patent 2253606, IPC[7] C 01 B 3/08. Filed 16 Feb 2004, published 10 June 2005 (in Russian)

36. Lukin YeG, Dunaev YuD, Kozin LF et al (1980) Aluminium-base protector alloy. USSR inventor's certificate785371, C 22C 21/00. Bull Invent 45 p112 (in Russian)
37. Kozin LF, Anikina NS (1984) Aluminium-base protector alloy. USSR inventor's certificate 1104896, C 22C 21/100. Bull Invent 27 (in Russian)
38. Lukin YeG, Dunaev YuD, Kozin LF et al (1979) Patent 4240829 USA, Int. Cl. C22C 21/06./ Aluminum base alloy used as material for galvanic protector/ – 25 Jan 1979
39. Lukin YeG, Dunaev YuD, Kozin LF et al. Patent 1142778 Canadian, Int. Cl. C22C 21/00./ Aluminum–base alloy used as material for galvanic protector/ Lukin YeG, Dunaev YuD, Kozin LF et al – APPLICATION No 320, 559
40. InfoLine M (2009) Survey of the aluminum deoxidizers market in Russia, 61p (in Russian)
41. Kozin LF, Gorodyskii AVVA, Sakharenko VA et al (1985) Method for the manufacture of beta-alumina. Bull Invent Inventor's certificate1358329, IPC C 01F 7/04. 45, 245 (in Russian)
42. Kozin LF, Gorodyskii AV, Sakharenko VA et al (1985) Method for the manufacture of beta-alumina. Bull Invent Inventor's certificate 1192280, IPC C 01F 7/04. 42: 258 (in Russian)
43. Haughton JL (1956) Constitution of alloys bibliography. Lond J Inst Metals p 236
44. Hultgren R, Orr RL, Anderson PhD, Kelley KK (1963) Selected values of thermodynamic properties of metals and alloys–John Wiley & Sons, Inc., 963 p
45. Kozin LF (1964) Physicochemical fundamentals of amalgam metallurgy. Nauka, Alma-Ata, p 361 (in Russian)
46. Kozin LF (1992) Physicochemistry and metallurgy of high-purity mercury and its alloys. Naukova Dumka, Kyiv, p 564 (in Russian)
47. Lukashev RV (2008) Carboniferous MgH_2-based hydrogen-storing and hydrogen- generating materials, Abstract of the thesis for the degree of candidate of chemical sciences. M: Moscow State University, 24 p (in Russian)
48. Handbook of electrochemistry (edited by Sukhotin AM), Leningrad: Khimiya 1981, 486 p. (in Russian)

Chapter 2
Materials Containing Carbon Nanoparticles for Hydrogen Power Engineering

E.M. Shpilevsky, S.A. Zhdanok, and D.V. Schur

Abstract Nanostructural date of the materials alters quantitatively and qualitatively their properties in comparison with traditional materials. The nanostructured materials often exhibit unusual combination of properties, which attracts more and more researchers and provides an intensive development of this direction. The materials containing carbon nanoparticles (fullerenes, carbon nanotubes and wires, nanodiamonds, thermally expanded graphite, graphenes, etc.) occupy the prominent place among the nanostructured materials. Despite its still high cost, the nanostructured materials have already found the practical applications. It may be the most effective in hydrogen energy, biomedicine, and as the active elements of sensors. The report are based on the analysis of published data of last 10 years, the prospects of the use of materials with carbon nanoparticles in hydrogen energy as: sorbents (hydrogen storage), membranes, catalysts, constructional, electrical and decorating materials, materials for friction, sensor elements, seals and coatings are considered.

Keywords Carbon nanoparticles · Cluster · Nanotube · Hydrogen storage · Lubricant · Electron microscopy · Grain size

2.1 Introduction

Nanostructural state of the materials alters their properties quantitatively and qualitatively in comparison with traditional materials. The values of the melting temperature, the solubility limit of components (chemical elements), the kinetic

E.M. Shpilevsky (✉) and S.A. Zhdanok
A.V. Luikov of Heat and Mass Transfer Institute, NAS of Belarus,
P. Brovki str. 15, Minsk, Belarus
e-mail: eshpilevsky@rambler.ru

D.V. Schur
Institute for Problems of Materials Science, NAS of Ukraine,
Kiev, Krzhyzhanovsky str. 3, Ukraine
e-mail: shurzag@materials.kiev.ua

S.Yu. Zaginaichenko et al. (eds.), *Carbon Nanomaterials in Clean Energy Hydrogen Systems - II*, NATO Science for Peace and Security Series C: Environmental Security 2,
DOI 10.1007/978-94-007-0899-0_2,

parameters of charge transfer processes in nanostructured materials are substantially different from the bulk samples. In nanostructures the uncharacteristic of traditional materials the structural states (metastable phases) are realized. For nanostructures the band theory of solid state is not acceptable. The properties of nanostructures are being actively studied and already it is known much. A number of reviews and monographs [1–4] is devoted to the exploration of nanomaterials and to the development of nanotechnology. It is shown that with the grain decreasing the electrical, mechanical, optical and magnetic properties of materials and also the phase transition temperature, elastic module are changed. Nanostructured materials often exhibit unusual combination of properties that attracts more and more researchers and provides an intensive development of this direction. The materials containing carbon nanoparticles (fullerenes, carbon nanotubes and wires, nanodiamonds, thermally expanded graphite, graphenes, etc.) occupy the prominent place among nanostructural materials. Despite its still high cost, the nanostructured materials have already found the practical applications. They may be the most effective in hydrogen energy, biomedicine, and as the active elements of sensors.

In this paper, on the base of an analysis of published data during last 10 years, the prospects of the use of materials containing carbon nanoparticles in hydrogen energy are submitted, such as: sorbents (hydrogen storage) and catalysts, constructional and electrical materials and also the materials for internal friction, seals, sensors.

2.2 Nanoparticles and Nanostructures

The fullerenes – carbon clusters with an even, more than 20, the number of carbon atoms, forming three bonds with each other. The atoms in molecules of fullerene are located on the surface of the spheroid at the tops of hexagons and pentagons.

The fullerene molecule C_{60} has the highest symmetry and the highest stability among all fullerenes. The carbon atoms in C_{60} molecule are placed on a spherical surface at the tops of 20 irregular hexagons and 12 regular pentagons and are linked by a covalent bond. The length of the C–C in the pentagon is 1.43 Å, the same length has the side of the hexagon, which is common to both figures, but the side which is common to two hexagons has a length of about 1.39 Å. The radius of the molecule C_{60} is 0.357 nm. Fullerenes have a high chemical inertness to the process of molecular decay: molecule C_{60}, for example, is stable in an inert atmosphere up to 1,700 K. At the oxidation, one molecule of C_{60} can attach 12 atoms of oxygen. Fullerene molecules can form crystalline phases – fullerites as endohedral and exohedral structure.

Fullerenes and fullerenelike particles – are the special large and diverse type of nanoscale clusters with a stable electronic bond between the atoms, they are much more stable compared to usual clusters (conglomerates of atoms). For fullerene-like particles are carbon nanotubes, aggregates of ultrafine carbon (“onion”

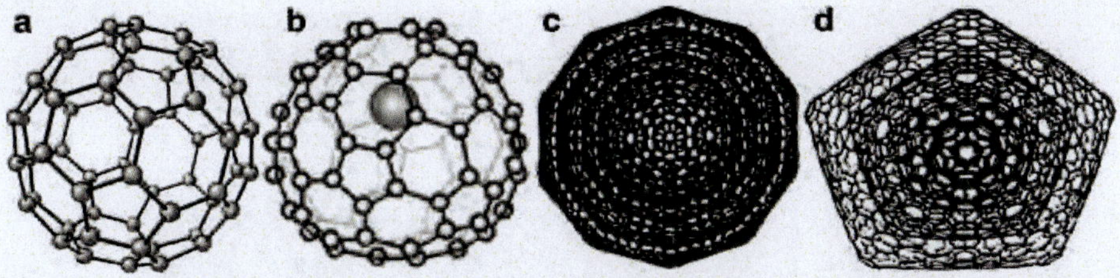

Fig. 2.1 Models of fullerenes and multilayer carbon clusters: **a** – fullerene $C_{60;}$ **b** – endofulleren; **c, d** – ultrafine carbon aggregates of multilayer type

structures, thermally expanded graphite, graphene, nanodiamonds). Ultrafine carbon aggregates (UFCA) are the association of carbon clusters.

Clusters can have a different structure: the chainlike, singlelayer, multilayer ("onion"), etc. Figure 2.1 shows the models of fullerenes and multilayer carbon clusters.

Although fullerenes are the weak acceptors (E for $C_{60} = -0.44$ v, E for $C_{70} = -0.41$ v) they form a series of molecular complexes with organic molecules of different classes. In which fullerene C_{60} acts as acceptor and organic molecule acts as donor. In these compounds the donor molecules and C_{60} are connected by Van-der-Waals bonds or π-π interactions. In crystals C_{60} molecules are packed in chains, layers or three-dimensional structures. The charge transfer from the organic molecule on the C_{60} is virtually absent. Photoexcitation of molecular complexes leads to a state with charge separation, the mechanisms of carrier generation are different.

The organization of associates from nanoclusters occurs on the same mechanisms as the formation of crystals from atoms. However, clusters have a real surface and the intercluster boundaries. Therefore, the formation of associates of the nanoclusters is accompanied by the emergence of a large number of defects and significant mechanical stresses. Defects and mechanical stresses lead to a significant change in the properties of UFCA.

Carbon nanotubes (CNT) are the graphite planes which are rolled into a cylinder, i.e. they form the surfaces from the regular hexagons with the carbon atoms in the tops. There are single-and multiwall nanotubes. The last differ from single-walled carbon nanotubes that they consist of several layers. The distance between the layers is 0, 34 nm, which corresponds to the distance between the layers in crystalline graphite. In addition, the properties of individual nanotube are determined by its chirality, i.e. by angle of orientation of the graphite plane relative to the tube axis. Figure 2.2 shows the electron microscopic images of multiwalled CNT obtained by the method of [5] and used by us to modify various materials [6].

Thermally expanded graphite (TEG) can be considered as a polymer layered graphite crystal with thickness of several atomic layers, which is characterized by the presence of active chemically "dangling" bonds. Figure 2.3a shows the structure

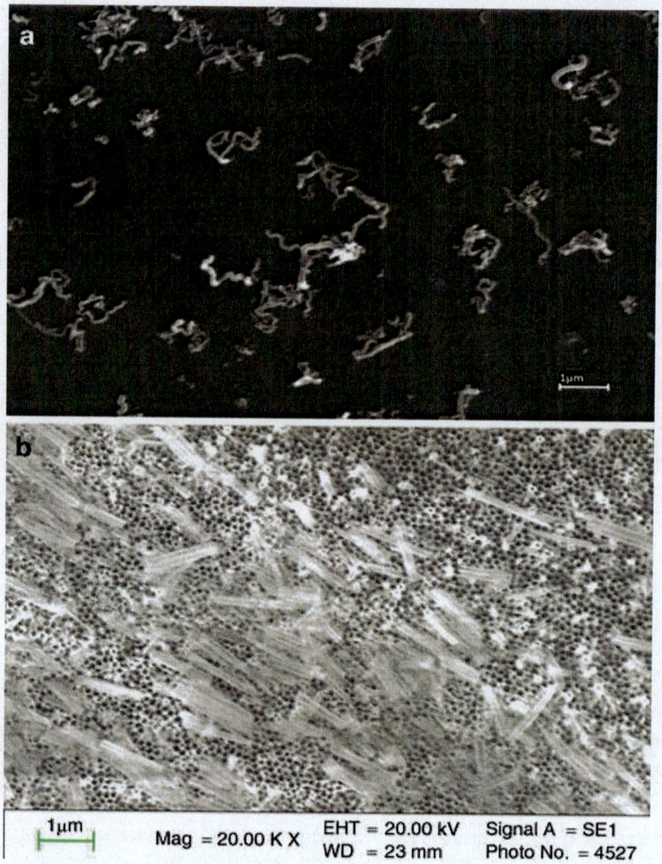

Fig. 2.2 Electron microscopy image of multiwalled CNTs used for the modification of metals

of TEG, which was received with a scanning electron microscope. TEG in our papers was used for modification of polymers [7].

In recent years the possibility of obtaining a single layer of graphite was revealed [8]. Such layers one atom thick, were called graphene. Graphene model are presented in Fig. 2.3b. The adsorbtion of the atoms of other elements on graphene surface can get new two-dimensional nanostructures. The authors of [9] managed to attach the hydrogen atoms to the graphen. This structure was called grafan. Figure 2.3c illustrates the graphan structure.

Nanostructures can be formed exposing the known materials by thermo-mechanical treatment, electrochemical treatment, by the condensation of evaporated substances in vacuum into the place of the condensation of nanoparticles, and by using other methods of action. Figure 2.4 shows the nanostructure of Al_2O_3, obtained by multistage anodization.

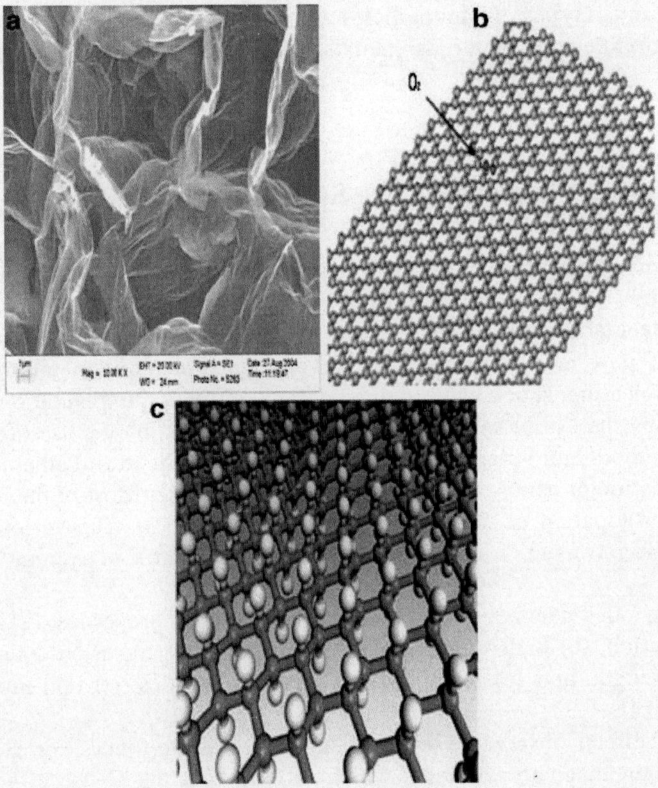

Fig. 2.3 Electron microscopy image of thermally expanded graphite (**a**) and the graphene (**b**) and grafan (**c**) models

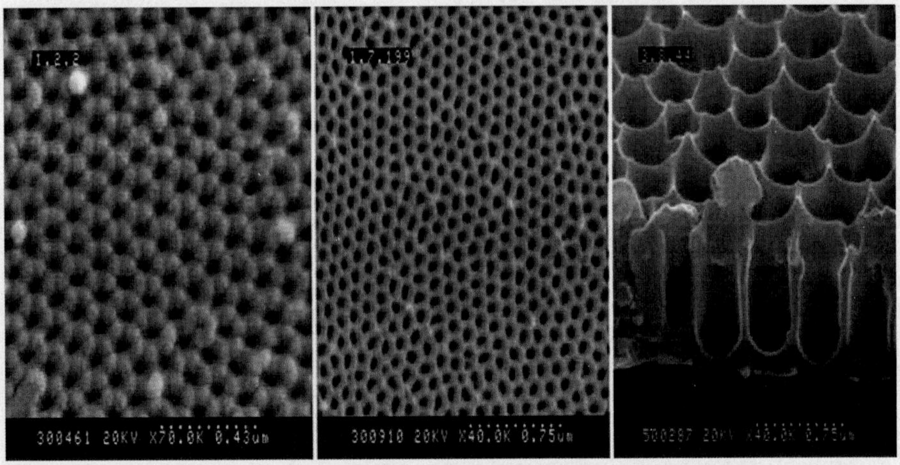

Fig. 2.4 SEM images of nanostructures Al2O3, obtained by multistage anodization

The presented types of nanoparticles and nanostructures are the convenient and effective modifiers of materials (metals, semiconductors, dielectrics, polymers, ceramics).

2.3 Key Achievements in the Study of Nanomaterials

To date, many methods for obtaining of nanomaterials and nanostructures are developed [10, 11]. We say only the most common: from the combined atomic-molecular beams, in electric arc, by the electrochemical deposition, with the help of powder metallurgy and conventional (vacuum) metallurgy, by the diffusion annealing, shock wave and dynamic loading, ion implantation, mechanical action, by the use of self-organized natural matrices, by the use of Langmuir-Blodgett method, multi-stage anodizing of aluminum, silicon and other substances, by the formation of nanoscale objects through solid mask. Most of them are tested with the participation of the author of this work [12, 13]. The various methods to ensure the introduction of nanoparticles into the matrix of various substances are tested.

It is shown that nanomaterials possess several unique properties: (a) low coefficient of friction, (b) high wear resistance, (c) increased corrosion resistance, and (d) high strength, and (e) increased the elastic properties; (f) non-linear optical properties [14].

The new effects observable only in nanostructures are found. For example, the effect of giant magnetoresistance, which differs from the Gause effect by size, sign and mechanism of carriers dispersion; the effect of the plasmon resonance absorption observed on the metal nanoparticles; the quantum Hall effect, and other dimensional effects [14–16]. The features of displaying of known physical effects in nanostructures (Hall, Gause, Seebeck, other galvanomagnetic, tenzoelectric, thermoelectric effects) are shown [17–19].

It is shown that fullerenes as a special type of carbon molecules can enter into chemical interaction with metals, even those that do not form carbides (Cu, Sn) [20]. Scientific instrumentation provided an opportunity for observation of nanoscale objects by creating a high-resolution electron microscopes, atomic force and tunneling microscopes. The technologies of industrial production of fullerene C_{60}, carbon nanotubes, TEG are developed [21, 22].

The ability of fullerite to polymerize at high temperatures and pressures, as well as the impact of radiation is established [23, 24]. Ferromagnetism of rhombohedral phase of C_{60} is discovered [25].

It is shown that the introduction of fullerenes in material even in small proportions (up to 1.0 wt.%) significantly (in some cases in several times) changes their physical and physico-chemical properties [26].

Experimentally proved that the diffusion processes in metal-fullerene structures have several features: (a) high partial diffusion coefficients of metal atoms, and (b) change in concentration of components in the surface layer during annealing,

(c) the formation of conglomerates of metal on the structure of fullerite defects, including vacancies (d) high rates of migration of fullerene molecules of C60 on metal surfaces (10^{-8} $sm^2 \cdot s^{-1}$) [27].

2.4 Materials Containing Carbon Nanoparticles for Hydrogen Power Engineering

To create new materials, fuel elements, environmentally harmless engines and power plants which are running on hydrogen fuel, the great scientific potential and significant financial investment are already exploited. In the search of new materials the materials containing carbon nanoparticles occupy a special place [28, 29].

2.4.1 *Sorbents, Accumulates, Catalysts, Membranes*

Carbon nanoparticles and structures since the discovery of ways to get them have attracted the attention of scientists as potential sorbents, batteries, catalysts, membranes. The particular interest have been shown to the fullerenes C60 and carbon nanotubes [30–34].

The process of adsorption is significantly affected by the electronic structure of surface atoms, which, in turn, depends on the size of the nanoparticles. Figure 2.5 shows the dependence of the rate constants of adsorption of nitrogen molecules (molecules/second) on the number of niobium atoms within the cluster. As can be seen from the figure, the rate of adsorption has a complex dependence on cluster size. This is due to the fact that the union of atoms in the nanoparticle is accompanied by a decrease in the interaction energy of the electrons and nuclei. The minimum energy of interaction is possible with a certain arrangement of atoms.

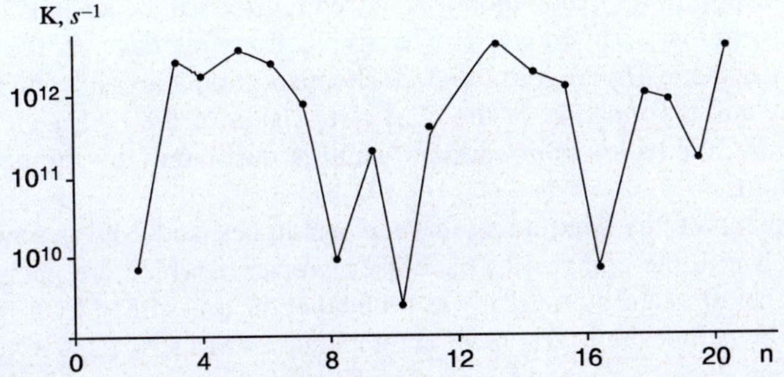

Fig. 2.5 The dependence of the rate constant of N2 adsorption (molecules/second) from the number of niobium atoms in the cluster

For this the size of the cluster is not important, but the redistribution of electrons at the changing of the number of atoms in the cluster and the surface area of the cluster.

Due to the high specific surface area and good sorption characteristics the carbon nanostructures can adsorb on their surface and enough strongly retain the metal nanoparticles, which are highly efficient catalysts. In [32] the methods of applying Pt clusters on the surface of carbon nanofibers have been developed. According to the data of transmission microscopy the size of Pt clusters, deposited on carbon nanofibers, are at an average of 5 nm. The obtained Pt/CNF catalysts were investigated in the membrane-electrode units of the hydrogen-air fuel elements. The measurements showed that the maximum power at the anode is 105 mW/cm^2, that indicates the prospects of using of these electrocatalysts.

The carrier for the electrocatalysts must satisfy the following requirements: to provide the high electrical conductivity and the availability of the reagents to the catalyst surface, to have the high corrosion resistance. At last time, the nanostructured forms of carbon are actively studied as a carrier of catalytic particles [32, 33].

The membrane properties of nanoporous materials are determined by both adsorption and diffusion processes. The study of these characteristics is especially important for creating a new generation of membranes in which the selective layer contains carbon nanotubes, nanostructured graphite and other carbon nanostructures. The contradictory of few published data related to the behavior of gases and liquids inside carbon nanotubes (CNT) and other nanostructures, are apparently conditioned by the fact that it does not take into account the important questions about the nature and mechanisms of adsorption and diffusion kinetics. The analysis for hydrogen, conducted in [34–37], shows that the unusual properties of membranes based on carbon nanotubes and other carbon nanostructures (the enhanced diffusion compared with the normal in the regime of Knudsen and the selectivity of 10 . . . 20 for a pair of methane – hydrogen which does not conform to this regime) which have been found by some researchers, it seems, can be explained by the diffusion models and characteristics, taking into account the possible chemisorption processes for hydrogen in graphite and related carbon nanomaterials with sp2 -hybridization [34]. “Anomalous” Knudsen regime can be attributed to monolayer chemisorption of hydrogen (for example from methane in the case of a mixture of methane, hydrogen) on the chemisorption carbon centers. In this case there is an amplification of the elastic repulsion of the hydrogen molecules (or, even more, methane) from carbon “walls”, “decorated” by the chemisorbed hydrogen atoms.

By treatment of the mixture of carbon nanotubes and MgH_2 powders in a planetary ball mill the composites have been prepared, which are perspective for hydrogen generation and storage. It was found that the pressure of hydrogen, which is separated by composite MgH_2-CNT at temperatures of 150°C, 250°C and 335°C, are higher than that of activated magnesium hydride. The process of hydrogenation of the composites after desorption proceeds more fast than for high dispersive magnesium. MgH_2-CNT composites have high activity in the reaction with water and acid solutions, which allows to use this reaction to create hydrogen accumulators.

Thus, the activity of the catalyst depends on the size of the cluster. A carbon nanostructures can be used as a carrier of metal catalysts and membranes, as components of hydrogen storage.

2.4.2 *Construction Materials*

At the joint condensation of metal and fullerene, the size and the shape of the alloy grains depend on the type of metal, its concentration and the temperature of the substrate. Thus, for the system Al-C60 the grains mainly take the form of pyramids, for the system Cu–C_{60} – elongated domes, and for the system Ti–C_{60} – hemispheres. Their linear dimensions are 30 ... 3,000 nm [13]. Homogeneous metal films depending on technological conditions may have a grain size of 80–3,000 nm. With the increasing of concentration of C_{60} molecules in the atomic-molecular flow the grain size initially decreases. By exceeding the volume concentration of fullerene 50%, the grain size of the alloy increases with the concentration of molecules of C60. The structure of the films is determined by the conditions of their formation (composition, substrate temperature, the density of an atomic cluster flow).

At the changing of the share of the components in the films the grain size decreases with increasing heterogeneity. Figure 2.6 shows the size dependence of the structural particles on the number of fullerene molecules per one metal atom.

The nanomaterials are characterized by the dependence of the mechanical properties on the grain size [38]. For example, the microhardness is proportional to the limit of the fluidity σy, whose dependence on the grain size (d) is determined

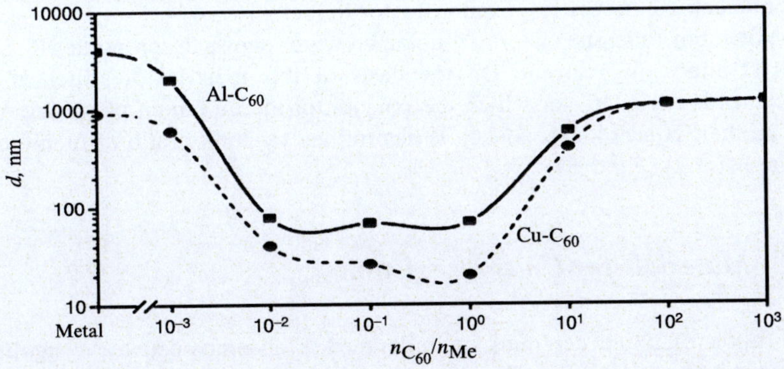

Fig. 2.6 The dependence of the average grain size on the number of fullerene molecules per 1 atom of metal

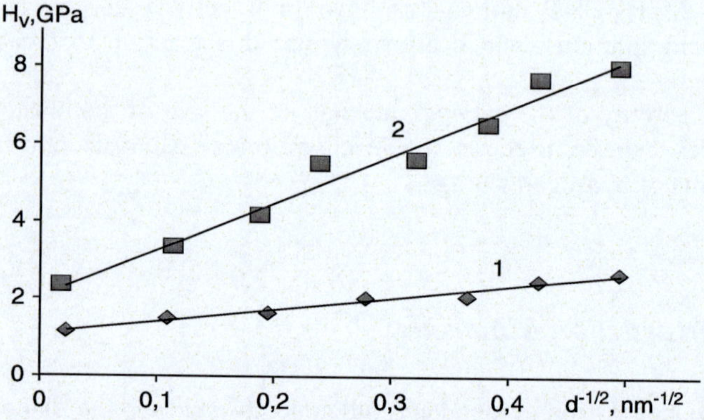

Fig. 2.7 The influence of grain size on the microhardness of metals: 1 – for copper, 2 – for Armco-iron

by the law of the Hall–Petch: σy = σo + kyd−1/2. Figure 2.7 shows the dependence of microhardness of copper and Armco-iron on grain size.

The significant increase in Young's modulus and strength of polymer material can be achieved by the introduction of nanoparticles (eg, fullerenes or CNTs) into polymer matrix. For example, the addition of 1% wt. CNTs in polystyrene composite leads to an increase in Young's modulus by 40%, and in the rupture strength – at 25% [39, 40].

2.4.3 *Electrotechnical Materials*

Any modern power plant, including running on hydrogen fuel, requires the equipment with electrotechnical devices, and consequently, the materials with different electrical characteristics. Metal-fullerene materials have the widest range of electrical properties: from insulator to metal [6, 14, 41]

The titanium-fullerene material is obtained which shows the properties R–C–L – chains at alternating current. On the basis of this material the filter of high frequencies is designed, for which the position of the minimum of electric resistance on the frequency dependence is determined by equity ratio of titanium and fullerite [41, 42] (Fig. 2.8).

2.4.4 *Materials for Units of Friction*

Modification of oxide ceramics by fullerenes is accompanied by a significant increase in its wear durability. The ceramics, containing 0.1–0.6 wt.% of fullerenes has the best anti-wear properties. In this case, the wear rate of oxide ceramics is reduced to about 15 times in comparing with the original unmodified state.

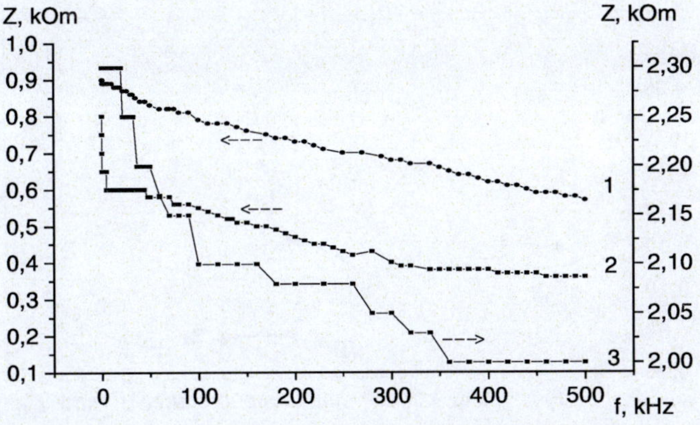

Fig. 2.8 Dependence of the changes in electrical resistance of the films on the frequency at different concentrations Ti/C_{60}: 1 – nTi/nC_{60} = 680, 2 – nTi/nC_{60} = 270, 3 – nTi/nC_{60} = 140

The increase of fullerene concentration in the ceramics up to 1.2 wt.% does not lead to such significant increase in its durability. Mass wear rate of the modified oxide ceramics in this case is 0.8 • 10–5 mg/m, that is only 3 times lower in comparison to the unmodified state. Thus, the modification of ceramics by fullerenes at the concentration 0.1–0.6 wt.% provides the marked improvement in wear resistance of oxide-ceramics coating. The reduction of tribological properties of oxide ceramics at content exceeding 0.6 wt.%, may be associated with the possibility of the formation in this case of the sufficiently large crystals of fullerite. Coming out the pores of the oxide ceramics and getting into the zone of frictional contact, such particles may have (under certain conditions) the properties inherent in abrasives [43], and, consequently, may increase the rate of the wear of the surface of triboconjugation.

The studies of elemental composition on the base of the intensity of characteristic X-rays have discovered in thin layers, formed on the surface friction of ceramic coating, carbon, iron and oxygen, and, surprisingly, the absence of aluminum. It turns out that the iron atoms not only actively diffuse into the modified oxide ceramics, but at the same time replace aluminum. The atomic ratio of elements in the formed layers is Fe: O: C = 1:2:3, or one molecule of C60 per 20 iron atoms and 40 oxygen atoms.

The appearance of these layers coincides with the sharp decrease in the coefficient of friction from 0.13 to 0.08 at the friction distance L = 300 m (see Fig. 2.9).

The stopping of the motion and the holding out the friction units under the pressure changes the friction coefficient, that indicates on diffusion processes and phase formation. The filling of the ceramic coating by fullerenes has also a significant influence on his wear strength.

Figure 2.10 shows a diagram of changes in the wear rate of modified and unmodified ceramics depending on pressure on the unit of friction.

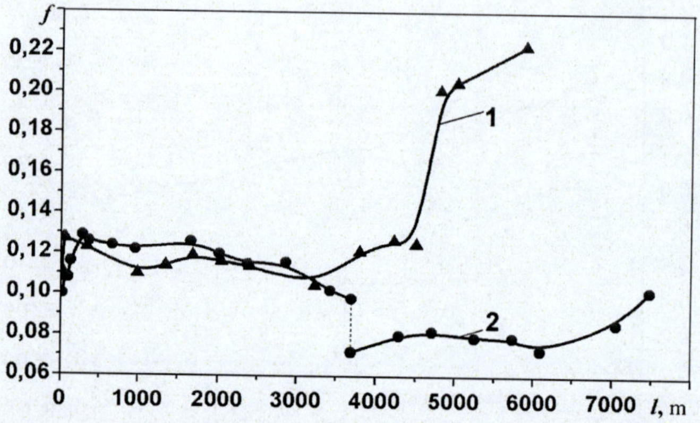

Fig. 2.9 Changes in the coefficient of friction from duration of a test at a pressure of 45 MPa

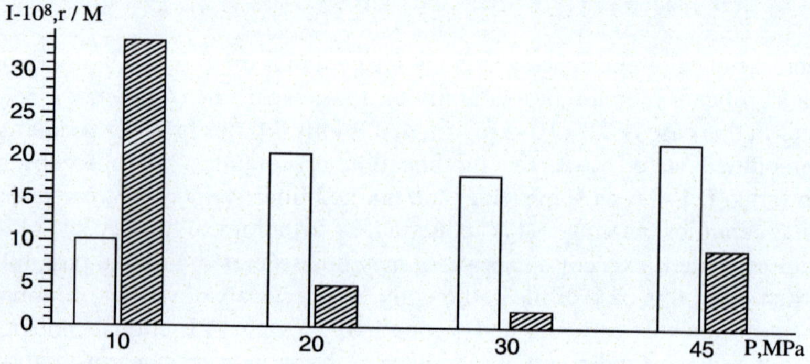

Fig. 2.10 The wear rate of the modified and unmodified ceramics depending on pressure on the unit of friction: P = 10, 20, 30, 45 MPa

The dependence of the wear for the coatings which were modified with C60 fullerenes on the contact pressure has the "extremum" character.

For the investigated pressure the minimum wear rate ($2.05{\cdot}10^{-8}$ g/m) is recorded at a pressure P = 30 MPa, which should be associated with the formation of iron-oxide fullerene surface layers (Fig. 2.2). The excess or reduction of this pressure is accompanied by the increasing wear, the most value of which equal to $33.6{\cdot}10^{-8}$ g/m and was recorded at a pressure of 10 MPa, which is associated with mechanical destruction of the layers and abrasive wear.

2.4.5 *Materials for the Sensitive Elements of Sensors*

On the base of materials which have been modified by carbon nanoparticles and metal carbon complexes the variety of sensors to determine the pressure, the flow rate of liquid or gas, temperature, optical properties, mass, pressure, strength, and also devices for determining various chemical substances can be created [44–46].

Sorption sensors. Metal-fullerene films are the good sorbents. Our studies of the electrical properties of thin films of Cu – C60 with different composition (the ratio of copper atoms per fullerene molecule NCu: NC60 have been changed) showed a high sensitivity of their electrical resistance to the adsorption of oxygen [45].

These changes in electrical resistance (in the tens of percentages) point out the good prospects for the use of such structures as sorption detectors.

Using tenzoelectrical effect the tensor sensors can be built on the base of metal-fullerene films. Metal-fullerene films have a high coefficient of tensor sensitivity, it is more than 10, while the highest for metals (platinum) is 1.6.

Fullerenes possess photoconductivity in the range of wavelengths from 280 to 680 nm. The probability of formation of electron-ion pair at the absorption of one photon is 0.9. On the basis of fullerene and metal particles structures of two types can be created: islet (i.e., with isolated inclusions of metal) and network (i.e., with interconnected metal inclusions). Such structures with periodicity much less than the wavelength of electromagnetic radiation behave as photonic crystals with forbidden photonic band. There is a significant change in plasmon frequencies in such structures.

At the study of the transmission spectra of nanostructures fullerite C60, copper and C60 – Cu in the visible and near infrared spectrum it is revealed that the spectral position and the intensity of the resonance plasmon absorption depend on the parameters of nanostructures C60–Cu, the conditions of their production and storage period on the air.

The formation of phases in the metal-fullerene structures [46] allows to achieve their desired characteristics and high selectivity of adsorption by technological methods. On these parameters it can be seen that the fullerene materials are perspective for the photoelectric sensor devices.

The external electric and magnetic fields change the electrical properties of the metal-fullerene films thanks to the interaction with electrons. This allows the use such films as sensors not only to determine the values of external influence, but to fixate the positions, deformation values and others. Fullerenes interspersed into the metal matrix, may serve as sensors of weak electronic and electromagnetic flows, deformations, force fields, in addition to other used materials. That expands the range of possible approaches and solutions.

Sensors for the determination of small external influences play an increasing role in various areas, which require precision, low cost, speed and efficiency of obtaining information. The use of nanoparticles as a specific electrochemical label is the new platform to enhance the effectiveness of sensors for determination of the small external influences.

Using different properties of nanostructures (electrical, tenzoelektrical, galvanomagnetic, thermoelectrical, photovoltaic, optical, adsorbtion, diffusion, chemical, biochemical) the sensitive elements of the sensors of new generation can be developed. The typical features of such sensors will be: the high localisation (from several to tens of nm), fast response time (at least an order of magnitude faster than the classical solid-state sensors), the highest sensitivity to virtually all physical, physicochemical, chemical and biochemical parameters, the reducing of the influence of it's presence on the object of the control (on the orders).

Nanostructures, particularly carbon nanotubes and metal nanoparticles have the excellent catalytic properties. Their introduction to electrochemical sensors reduces the strain of many important for the analysis electrochemical reactions, and even allows to carry out some redox reactions in the opposite direction, that it is impossible for classical non-modified electrodes [45]. Such properties of metal-carbon complexes at their use in sensors increase the sensitivity and selectivity, and also reduce the response time [46, 47].

2.4.6 Lubricants

To modify the lubricants the carbon nanoparticles represent the considerable interest, thanks to their unique complex of physical and mechanical properties [6, 48]. The use of lubricants, modified with carbon nanoparticles, reduces the wear rate of the test material in 1.4–1.8 times. At using lubricants modified with carbon nanoparticles the wear resistance of materials increases with increasing of specific load in the friction zone. The modification of oils by carbon nanoparticles allows to stabilize the friction coefficient with increasing temperature and raise to 20–30% upper the limit of working temperatures for modified oils [49, 50].

2.5 Conclusions

The research area related to the obtaining of new substances and materials in the base of fullerenes and other carbon nanostructures is successfully developing. The effective methods to obtain nanostructured materials containing carbon nanoparticles and nanostructures are developed. The structure, optical, magnetic properties, photoconductivity, processes of photoseparation of charges, the possibility of practical use of these materials are investigated. The introduction of carbon nanoparticles or nanostructures into the materials, even in small proportions (up to 1.0 wt.%) significantly (in some cases in several times) changes their physical and physico-chemical properties.

The use of carbon nanoparticles and nanostructures in multifunctional materials in engineering (including the hydrogen power engineering) often requires the chemical modification of their surface. To improve the interaction and compatibility of nanoparticles and nanostructures with the matrix material it is efficiently to apply the fluoridation.

Various methods for modification of carbon nanoparticles and nanostructures – oxidation, fluorination, hydrogenation, and the addition of free radicals and other highly reactive molecules are developed.

References

1. Poole C Jr, Owen F (2006) Nanotechnology. Technosphere, Moskva, 336 p (in Russian)
2. Grechihin LI (2004) Physics nanoparticles nanotechnology. Tehnoprint, Minsk, 399 p (in Russian)
3. Chaplygin YA (ed) (2005) Nanotechnology in electronics. Technosphere, Minsk, Collective monograph 448 p (in Russian)
4. Borisenko VE, Tolochko NK (eds) (2008) Nanomaterials and nanotechnology. Izd. Center BSU, Minsk, Collective monograph 375 p (in Russian)
5. Drozd AS, Matyushkov VE, Stelmach VF, Shpilevsky EM (2001) Arc plant for the production of fullerene-containing product. In: Fullerenes and fullerene-containing materials. Tehnoprint, Minsk, pp 143–149 (in Russian)
6. Shpilevsky ME, Shpilevsky EM, Stelmach VF (2001) Fullerenes and fullerene structure the basis of promising materials. J Eng Phys 74(6):106–112
7. Dlugunovich VA, Zhumar AYu, Shpilevsky EM, Lisovskaya GB (2007) Electron beam irradiation influence on polarization characteristics of He-Ne laser radiation scattered by polystyrene films with carbon nanoparticles. P SPIE 6732:673216–673220
8. Novoselov KS et al (2004) Electric field effect in atomically thin carbon films. Science 306:666–671
9. Elias DC, Nair RR, Mohiuddin TMG, Morozov SV, Blake P, Halsall MP, Ferrari AC, Boukhvalov DW, Katsnelson MI, Geim AK, Novoselov KS (2009) Control of graphene's properties by reversible hydrogenation: evidence for graphene. Science 323:610–613
10. Siegel RW, Fougere GE (1995) Mechanical properties of nanophase metals. Nanostr Mat 6 (1–4):205–216
11. Liakishev NP, Alymov MI, Dobatkin SV (2002) Nanomaterials for constructional purposes. Convers Mech Eng (6): 56–62 (in Russian)
12. Shpilevsky EM, Zhdanok SA, Shpilevsky ME (2007) E Designing metal-fullerene materials. In: Thin film electronics. Open TsNITI "Technomash, M, pp 114–120 (in Russian)
13. Shpilevsky EM, Zhdanok SA (2007) Methods of forming metal-fullerene materials. In: Modern methods and technologies of creation and processing of materials, vol 1. Ekoperspektiva, Minsk, pp 9–16 (in Russian)
14. Shpilevsky EM, Zhdanok SA (2008) Fullerenes and carbon nanotubes in modern materials science / nanotechnology in condensed matter. In: Izd. Center BSU, Minsk, Proc. SPIE. pp 231–236 (in Russian)
15. Vityaz PA, Shpilevsky EM, Shpilevsky ME (2009) Fullerene materials and functional elements based on them. NanotechnologySci Prod 2:12–16
16. Zhdanok SA, Shpilevsky EM, Shpilevsky EM, Baran LV (2009) The properties of metal-fullerene materials. Hydrogen materials and chemistry of carbon nanomaterials, XI ICHMS'2009, Yalta, 25–31 Aug 2009. IHSE, Kiev. pp 434–437
17. Shpilevsky EM (2010) Nanomaterials and nanotechnology: successes, hopes and fears. In: Modern Methods and tehnologiisozdaniya and materials processing, vol 3. PhTI of NAS of Belarus, Minsk, pp 301–308 (in Russian)
18. Fedosyuk VM (2000) Multilayer magnetic structure. BSU, Minsk, 197 p (in Russian)
19. Borisenko VE, Vorob'eva AI, Utkina EA (2004) Nanoelectronics: a manual for students. In: Transfer of charge carriers in low-dimensional structures, vol 3. Belarusian state university, Minsk, 88 p (in Russian)
20. Baran LV, Shpilevsky EM, Ukhov VA (2004) The formation of phases in the layers of copper-fullerite during annealing in vacuum. Vac Techn Technol 14(1):41–46 (in Russian)
21. Tkachev AG (2007) The eguiment and manufacture technology of nanostructured carbon materials. Hydrogen materials science and chemistry of carbon nanomaterials, X International Conference 22–28 Sept 2007. Sudak-Crimea-Ukraine. AHEU, Kiev pp 418–421

22. Zhdanok SA, Buyakov IF, Chernukho AP, Krauklis AV, Solntsev AP, Shashkov AE (2005) Fullerenes and fullerenelike structures. Heat and Mass Transfer Institute NAS of Belarus, Minsk, pp 32–40
23. Gu Z, Khabashek VN, Davydov VA, Rakhmanina AV, Agafonov VI (2006) Fluorination of crystalline polymerized phases of C_{60} fullerene. Fullerenes Nanotubes Carbon Nanostructures 14(2–3):303–306
24. Dmytrenko OP, Kulish NP, Popenko VI, Stashchuk VS et al (2008) Polimerization of the C_{60} fullerene films doped by the copper and titanium atoms. Metallophisika New Technol 30 (7):925–932
25. Vityaz PA, Shpilevsky EM, Shpilevsky ME (2009) Fullerene materials and functional elements based on them. Nanotechnology Science Prod 2:12–16
26. Davydov V (2002) Magnetically ordered state of carbon based polifullerenov C_{60}. UFN 172 (11):1295–1299
27. Bulavenko AL (2003) Nanochemistry - a direct path to high technology of the new century. Uspekhi himii 72(5):419–437
28. Shpilevsky EM (2009) Mass transfer in THE metall-fullerene structures. Fullerenes and atomic clasters, 9th Biennial International workshop. St. Petersburg, Russia, 6–10 July 2009. A.F. Ioffe PhTI RAS, St. Petersburg p 267
29. Schur DV, Matysina ZA, Zaginaichebko SYu (2007) Carbon nanomaterials and phase transformations in these materials. Nauka i obrazovanie, Dnepropetrovsk, 678 p (in Russian)
30. Rakov EG (2007) Preparation of thin carbon nanotubes by catalytic pyrolysis on a support. Russ Chem Rev 76(1):1–22 (in Russian)
31. Schur DV, Matysina ZA, Zaginaichebko SYu (2004) Hydrogen solubility in fcc fullerene. Hydrogen materials science and chemistry of carbon nanomaterials. NATO Science Series. Kluwer 172: 25–44
32. Schur DV, Zaginaichebko SYu, Matysina ZA (2007) Effects of stimulation of hydrogen solubility in fullerene C_{60}. Nanosystems Nanomaterials Nanotechnologies 5(2):371–384
33. Alekseeva DC, Kotenko AA, Chelyak MM (2007) High-temperature filters and gas separation membranes, obtained under controlled carbonization of polymers. Membranes 36(4):3–16 (in Russian)
34. Nechaev YuS, Alekseeva DC (2007) Ways of solving the urgent problem of sorption of hydrogen storage on board the car with fuel cells. Altern Energy Ecol (ISJAEE) 3:32–35
35. Lukashev RV, Klyamkin SN, Tarasov BP (2006) Properties of hydrogen-accumulating composites in the system MgH2-C. Neorg Mater 42(7):803–810 (in Russian)
36. Gerasimova EV, Volodin AA, Archangel IV, Dobrovolsky YuA, Tarasov BP (2007) Platina nanocarbon electrocatalysts for hydrogen-air fuel cells. Altern Energy Ecol 7:92–96
37. Li X, Hsing I-M (2006) The effect of the Pt deposition method and the support on Pt dispersion on carbon nanotubes. Electrochimica Acta 51:5250–5258
38. Ocampoa AL et al (2006) Characterization and evaluation of Pt-Ru catalyst supported on multi-walled carbon nanotubes by electrochemical impedance. J Power Sources 160:915–924
39. Gusev AI (2005) Nanomaterials, nanostructures, nanotechnology. Fizmatlit, Moskva, 416 p
40. Vityaz PA, Shpilevsky EM (2007) Carbon nanoparticles as active modifiers of materials. Materials, equipment and technologies in the production, maintenance, repair and modernization of machines, vol 1. PSU, Navapolatsk, pp 54–57 (in Russian)
41. Shpilevsky EM (2009) Polymers and ceramics, modified fullerenes. Mech Eng 1:10–13
42. Zhdanok SA, Shpilevsky EM (2009) Metal-fullerene materials. In: Third All-Russian Conference on Nanomaterials NANO-2009. Ural Publishing, Ekaterinburg, pp 144–146 (in Russian)
43. Shpilevsky EM, Zhdanok SA, Shpilevsky ME (2007) Modification of carbon-metal nanoparticles. In: Vacuum nanotechnology and equipmen. Contrast, Kharkov, pp 311–316 (in Russian)
44. Komarov AI, Komarova VI, Shpilevsky EM, Kovaleva SA (2008) Effect of concentration included in the ceramics of fullerenes on its tribological properties. In: Nanoparticles in Condensed Matter. Sat scientific. articles. Izd. Center BSU, Minsk, pp 30–36 (in Russian)

45. Matsumoto T et al (2004) Reduction of Pt usage in fuel cell electrocatalysts with carbon nanotube electrodes. Chem Commun 840–841
46. Shpilevsky EM, Zhdanok SA, Prokoshin VI (2004) Prospects for the use of carbon nanoparticles and related materials in the sensory elektronikike. Sensor Electronics and Microsystem Technology. Abstracts. Proceedings. Interernational Conference Sems-1, Odessa p 247–248 (in Russian)
47. Merkoci A (2008) Carbon nanotube PVC based matrix modified with glutaraldehyde suitable for biosensor applications. Electroanalysis 20:603–610
48. Angels GA, Lopez BP (2008) Enhanced host-guest electrochemical recognition of dopamine using cyclodextrin in the presence of carbon nanotubes. Carbon 46:898–906
49. Vityaz PA, Zhornik VI, Kukareko VA, Ivakhnik AV (2008) Improving tribological properties of materials modefitsirovaniem solid nanoscale components Proceedings of the National Academy of Sciences. Ser. Sci. nauk. 2008, 4: 58–62 (in Russian)
50. Vityaz PA (2009) Nanomaterials and Nanotechnology: achievements and prospects. Nanostructured Materials: Synthesis, properties, application. Minsk: a Belarusian nauka pp 5–51 (in Russian)

Chapter 3
About Fe-Graphite-H Phase Diagram Like a Scientific Base of Hydrogen Storages and Hydrogen Membranes

V.I. Shapovalov

Abstract Hydrogen-metal interaction phenomena belong to the most exciting challenges of today's physical metallurgy and physics of solids due to the uncommon behavior of hydrogen in condensed media and to the need for hydrogen power engineering and understanding hydrogen's strong negative impact on properties of some high-strength steels and alloys. The paper cites and summarizes research data on fundamental thermodynamic characteristics of hydrogen in iron-graphite system at elevated pressures and temperatures. It was shown that the iron-graphite-hydrogen system can serve very effective hydrogen storages and hydrogen membranes.

Keywords Hydrogen · Iron · Graphite · Phase diagrams · Solubility · Transformation temperature · High pressures

3.1 Introduction

A phase diagram not only represents fundamental thermodynamic characteristics of the interacting elements and phases but also provides basic information needed to engineer processes of melting, solidification and heat treatment and to determine alloy service conditions. Research in this important area has been hindered by the well-known difficulties involved in experimentation with hydrogen at elevated temperatures. With the limited amount of information available on high-temperature interactions of hydrogen with Fe-C (iron-graphite) system, a matter of great practical importance, a task was set to determine how high-pressure hydrogen can transform iron-graphite diagram.

V.I. Shapovalov (✉)
Materials & Electrochemical Research Corporation, 79060,
S. Kolb Road, Tucson, AZ 85706, USA
e-mail: vshapovalov@mercorp.com

S.Yu. Zaginaichenko et al. (eds.), *Carbon Nanomaterials in Clean Energy Hydrogen Systems - II*, NATO Science for Peace and Security Series C: Environmental Security 2,
DOI 10.1007/978-94-007-0899-0_3, © Springer Science+Business Media B.V. 2011

3.2 Experimental

A number of techniques were developed by the present writer that enable investigations into metal-hydrogen systems over broad ranges of temperature and pressure. Experimental apparatuses were made in several configurations/ modifications. The basic units (Figs. 3.1–3.4) enable heating, cooling, holding and quenching in a vacuum or in an atmosphere of hydrogen, helium, argon, nitrogen or their mixtures, together with thermal analysis and DTA at temperatures up to 2,300 K and pressures as high as 100 MPa. A refined technique for hydrogen determination was developed that improved the accuracy of hydrogen solubility measurements by an order of magnitude.

The materials used (iron and graphite) were at least 99.98% pure. Hydrogen or helium or argon in control or special runs – was pressurized above 20 MPa with a membrane pump specially designed to prevent gas contamination during pressurizing. The control runs were performed in ultra-pure hydrogen prepared by filtration of commercial hydrogen through palladium membranes.

Instruments from LECO, Balzers, Niaphot, and Cameca were used in the experiments. Also employed were conventional methods of quantitative metallurgy, hydrostatic weighing, and selective etching by thermal, chemical and electrochemical techniques, microhardness measurements and tensile tests.

The experimental procedures were modified depending on the chemical nature of the metal at hand. For this reason, sections dedicated to individual systems may contain information on features of the experimental techniques.

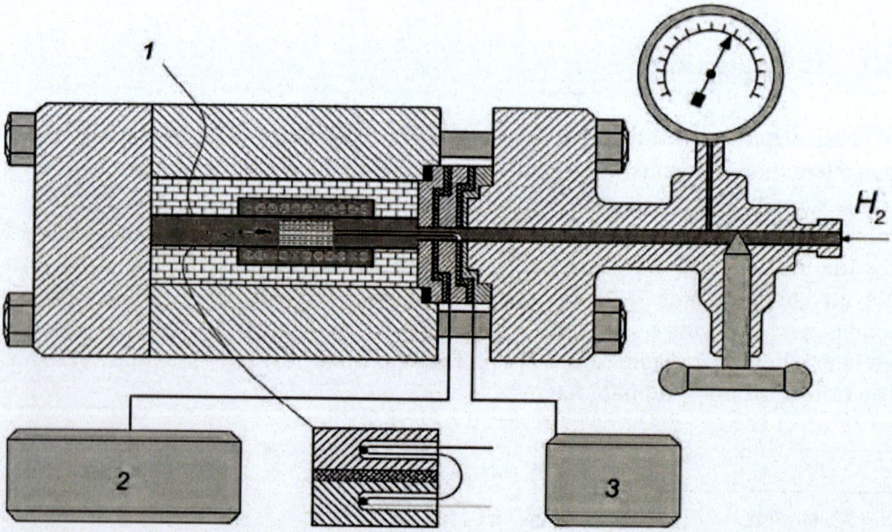

Fig. 3.1 Apparatus for studies into the effects of hydrogen on metal transformation temperatures: (1) specimen; (2) voltage regulators; (3) balance galvanometer

Fig. 3.2 General view of main experimental device

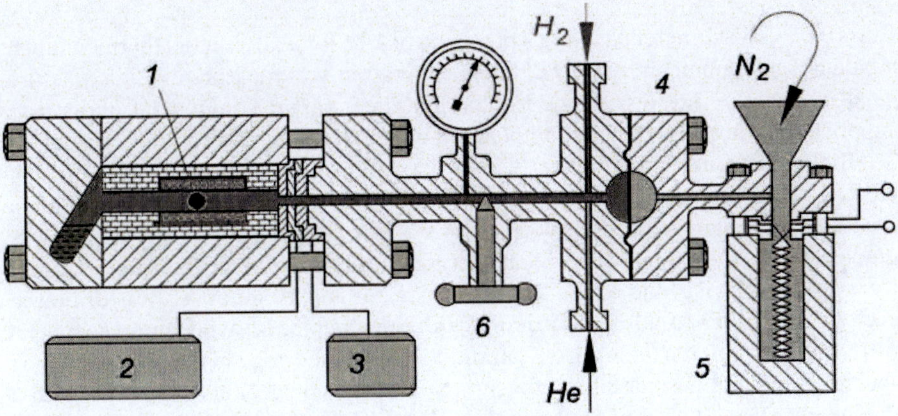

Fig. 3.3 Apparatus for charging metals with hydrogen and quenching in water: (1) specimen; (2) voltage regulator; (3) potentiometer; (4) high pressure pump; (5) liquid argon (or nitrogen) evaporator; (6) high pressure valve

Figures 3.1–3.4 show the experimental apparatuses used in the study. The high-pressure chamber made of austenitic stainless steel had a water-cooled casing and covers. Resistance furnaces with molybdenum or tungsten heating elements were used. The chamber walls had a thermal insulation of quartz washers and alumina tubes. Thermal analysis and DTA were used to determine the transformation temperatures in heating. W-Re thermocouples in alumina sheaths served as sensors. The electromotive force of a plain thermocouple was measured with a

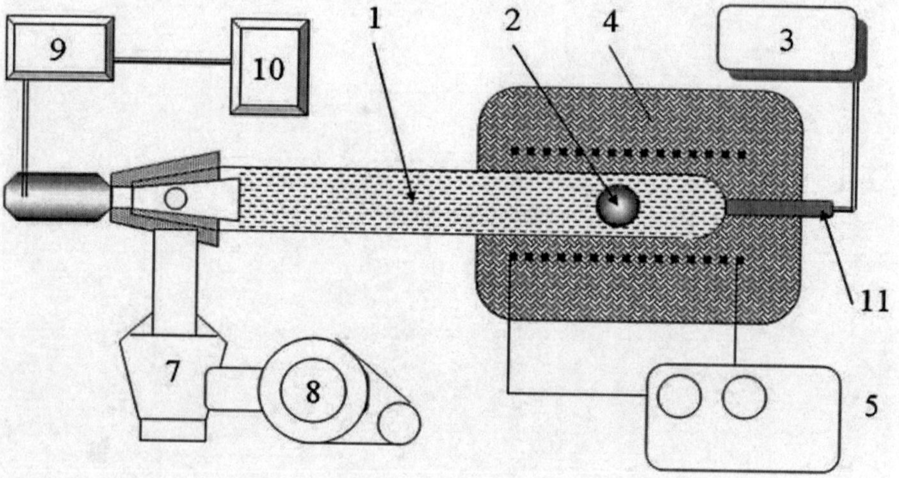

Fig. 3.4 Apparatus for hydrogen determination by cyclic vacuum extraction: (1) reaction tube; (2) specimen; (3) galvanometer; (4) furnace; (5) voltage regulator; (6) pressure sensor; (7) diffusion pump; (8) roughing pump; (9) vacuum gage; (10) recorder; (11) thermocouple

Class 0.05 galvanometer enabling a precision of 1 K; for a differential thermocouple, a balance galvanometer was used. The gas pressure was measured with a pressure gage to accuracy not less than 2%. The hydrogen gas of commercial purity was supplied in cylinders. The gas composition was checked with a chromatograph.

High-pure iron-graphite alloys with a maximum total amount of impurities at 0.002% were prepared for the experiments by induction melting in an alumina crucible in an atmospheric vacuum. An ingot about 350 g in weight was graphitizated. Then they were cut and forged to rods 5–15 mm in diameter. These were cut lengthwise, and a hole 2 mm in diameter and 10 mm deep was drilled in each of the halves to allow insertion of a thermocouple. Molybdenum specimens were used for reference. An iron sample was degassed together with a reference specimen in a vacuum at 800°C for 5 h. Next, the two specimens together with a thin quartz plate interposed for thermal insulation were placed into the experimental chamber. In order to attain the closest possible proximity to the gas-metal equilibrium, the iron specimens for some runs were composed of stacked plates 0.5mm thick made by rolling at room temperature.

The saturation time at 900°C was 300 s for a solid specimen and not longer than 10 s for a stacked specimen. At higher temperatures, the saturation time lengths were even shorter. With this in mind, the rates of heating and cooling were varied in the range from 1 to 100 K/min.

The experimental data were practically the same for the two specimen types because preliminary saturation with hydrogen was carried out in the vicinity of phase transformation temperatures. The heating and cooling curves were obtained simultaneously with differential curves. Several replications were run for each experimental point (10–15). No cooling curves were used to determine phase diagrams because of marked supercooling in phase transformations.

Hydrogen solubility was determined via quenching – so far the only technique available for high temperatures and hydrogen pressures. Figure 3.3 gives a schematic of a unit for holding in hydrogen gas and subsequent quenching. The holding temperature was maintained to a precision of 2 K. The saturation time was selected for each temperature by trial and error. When the equilibrium was attained, the furnace was rotated, so that the specimen was abruptly dropped into the quenching section filled with ice water. When the experimental temperature was above the melting point, the molten iron flowed into the quenching section via a quartz tube. The average cooling rate in water was about 300 K/s.

The hydrogen content was measured by vacuum extraction directly after quenching. In order to improve the accuracy of measurements, a special apparatus was designed, Fig. 3.4. It eliminated the main disadvantage of Sieverts method, namely equilibration between the solute and the gaseous hydrogen during the vacuum extraction. This disadvantage was corrected via cyclic degassing. On reaching an internal pressure of 10 Pa, the reaction tube was connected to the vacuum system preevacuated to 0.001 Pa. The hydrogen that released was removed in a few seconds, the reaction chamber was again isolated from the vacuum system, and the release of hydrogen was resumed under a high vacuum. The absolute accuracy of hydrogen measurement for a 450 cm^3 reaction tube was 0.03 cm^3. The corrections for possible hydrogen release during specimen quenching were negligible at less than 1%. The interval between the quenching and the degassing step was not longer than 5 min. An estimated 0.5% hydrogen could release from a spherical specimen 15 g in weight at room temperature. The bake-out was carried out stepwise: first at 300°C, next at 800°C. At least 80% of the hydrogen evolved at 300°C. The correction for residual hydrogen due to vacuum melting was 0.05 cm^3. To estimate the side effects of quenching, namely possible iron charging with hydrogen, oxygen and nitrogen during contact of hot specimen with water, the unit was filled with helium at a similar pressure. The amount of these gases was found to range from 0.1 to 0.3 cm^3 per 100 g iron.

3.3 The Iron-Graphite System

Previous research on the Fe-C-H system was mostly focused on hydrogen solubility in at relatively low pressures. On the other hand, data on how hydrogen influences the melting temperature and the allotropy are very scarce. Geller and Tak-Hosun tried to assess the Fe-H diagram for a hydrogen pressure of 0.1 MPa, using the existing data on hydrogen solubility and the general thermodynamic relationships [1]. It was concluded that hydrogen reduces the melting temperature by 1.8 K and extends the γ-region. It was observed even earlier that the α-γ equilibrium temperature drops by 4 K in the presence of hydrogen. The present

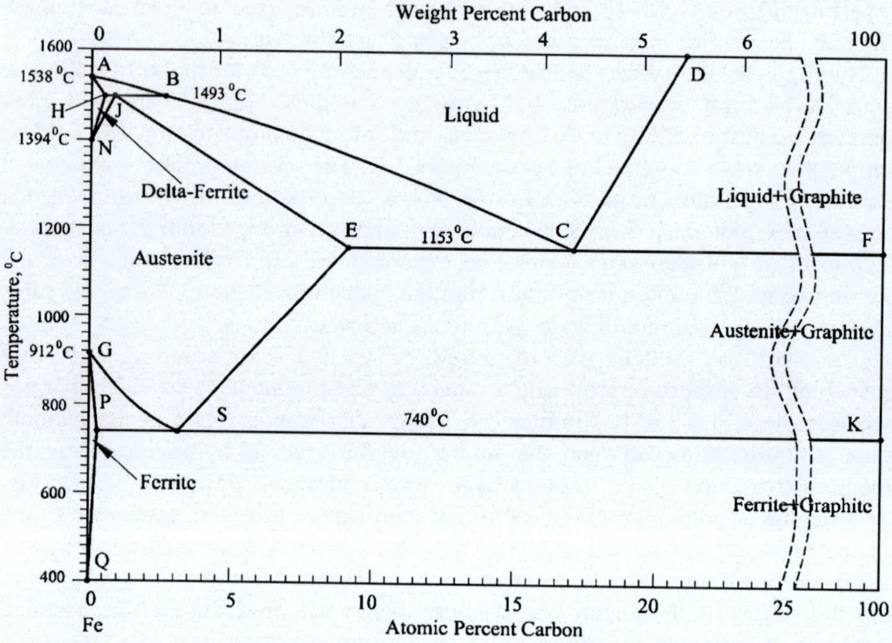

Fig. 3.5 Modern iron-graphite phase diagram

writer investigated the effects of hydrogen on iron-graphite system transformation at pressures up to 90 MPa and temperatures ranging from 500 to 1,600°C [2, 3].

Modern iron-graphite phase diagram at low hydrogen pressure is well known (Fig. 3.5). Line and point positions are almost same without hydrogen and at hydrogen pressure up to 2.0 MPa [4].

Hydrogen presenting at high temperature and high pressure dramatically decrease graphite solubility (Figs. 3.6 and 3.7). And finally at pressure upper 30.0 MPa hydrogen totally blocks interaction between iron and graphite.

So the iron-graphite phase diagram (Fig. 3.5) fully transforms under hydrogen influence (Fig. 3.8).

Graphite particles absorb hydrogen at high pressures in large amounts (Fig. 3.9) and upper 300 MPa the absorption is reaching maximum. Smaller particles absorb more hydrogen. Hydrogen %-temperature functional dependence has maximum at 800°C.

At low pressures hydrogen can escape graphite-iron composition relatively quickly at elevated temperatures and very slowly at room temperature (Fig. 3.9). So the composite material is a very promising safe hydrogen storage.

The results can be explained by very unusual graphite atomic structure and very particular hydrogen atomic size (Fig. 3.10) as well as the abnormally high energy of the C-H bond.

Fig. 3.6 Dissolution kinetics of graphite in austenite at different hydrogen pressures and 1,000°C: (1) argon 20 MPa; (2) hydrogen 10 MPa; (3) hydrogen 20 MPa; (4) hydrogen 30 MPa

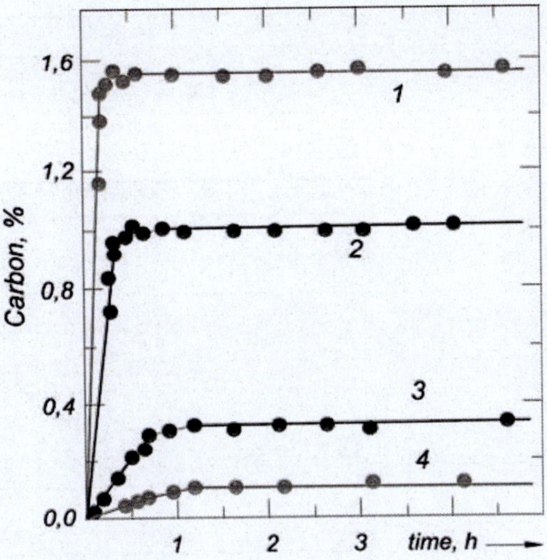

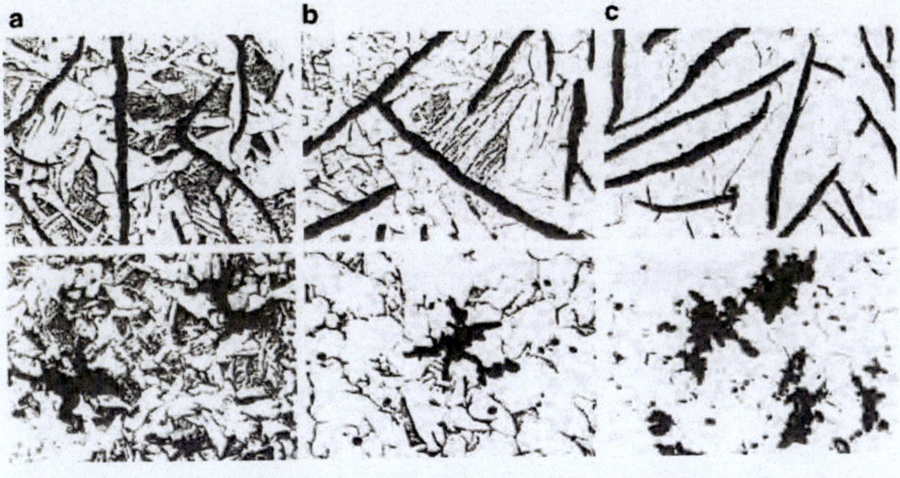

Fig. 3.7 Iron-graphite structures after 4 h at 1,000°C and fast cooling in hydrogen atmosphere: (**a**) 10 MPa; (**b**) 20 MPa; (**c**) 30 MPa (*upper row* – laminar graphite crystals; *lower row* – compact graphite crystals; *light fields* – ferrite; *black* – graphite; *other* – pearlite)

Hydrogen can interact with graphite by many ways (Figs. 3.10–3.12):

- Atomic chemical adsorption (on graphite surface)
- chemical interaction (inside micro porosity and making
- molecular adsorption (on graphite surface)
- molecular absorption (dissolution inside graphite particles)
- atomic dissolution inside graphite particles

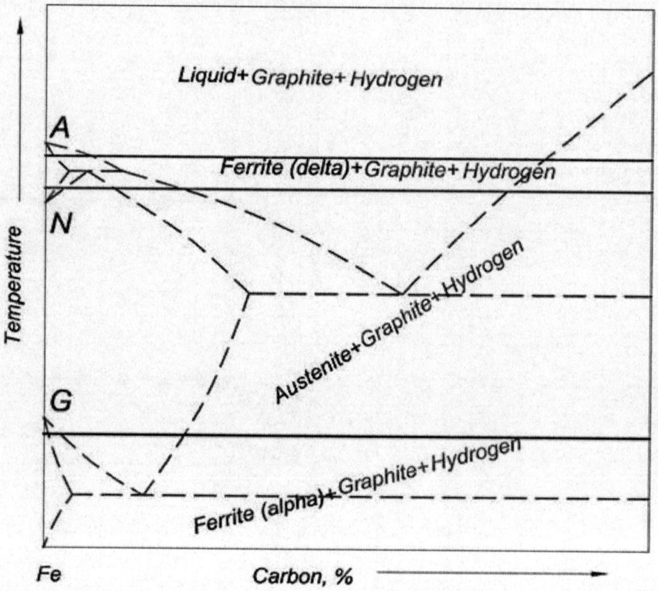

Fig. 3.8 Iron-graphite phase diagram transformation at high pressure hydrogen condition: *dotted lines* are the diagram without hydrogen; *firm horizontal lines* are the diagram at hydrogen pressure 600 MPa

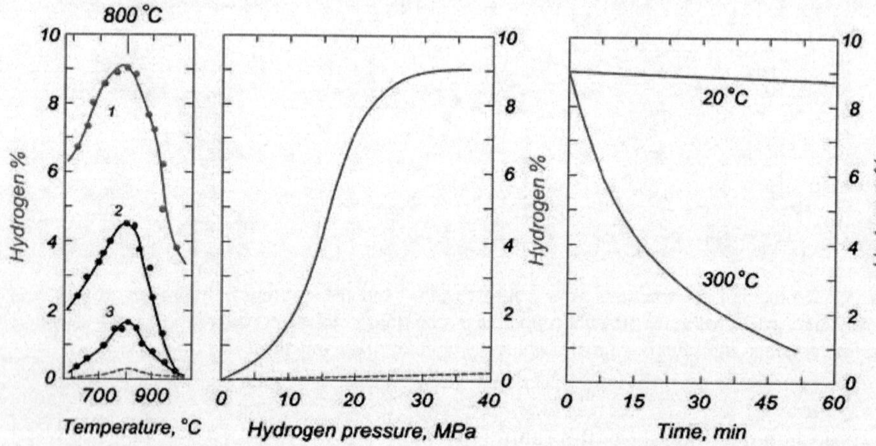

Fig. 3.9 Hydrogen absorption and desorption at different hydrogen pressures and temperatures: (1) graphite particles 1–3 μm; (2) particles 50–300 μ; (3) 250–600 μ; *dotted line* is solid graphite samples

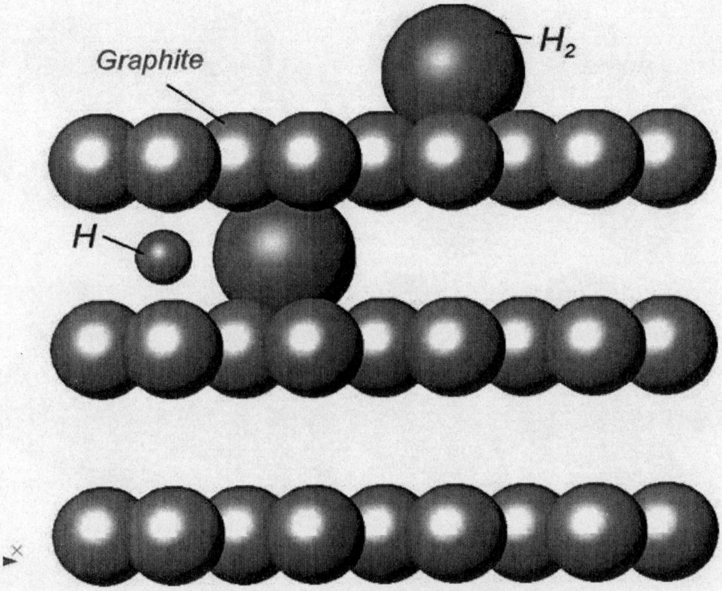

Fig. 3.10 Correspondence of atomic size in hydrogen graphite system

3.4 Conclusions

1. Hydrogen can completely block graphite crystals from iron matrix at high pressures and temperatures.
2. Hydrogen-graphite interaction is complicated and can occur through:
   - chemical adsorption
   - chemical interaction
   - molecular adsorption
   - molecular absorption (dissolution)
   - atomic dissolution
3. Hydrogen absorption by graphite crystals (in graphite iron system) depends on graphite particle size and graphite structure and can reach 8% (weight).
4. Hydrogen diffusion through iron-graphite compositions occurs two times faster than through palladium membranes with the same thickness.
5. Iron-graphite composition can serve for hydrogen storage and hydrogen purifying membranes as well.

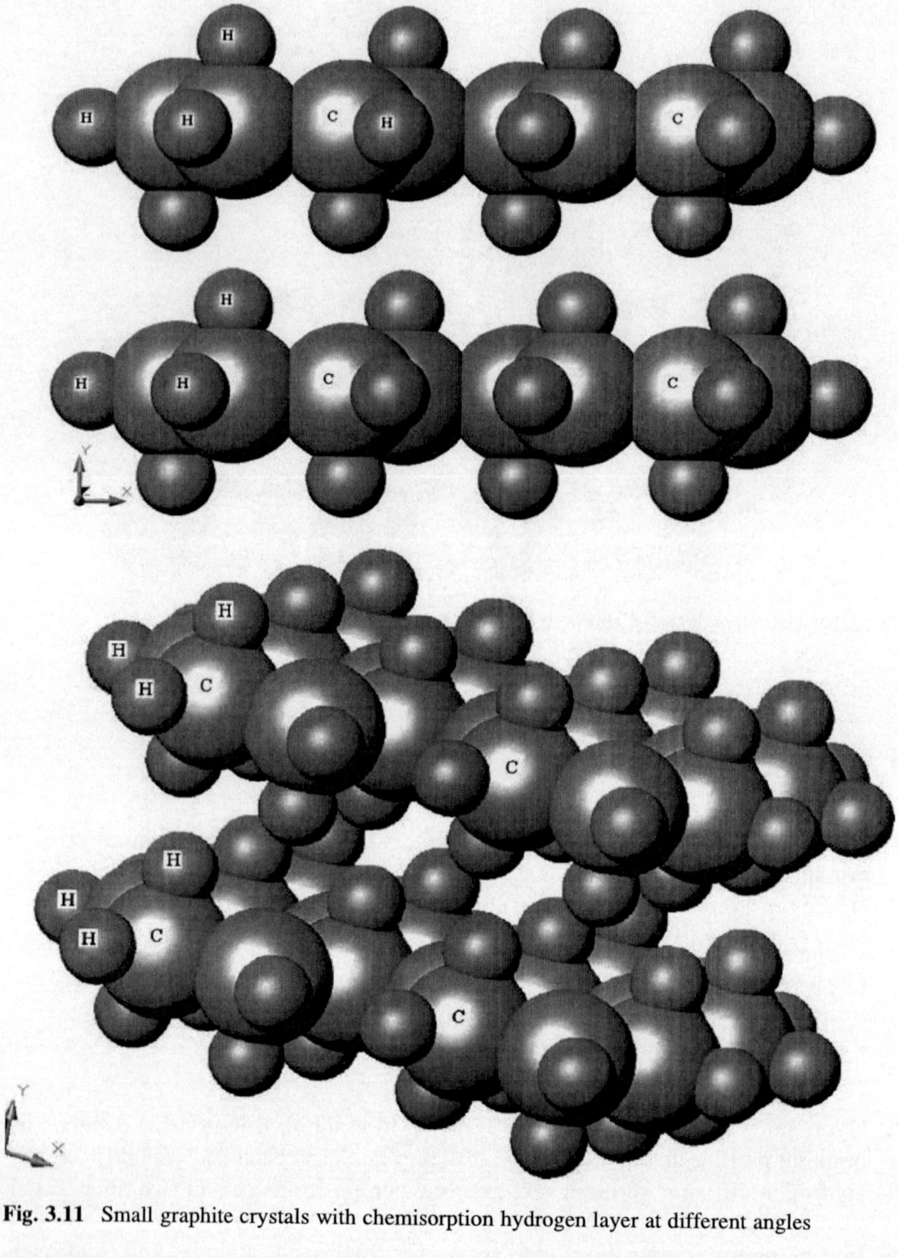

Fig. 3.11 Small graphite crystals with chemisorption hydrogen layer at different angles

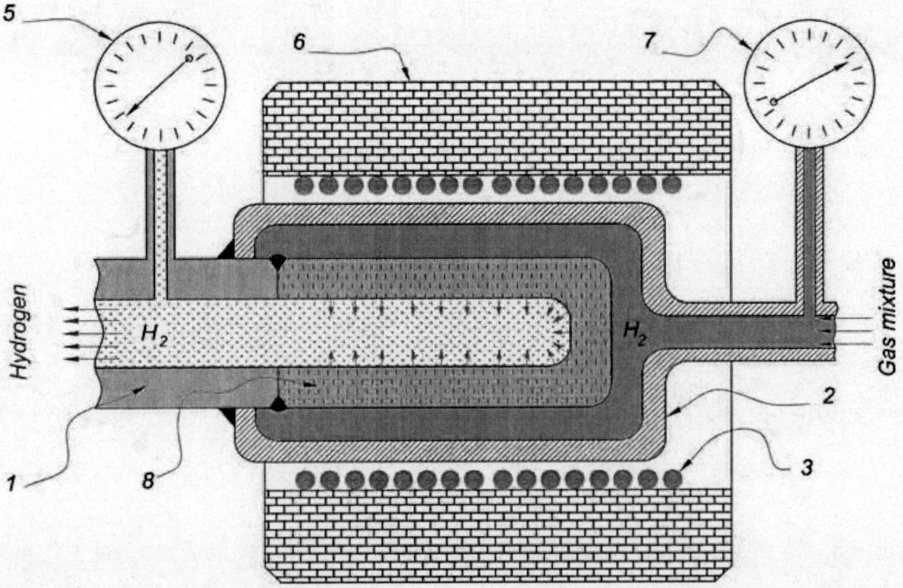

Fig. 3.12 Experimental cell for hydrogen purification: (1) stainless steel tube; (2) stainless steel shell; (3) heater; (5) low pressure manometer; (6) thermo insulation; (7) high pressure manometer; (8) iron-graphite tubular membrane

References

1. Geller W, Tak-Hosun (1950) Arch. eisenhuettenwesen, vol 21., pp 423–431
2. Shapovalov VI (1982) Effects of hydrogen on structure and properties of Fe-C alloys. Metallurgiya Publishing House, Moscow, p 235, Russian
3. Shapovalov VI (1978) Constitution diagram Fe-C-H, *Izvestiya Vuzov*. Chernaya Metallurgiya 6:117, In Russian
4. Venkatraman M, Neumann JP (1991) The Cr-H (chromium-hydrogen system). J Phase Equil 12(6):672–677

Chapter 4
Special Features and Regularities of Interaction Between Fullerene Molecules and Aromatic Solvents

N.S. Anikina, O.Ya. Krivuschenko, D.V. Schur, S.Yu. Zaginaichenko, and E.A. Kamenetskaia

Abstract The reaction of C_{60} fullerene dissolution in monosubstituted benzenes has been studied. The correlation method of physical properties of solvent molecules with reaction characteristics has been applied under the assumption of their donor-acceptor mechanism. The effect of electron-donor substituents of aromatic ring has been found.

Keywords Fullerene · Aromatic hydrocarbons · Electronic effects of substituents · Ionization potential · Solubility · Dipole moment · Electronegativity · Mesomeric effect

Nomenclature

QSPR	quantitative structure-property relationship
QSAR	quantitative structure activity relationship
HOMO	high occupied molecular orbital
LFMO	lower free molecular orbital
LCM	liquid-crystal media
DAC	donor-acceptor complex
DAI	donor-acceptor interaction
CCT	complex with charge transfer
HB	hydrogen bond
C_s	concentration of solvent
C_{CCT}	concentration of CCT
$S_{C_{60}}$	solubility of C_{60}
μ_{HB}	dipole moment of halogen benzene

N.S. Anikina (✉), O.Ya. Krivuschenko, D.V. Schur, S.Yu. Zaginaichenko, and E.A. Kamenetskaia
Institute for Problems of Materials Science of NAS of Ukraine, Krzhyzhanovsky str. 3, 03142 Kiev, Ukraine
e-mail: shurzag@materials.kiev.ua

S.Yu. Zaginaichenko et al. (eds.), *Carbon Nanomaterials in Clean Energy Hydrogen Systems - II*, NATO Science for Peace and Security Series C: Environmental Security 2,
DOI 10.1007/978-94-007-0899-0_4,

M_{HB}	mesomeric effect of halogen benzene
μ_{FB}	dipole moment of fluorine benzene
HC	hyperconjugation
φ_{AB}	ionization potential of alkyl benzene
χ	electronegativity

4.1 Introduction

The mechanism of fullerene interaction with molecules of solvents is one of the essential lines of investigations in the chemistry of fullerenes. The interest to the chemistry of C_{60} fullerene dissolution, as to the molecule with unique structure, has been generated both by the applied and fundamental significance of this problem.

4.1.1 Approaches and Physical Parameters of Solvents Using at Interpretation of Dissolution Effects of Fullerene Molecule

The process of C_{60} dissolution is considered predominantly from the positions of similarity like "similar dissolve similar" and also from the widely applied method "conceptions of molecular similarity" (CMS). The enthalpy of evaporation, polarity, volume of solvent molecule [1, 2] and others are used as the similarity factors. The capability of C_{60} molecule to be dissolved in aromatic solvents is often explained on the basis of magnetic interaction of ring currents of aromatic rings of solvent and hexahedrons of fullerene molecules [3].

An active search for physicochemical universal (multi-purpose) parameter, affecting immediately on the value of C_{60} solubility, has been carried out and polarizability – $(n^2 - 2)/(n^2 + 2)$, polarity – $(\varepsilon - 1)/(2\varepsilon + 1)$, molar volume, Hildebrand solubility parameter are seen as such parameters [1–3]. However as these investigations have shown, none of considered properties of solvent cannot predict adequately the C_{60} solubility [1] in one or another solvent. On this basis it has been proposed [1] that C_{60} dissolution is dictated at once by the going on processes, the common energetic effect of which determines the C_{60} solubility. The solution of problems of such type has been proposed within the limits of CMS conception.

The Famini-Wilson approach was used in the paper [4] with descriptors: molar volume V_m, dipole moment μ and capability to the formation of hydrogen bonds (HB) and this way of looking showed that the greater is V_m and μ of solvent molecule, the smaller is solubility, but increase of capability to give up the electron pair and to be polarized cause a rise in solubility.

In the paper [5] the equation with four descriptors was solved. Analysing the polarizability by Lorents-Lorenz $(n^2 - 1)/(n^2 + 2)$, dipole moment μ, bipolarity parameter π of Kamlet-Taft and index of solvent polarity the authors came to conclusion that solvent polarizability has a maximum influence on C_{60} solubility and dipole moment of solvent molecule has very slight effect on its. The application of combination of another parameters, as capability to formation of hydrogen bond by Kamlet-Taft, Hildebrand solubility parameters gave the better results in the paper [6].

The attempt to use the extended equation of Compel-Palm [7] was made. After elimination of some solvents differing by their physicochemical characteristics (polarity $(\varepsilon - 1)/(2\varepsilon + 1)$, basicity by Palm, electrophilicity by Reinhard and also by parameters of specific solvation) that by the authors opinion are slightly important the two-parameter equation was proposed for data generalization on C_{60} solubility. This equation involves only polarizability and density of cohesion energy defined by the square of Hildebrand parameter δ^2.

In addition to the empirical characteristics of solvent, as volume of saturated surface and average polarizability, the quantum-chemical parameters (HOMO and LFMO energies) and a number of anothers were taken into account in investigation [8]. As calculation showed, the molar volume and polarizability are parameters determined the C_{60} solubility.

The five-parameter equation of regression of Kamlet-Taft-Abraham [9] was used for the extension of results on C_{60} solubility in 20 solvents. In this case besides the solvation parameters investigators took into consideration the excessive molar refraction and coefficient of fullerene C_{60} distribution between water and solvent. These calculations showed that excessive refraction, bipolarity and capability to manifest itself as acceptor of hydrogen bond are the basic parameters of solvent.

In the paper [10] topology indexes and parameter of polarizability of solvent were used as descriptors with application of the repeated linear regression and division on groups of compounds according to their chemical nature.

An active search was performed also for universal parameter for prediction of C_{60} solubility with application of three-parameter approach of Hansen [11] taking into consideration three types of molecular interactions: dispersive, direct electrostatic and capability to form the hydrogen bond. As the result of these calculations the compositive parameter of affinity, as RED number, was proposed for qualitative assessment of C_{60} solubility. The material for which RED <1 is believed to be the good solvent of fullerenes.

4.1.2 *Conception of Dissolution Mechanism of Fullerene Molecules Proposed for Discussion*

In the present paper the process of fullerene C_{60} dissolution has been considered within the limits of laws of intermolecular donor-acceptor interaction (DAI). Such approach has been chosen on the basis that substances, dissolved C_{60}

substantially, share one common property – their molecules are donors of π-electrons. The aromatic hydrocarbons and also aliphatic hydrocarbons containing atoms with nondivided pair of p-electrons are such material. In its turn fullerene molecule C_{60} wilt unique structure, that lies in the nonsaturation of conjugated bonds and the absence of substituents, is the acceptor of π-electrons. They are restored reversibly in the fundamental and excited states [12–15]. The donor-acceptor complexes (DAC) are known, fullerene molecules C_{60} fulfil the role of acceptors in these complexes [16–20]. Because of this, in this case the main conditions necessary for the initiation of intermolecular donor-acceptor interaction are fulfilled. The existence of correlation between C_{60} solubility and electron-donor capability (EDC) of solvent molecules can serve as the weighty argument in support of the validity of selected approach.

In our opinion the process of molecule C_{60} dissolution occurs in the following manner. As the result of interaction between solvent molecules and fullerene molecules the intermolecular redistribution of electronic density has been going on that leads to its intermolecular redistribution between atoms forming both of interacting with each other molecules. As this takes place, the electronic density and consequently the chemical activity of interacting molecules of solvent will be different from the initial. The solvent molecule forming the complex will fullerene molecule will be in a position to come into interaction with the free molecule of solvent but on new conditions because it will possess the properties of the formed complex, carried in itself the influence of fullerene molecule. The process will equilibrate in the case when electron-deficient properties of fullerene molecule will be not compensated by the electron-donor properties of solvent molecules. The electron-donor properties of monosubstitutes of benzene depend on the structure and chemical composition of substituents. It is profitable to consider the influence of substituents on fullerenes solubility in monosubstituted benzene for the verification of our conception.

In the present-day chemistry the series of physicochemical and thermodynamic characteristics are used as a measure of electromotive force EMF. The constant of equilibrium (C_{eq}) of DAI reaction as well as ionization potentials (φ) of donor-molecules fit the process most adequately.

At the mechanism of reaction of fullerene C_{60} dissolution with formation of complex with charge transfer (CCT):

$$^{1}C_{60} + {}^{1}C_{p} \rightleftarrows C_{CCT} + {}^{2}C_{60} + 2C_{p}, \tag{4.1}$$

the fullerene C_{60} solubility ($S_{C_{60}}$) is numerically equal to the equilibrium constant C_{eq} of reaction, as exemplified by Eq. 4.2

$$C_{eq} = C_{CCT} \cdot {}^{2}C_{60} \cdot {}^{2}C_{s} / {}^{1}C_{60} \cdot {}^{1}C_{s}, \tag{4.2}$$

where

$$^{1}C_{60} = {}^{2}C_{60} \tag{4.3}$$

because crystalline substance is dissolved and

$$^{1}C_s = {}^{2}C_s \tag{4.4}$$

are constants owing to the low C_{60} solubility. It follows that

$$C_{eq} = C_{CCT}. \tag{4.5}$$

Here the following designations are inserted: $^{1}C_{60}$ and $^{1}C_s$ are initial concentrations of crystalline fullerene and solvent respectively; C_{CCT} is equilibrium concentration of complex with charge transfer; $^{2}C_{60}$ and $^{2}C_s$ are concentrations of crystalline fullerene and solvent in the reaction product.

At equilibrium

$$C_{CCT} = S_{C_{60}}, \tag{4.6}$$

Then

$$C_{eq} = S_{C_{60}}. \tag{4.7}$$

This provides a way of correlation between published values of $S_{C_{60}}$ (as measures of electron – donor capability of solvent molecule) and already known measures received by independent methods as well as physical parameters of solvent molecules representing their capability to give up electrons; in this case ionization potential φ(eV), dipole moment μ(D) of solvent molecules and also electronegativity χ(eV) and mesomeric effect M(D) of substituents can be such parameters.

The basis of experimental data on C_{60} solubility has been taken from works [1, 21–23].

4.2 Experimental Data

4.2.1 C_{60} Dissolution in Halogen Benzenes

Mush research is devoted to investigation of molecule C_{60} solubility in halogen benzenes that allow to make sufficiently realistic conclusions about the process course, but results of different authors are somewhat distinguished and we have used the research results of paper [23] for our investigations.

Experimental data acquired by various researchers testify that the fullerene C_{60} solubility in halogen benzenes grows in the series of substituents:

$$-F < -Br < -Cl, \tag{4.8}$$

sharply decreasing in fluorine benzene.

Table 4.1 The physicochemical parameters for monosubstituted halogen benzenes and halogen substituents [23]

No	Substituents	C_{60} Solubility S (mg/ml)	Mesomeric effect M (D)	Dipole moment μ (D)	Ionization potential φ (eV)	Electronegativity χ (eV)
1	*–F*	1.2	+0.35	1.57	9.20	4.00
2	*–Cl*	5.7	+0.32	1.73	9.07	3.00
3	*–Br*	4.6	+0.28	1.72	8.98	2.85
4	*–I*	2.1	+0.20	1.70	8.73	2.44

We have used the difference of dipole moments (μ_{HB}) of halogen benzenes (C_6H_5-X) and aliphatic (μ_{AL}) halogens (C_2H_5-X), for which the mesomeric effect is impossible [24], as the measure of halogen substituents in benzene

$$M_{HB} = \mu_{HB} - \mu_{AL}. \tag{4.9}$$

The mesomeric effect of halogen substituents, defined in such a manner, grow from iodine up to fluorine (Table 4.1) according to the sequence

$$-\mathrm{I} < -\mathrm{Br} < -\mathrm{Cl} < -\mathrm{F}, \tag{4.10}$$

whereas dipole moments of halogen benzenes, in values of which the mesomeric effect of substituent [25] is bound to manifest itself, are arranged in an orderly sequence corresponding to the increase of C_{60} molecules solubility in them, i.e. according to the sequence (4.8). The insubordination of dipole moment μ_{FB} of fluorine benzene to the sequence of mesomeric effects (4.10) that has the low value $\mu_{FB} = 1.57$ D, can be assumed as the exception from the rule of mesomeric effects (4.10). The graphical representations of dependences $S_{C_{60}} = f(M)$ and $S_{C_{60}} = f(\mu_{FB})$, illustrated in Figs. 4.1 and 4.2, show respectively that C_{60} solubility in the series of substituents

$$-\mathrm{I} < -\mathrm{Br} < -\mathrm{Cl}, \tag{4.11}$$

grows directly with the values of M_{HB} of substituents and μ of halogen benzenes. Furthermore, the dependence, corresponding to the series (4.11), is observed between the rise of C_{60} solubility and ionization potentials of halogen benzenes (Fig. 4.3) and values of electronegativity of halogen atoms (Fig. 4.4).

The character of dependences shown in Figs. 4.1 and 4.2 confirms the donor – acceptor mechanism of dissolution of C_{60} fullerene because according to the regularities of intermolecular donor – acceptor interaction, leading to the formation of charge transfer complexes (CCT) of π-type, the complex stability and hence also the solubility are the greater, the greater is the density of π-electron system of aromatic ring of solvent molecule.

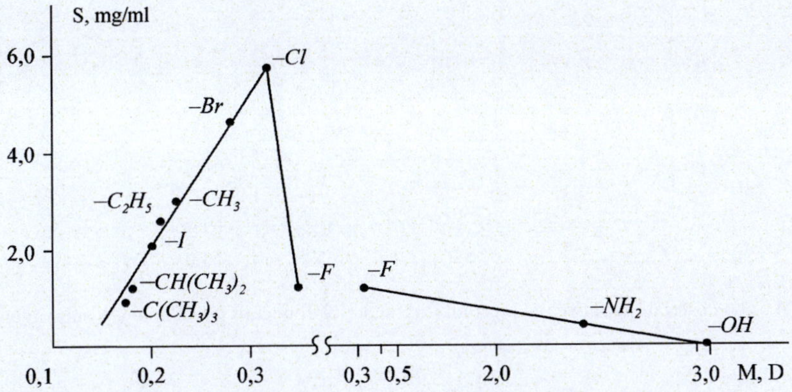

Fig. 4.1 The C_{60} solubility dependence on positive M. Substituents

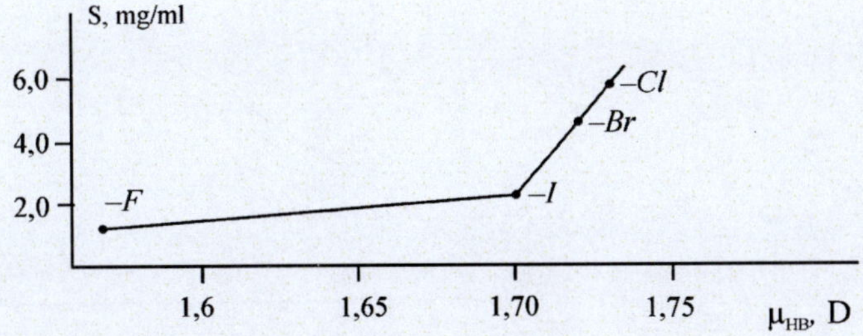

Fig. 4.2 The relationship between C_{60} solubility and dipole moments of halogen benzenes

Fig. 4.3 The correlation between C_{60} solubility and ionization potentials of halogen benzenes

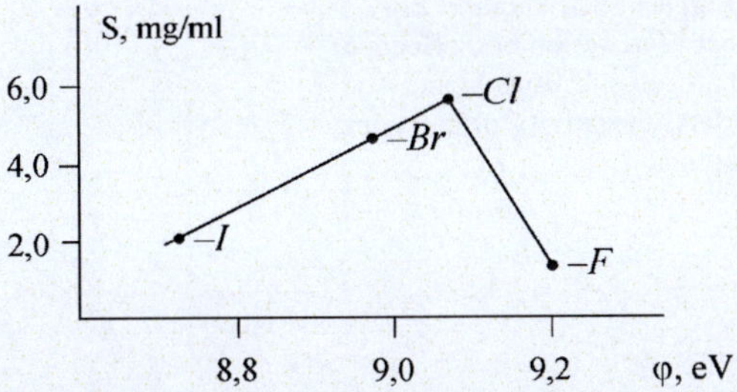

At the same time the direct proportional relationship between solubility $S_{C_{60}}$ and ionization potentials φ_{HB} as well as between $S_{C_{60}}$ and electronegativity χ_{HB} (Figs. 4.3 and 4.4) are in contradiction with the concept of donor – acceptor interactions. However, the same conflict (we shall call it as effect of ionization potential – "effect PI") are the linear relationships between mesomeric effects and ionization potentials $M_{HB}/\ \varphi_{HB}$ as well as between M_{HB}/χ_{HB} (Figs. 4.5 and 4.6 respectively).

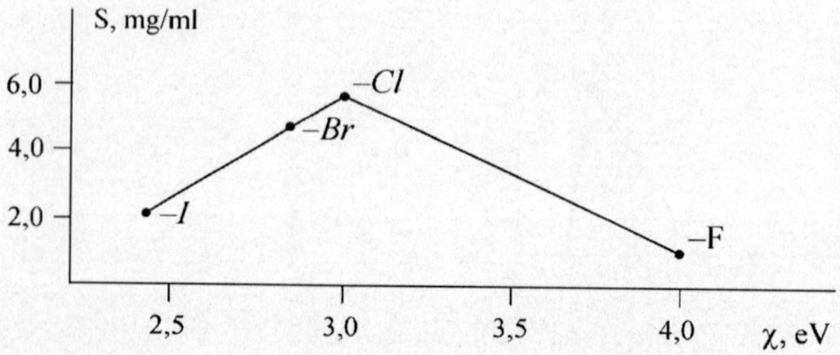

Fig. 4.4 The correlation between C_{60} solubility and electronegativity of halogen substituents

Fig. 4.5 The correlation between mesomeric effects of halogen substituents and ionization potentials of halogen benzenes

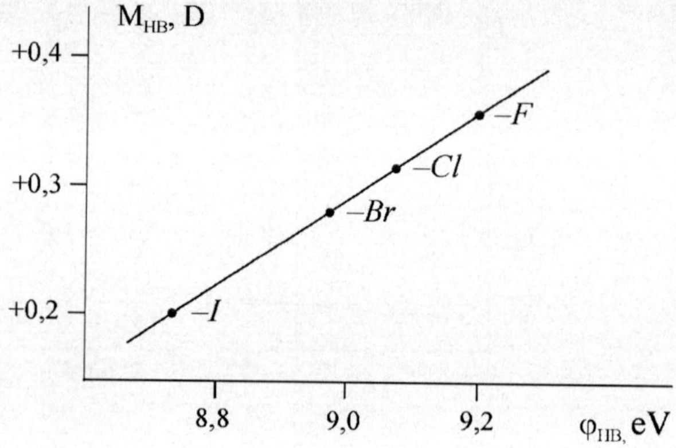

Fig. 4.6 The relationship between mesomeric effects of halogen substituents and electronegativity of halogens atoms

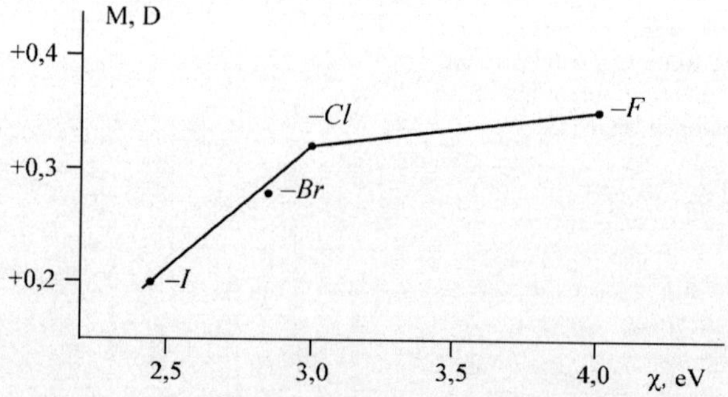

This means that the observed anomalous from the view point of regularities of donor-acceptor intermolecular interaction, the directly proportional dependences between C_{60} solubility and ionization potentials of monohalogen benzenes $S_{C_{60}}/\phi_{HB}$ and between solubility and values of electronegativity of halogen atoms $S_{C_{60}}/\chi_{HB}$ are not the result of "singularity" of molecular structure of fullerene

C_{60} and its chemical properties. On the contrary, in this case molecule C_{60} fall harmoniously into the existing, poorly investigated system of chemical and quantum – chemical processes, associated with interaction between unshared pair of atoms ofp-electrons substituents and the π-electron system of benzene ring. Several effects of these processes have been identified in the study of reaction mechanism of C_{60} dissolution due to its high sensitivity to the electronic effects of substituents of aromatic ring. For example, the effect of ordering of meta – nitroisomers [26] was discovered at the correlation of C_{60} solubility with composition of mononitroisomers in the product of nitration of monosubstituted benzene; the effect of metaposition of polysubstituted benzene [27] and several other previously unstudied effects – at the investigation of C_{60} solubility in the methyl – polysubstituted benzenes.

4.2.2 Dissolution of C_{60} in Substituted Benzenes with Substituents That Have High Values of Positive Mesomeric Effects

The increase of C_{60} solubility with increase of mesomeric effect of substituents (Fig. 4.1) is replaced on the inversely proportional dependence for –F, $-NH_2$ and –OH.

The sharp decrease of C_{60} solubility in fluorobenzene, that has the highest value + M in the series of halogen substituents, is also observed with increasing positive mesomeric effects for $-NH_2$ and –OH in the series of substituents

$$-F > -NH_2 > -OH. \tag{4.12}$$

It can be seen from Table 4.2 that ionization potential of aniline is much smaller than that of fluorobenzene. However contrary to all expectations, the phenol has the greater this potential than aniline. Consequently in going from aniline to phenol the PI-effect is observed: the positive M grows with increase of ionization potentials of these compounds. This example clearly demonstrates that the evaluation of electron – donor power of compound is not always legitimate on the value of its ionization potential.

For example, it is noted [28] that the amino group, that is the substituent of the first kind, has a stronger electron – donor properties than the phenolic group since the experimental value of ionization potential of phenol is greater than that of aniline, losing sight in this case of values of their positive mesomeric effects

Table 4.2 The physicochemical parameters of phenol, aniline and fluorobenzene

No	Substituents	C_{60} Solubility S (mg/ml)	Mesomeric effect M (D)	Ionization potential φ (eV)
1	*–OH*	≈0.0	+3.06	8.50
2	*$-NH_2$*	<0.40	+2.47	7.70
3	*–F*	1.20	+0.35	9.24

(Table 4.2). The same principle is used to estimate the electron – donor properties of halogen benzenes [28]. The electron – donor properties of benzene and fluorobenzene are considered equal on the grounds that their ionization potentials are almost identical – 9.24 eV and 9.20 eV respectively. If this is true, then the dative property of iodobenzene, the ionization potential of which is equal to 8.73 eV, should be higher than, for example, of chlorobenzene, which it is not. The same especially is true for mesitylene that possess high electron – donor ability. For example in the paper [11] the calculated affinity parameter RED for mesitylene equals 1,182 which is meant, according to the accepted situation, the low solvent capacity with respect to the C_{60} molecule that corresponds to the reality: the C_{60} solubility in mesitylene is equal to 1.5 mg/ml, i.e. it is the same as in benzene. However mesitylene has been assessed positively "Good", whereas toluene, which parameter of affinity RED is 0.970, that means its more high solubility of C_{60} – has the assessment "Bad", although the C_{60} solubility in toluene is two time higher than in mesitylene and is – 3.0 mg/ml. In this case the ionization potentials of mesitylene and toluene are equal to 8.39 and 8.82 eV respectively.

The detailed discussion of the character of linear correlations SC_{60} and M with ionization potentials of substituted benzenes is beyond the scope of this work and is the subject of separate study, during which the similar "effect – PI" is found. The essence of it and also of the low solubility of C_{60} in phenol, aniline and fluorobenzene are revealed through the use of bonds polarity. The preliminary results of investigation in brief form were published in [26, 29, 30].

4.2.3 Dissolution of C_{60} in Alkyl Benzenes

The solubility of C_{60} fullerene in alkyl benzenes (Table 4.3) grows in the series of alkyl substituents:

$$\text{grinds} - \text{butyl} < - \text{isopropyl} < - \text{ethyl} < - \text{methyl}. \qquad (4.13)$$

The positive mesomeric effect increasing in the series (4.13) and positive inductive effect (+I) growing in the opposite direction are ascribed to the alkyl substituents of benzene. The last is considered the dominant.

Table 4.3 The physicochemical parameters for alkyl benzene

No	Substituents	C_{60} Solubility S (mg/ml)	Mesomeric effect M (D)	Dipole moment μ (D)	Ionization potential φ (eV)	Electro-negativity χ (eV)
1	$-CH_3$	3.0	+0.227*	–	8.82	2.580*
2	$-C_2H_5$	2.6	+0.204*	–	8.76	2.501*
3	$-CH(CH_3)_2$	1.2	+0.178*	–	8.69	2.336*
4	$-C(CH_3)_3$	0.9	+0.175*	–	8.68	2.310*

*Arithmetical average values for mesomeric effect and electronegativity

By their nature alkyl substituents can be neither donors nor acceptors of π-electrons. Their + M can be called a quantum – chemical value because its theoretical explanation is given by Milliken on the basis of MO method and it reduces to that the pairing integration analogous to π- π- interactions exists between the alkyl group and π- electronic system of benzene ring. Milliken have called this integration as hyperconjugation (HC).

The HC concept is not shared by all scientific workers. Often at the interpretation of effects similar to the mesomeric effect, the concept of inductance or steric and solvation effects are preferred the HC concept. Nevertheless, there is a large number of reactions in which the alkyl substituents behave as substituents – donors of π-electrons. In this case for comparative study as an independent measure of donor ability of the alkyl substituents we used the values of σ_C constants of the aliphatic substituents and the relative rates (K_{rel}) of monohydrolysis reaction of n-alkylbenzhydrilchloride.

Quantitatively the value of σ_C constant is the difference

$$\sigma_C = \sigma - \sigma_1, \tag{4.14}$$

where σ is the Hammett constant, σ_1 is the inductive constant of substituent [31].

The unit M of alkyl substituents was observed firstly in the study of the hydrolysis reaction of n- alkylbenzhydrilchloride [24], which is bound to undergo more easily in the case when in the aromatic cycle of molecule there are groups increasing the electron density in the reaction centre. It has been found that the relative rate of hydrolysis K_{rel} increases in the series of substituents from the

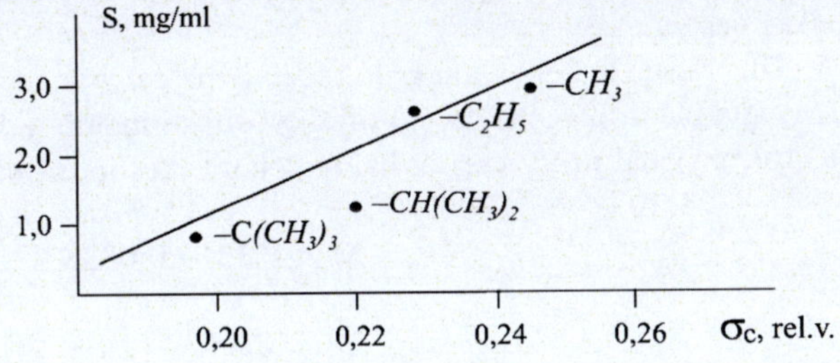

Fig. 4.7 The relationship between C_{60} solubility in alkyl benzenes and σ_c constants of alkyl substituents

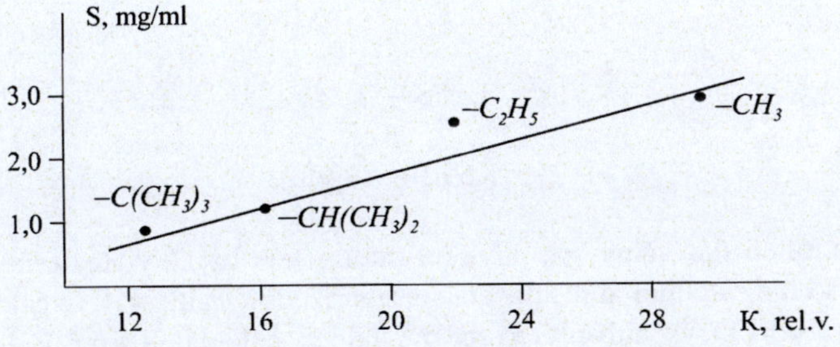

Fig. 4.8 The correlation between C_{60} solubility and values of relative rates of monohydrolysis reaction of *n*-alkylbenzhydrilchloride

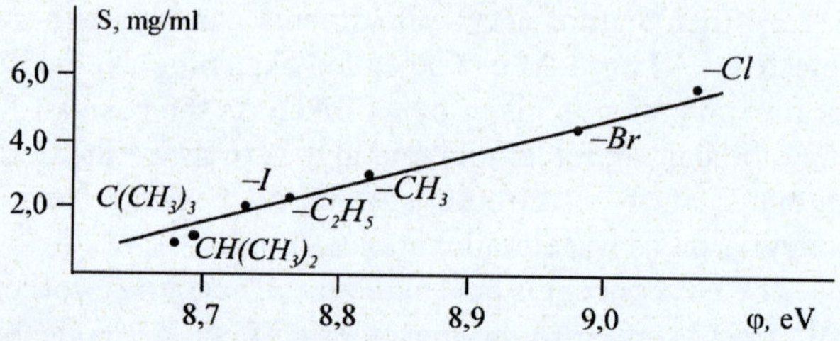

Fig. 4.9 The correlation of C_{60} solubility in alkyl- and halogen benzenes with ionization potentials of these solvents

grinds-butyl to the methyl group, i.e. in opposite direction expected by the inductive direction of substituents action. The rise of C_{60} solubility occurs in the same order: directly proportional to the values of σ_C – constants (Fig. 4.7) and values of K_{rel} (Fig. 4.8).

This means that C_{60} solubility in monoalkyl benzenes is proportional directly to the density of π- electronic system of aromatic ring of alkyl benzenes, which is the weighty argument in favour of donor – acceptor mechanism of C_{60} dissolution in the alkyl benzenes. But even here, as in the case of C_{60} dissolution in halogen benzenes, the C_{60} solubility is directly proportional to the ionization potentials of monoalkyl benzenes (φ_{AB}) in Fig. 4.9, i.e. in this system also the PI effect is observed. The values ofσ_C- constants and K_{rel} are also directly proportional to the values φ_{AB}, so there is the complete qualitative and quantitative analogy of relations of physical parameters of alkyl benzenes and halogen benzenes molecules. As is seen from Fig. 4.9, the values of C_{60} solubility in halogen benzenes and alkyl benzenes fall on the one straight line. This means that linear correlations is in existence between all physical properties of molecules of halogen benzenes and alkyl benzenes.

This raises the question as to receive the calculated values of mesomeric effects (M) and electronegativity (χ) of alkyl substituents by Eqs. (4.15–4.19) derived by us for physical properties of halogen benzenes in the series Cl, Br, I

$$M = 0.13 + 0.033S \tag{4.15}$$

$$M = 0.35\varphi - 2.83 \tag{4.16}$$

$$M^* = 0.22\chi^* - 0.348 \tag{4.17}$$

$$\chi = 1.15S + 2.146 \tag{4.18}$$

$$\chi = 1.65\varphi - 11.98. \tag{4.19}$$

We substituted the values of physical parameters of alkyl benzenes in these equations and the calculation results are summarized in Table 4.4. In calculation of M value by Eq. 4.17 we substituted the χ* values, calculated from Eq. 4.19. The Figs. 4.10–4.12 demonstrate the linear correlations of parameters respectively S, K_{rel}. and σ_C with the average values M*.

Table 4.4 The design values for mezomeric effect and electronegativity of alkyl substituents of benzene

		Calculated values of M(D) for alkyl substituents by equations:			Design values of χ(eV) by equations:	
No	Substituents	(4.15)	(4.16)	(4.17)	(4.18)	(4.19)
1	$-CH_3$	0.228	0.229	0.225	2.587	2.573
2	$-C_2H_5$	0.207	0.201	0.204	2.528	2.474
3	$-CH(CH_3)_2$	0.182	0.176	0.176	2.322	2.350
4	$-C(CH_3)_3$	0.179	0.173	0.174	2.278	2.342

Table 4.5 The physicochemical parameters for monosubstituted benzene containing electron-acceptor substituents

No	Substituents	C_{60} Solubility S (mg/ml)	Mesomeric effect M (D) [24]	Ionization potential φ (eV)
1	$-CHO$	0.41	−0.22	9.53
2	$-C\equiv N$	0.40	−0.39	9.70
3	$-NO_2$	0.80	−0.69	9.92

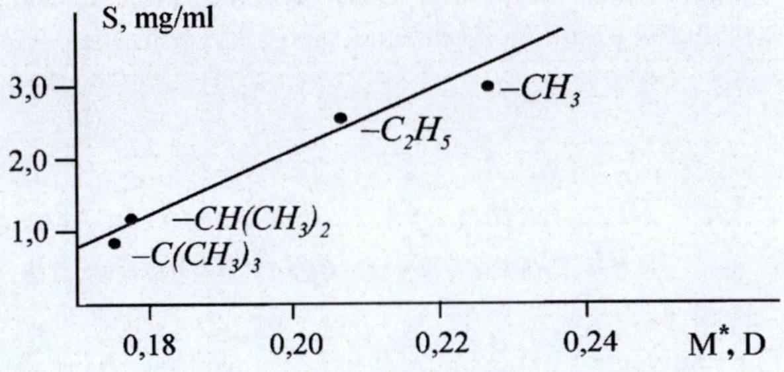

Fig. 4.10 The interrelation between C_{60} solubility in alkyl benzenes and M. calculated values of alkyl substituents

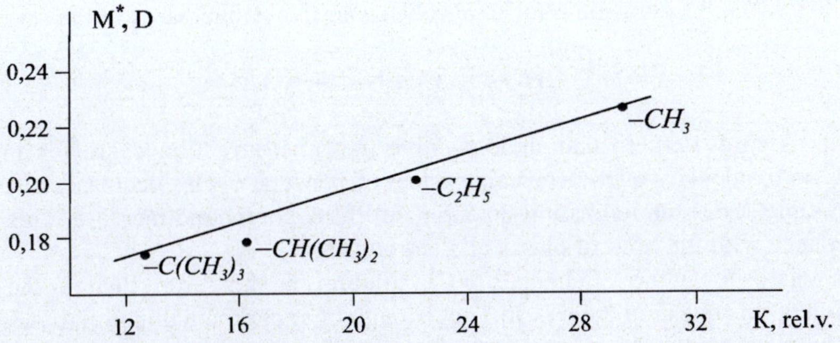

Fig. 4.11 The correlation of calculated values of mesomeric effects of alkyl substituents with relative rate of monohydrolysis of *n*-alkylbenzhydrilchloride

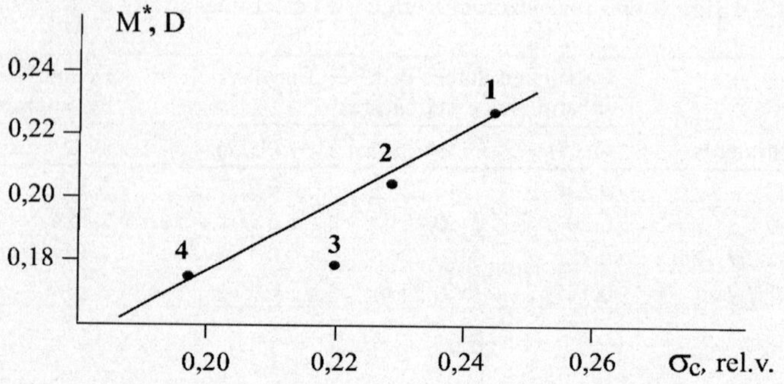

Fig. 4.12 The linear relationship between calculated values of mesomeric effects of alkyl substituents of benzene and σ_c constants

The linear correlation of these parameters with average values χ^* in graphical form are not presented here.

The results of comparative study given in this paper provide support for the donor-acceptor mechanism of dissolution of C_{60} fullerene in monosubstituted benzenes. In addition at the use of this method it has been firstly revealed that electronic effects of halogen and aliphatic substituents in the reactions of intermolecular interactions with charge transfer follow the same laws that indicates on the same mechanism of C_{60} dissolution in halogen benzenes and alkyl benzenes.

4.2.4 *Dissolution of C_{60} Fullerene in Monosubstituted Benzenes with Electron-Acceptor Substituents*

The analysis of experimental data shows that effect –PI is observed not only in the case when substituents of benzene rings are π-electron-donors. For example, the negative mesomeric effect of electron-acceptor substituents of compounds benzaldehyde (No 1), benzonitrile (No 2), nitrobenzene (No 3) increases in the series:

$$-\mathrm{CHO} < -\mathrm{C} \equiv \mathrm{N} < -\mathrm{NO_2}.$$

This is equivalent to that their positive mesometric action decreases in this series, so the density of π-electronic system of benzene cycles decreases also and in the same series the ionization potentials of these compound increase. This is in accordance with the laws of classical electron theory.

As regards the nature of changes of C_{60} solubility in this series, then for the final solution of this question there is still not enough of experimental material. Account must be taken of the dependence of reactivity of π-electronic system of benzene ring in the reactions of donor-acceptor interactions not only on its density, but also on its mobility and affinity to the electron of reagent. Besides, it is necessary to take into account that according to literature data [28] the dipole moments of these compounds are directed toward the substituents that carry consequently the negative charge and C_{60} molecules can interact directly with the substituents.

4.2.5 *Dissolution of C_{60} in Aliphatic Hydrocarbons Containing the Halogen Atom*

The fact that C_{60} fullerene, which is practically insoluble in the aliphatic and alicyclic hydrocarbons, begins to dissolve at the substitution of hydrogen in compound by halogen atom is the confirmation of donor - acceptor mechanism of C_{60} dissolution in the aliphatic hydrocarbons. It is evident from the data of Table 4.6 that in the saturated hydrocarbons involving halogen atoms with the bond R-C (sp^3)-X, where X = Cl, Br, I, the reaction of C_{60} dissolution proceeds in accordance wilt the requirements of classical electron theory regardless of the nature of the hydrocarbon residue "R-" (is it aliphatic or alicyclic saturated hydrocarbon). The C_{60} solubility increases in the series of substituents: –Cl < –Br < –I, i.e., inversely proportional to the electronegativities of halogen atoms and to the ionization potentials of solvent molecules.

The overview of literature data has showed that electron-donor properties of halogen derivatives of hydrocarbons with the bond C (sp^3) – X are not studied systematically. In this regard the substituted benzenes are of particular interest, the halogen atom in them is in the side chain. This means that molecule has two reaction center, as π-electronic system of benzene ring and the non-divided pair of p-electrons of halogen substituents. Until the present time there is still a problem requiring the answer to the question, as – the activity of what center prevails in the reactions of donor-acceptor interaction.

From the above discussion it follows that the form of relations "C_{60} solubility" – "physical parameters" of solvent molecules depend on the type of hybridization of carbon atom bound with the substituent. Therefore, using our developed method by the character of C_{60} solubility changes at variation of substituents it is possible to

Table 4.6 The values of fullerene C_{60} solubility S, ionization potential φ of solvents and elecrtonegativity χ of substituents

No	Substituents	S (mg/ml)	φ (eV)	χ (eV)
1	Cyclohexile chloride	0.53	–	3.00
2	Cyclohexile bromide	2.20	10.5	2.85
3	Cyclohexile iodide	8.06	9.5	2.44
4	1–Chloropropane	0.02	–	3.00
5	1–Bromopropane	0.05	–	2.85
6	1–Iodopropane	0.17	–	2.44
7	Bromomethyl	–	10.54	2.85
8	Iodomethyl	–	9.54	2.44

define what is involved in the charge transfer from donor to acceptor-unshared pair of p-electrons of halogen located in the side chain or the π-electronic system of benzene ring. The proposed method is based on the comparative investigation of C_{60} solubility with the physicochemical parameters of solvent molecules in combination with experimental method of C_{60} solubility determination in compounds for which it is still unknown.

4.3 Discussion of Final Research Results

The obtained information discussed in the Tables 4.1–4.6 has been accumulated in the Summary Table 4.7 and this has made it possible to receive a convincing picture. The obviousness of coherent, decisive role of electron-donor substituents of the aromatic ring in the chemical activity of solvent molecules is evident.

Since the electron – donor ability of solvent molecules is defined by the chemical composition and molecules structure and expressed by parameters entered in the Table 4.7, then the analysis of these data can shed light on the peculiarities of interactions passing and their mutual influence.

From the presented above information it should be noted that fullerene C_{60} solubility in the monosubstituted benzenes is determined by the density of π-electronic system of benzene ring. This conclusion is confirmed by the existence of linear dependence between C_{60} solubility and values of positive mesomeric effects of substituents, that reveal themselves in parameters σ_c, M, $K_{rev.}$

Table 4.7 The summary table for values of C_{60} solubility and physicochemical parameters of solvents

No	Substituents	C_{60} Solubility S (mg/ml)	Mesomeric effect M (D)	Dipole moment μ (D)	Ionization potential φ (eV)	Electro-negativity χ (eV)
1.	$-OH$	–	+3.06	–	8.50	–
2.	$-NH_2$	<0.40	+2.47	–	7.70	–
3.	$-F$	1.2	+0.35	1.57	9.20	4.00
4.	$-Cl$	5.7	+0.32	1.73	9.07	3.00
5.	$-Br$	4.6	+0.28	1.72	8.98	2.85
6.	$-I$	2.1	+0.20	1.70	8.73	2.44
7.	$-CH_3$	3.0	+0.227*	–	8.82	2.580*
8.	$-C_2H_5$	2.6	+0.204*	–	8.76	2.501*
9.	$-CH(CH_3)_2$	1.2	+0.178*	–	8.69	2.336*
10.	$-C(CH_3)_3$	0.9	+0.175*	–	8.68	2.310*
11.	$-CHO$	0.41	−0.22	–	9.53	–
12.	$-C{\equiv}N$	0.40	−0.39	–	9.70	–
13.	$-NO_2$	0.80	−0.69	–	9.92	–

*Arithmetical average values for mesomeric effect and electronegativity

Also it has been demonstrated that the greater is the density ofπ-electronic system of benzene ring of compound, the higher is C_{60} solubility in this compound. In this case the ionization potential of solvents should decrease in accordance with the laws of classical electron theory.

However, as seen from the Table 4.7, the ionization potential increases in direct proportion to the density of π-electronic system of aromatic ring of solvent molecule. This effect is called by us as "Effect of ionization potential" (Effect-PI).

Using the method of comparison it has been established that Effect-PI is observed only in substituents of first kind.

The explanation to this effect can be offered, for example, by the assumption that positive mesomeric effects of halogen and aliphatic substituents prevail their inductive effects. Then, in the case of halogen substituents the halogen atoms are bound to be charged positively, but not negatively, according to the modern concepts. As regards the aliphatic substituents, their inductive and mesomeric effects are positive. In such situation the energy field of positive charges of substituents should be partially passivate the mobility of π-electronic system of benzene ring proportionally to the mezomeric effect of substituent.

It is necessary to stress that in the reaction of intermolecular donor-acceptor interaction with charge transfer, leading to the CCT formation, a charge that is transmitted from donor to acceptor does not lose bond with donor molecule, i.e. ionization of donor molecule as such is not happen. Conceivably this process does not depend at all on ionization potentials of the reacting molecules. In particular the following point indicates on that as: the ionization potential of fullerene C_{60} molecule, equal to 7.6 eV, is significantly smaller than the ionization potentials of compounds whose substituents have positive mesomeric effects. Their values are given in Table 4.8. More detailed examination of the complicated relationship between the C_{60} solubility and physico-chemical properties of solvent molecules is beyond the scope of this work and is the subject of separate investigation.

The obtained results of this study confirm the assumption advanced by us about the donor-acceptor mechanism of fullerene C_{60} dissolution in organic hydrocarbons. Due to the method developed by us it has been opened the effect of electron-donor substituents, the manifestation of which is also observed in other systems and which discussion we could not find in the literature.

As an example we shall give some problematic situations in which the devised method of evaluation of the donor ability of hydrocarbons and their derivatives can be used.

One of the promising areas of practical application of fullerenes is their use in the systems of liquid crystal materials, that are used in optoelectronics, display technique, perhaps in the future they shall find in solar energetic. The most central problem of these systems is the contraction of times of change-over of liquid-crystal media (LCM). It is assumed that one of possible ways of increase of fast activity would be the creation of additional polarization of LC system by introducing into system of new compounds with high dipole moments, such as charge-transfer complexes. Experimentally this assumption is confirmed by the research results of paper [19], in which fullerenes C_{60} and C_{70} were used as an electron acceptor in CCT. It is believed that the greater is polarization of CCT (donor-fullerene), the greeter is its

Table 4.8 The summary table of descriptors and their qualitative effect on C_{60} solubility

References	No	Descriptors	Qualitative influence of parameters increase on C_{60} solubility
	1	Molar volume V_m	Decrease
	2	Dipole moment μ	Decrease
	3	Capability to *HB*	Increase
[4]	4	Polarizability f_n	Increase
	5	Polarizability f_n	Increase
	6	Bipolarity π^*	Slight influence
[5]	7	Dipole moment μ	Scarcely influence
	8	Capability to *HB*	
[6]	9	Square of Hildebrand parameter δ^2	Give worse results
	10	Polarizability f_n	Increase
	11	Cohesion δ^2	Decrease
	12	Polarity f_ε	Slight influence
	13	"Basicity" by Palm	Slight influence
[7]	14	"Electrosensitivity"	Slight influence
	15	Polarizability f_n	Increase
	16	Molar volume V_m	Decrease
	17	LFMO Energy	Little sizable
[8]	18	"Volume of saturated surface"	Little sizable
	19	Division of compounds on groups "Topology indexes"	Acceptable prediction of C_{60} solubility in each subgroup
[5]	20	"Parameter of polarizability"	Except in subgroups of aliphatic and aromatic hydrocarbons
	21	"Solvation parameters"	Weak effect
	22	"Coefficient of C_{60} distribution between water and solvent"	Weak effect
	23	"Bipolarity"	Basic parameter
	24	"Capability to *HB*"	Basic parameter
[9]	25	"Excessive molar refraction"	Basic parameter

ability to the further polarization of LC system and to the orientation of the LC director along the dipole vector of the complex that leads to a significant increase of system fast activity.

The efficient transfer of charge between the organic donor and acceptor depends not only on the affinity to the acceptor electron, but also on the dative properties of the donor. As in the work [19] in the complexes, introduced in LCM, only donor was varied and the derivatives of pyridine and aniline were used as donor, the change of LCM fast activity can be attributed to its influence. However it was found that the donor replacement did not affect the time characteristics of LCM. On this basis conclusion has been reached that the fast activity is almost independent on the nature of organic donor.

Nevertheless, taking into account the obtained results in this research work, this conclusion cannot be unambiguous. Within the boundaries of developing concept about electron-donor activity of organic compounds the donor activity of pyridine and aniline derivatives, as shown in the present paper, is very low in the reactions of intermolecular interaction with charge transfer. The fullerene C_{60} fulfils the acceptor role in these reactions. In the aniline the C_{60} solubility is virtually absent, while in the pyridine it is lower than in benzene. In this connection, if the increase in fast activity of LCM is associated with the dipole moment of introduced CCT, the role of donor nature in the studied process should be investigated with more active organic donors of π-electrons which substantially dissolve the fullerene C_{60}.

The main direction of publications, as is seen from the papers review on the dissolution of fullerene C_{60}, is the search for linear multiparameter equations with aim of their use to predict the solubility of fullerene C_{60} in the compounds for which there is no experimental data at present time.

From the analysis of publications it is clear that investigations differ from one another by the quantity and composition of compounds in samples and by combinations of using descriptors given in Table 4.8.

The parameters that accounts in one or another extent the distribution of electron density in the solvent molecule are not used in papers as discriptors. These parameters include the ionization energy of molecule, inductive and mesomeric effects of substituents of aliphatic and aromatic hydrocarbons, distinguished by the highest solvent power with respect to fullerene C_{60}, the quantity and distribution of substituents in the solvents molecules. In this case some studies excluded from the consideration of parameters related to the electron-donor activity of compounds, for example, basicity by Palm and polarity as unimportant factors [7].

In the paper [32], in distinction to the prevention to the previous works, theQSPR-QSAR analysis was used with optimal descriptors calculated using a system of notations SMILES. Because of rather complicated coordination between the fullerene C_{60} solubility and the solvent structure for the analysis GSPR it has been selected the representative series of benzene derivatives. The great difference between the experimental and calculated values of C_{60} solubility in relatively simple structures was received by this method. It has been (as 1,2-dimethyl benzene, bromobenzene and benzal dehyde) suggested that this difference is due to the formation of specific compounds – clusters of solvents near the molecules C_{60}. Also it has been concluded that fullerenes and other nanostructures can hardly be represented by the classical molecular circuits.

We believe, taking into account the results of present paper, that the last conclusion does not correspond to reality. On the contrary considering the fullerene C_{60} molecule as molecule-standard, the major opportunity of studying the electronic effect of substituents has been provided.

The presented in this paper results on the mechanism of fullerene C_{60} dissolution strongly provide support for the view expressed by us [26, 27, 29, 30] about the donor-acceptor mechanism of dissolution of fullerene C_{60}. One of supporting factors is the high sensitivity to electronic effects of solvent molecules. Therefore, the use as descriptors parameters (representing the electronic state in reaction centers of solvent molecule) to make the choice of descriptors (with the use of CMS method) targeted and therefore more productive.

It is necessary to note also that fullerene C_{60} solubility in the aromatic hydrocarbons depends not only on the density of π-electronic system of benzene ring, but also on its mobility. Therefore at the completeness of sample it is necessary to separate the compounds with different degrees of hybridization of the carbon atom bound with the substituent.

4.4 Conclusions

The main results of the carried out comparative analysis of the values of fullerene C_{60} solubilities with the physicochemical parameters of the solvents molecules are as follows:

1. the comparative analysis of the values of fullerene C_{60} solubilities with physicochemical parameters of solvent molecules for the purpose of revealing of reaction mechanism of C_{60} dissolution;
2. it has been established that the fullerene C_{60} dissolution is the reaction of intermolecular donor-acceptor interaction with charge transfer, leading to the formation of charge transfer complexes;
3. it has been determined that the equilibrium constant of the C_{60} dissolution reaction is numerically equal to the C_{60} solubility that makes it possible to use the values of the C_{60} solubility as a measure of π-electron force of solvent molecule;
4. it has been discovered the effect of direct proportionality between ionization potentials of the monosubstituted benzene molecules containing the substituents of the first king with a density of π-electronic system of the benzene nucleus that we call as the effect of the ionization potential – "Effect-PI";
5. it has been suggested the proposal that the intramolecular charge transfer of the nondivided pair of p-electrons of substituent into π-electronic system of benzene ring does not depend on the electronegativity of substituent, as well as intermolecular charge transfer from the donor-molecule to the acceptor-molecule does not depend on the ionization potentials of interacting molecules;
6. it has been obtained for the first time the values of electronegativities and mesomeric effects of the alkyl substituents of benzene.

References

1. Ruoff RS, Tse DS, Malhotra P, Lorents DC (1993) Solubility of fullerene C_{60} in a variety of solvents. J Phys Chem 97(13):3379–3383
2. Bezmel'nitsyn VN, Eletskii AV, Okun MV (1998) Fullerenes in solutions. Phys Usp 41:1091–1114
3. Eletskii AV, Smirnov BM (1995) Fullerenes and carbon structures. Phys Usp 38:935–964

4. Smith AL, Wilson LY, Famini GR (1996) Quantitative structure - property relationship study of C_{60} solubility, recent advances in the chemistry and physics of fullerenes and related materials. Los Angeles, CA
5. Marcus Y (1997) Solubilities of buckminsterfullerene and sulfur hexafluoride in various solvents. J Phys Chem B 101(42):8617–8623
6. Kamlet MJ, Abboud JLM, Taft RW (1981) An examination of linear solvation energy relationships. Progr Phys Org Chem 13:485–630
7. Makitra RG, Pristanskii RE, Flyunt RI (2003) Solvent effects on the solubility of C_{60} fullerene. Russ J Gen Chem 73(8):1227–1232
8. Kiss IZ, Mandi I, Beck MT (2000) Artifical neural network approach to predict the solubility of C_{60} in various solvents. J Phys Chem A 104(20):8081–8088
9. Abraham MH, Green CE, Acree WE (2000) Correlation and prediction of the solubility of buckminsterfullerene in organic solvents, estimation of some physico-chemical properties. Chem Soc Perkin Trans II 2:281
10. Sivaraman N, Srinivasan TG, Rao PRV, Natarajan RN (2001) QSPR modeling for solubility of fullerene C_{60} in organic solvents. J Chem Inf Comput Sci 41(4):1067–1074
11. Hansen CM, Smith AL (2004) Using Hansen solubility parameters to correlate solubility of C_{60} fullerene in organic solvents and in polymers. Carbon 42(8–9):1591–1597
12. Dubois D, Kadish KM, Flanagan S, Wilson LJ (1991) Electrochemical detection of fulleronium and highly reduced fulleride (C_{60}^{5-}) ions in solution. J Am Chem Soc 113:7773–7774
13. Arbogast JW, Foote CS, Kao M (1992) Electron-transfer to triplet C_{60}. J Am Chem Soc 114:2277
14. Hirsch A (1993) The chemistry of the fullerenes: an overview. Angew Chem 32(8):1138–1141
15. Tayior R, Walton DMR (1993) The chemistry of the fullerenes. Nature 363:685–693
16. Wang Y (1992) Photophysical properties of fullerenes and fullerene/N, N-diethylaniline charge-transfer complexes. J Chem Phys 96(2):764–767
17. Caspar JV, Wang Y (1994) Excited state electron transfer of fullerenes. Singlet states versus triplet states. Chem Phys Lett 218(3):221–228
18. Kotov AI, Konovalikhin SV, Pisarev PV, Shilov GV et al. (1994) Preparation and X-ray study of a molecular complex of C_{60} with a new organic cyclotetrasulfide, $4C_{60}$ 3 (twin TDAS). Mendeleev Commun 4(5):180–182
19. Kamanina NV (2005) Fullerene-dispersed nematic liquid crystal structures: dynamic characteristics and self-organization processes. Phys Usp 48(4):419–427
20. Evlampieva NP, Yakimanskii AV, Dobrodumov AV et al (2005) Synthesis, polar and electrooptical properties of butylaminoderivative of fullerene C_{60}. J Gen Chem 75 (5):795–802, in Russian
21. Beck MT, Mandi G, Keki S (1995) Fullerene Sci Technol 2:1510
22. Beck MT, Mandi G (1996) Fullerenes Sci Technol 3:32
23. Scrivens WA, Tour JM (1993) Potent solvents for C_{60} and their utility for the rapid acquisition of 13 C-NMR data for fullerenes, Phys Inorg Chem 24(44):1207
24. Bekker G (1965) Introduction to the electronic theory of organic reactions, Moscow: Mir, p 576 (in Russian)
25. www.encyclo.co.uk/define/mesomerism
26. Schur DV, Zaginaichenko SYu, Anikina NS, Matysina ZA, Krivuschenko OYa, Skorokhod VV (2009) The discovery of "ordering effect" in meta-isomer, the product of benzene mono-derivative nitration and relation between this effect and reactivity of monosubstituted benzenes in the reactions of intermolecular donor-acceptor interaction with C_{60} fullerene. In: Book of abstracts of Proceeding of 11th international conference "hydrogen materials science and chemistry of carbon nanomaterials", Yalta, Crimea, 25–31 Aug 2009, pp 604–605
27. Anikina NS, Schur DV, Zaginaichenko AD, Zolotarenko AD et al. (2007) On the donor-acceptor mechanism of fullerene C_{60} dissolution in aromatic hydrocarbons. In: Book of abstracts of 10th international conference "hydrogen materials science and chemistry of carbon nanomaterials", Sudak, Crimea, 22–28 Sept 2007, pp 672–677

28. Neiland O Ya (1990) Organic chemistry. Moscow: Vysshaya shkola, p 750 (in Russian)
29. Anikina NS, Schur DV, Zaginaichenko AD, Golovchenko TN et al. (2009) About physical nature of ionization potentials of benzene mono-derivatives and C_{60} fullerene solubility in these compounds. In: Book of abstracts of 11th international conference "hydrogen materials science and chemistry of carbon nanomaterials", Yalta, Crimea, 25–31 Aug 2009, pp 796–799
30. Schur DV, Anikina NS, Krivuschenko OYa, Voronaya TV et al. (2009) About fullerene C_{60} solubility in mono- and polysubstituted benzenes, fullerene constantans of substituents and ionization energy of substituents. In: Book of abstracts of 11th international conference "hydrogen materials science and chemistry of carbon nanomaterials", Yalta, Crimea, 25–31 Aug 2009, pp 1166–1167
31. Toporov AA, Leszczynska D, Leszczynski J (2007) QSPR study on solubility of fullerene C_{60} in organic solvents using optimal descriptors calculated with SMILES. Chem Phys Lett 441 (1–3):119–122
32. Knunyants IL (1992) Chemical encyclopedia, Moscow: Solvent encyclopedia, 3:641 (in Russian)

Chapter 5
Fullerene Molecule as Catalyst of Synthesis of Carbon Nanotubes

D.V. Schur, S.Yu. Zaginaichenko, V.A. Bogolepov, V.M. Adeev, E.P. Rudakova, A.V. Kotko, V.V. Skorohod, and Yu.M. Shulga

Abstract Advantage was taken of new materials produced by original (developed by authors) technology with the use of non-metallic catalysts. It is common knowledge that metals of iron group (Fe, Ni, Co and their mixtures) are most often used as an catalysts for synthesis of carbon nanostructures. After synthesis the catalyst which content amounts to as much as 30 wt.% is dissolved in mineral acids by boiling.

The carbon nanostructures prepared by pyrolysis of hydrocarbons on the non-metal catalysts don't require the harmful and power-intensive stage of chemical treatment and they contain on their surface the molecules of inorganic acids in adsorbed form and another associated impurities.

In the present work the carbon fibres were impregnated by fullerene solution in toluene that catalyzed the process of carbon nanotubes growth at the fibres surface.

Keywords Fullerene · Catalyst · Carbon · Nanostructure synthesis · Carbon microfiber · Pyrolysis

5.1 Introduction

Pyrolytic method is the most widespread technique of carbon nanostructures synthesis. Simplicity of technological equipment, significant deposition rate, low fabrication cost, control over properties and chemical flexibility are a driving

D.V. Schur (✉), S.Yu. Zaginaichenko, V.A. Bogolepov, V.M. Adeev, E.P. Rudakova, A.V. Kotko, and V.V. Skorohod
Institute for Problems of Materials Science, The National Academy of Sciences of Ukraine, Krzhyzhanovsky str. 3, Kyiv 03142, Ukraine
e-mail: shurzag@ipms.kiev.ua; shurzag@materials.kiev.ua

Yu.M. Shulga
Institute of Problems of Chemical Physics, Russian Academy of Sciences, Chernogolovka, Moscow District 142432, Russia
e-mail: shulga@icp.ac.ru

S.Yu. Zaginaichenko et al. (eds.), *Carbon Nanomaterials in Clean Energy Hydrogen Systems - II*, NATO Science for Peace and Security Series C: Environmental Security 2, DOI 10.1007/978-94-007-0899-0_5,

force of this method. Synthesis of various nanoproducts is successfully performed by the pyrolytic method.

The size and shape of synthesized particles are of essential fundamental and applied interest as they define micro (electron and quantum) as well as bulk behavior of the system. Dependence of these parameters on the process conditions is crucial in optimization of the generated structures. Critical parameters of the process are profiles of chemical, dynamical and thermal properties of reacting gases (type of gas, composition, flow rate, gas pressure, direction of convective streams, angle of the reactor axis relative to horizon and temperature gradient in the reactor) and chemical composition of a catalyst. The latter affects chemical composition of the product. The product purification from metallic catalysts is a serious problem.

In our previous papers on the use of fullerene solutions for advantage was taken of new materials produced by original (developed by authors) technology with the use of nanostructures synthesis, it was mentioned that at the arc evaporation of nickel in the fullerene solution in toluene the net from carbon nanotubes was formed on the surface of nickel particles [1]. At the evaporation of $LaNi_5$ alloy in the fullerene solution in toluene (compound – 2355) the nanotubes were identified on the surface of metallic particles [2].

In consequence of these research works we made a conclusion that fullerene molecules in the reaction zone can serve not only as a source of carbon pair (arc discharge) but as the catalyst of synthesis of carbon nanostructures (pyrolysis of hydrocarbons) [3–5].

The majority of hydrocarbons as a sources of carbon atoms at the pyrolytic synthesis of nanostructures are sublimated and decompose in the temperature range from 200°C to 500°C. For this reason the thermostable catalyst is required over this interval. The carbon fullerene molecules that are coming into sublimation at the temperature higher 600°C can go to work as such catalysts.

In the present work authors have synthesized carbon nanotubes (CNT) on the non-metallic substrate by the pyrolytic method [6, 7] using nonmetallic catalysts.

5.2 Results and Discussion

The studies were performed on a pyrolytic setup with a vertical reactor. This installation makes it possible to install the reactor at any angle to horizon.

The change in the angle of the reactor position (γ) allows the change in the effect that processes conditioned by both convection and gravitation has on the nanostructures formation.

Carbon microfibers 5–15 μm in diameter were used as a nonmetallic substrate. Before synthesis these fibers were saturated with metal-free catalysts. Gas mixture of acetylene and helium was used as a precursor. The synthesis was carried out in the temperature range from 350°C to 800°C.

Different organic compounds, their mixtures and fullerene solutions were used as nonmetallic catalysts. The mixture 2355 containing up to 2 g of fullerenes per liter of toluene with other admixtures proved to be optimal. The use of this composition did not cause the fibers erosion (Fig. 5.1) and allowed the control over geometric sizes (Fig. 5.2) of carbon nanotubes and fibers (Fig. 5.3) using technological conditions.

The prolonged process of synthesis brings about the formation of nanofibers up to 100–200 nm in diameter and tens of micrometers in length.

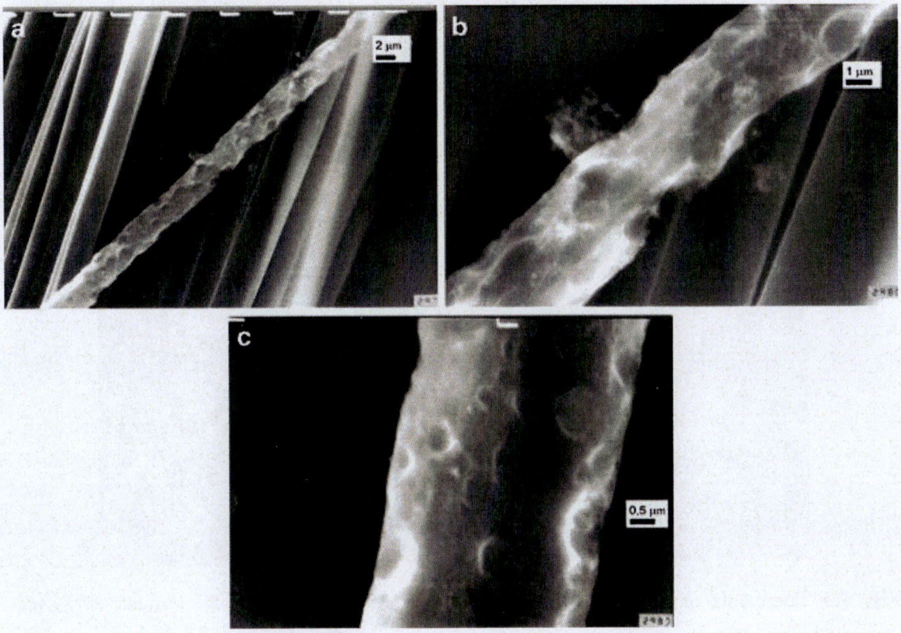

Fig. 5.1 Carbon fibers with surface defects formed as a result of action of unfortunate choice catalyst (**a**) ×5,000; (**b**) ×10,000; (**c**) ×20,000

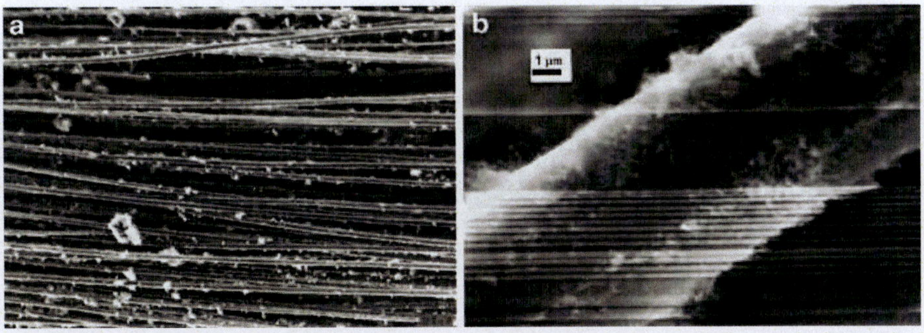

Fig. 5.2 Carbon microfibers after synthesis of carbon nanotubes and nanofibers on them: (**a**) fibers bundle; (**b**) individual fiber with carbon nanostructures on the surface

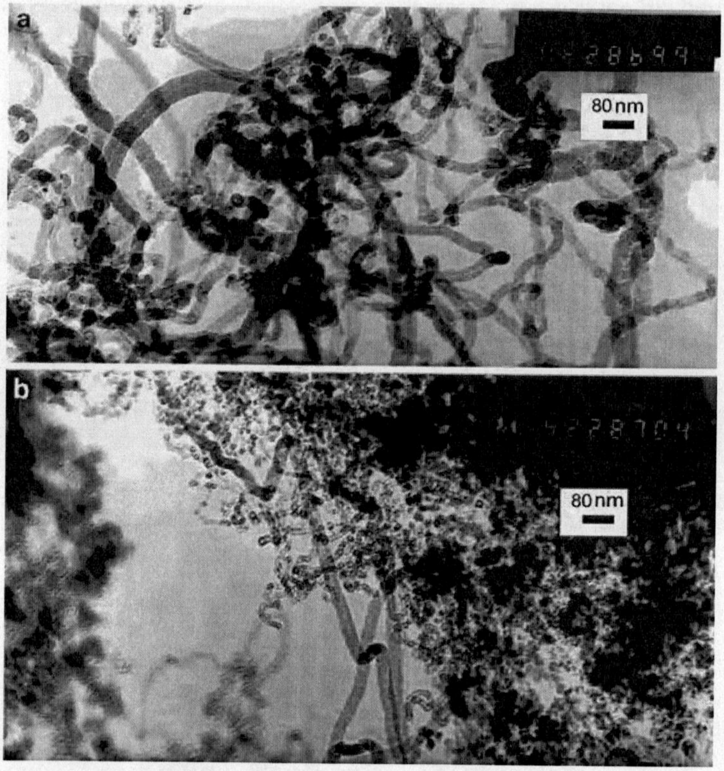

Fig. 5.3 Fragments of bundles of carbon nanotubes synthesized on the surfaces of carbon microfibers

Our design of the reactor makes it possible to pass continuously a fiber through the reaction zone at a specified velocity. In this case a bunch of fibers is rewound from one cassette to another.

5.3 Conclusions

Carbon microfibers synthesized under optimal conditions ($T = 630°C$, $\gamma = 50°$) have carbon nanotubes 5–25 nm in diameter and up to 1–2 μm in length present on their surface.

It has been shown that fullerene molecules are possible to use as a nonmetallic catalyst.

The pyrolytic setup has been designed for continuous synthesis of carbon nanotubes on the surface of carbon fibers.

References

1. Hu J, Odom TW, Lieber CM (1999) Chemistry and physics in one dimension: synthesis and properties of nanowires and nanotubes. Acc Chem Res 32:435–445
2. Maruyama S, Miyauchi Y, Edamura T et al (2003) Synthesis of single–wall carbon nanotubes with narrow diameter – distribution from fullerene. Chem Phys Lett 375(5–6):553–559
3. Lee CJ, Park J (2000) Growth model of bamboo-shaped carbon nanotubes by thermal chemical vapor deposition. Appl Phys Lett 77(21):3397–3399
4. Huczko A (2002) Synthesis of aligned carbon nanotubes. Appl Phys A A74:617–638
5. Mothioux M, Serp P, Flahout E (2007) Introduction to carbon nanotubes. In: Bushan B (ed) Handbook of Nanotechnology. Springer, NewYork, pp 43–112
6. Shaijumon MM, Ramaprablu S (2003) Synthesis of carbon nanotubes by pyrolysis of acetylene using alloy hydride materials as catalysts and their hydrogen adsorption studies. Chem Phys Lett 374:513–520
7. Mahanandia P, Vishwakarma PN, Nunda KK et al (2008) Synthesis of multi – wall carbon nanotubes by simple pyrolysis. Solid State Commun 145(3):143–148

Chapter 6
Formation and Properties of Magnetic Nanocrystallites Embedded in Carbon Beads

M. Leonowicz, M. Izydorzak, A.D. Pomogailo, and G. Dzhardimaleva

Abstract Ferromagnetic or superparamagnetic nanocrystallites, stabilized in carbon matrix, were prepared by the procedure comprising formation of the appropriate metal acrylamide complex, followed by frontal polymerization and pyrolysis of the polymer. The pyrolysis products were in a form of beads, which contained in their volume randomly distributed nanocrystallites. The nanocrystallites had various composition, magnetic properties and sizes ranging from few to tens of nanometers, depending on the element type. Application of this procedure stabilizes the nanostructure and enables processing of nanoparticles within a narrow window of sizes. The magnetic parameters depend on the crystallite size, determined by their chemical composition. The nanocrystallites pyrolysed at 773 K exhibited ferromagnetic properties for Co and Fe, and superparamagnetic behaviour for the Ni.

Keywords Magnetic nanocrystallite · Pyrolysis · Polymerization · TEM microscopy · Composition · Superparamagnetism · Hysteresis

6.1 Introduction

Processing of composites containing nanodispersed particles focuses continuously substantial scientific and technological interest. Individual nanoparticles exhibit unique physical properties resulting from size effects and highly developed surface. Such materials can exhibit either superparamagnetic or ferromagnetic properties. They can find medical applications in targeted drug delivery therapy and as contrast in NMR body screening, as well as at high density magnetic recording systems.

M. Leonowicz (✉) and M. Izydorzak
Faculty of Materials Science and Engineering, Warsaw University of Technology, Woloska 141, 02-507 Warsaw, Poland
e-mail: mkl@inmat.pw.edu.pl

A.D. Pomogailo and G. Dzhardimaleva
Institute of Problems of Chemical Physics 142432 Chernogolovka, Moscow Region, Russia

S.Yu. Zaginaichenko et al. (eds.), *Carbon Nanomaterials in Clean Energy Hydrogen Systems - II*, NATO Science for Peace and Security Series C: Environmental Security 2, DOI 10.1007/978-94-007-0899-0_6, © Springer Science+Business Media B.V. 2011

A significant problem, arising in the preparation and application of the nanoparticles, is their stabilization. Among the methods ensuring stabilization and conservation of the physicochemical properties is synthesis in a polymer matrix. Frontal polymerization, which enables conversion of the monomer to polymer in localized reaction zone, was first mentioned in 1970s [1, 2]. Application of the frontal polymerization and study of its kinetics and dynamics is widely exploited [3–6]. Among the advantages of the frontal polymerization one can mention: high reaction rate, short reaction time, low energy consumption and the possibility of conducting polymerization in ecologically friendly way without addition of a solvent. The frontal regime is formed and stabilized by the chemical heat release and its transfer by heat conductivity.

In the current study the nanocomposite beads containing Co, Fe and Ni nanocrystallites were prepared by frontal polymerization of metal acrylamide (AAm) monomer and subsequent pyrolysis of the polymer. Structure and magnetic properties of the nanocomposites were evaluated and discussed.

6.2 Experimental

The acrylamide metal complexes (MeAAm) (Me = Co, Ni, Fe) were obtained by substitution reaction of inorganic hydrates by the acrylamide. Frontal polymerization was carried out at atmospheric pressure in self-generated atmosphere. The nanocomposites were formed by further pyrolysis. The process of frontal polymerization is accompanied by melting of the monomer. The melting and formation of the polymer matrix promote encapsulation of the nitrite groups and prevent their further decay. The metal containing nanocomposite powder particles were formed in the process of pyrolysis of the polymeric matrix. The pyrolysis was carried out without evacuation in a self generated atmosphere at a temperature 773 K. The pyrolysis products were cooled down slowly with the reactor to the room temperature. The solid products, formed during the pyrolysis, consisted of metal nanocrystallites in amorphous matrix which protects them from oxidation. The EDS analysis of the amorphous matrix revealed the presence of carbon, nitrogen and oxygen atoms. More detailed description of the processing route, as well as the products, can be found in [7].

Structure of the material, after all stages of preparation, was studied by X-ray diffraction (Philips X-pert, Cu-K_{α}). Microstructure was analyzed using High Resolution Electron Microscope JEOL JEM 3010. The hysteresis loops were recorded in a temperature range 90–300 K, in external magnetic field ±1,600 kA/m. Zero-field-cooled (ZFC) and field-cooled (FC) magnetization curves were studied versus temperature in a range of 4–300 K, in a field of 8 kA/m.

6.3 Results and Discussion

The monomers had either crystalline (cobalt) or amorphous (iron and nickel) structure. The frontal polymerization products were amorphous and exhibited paramagnetic properties. Because the nanocomposite material is formed in the course of pyrolysis, the first two stages - monomer formation and frontal polymerization are not discussed in this paper in more detail.

All the pyrolysis products were in a form of beads exhibiting broad size distribution, ranging from few up to 500 μm and had high porosity with pores size within a range 5–50 μm (Fig. 6.1). The magnetic nanocrystallites were distributed within the volume of the powder particles.

After pyrolysis distinct crystalline reflections, which appeared in the pattern, matched well the characteristic lines of metallic Co, Fe and Ni, respectively (β-Co, FCC; αFe, BCC and Ni, FCC). Broad reflections suggested nanocrystalline structure. The Scherrer assessment gave the mean crystallite size 7–30 nm (Fig. 6.2). In fact the transmission electron microscopy (TEM) observations confirmed that the crystallites are very fine.

The TEM images (Fig. 6.3) proved that the Co crystallites obtained at the pyrolysis temperature 773 K had uniform size, close to 7 nm, which well corresponded to the value obtained from the X-ray pattern. For the Fe and Ni crystallites their size was in the ranges 10–35 and 8–14 nm, respectively. The Co and Ni crystallites had spherical shape, whereas the Fe crystallites were in a form of floccules.

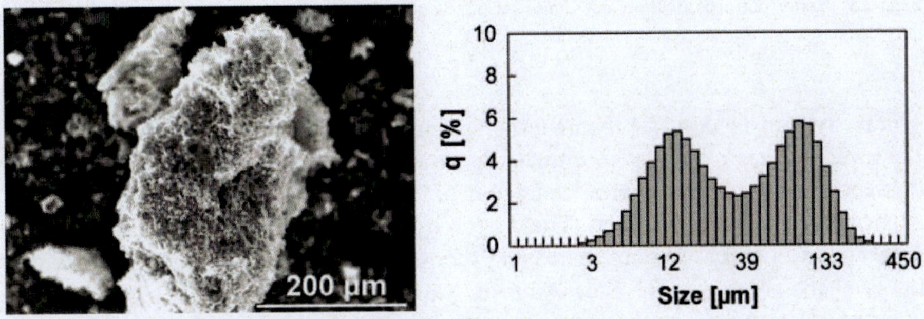

Fig. 6.1 Particles morphology after pyrolysis (*left*) and particle size distribution (*right*) for beads containing magnetic nanocrystallites in their volume

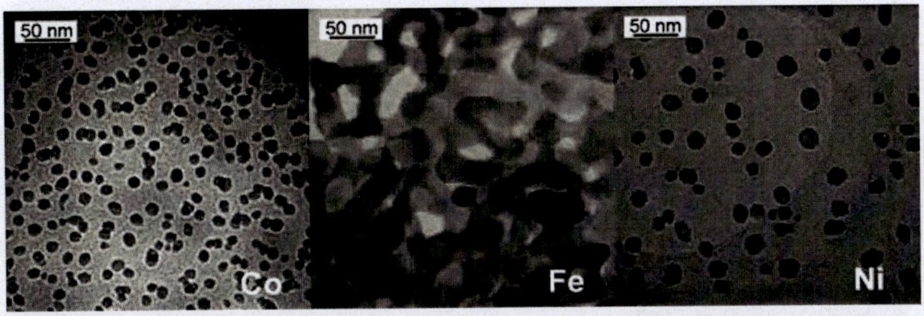

Fig. 6.2 TEM images of nanocrystallites obtained by pyrolysis of materials containing Co, Fe and Ni

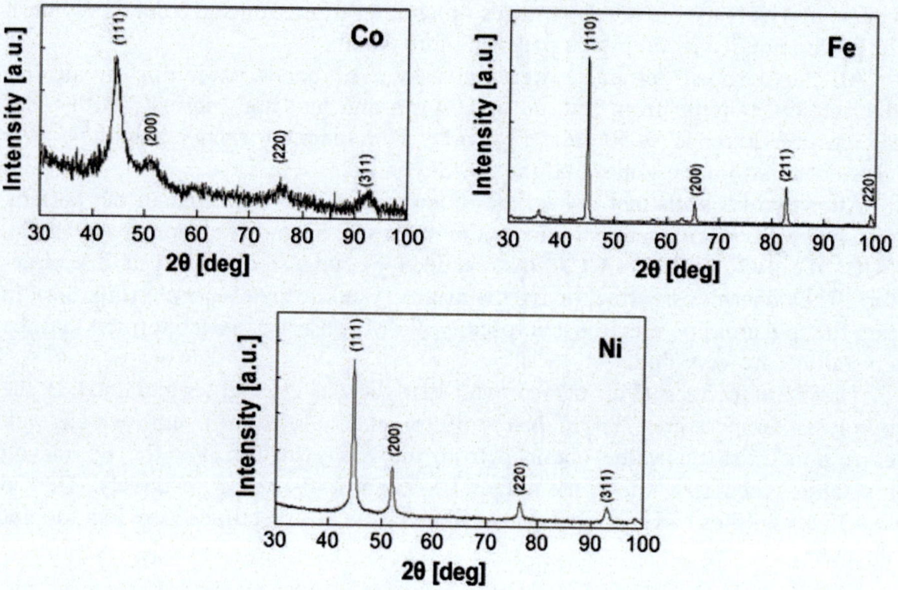

Fig. 6.3 X-ray patterns of the Co, Fe and Ni beads after pyrolysis

The hysteresis loops for the material containing cobalt were closed and symmetrical versus origin of the coordinate system (Fig. 6.4). The shape of the loops evidenced the ferromagnetic character of the material. No evidence for the superparamagnetism was seen. The coercivity depended on the measurement temperature and decreased with increasing the temperature. At room temperature the material pyrolysed at 773 K showed the saturation magnetization 33 emu/g and coercivity 10 kA/m. The FC and ZFC curves did not show any evidence of blocking temperature, confirming the conclusion from the hysteresis measurements. Taking into account that the saturation magnetization of cobalt is 165 emu/g, the amount of metallic Co in the composite powder can be assessed at a level of 20 wt%. For the nanocomposite beads containing iron, room temperature saturation magnetization and coercivity were 100 emu/g and 17.3 kA/m, respectively. The coercivity slightly depended on the temperature. However, this rather high coercivity, about twice the value measured for Armco iron, suggests that some amount of carbon went into solution with Fe. The loops were closed and symmetrical versus origin of the coordinate system. Shape of the loops evidenced the ferromagnetic character of the material. No evidence of superparamagnetism was seen. The magnetization versus temperature curves, taken in a field 8 kA/m are shown in Fig. 6.4. The magnetization for the ZFC curve increased with growing temperature, however, no distinct maximum, representing the blocking temperature, was seen. As the saturation magnetization of Fe is 217 emu/g, hence the percentage of iron nanocrystallites in the composite powder amounts to 46 wt%.

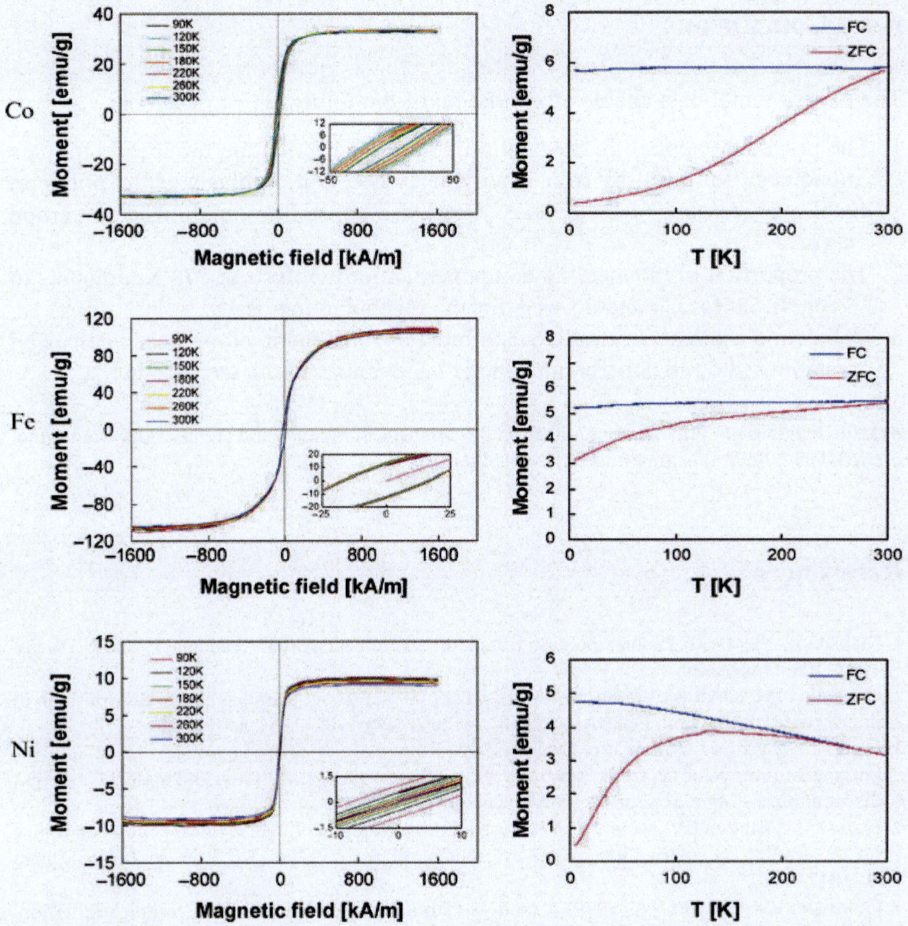

Fig. 6.4 Hysteresis loops, FC and ZFC curves versus temperature for beads containing Co, Fe and Ni. Insets show parts of hysteresis loops close to zero field

The magnetic studies for the beads containing nickel are also shown in Fig. 6.4. The saturation magnetization was 10 emu/g, however, the coercivity decreased from 8 kA/m at 90 K, down to zero at room temperature. This suggests that substantial proportion of the nanocrystallites exhibited superparamagnetic properties. This was confirmed by studies of ZFC and FC magnetization versus temperature. The broad maximum on the ZFC curve suggests existence of the nanocrystallites having broad spectrum of sizes and thus blocking temperatures. The material becomes superparamagnetic at around 200 K. Assuming the saturation magnetization of nickel equal to 54 emu/g, the amount of Ni in the material is 18 wt%. More detailed studies revealed that the proportion of the metallic component can be altered by changing the pyrolysis temperature [8].

6.4 Conclusions

The experimental data enables formulation of the following conclusions:

1. The procedure applied in the studies, comprising formation the metal complex (monomer), followed by frontal polymerization and pyrolysis of the polymer, enables processing of elementary metal nanocrystallites, stabilized in carbon matrix.
2. The proportion of the metallic component, after pyrolysis at 773 K, amounts to 18 for Ni, 20 for Co and 46 wt% for the Fe containing beads.
3. Cobalt and iron nanocrystallites exhibited ferromagnetic properties, whereas Ni crystallites showed superparamagnetic behaviour at room temperature.

Acknowledgement Financial support from the Ministry of Science and Higher Education (grant 2761/B/T02/2009/36) is gratefully acknowledged.

References

1. Maksimov EI (1970) Frontal polymerization of monomers. Dokl Akad Nauk SSSR 191(5): 1091–1093, in Russian
2. Chechilo NM, Khvilitskij R Ya, Enikolopyan NS (1972) About phenomenon of transmission of polymerization reaction. Dokl Akad Nauk SSSR 204(5):1180–1181, in Russian
3. Pojman JA, Willis J, Fortenberry D et al (1995) Factors affecting propagating fronts of addition polymerization: velocity, front curvature, temperature profile, conversion and molecular weight distribution. J Polym Sci Part A: Polym Chem 33:643–652
4. Ivanov VV, Stegno EV, Melnikov VP, Pushacheva LM (2002) Influence of reaction conditions on the frontal polymerization of methyl methacrylate. J Polym Sci Part A: Polym Chem 40:1017–1022
5. Goldfeder PM, Volpert VA (1999) A model of frontal polymerization using complex initiation. Math Probl Eng 5(2):139–160
6. Goldfeder M, Volpert VA, Ilyashenko VM et al (1997) Mathematical modeling of free-radical polymerization fronts. J Phys Chem B 101(18):3474–3482
7. Pomogailo AD, Dzhardimaleva GI (2004) Frontal polymerization of metal-containing monomers: achievements and problems. Polym Sci Ser A+ 46:250–263
8. Sówka E, Leonowicz M, Pomogailo AD, Dzhardimalieva GI, Kaźmierczak J, Ślawska-Waniewska A, Kopcewicz M (2008) Tailoring of the magnetic properties of Co, Fe and Ni nanocrystallites. Rev Adv Mater Sci 18:654–657

Chapter 7
Hydrogenation of Fullerite C_{60} in Gaseous Phase

D.V. Schur, S.Yu. Zaginaichenko, A.F. Savenko, V.A. Bogolepov, N.S. Anikina, A.D. Zolotarenko, Z.A. Matysina, T. Nejat Veziroglu, and N.E. Skryabina

Abstract The fullerene is the fourth allotropic modification of carbon and its properties, as volume, gravimetric and electrochemical capacities, are in excess of many similar properties of metal hydrides and hydrocarbons. The solution of the problem of the reversible hydrogenation of each carbon atom in the frame of fullerene molecule will allow to create the hydrogen storage with the capacity up to 7.7 wt.% H.

A series of experiments have been conducted to evaluate the full hydrogenation of fullerite C_{60}; hydrofullerenes have been produced experimentally with the variable content of hydrogen. The optimum regime of C_{60} hydrogenation has been determined resulting in the full hydrogenation of fullerene molecule C_{60}. As was apparent after the tests, the sequence of formation of hydrogenated fullerene molecule $C_{60}H_{60}$ in fullerite has been going in the following order: the molecular hydrogen dissolution in octahedral interstices of fcc lattice of fullerite, the dissociation of hydrogen molecules in going from octa- to tetrahedral interstices, the interaction of hydrogen atoms with fullerene molecule. It has been demonstrated that chemisorption process of hydrogen by molecule C_{60} is limited by diffusive processes in fullerite after hydrogen concentration conformed to $C_{60}H_{36}$. The spectral analysis have shown that the second stage process of chemisorption follows the compressive shell model. The suggestion of the model of processes going on at the interaction between H_2 and fullerite C_{60} has been

---

D.V. Schur (✉), S.Yu. Zaginaichenko, A.F. Savenko, V.A. Bogolepov, N.S. Anikina, and A.D. Zolotarenko
Institute for Problems of Materials Science of NAS of Ukraine, Krzhyzhanovsky str. 3, 03142 Kiev, Ukraine
e-mail: shurzag@ipms.kiev.ua

Z.A. Matysina
Dnepropetrovsk National University, 72 Gagarin str, Dnepropetrovsk 49000, Ukraine

T.Nejat Veziroglu
International Association for Hydrogen Energy, 5794 SW 40 St. #303, Miami Fl 33155, USA
e-mail: verzioglu@iahe.org

N.E. Skryabina
Perm State University, Bukireva str., 15, Perm 614990, Russia

S.Yu. Zaginaichenko et al. (eds.), *Carbon Nanomaterials in Clean Energy Hydrogen Systems - II*, NATO Science for Peace and Security Series C: Environmental Security 2, DOI 10.1007/978-94-007-0899-0_7, © Springer Science+Business Media B.V. 2011

made. The mechanism for the definition of hydrogenation degree of molecule C_{60} has been proposed in the present paper.

Keywords Fullerene C_{60} · Hydrogen · Fullerite · Absorption · Chemisorption · Hydrogen capacity · Hydrogen energetics

7.1 Introduction

It is well known that, fossil fuels resources – headed by petroleum and natural gas – are being depleted rapidly and the environmental problems (climate change, acid rains, ozone layer depletion, pollution, oil spills, etc.) caused by their utilization are increasing. Rising energy prices and the growing environmental damage are hurting the economies of the developing countries mainly, and the income per capital gap between industrial countries and developing countries is widening. This is causing many problems between the developing countries and the industrial countries, including ever-growing illegal immigration to industrial countries such as EU and USA.

The Hydrogen Energy System is the permanent answer to these interrelated global problems. Under this system, each country will be able to produce the fuel it needs, i.e., hydrogen, to power its economy, using the primary energy sources available. Therefore, they will not be affected by the rising energy prices and will not spend foreign exchange to import petroleum and/or natural gas. This will speed up their economic growth and at the same time eliminate the environmental damage. Consequently, the Hydrogen Economy will help developing countries develop faster, protect the environment, and eliminate the tension between the developing countries and the industrially developed countries.

The hydrogen safe storage and transportation are the problems restrained the introduction of hydrogen energetics. Solving this problem can contribute considerably to the fast progress of hydrogen at the power market. The hydrogen storage in solids meets the requirements and supply most of the needs of life. The hydrogen density in the volume unit for the majority of hydrides exceeds the density of liquid hydrogen. The bulk of such materials are not explosive, containers for their storage are compact and cheap and it is less of a problem for these containers transit than that during liquid and gaseous hydrogen transportation.

The recently found fourth allotropic modification of carbon, as fullerene, can be one of such materials. As pointed out in [1], the fullerite properties, as volume ($8.25 \cdot 10^{22}$ H atoms/cm^3), gravimetric (7.7 wt.% H) and electrochemical (2,000 mA/h·g) capacities, exceed many similar properties of metal hydrides and hydrocarbons. The solution of the problem of the reversible hydrogenation of each carbon atom in the frame of fullerene molecule will allow to create the hydrogen storage with the capacity up to 7. 7 wt.% H.

Over the last two decades, a great number of papers were devoted to the solution exactly of this problem [2–17]. In our papers [1, 18] we specified that hydrogen in fcc lattice of fullerite (fullerene in the solid state) can be in the lattice (absorbed) and fullerinated (chemisorption) states. As shown in the paper [19], the fcc fullerite lattice can contain 18 hydrogen atoms in the absorbed state per one molecule C_{60} at the temperature of T → 0 K. However at the normal conditions only octa- and tetrahedral interstices, as positions of possible location of hydrogen atoms, are worth consideration. The theoretical justification of possibility of all carbon atoms hydrogenation in the fullerene structure is given also in this work.

The conditions of experiments and discussion of experimental results, received during the investigation of special features of processes of gaseous hydrogen interaction with fullerite, are being given in the present paper.

7.2 Experimental

7.2.1 Choosing the terms of experimental investigations

Considerable recent attention has been focused on the study of possible execution of full hydrogenation of fullerene molecule C_{60} [1–18]. However the final objective has not been achieved by none of these works. Only the scientific group of Bashkin [20] has succeeded the setted purpose not only for fullerite C_{60}, but also for CNT, nanofibres [21] and even as for graphite using hydrogen pressure about 7–9 GPa. Each carbon atom, formed one or another structure, has been hydrogenated in these experimental conditions, as it was reported in these papers [22].

Using high pressures for hydrogenation of fullerene molecules, researchers have mentioned that the process of hydrogenation causes often the destruction and fragmentation of fullerene molecules. Therefore, the optimum pressure of hydrogen is necessary for the performance of the experiment and this pressure must not break down the fullerene structure and at the same time must initiate the flowing of reaction between hydrogen and carbon atoms formed the fullerene frame.

The hydrogenation of fullerene molecule C_{60} on the Ru catalyst in the liquid phase up to the state of $C_{60}H_{48}$ at the pressure of 12.0 MPa has been gained by the authors of paper [4]. In this case, neither destruction nor fragmentation of fullerene molecules have not been noted.

7.2.2 Determination of optimum temperature of interaction

The 75-h experiment has been conducted for determination of the optimum regime of reaction carrying out on the fullerene molecule hydrogenation at the hydrogen pressure of 12 MPa. In the course of this experiment it has been elucidated the temperature at which the rate of hydrogen interaction with fullerene molecule is at its maximum. The construction and the operation of experimental installation were described in our paper [23]. The fullerite C_{60} sample of mass 0.5–1 g has been used in the experiment. The test specimen has been placed into the working ampoule in the experimental reactor from stainless steel. The reactor has been vacuumized during 1 h for the removal of impurities from fullerite. Thereafter the hydrogen has been supplied to reactor from metal-hydride accumulator up to the pressure of 12 MPa and conditioned for 4 h (Fig. 7.1). Then, the temperature in reactor has been raised up

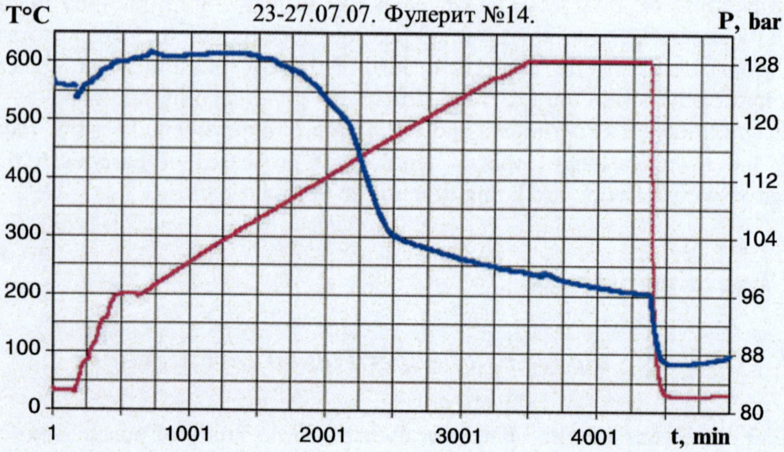

Fig. 7.1 The pressure change in reactor by varying the temperature at the rate of 0.125°C/min.: *1* – curve at pressure change in reactor; *2* – curve at temperature change in reactor

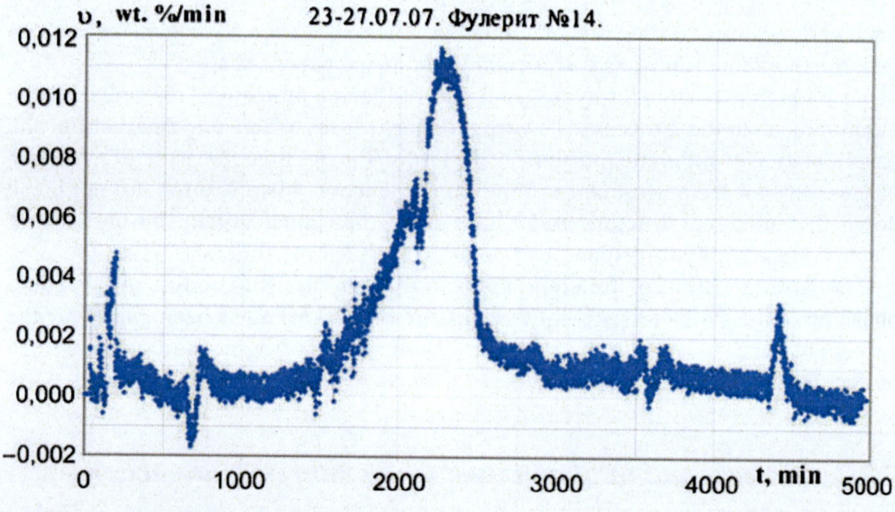

Fig. 7.2 The time-variation of the rate of fullerite C_{60} interaction (the rate of temperature growth from 200°C to 600°C is 0.125°C/min)

to 200°C at the rate of 1°C/min. In doing so the pressure in reactor has been increased on 0.3 MPa, reached 12.8 MPa. This can be explained by the decomposition of solvates formed at the fullerite crystallization. For stabilization of the bar-temperature conditions the specimen has been kept for 4 h more at the temperature of T = 200°C and the pressure of P = 12.8 MPa. At the 625 min the rise of temperature in reactor has been continued from 200°C to 600°C at the rate of 0.125°C per min.

During this experiment it has been found that the maximum rate of interaction has been reached at the 2,250 min (Fig. 7.2) that corresponds to the temperature of

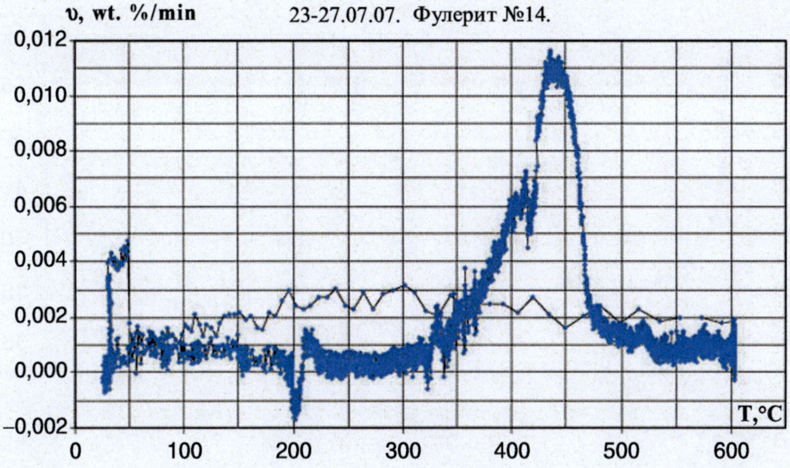

Fig. 7.3 The variation of the rate of interaction between fullerite C_{60} and hydrogen with the temperature

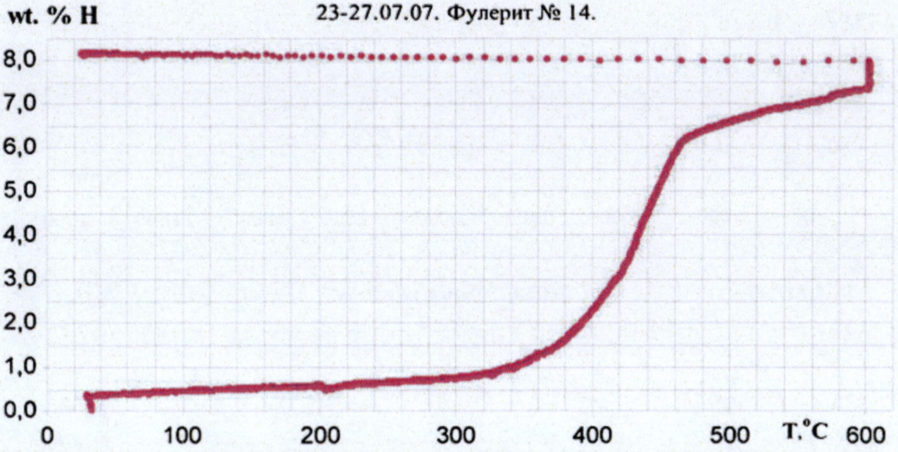

Fig. 7.4 The variation of specimen mass of fullerite with temperature in the course of experiment

433°C (Fig. 7.3). It should be pointed out that interaction rate has been sufficiently high over the interval between 2,181 and 2,410 min (about 4 h) (Fig. 7.2). This rate has been at the temperature change from 425°C to 455°C (Fig. 7.3).

On further increase of temperature the interaction rate begins to reduce. In this experiment after 10 h of exposure at T = 200°C and P = 13 MPa the specimen mass has been changed on 0.6 wt.% (Fig. 7.4). The total amount of hydrogen absorbed by specimen has been fixed at the level of 8.2 wt.% H.

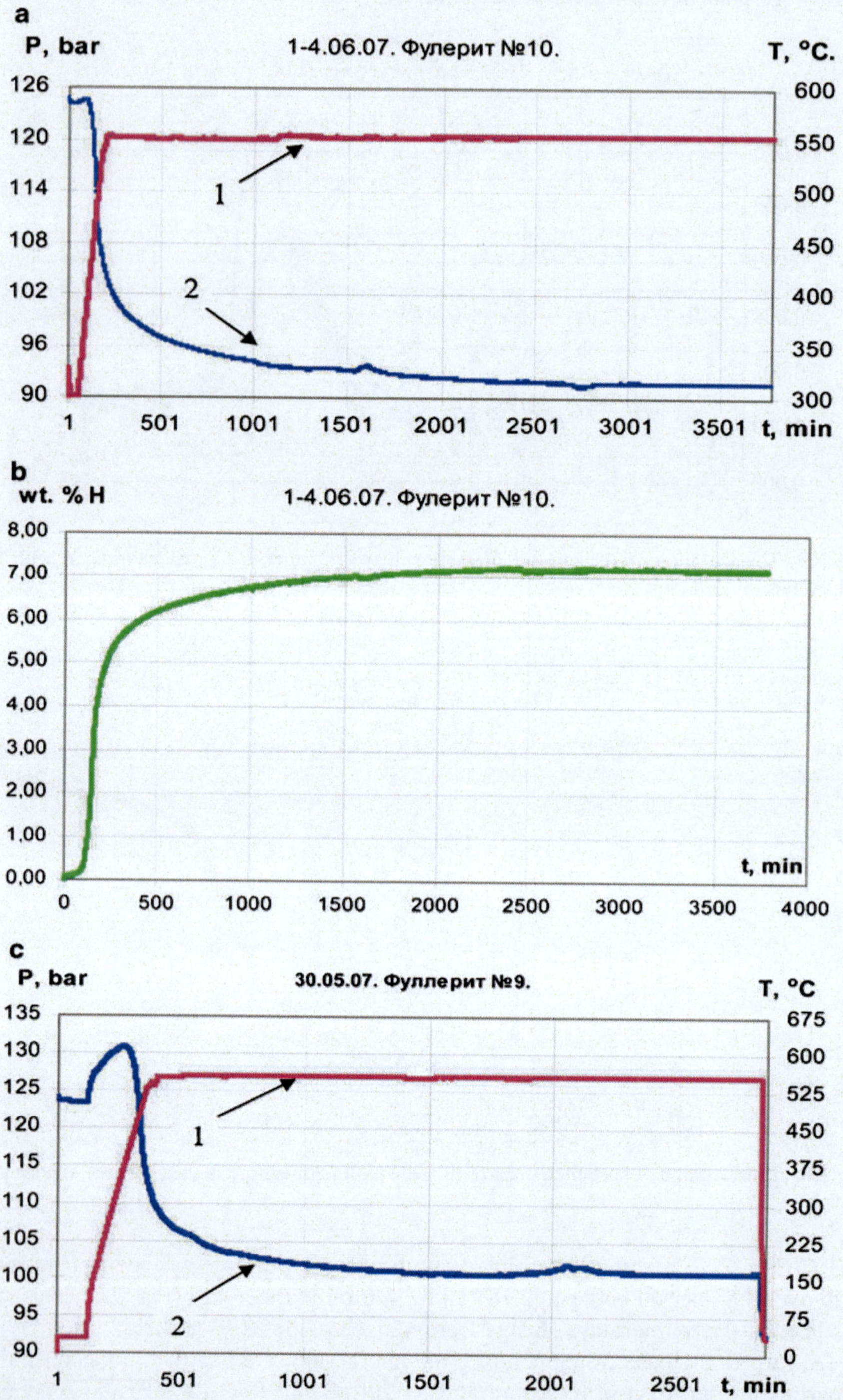

Fig. 7.5 The experimental curves represented the process of interaction of gaseous hydrogen with fullerite C_{60} (starting pressure $P_{st.} = 12.5$ MPa, final temperature $T_{fin} = 550°C$, the heating rate of 1.91°C/min.): (**a**) the temporal changes in reactor temperature (*1a*) and pressure (*2a*) and mass of specimen No x10 (**b**); (**c**) – the temporal changes in reactor temperature (*1*) and pressure (*2*) for specimen No 9

Another experiment (specimen No 10) has shown that the prior annealing of fullerite in vacuum at T = 350°C brings about the more intensive interaction with gaseous hydrogen (Fig. 7.5)

A quick-heating of specimen (over 3 h) up to the temperature of 550°C at hydrogen pressure of 12.5 MPa give rise to specimen saturation with hydrogen up to 7.3 wt.% H (Fig. 7.5b) over this period. The further exposure atT = 550°C more than 65 h has not increased the hydrogen volume in specimen. Conceivably the portion of hydrogen has been absorbed by specimen in the initial stage (T = 350°C) at the hydrogen-filling of reactor.

The results of our experiments for specimen No 9 (Fig. 7.5c) are given for sample that was not annealed tentatively in vacuum. The characteristic splash of pressure in reactor initiated by hydrocarbons ejection at the heating up to 250°C has been seen in the curve "2". The experiments have shown that for the removal of solvated hydrocarbon impurities from fullerite lattice it is sufficiently to heat up fullerite in vacuum over the temperature of 250°C.

By virtue of the fact that initial stage of fullerite interaction with gaseous hydrogen remains the elucidated not to the end the necessity of more careful study of this interaction period arises.

7.2.3 *Study of initial stage of interaction between gaseous hydrogen and fullerite*

Since the foregoing investigations have determined that the optimum temperature of gaseous hydrogen interaction with fullerite (at P = 12–13 MPa) is the temperature range 425–455°C, all further study has been carried out under this conditions.

Before testing, the sample of fullerite (1 g) has been annealed in vacuum at the temperature T = 449°C over 1 h. Thereafter hydrogen bleeding-in to reactor has been performed from metal-hydride accumulator up to the pressure P = 13 MPa.

In the course of experiments, we have studied main dependences (Fig. 7.6):

1. the time-dependence of fullerite mass change (Fig. 7.6a, d);
2. the fullerite temperature modification in time (Fig. 7.6b, e);
3. the time-change of hydrogen pressure in reactor (Fig. 7.6c, f).

The data measurements have been conducted by computer program each minute. The supplementary lines "nm" and "kl" have been entered on the experimental curves of Fig. 7.6d–f and this allows to ascertain the special features of processes under test.

The following peculiarities have been noted in studies of obtained data:

1. the mass change (OA) on 0.085 wt.% can attests the hydrogen adsorption on the surface of fullerite powder (Fig. 7.6a);

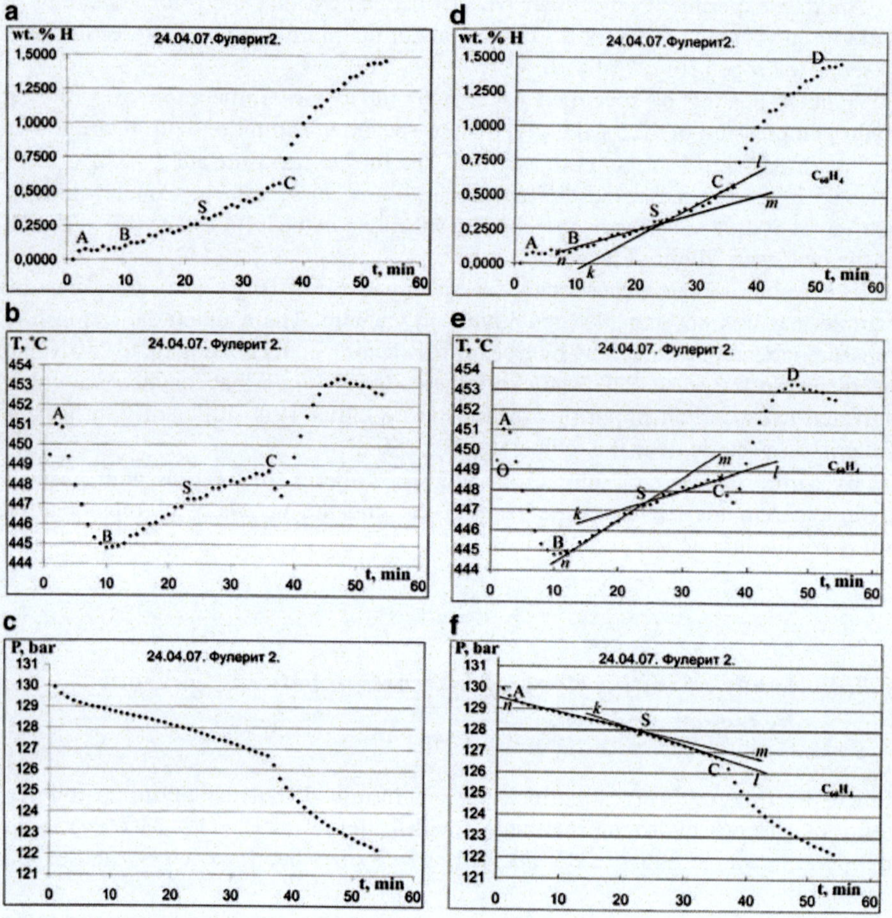

Fig. 7.6 Experimental plots on investigation of interaction processes between hydrogen and fullerite, presented the formation of hydrogen solid solution in fullerite: (**a**, **d**) dependences of fullerite mass change with time; (**b**, **e**) fullerite temperature modifications in time; (**c**, **f**) hydrogen pressure changes in reactor. The neighbouring points of diagram curves are spaced out at equal intervals of 1 min. (**d**, **e**, **f**) are supplemented by subsidiary straight lines "nm" and "kl". Section BS – absorption of molecular hydrogen, Section SC – dissociative absorption of hydrogen

2. exoeffect at the expense of pressure increase in reactor up to 130 MPa and H_2 absorption (section OA (3 min) of Fig. 7.6b); lowering the temperature on 6°C at the unchanged mass of specimen (section AB (7 min) of Fig. 7.6a);
3. increase of specimen mass of 0.5 wt.% corresponding to the introduction of four hydrogen atoms per one molecule C_{60} (section BSC (27 min) of Fig. 7.6a);
4. the break in the point S indicated that interaction changes its character (Fig. 7.6d, e);
5. the total process on section SC is more energy-intensive than on section BS (Fig. 7.6e);

6. the interaction rate on section SC is greater than on section BS (Fig. 7.6d);
7. point C is the site where the interaction nature is radically changed.

7.3 Results and discussion

7.3.1 Lattice hydrogen

On examination of experimental data and theoretical calculations by molecular-kinetic theory, we have a good chance of interpretation of effect existing on experimental curves.

Within the first 3 min the absorption process of hydrogen molecules has proceeded on the surface of fullerite powder and their diffusion is started into the volume of crystallites. In the course of absorption the hydrogen diffuses on octa- and tetrahedral interstitial sites.

In going from octa- to tetrahedral interstice the hydrogen molecule is subject to dissociation process (consuming energy in this case). On further moving into the fullerite volume hydrogen atoms arrive at octahedral interstices from tetrahedral, where they recombinate forming the molecule (releasing energy in this case). The dissociation and recombination processes runs sequentially up to the full occupation of octahedral interstitial sites and formation of molecular fcc sublattice of hydrogen (Fig. 7.7) occurs.

The activation energy of this process is combined from activation energy of elementary acts, making this sequence. As long as octahedral interstices exist, the quantity of dissociation events will be equal to the amount of recombination acts. As the process of filling of octahedral interstices of fcc sublattice of fullerite by hydrogen is completed, the both molecular sublattices will contain the equal quantity of molecules in the elementary cells. The formed solid interstitial solution can be written as follows:

$$C_{60} + H_2 = C_{60}H_2 \qquad (7.1)$$

The section "BS" on experimental curves corresponds to this process. The break of experimental curves is presented in the "S" indicating that the hydrogen interaction with fullerite changes its character.

It can be assumed that further process of hydrogen absorption follows by the way of occupation of tetrahedral interstices by atomic hydrogen (Fig. 7.6e, section SC). However in this case, the energy provision of dissociation would not be compensated by the subsequent exoeffects of recombination reactions. Because of this, the total energy consumption of this process will be more than at the earlier sections and it is defined only by the value of activation energy of the process of hydrogen molecules dissociation.

In the case of fully filling of tetrahedral interstitial sites, the hydrogen content in fullerite is doubled, because for every octahedral interstitial site there are

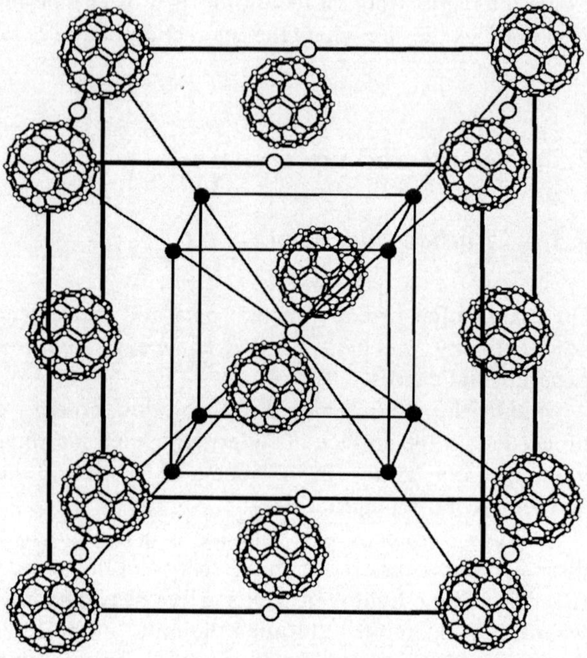

Fig. 7.7 Elementary cell of fullerite with fcc structure: ○ – octahedral (O) interstitial sites, in which molecules of hydrogen are located; ● – tetrahedral (T) interstitial sites, in which atoms of hydrogen are placed. (fullerene symbol) – sites of crystalline lattice, in which fullerene molecules are distributed

two tetrahedral interstices. The formation of solid solution can be expressed by equation:

$$C_{60} + H_2 + 2H \;= C_{60}H_4 \tag{7.2}$$

The atomic-molecular sublattice of hydrogen contains three hydrogen particles (one molecule and two atoms) per one molecule of fullerene.

At this stage the process of physical absorption of hydrogen by fullerite is completed. This process of absorption is changed by desorption process and the overall process goes in the reverse direction with decreasing hydrogen pressure.

From the view-point of molecular-kinetic theory such mechanism of hydrogen absorption by fullerite is quite possible. As interacting atoms and molecules can be considers as rigid spheres, it can be noted that radiuses R_O, R_T of sphere entered into octa- and tetrahedral interstices are determined correspondingly by formulae:

$$R_O = 0.41\ r;\ R_T = 0.22\ r,$$

where r is the radius of sphere, located in the sites of elementary cell (in the case of fullerene C_{60} r = 0.35 nm).

For fullerite we have:

$$R_O = 0.41 \times 0.35\ \text{nm} = 0.1435\ \text{nm};$$

$$R_T = 0.22 \times 0.35 \text{ nm} = 0\ 077 \text{ nm}.$$

The volumes of these interstices are equal to:

$$V_O = 12.37\ \text{Å}^3;\quad V_T = 1.91\ \text{Å}^3$$

with diameters and volumes of hydrogen molecule and atom taken as

$$D_{H_2} = 0.212 \text{ nm} \quad V_{H_2} = 4.99\ \text{Å}^3,$$
$$D_H = 0.092 \text{ nm}; \quad V_H = 0.523\ \text{Å}^3,$$

it is necessary to point out that hydrogen molecule is freely placed in octahedral interstices and hydrogen atom – in tetrahedral interstices.

Hence, it should be mentioned that in the course of fullerite lattice saturation by hydrogen the following processes of physisorbing hydrogen formation are proceeded (Fig. 7.6a, d): (1) OA – adsorption of H_2 on the surface of crystals; (2) AS – absorption of H_2; (3) SC – dissociative absorption of hydrogen; (4) AC – two-stage absorption of hydrogen by fullerite lattice; and (5) CK – chemisorption of hydrogen by fullerene molecule.

7.3.2 *Fullerinated hydrogen*

The performed experiments have shown that after pressure release the absorbed hydrogen by fullerite is desorbed gradually at normal conditions. Such mobility of hydrogen can signify that the process of hydrogen chemisorption by all molecules of fullerene C_{60} will undergo evenly with the formation of molecules of equal energy or only fullerene molecules with the equal quantity of hydrogen atoms over all lattice volume. In this case topochemistry of reaction of fullerite interaction with hydrogen corresponds to the volume formation of hydride.

The sets of experiments have been pursued for the purpose of elucidation of topological model of hydrogen interaction with fullerite. The method of express-analysis of definition of fullerene molecules C_{60} presence in the solution developed by authors has been used in the experiments. This method has been described in the paper [24]. The experiments have demonstrated that topochemistry of process of hydrogen chemisorption by fullerene molecule in fullerite up to concentration of $C_{60}H_{36}$ follows the topochemical model of the volume formation of product and after this note the model of compressive shell is operating (Fig. 7.8).

Analyzing experimental data on chemisorption of hydrogen, it should be emphasized that this process also runs in two stages. The uniform hydrogenation of fullerene molecules over the fullerite volume up to 36–40 atoms of hydrogen is related to the first stage. The second stage on the kinetic curves is expressed by the

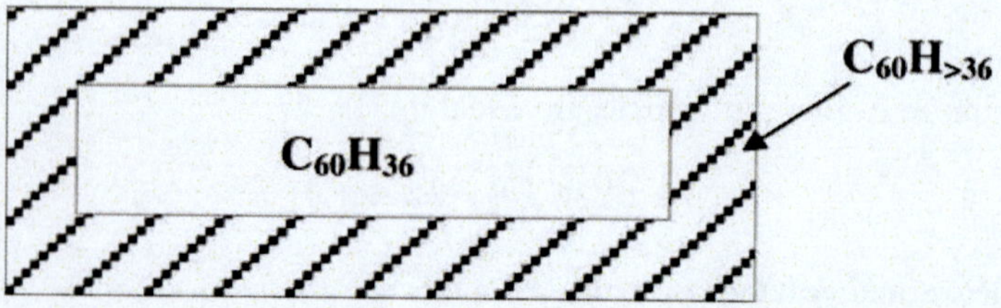

Fig. 7.8 The topochemical model of compressive shell typical for hydrogenation of fullerite more than concentration $C_{60}H_{36}$

parabolic dependence. This testifies that the process has diffusive-limitative character (Figs. 7.4 and 7.5).

The succession of experiments has been carried out to study the first stage of chemisorption and their findings are given in this work (specimens No 9 and No 10 – Fig. 7.5; No 12 and No 13 – Fig. 7.8 and No 14 – Figs. 7.1–7.4). The results of experimental data shown in these figures have been accumulated in the Table 7.1.

The results analysis has shown that the value of temperature range of chemisorption stage I ($\Delta T = T_I - T_{II}$) does not depend on the rate of temperature raise and corresponds to the range near 115–125°C, whereas the change of this temperature range may last from 60 to 2,000 min. However, it should be mentioned that the rise of temperature rate (υ_H) considerably affects the initiation of interaction with its shift into the more high-temperature field (see T_I in Table 7.1). At the low rates of heating the reaction proceeds more completely (Δm).

The change of average rate of fullerite mass in time ($\upsilon^t_{\Delta m}$) increases with increase in the rate of temperature rise. In this case the specific volume of absorbing hydrogen on 1°C of temperature increase remains constant for any rate of heating ($\upsilon^t_{\Delta m}$= const).

From the aforesaid and focusing attention on the behaviour of the last two parameters ($\upsilon^t_{\Delta m}$ and $\upsilon^T_{\Delta m}$), it might be assumed that hydrogenation of fullerene molecule is the energy-controlled process in which the system temperature determines the quantity of hydrogen atoms on the fullerene molecule shell. This conclusion has a vital importance for further investigation of fullerene-hydrogen system and specifically for its applied utilization.

Thus, from the results obtained it may be concluded that hydrogen in fullerite lattice at T = 450°C conserves its high mobility up to its total concentration 5.5–6.0 wt.% that correspond to >40 hydrogen atoms chemisorbed by each molecule C_{60}. Thereafter the process of interaction is limited by diffusive processes (Fig. 7.9).

Because on the second stage the interaction process between hydrogen and fullerite follows the model of compressive shell, the uniform distribution of hydrogen between all molecules of fullerenes is not to be expected in this case (i.e. the formation of molecules with equal amount of hydrogen atoms will not exist). The process of second stage of chemisorption in fullerite differs from the previous processes just by this.

Table 7.1 The summary table of experimental and design data obtained in studies of chemisorption stage I of hydrogen by fullerite

No Experiment	υ_H (°C/min) The rate of specimen heating	T_I (°C) Temperature of chemisorption initial stage I	T_{II} (°C) Temperature of chemisorption initial stage II	ΔT(°C) Temperature range of realization of chemisorption stage I ($T_I - T_{II}$)	Δm (wt. %) Mass change in the course of chemisorption stage I	t_I (min) Duration of chemisorption stage I	$\upsilon^t_{\Delta m}$ (wt.%/min) Average rate of mass change in time ($\frac{\Delta m}{t_I}$)	$\upsilon^T_{\Delta m}$ (wt.%/°C) Specific mass of chemi-sorbed hydrogen with temperature rise of 1°C ($\frac{\Delta m}{\Delta T}$)
12	0.061	314	436	122	5.9	2,000	0.0030	0.048
14	0.125	340	460	120	5.8	1,100	0.0052	0.048
13	0.139	346	465	119	5.8	1,000	0.0058	0.049
9	1.34	380	503	123	5.5	92	0.060	0.045
10	1.91	400	516	116	5.38	61	0.088	0.046

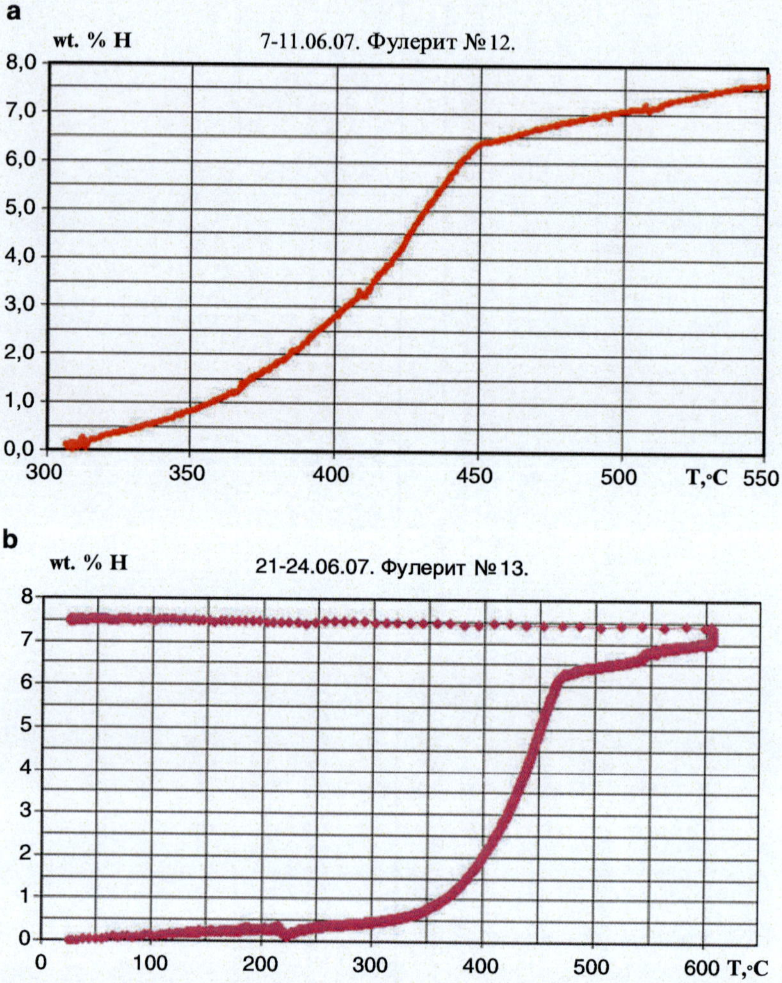

Fig. 7.9 The temperature dependence of mass change of fullerite specimen: (**a**) specimen No 12, (**b**) specimen No 13

7.3.3 *Model of gaseous hydrogen interaction with fullerite*

Using the above discussion data, it can be proposed the model of processes occurred at the interaction between fullerite and hydrogen (Fig. 7.10). It is evident from this model that the interaction process involves several activated stages. The surface stage involves adsorption and hydrogen molecules transition from the adsorbed state to the absorbed one. The diffusive processes involve molecular diffusion, atomic diffusion and hydrogen atoms transfer through the layer of hydrogenated molecules of fullerene. They are controlled by the volume and grain boundaries

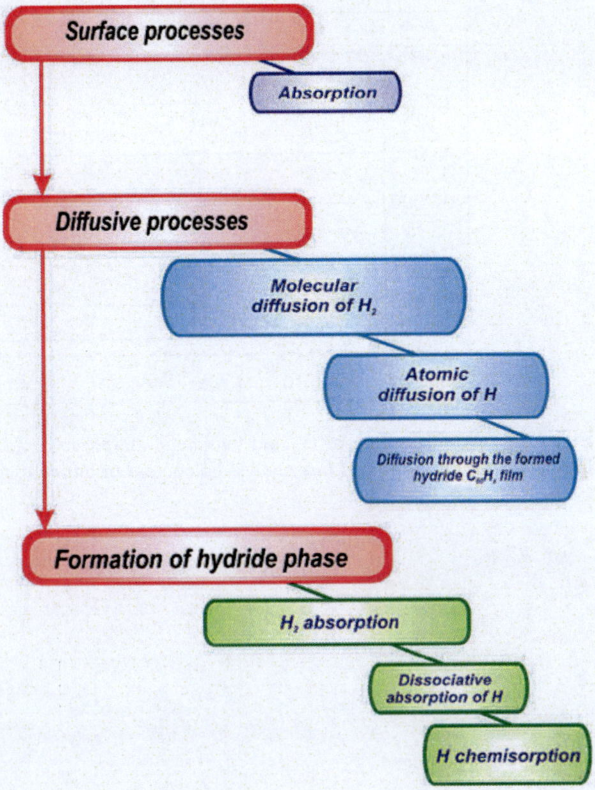

Fig. 7.10 Hypothetical model of processes at the gaseous hydrogen interaction with fullerite

diffusion. The hydride phase formation consists of the chain of elementary acts involving absorption of molecular hydrogen, dissociative absorption and chemisorption of hydrogen atoms by fullerene molecules.

The relative contribution of each stage to the interaction rate during the course of reaction has undergone a change. As this takes place, the limiting effect of the surface and diffusive processes on the reaction rate does not permit to study directly the process of interaction between fullerene molecule and hydrogen atom.

Taking into consideration the proposed model and experimental data obtained during hydrogenation of fullerite, it should be noted that in practice both components (lattice H_L (Fig. 7.11) and fullerinated H_F (Fig. 7.12) hydrogen) should be taken into account. The total content of hydrogen in fullerite (H_{Tot}) can reach the mass of 8.2 wt.% H, i.e.

$$H_{Tot} = H_L + H_F = H_4 + H_{60} = H_{64}, \text{ i.e. } 0.5 + 7.7 = 8.2 \text{ wt.\% H.} \qquad (7.3)$$

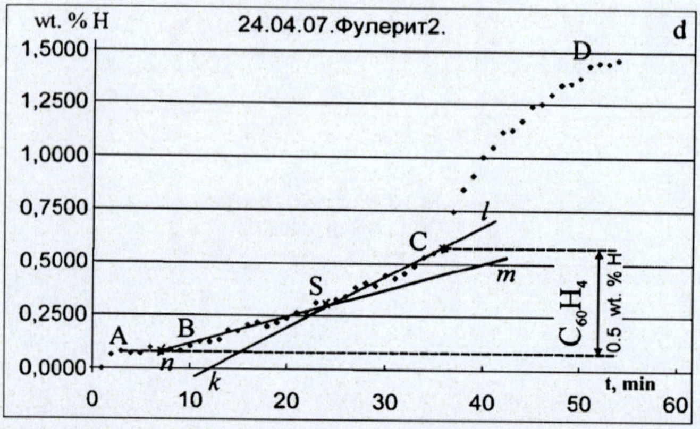

Fig. 7.11 The adsorption of H_2 and two-stage atomic-molecular absorption of hydrogen in fcc lattice of fullerite. *Arrowed line* shows the content of lattice hydrogen (H_L)

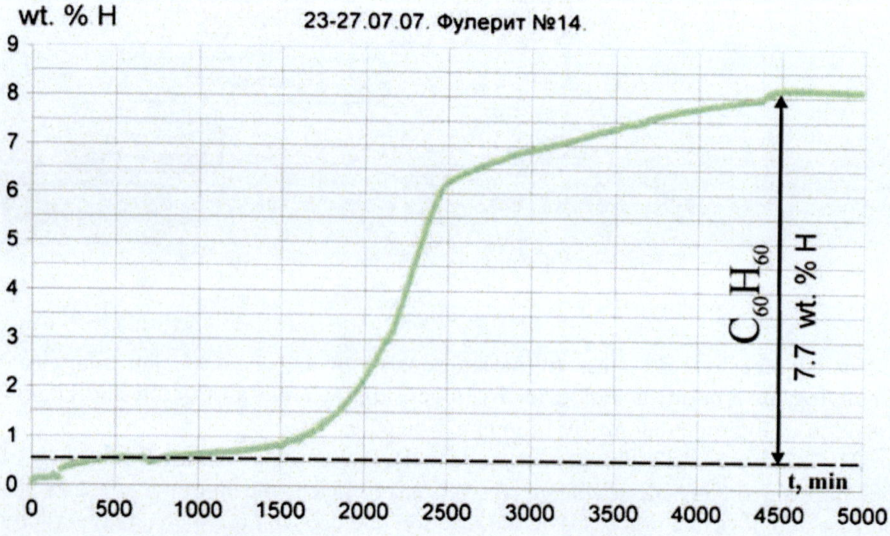

Fig. 7.12 Kinetics of interaction of gaseous hydrogen with fullerite. *Arrowed line* shows the fraction of fullerinated hydrogen (H_F)

In the course of saturation process of fullerite by gaseous hydrogen the total content of hydrogen consists for all time from these two components (Table 7.2). After pressure release and specimen removal from reactor the lattice hydrogen is desorbed spontaneously from fullerite lattice at the normal conditions. Only chemisorbed hydrogen remains predominantly in fullerite. For this reason, in practice it is desirable to use the content of Table 7.2 for investigation of fullerite-hydrogen system.

Table 7.2 Relation between the content of lattice and fullerinated hydrogen in fullerite in the course of saturation process

	Total content H_{Tot} of hydrogen		Lattice hydrogen H_L		Fullerinated hydrogen H_F in compound $C_{60}H_x$	
No	Wt.% H	Atoms H/C_{60}	Wt.% H	Atoms H/C_{60}	Wt.% H	x
1	1.346	10	0.519	4	0.826	6
2	2.158	16	0.519	4	1.639	12
3	2.958	22	0.519	4	2.439	18
4	3.745	28	0.519	4	3.225	24
5	5.281	40	0.519	4	4.761	36
6	6.769	52	0.519	4	6.250	48
7	8.211	64	0.519	4	7.692	60

7.4 Conclusions

A series of tests were conducted to systematically investigate the fundamental property of fullerite C_{60} on hydrogen capacity.

The main results are summarized as follows:

1. Hydrofullerenes have been produced experimentally with the dissimilar content of hydrogen.
2. The optimum regime of C_{60} hydrogenation has been determined resulting in the full hydrogenation of fullerene molecule C_{60}.
3. The pressure $P = 12$ MPa and temperature $T = 425/455°C$ are the optimal conditions of the performance of process of fullerite C_{60} hydrogenation.
4. As our experimental study has revealed, the induced period, caused by dissociative adsorption of molecular hydrogen, is absent at the interaction between molecular hydrogen and fullerite.
5. The sequence of formation of hydrogenated fullerene molecule $C_{60}H_{60}$ in the fullerite lattice has been going in the following order: the molecular hydrogen dissolution in octahedral interstices of fcc lattice of fullerite, the dissociation of hydrogen molecules in going from octa- to tetrahedral interstices, the interaction of hydrogen atoms with fullerene molecule.
6. It has been demonstrated that hydrogenation of fullerene molecule is the energy-controlled process in which the system temperature determines the quantity of hydrogen atoms on the shell of fullerene molecule especially for its applied use. The chemisorption process of hydrogen by molecule C_{60} is limited by diffusive processes in fullerite after hydrogen concentration conformed to $> C_{60}H_{36}$.
7. The results of our investigation into the application of developed method of spectral analysis have shown that the second stage of process of chemisorption follows the compressive shell model.
8. The model of processes going on at the interaction between H_2 and fullerite C_{60} has been proposed.
9. The mechanism for the definition of hydrogenation degree of molecule C_{60} in fullerite lattice has been suggested in the present paper.

Acknowledgements This work was financially supported by International Atomic Energy Agency (contract No 15895/RO).

References

1. Schur DV, Tarasov BP, Shul'ga YM, Zaginaichenko SYu, Matysina ZA, Pomytkin AP (2003) Hydrogen in fullerites. Carbon 41(7):1331–1342
2. Jin C, Hettich R, Compton R, Joyce D, Blencoe J, Burch T (1994) Direct solid-phase hydrogenation fullerenes. J Phys Chem 98(16):4215–4217
3. Lobach AS, Tarasov BP, Shul'ga Yu M, Perov AA, Stepanov AN (1996) The D_2 reaction with palladium fulleride $C_{60}Pd_{4,9}$, *Izv. RAN., Ser. khim.*, 483–484 (in Russian).
4. Shigematsu K, Abe K, Mitani M, Tanaka K (1992) Catalytic hydrogenation of fullerene C_{60}. Chem Express 7(12):37–40
5. Attalla MI, Vassallo AM, Tattam BN, Hanna JV (1993) Preparation of hydrofullerenes by hydrogen radical induced hydrogenation. J Phys Chem 97:6329–6331
6. Haufler LE, Conceicao J, Chibante LPF, Chai Y, Byrne NE, Flanagan S et al (1990) Reduction with tithium in ammonia in the presence of t-BuOH. J Phys Chem 94(24):8634–8636
7. Henderson CC, Cahill PA (1993) Synthesis of the simplest C_{60} hydrocarbon derivative. Science 259:1885–1887
8. Bashkin IO, Antonov VE, Kolesnikov AI, Ponyatovsky EG, Mayers J, Parker SF, Tomkinson J, Moravsky AP, Shul'ga YM (2000) Hydrogen in the vibrational spectra of high-pressure hydrofullerite. Mol Mater 13(1–4):251–256
9. Gerst M, Beckhaus H-D, Ruchardt C, Campbell EEB, Tellgmann R (1993) [7H] Benzanthrone, a catalyst for the transfer hydrogenation of C60 and C70 by 9, 10-dihydroanthracene. Tetrahedron Lett 34(48):7729–7732
10. Darwish ADM, Taylor R, Loutfy R (2000) In: Proceedings of 197th Meeting of Electrochemical Society, Toronto, Canada, 14–18 May 2000, Abstract. No 693
11. Nozu R, Matsumoto O (1919) Electrochemical hydrogenation of fullerenes in a 30%KOH solution. J Electrochem Soc 1996:143
12. Tarasov BP, Fokin VN, Moravsky AP, Shul'ga YM (1997) Hydrogenation of fullerite in the presence of intermetallic compounds. Izv RAN Ser khim 4:679–683 [in Russian]
13. Tarasov BP, Fokin VN, Moravsky AP, Shul'ga YuM, Yartys' VA (1997) Hydrogenation of fullerenes C-60 and C-70 in the presence of hydride-forming metals and intermetallic compounds. J Alloy Comp 25:253–254
14. Tarasov BP (1998) Mechanism of hydrogenation of fullerite-metallic compositions. Zhurn obshchei khimii 68:1245–1248 [in Russian]
15. Tarasov BP, Fokin VN, Moravsky AP, Shul'ga Yu M, Yartys' VA, Schur DV (1998) Promotion of fullerene hydride synthesis by intermetallic compounds. In: Proceedings of 12th World Hydrogen Energy Conference, Vol 2. Buenos Aires, Argentina, 21–26 June 1998, pp 1221–1230
16. Goldshleger NF, Tarasov BP, Shul'ga YM, Perov AA, Roschupkina OS, Moravsky AP (1999) Interaction of platinum fulleride $C_{60}Pt$ with deuterium. Izv RAN Ser khim 5:999–1002 [in Russian]
17. Goldshleger NF, Tarasov BP, Shul'ga Yu.M, Roschupkina OS, Perov AA, Moravsky AP (1999) In: Kadish KM, Kamat PV, Guldi D (eds) Recent advances in the chemistry and physics of fullerenes and related materials, Vol 7. The Electrochemical Society, Pennington, NJ, p 647
18. Matysina ZA, Schur DV (2002) Hydrogen and solid phase transformations in metals, alloys and fullerites, Dnepropetrovsk: Nauka i obrazovanie, 420p (in Russian)
19. Schur DV, Zaginaichenko SYu, Veziroglu TN (2008) The peculiarities of hydrogenation of pentatomic carbon molecules in the frame of fullerene molecule C_{60}. Int J Hydrogen Energy 33 (13):3330–3345

20. Bashkin IO, Antonov VE, Bazhenov AV et al (2004) Thermostable compounds of hydrogen based on carbon nanotubes and nanofibres produced under high pressure. JETP Lett 79 (5):280–285
21. Bashkin IO, Antonov VE, Bazhenov AV et al (2003) Carbon materials hydrogenated under high pressure. In: Proceedings of 8th International Conference "Hydrogen materials science and chemistry of carbon nanomaterials", Sudak, Crimea, Ukraine, 14–20 Sept 2003, pp 796–799
22. Bashkin IO, Antonov VE, Bazhenov AV et al (2007) High-pressure hydrogenation of graphite. In: Proceedings of 10th International Conference "Hydrogen materials science and chemistry of carbon nanomaterials", Sudak, Crimea, Ukraine, 22–28 Sept 2007, pp 686–689
23. Savenko AF, Bogolepov VA, Meleshevich KA, Zaginaichenko SYu, Schur DV, Lototsky MV, Pishuk VK, Teslenko LO, Skorokhod VV (2007) Structural and methodical features of the installation for investigations of hydrogen-sorption characteristics of carbon nanomaterials and their composites. In: Proceedings of NATO ARW on HMSCCN, Sevastopol; 2005. Published by Springer, The Netherlands, pp365–382
24. Schur DV, Matysina ZA, Zaginaichenko SYu (2007) Carbon nanomaterials and phase transformations in these materials, Dnepropetrovsk: Nauka i obrazovanie, 678 p. (in Russian).

Chapter 8
Carbon Nano/Microstructures for Hybrid Hydrogen Storage Based on Specially Treated Carbon Fibers

Zh.A. Mileeva, I.L. Shabalin, D.K. Ross, V.A. Bogolepov, S.Yu. Zaginaichenko, D.V. Schur, V.A. Begenev, and Z.A. Matysina

Abstract The fabrication of carbon 3D-nano/microstructures based on the nanostructure deposition from gas phase on the surface of specially treated carbon fibres is proposed as an initial preparative stage to produce a carbonaceous scaffold for hybrid (adsorption-absorption) hydrogen storage materials. This materials design approach is focused toward the hybrids/composites, which could combine the capacity of compounds consuming hydrogen chemically with high specific surface area of systems adsorbing hydrogen intensively by physisorption. The fullerene molecules in the reaction zone can serve not only as a source of carbon pair (arc discharge) but as the catalyst of synthesis of carbon nanostructures (pyrolysis of hydrocarbons). In the present work the carbon fibres were impregnated by fullerene solution in toluene that catalyzed the process of carbon nanotubes growth at the fibres surface.

Keywords Hydrogen storage · Pyrolytic synthesis · Arc discharge · Fullerene · Carbon nanostructures

8.1 Introduction

Hydrogen is a significant and progressive energy carrier, as energy release can be achieved without producing polluting and harmful by-products, as would be emitted by the combustion of fossil fuels [1]. Hydrogen storage is regarded as one of the most

---

Zh.A. Mileeva, I.L. Shabalin (✉), and D.K. Ross
Materials & Physics Research Centre, University of Salford, Maxwell Building, The Crescent, Salford, Greater Manchester M5 4WT, UK
e-mail: i.shabalin@salford.ac.uk

V.A. Bogolepov, S.Yu. Zaginaichenko, D.V. Schur, and V.A. Begenev
Institute for Problems of Materials Science of NAS of Ukraine, Krzhyzhanovsky str. 3, Kiev 03142, Ukraine

Z.A. Matysina
Dnepropetrovsk National University, 72 Gagarin str, Dnepropetrovsk 49000, Ukraine

S.Yu. Zaginaichenko et al. (eds.), *Carbon Nanomaterials in Clean Energy Hydrogen Systems - II*, NATO Science for Peace and Security Series C: Environmental Security 2, DOI 10.1007/978-94-007-0899-0_8, © Springer Science+Business Media B.V. 2011

important problems, which have to be solved in the developing area of hydrogen technology and economy, because the common storage systems based on liquefied and pressurized hydrogen exhibit principal drawbacks. According to the technical targets of the US Department of Energy for 2010–2015, the containment parameters of 6–9 wt.% and/or system volumetric capacity of 0.028–0.040 kg/l for hydrogen storage materials should be achieved [2]. Thus, many research projects and programs all over the world are concentrating on the search for hydrogen solid store solutions [3].

There are two principal mechanisms of hydrogen storage in solids connected with: (i) intermolecular forces, which do not cause any change in the electronic patterns (adsorption or physisorption), and (ii) interatomic electron exchange, which leads to formation of different hydrogen containing chemical compounds (absorption or chemisorption). The comprehensive research in these areas has largely concluded that physisorption systems cannot achieve the required capacities without maintaining cryogenic temperatures, whereas chemical absorbers store more hydrogen, but have limited kinetics and high desorption temperatures. Recently declared materials design approach is transferring research focus toward the hybrid hydrogen storage, which could combine capacity of chemisorbing materials with kinetics of physisorption systems [4]. One of the alternatives of such hybrid (composite) system could be: (i) dispersion of nanoparticles of the metals or metal compounds, which chemically interact with hydrogen to form hydrides or other hydrogen containing compounds (amides, imides, etc.) with high gravimetric/volumetric hydrogen uptake, inside of (ii) nanoporous carbonaceous medium (matrix) with high specific surface area, which is active for physisorption of molecular hydrogen and provided by the precise porosity and pore size distribution. For effective physisorption, surfaces suitable for adsorption must be maximized and activated. This could be done by special materials design, as in activated carbons, exfoliated or expanded graphite, carbon nanotubes, carbon pyrolyzed from polymers and some other carbonaceous media including some nanoporous crosslinked or hyper-crosslinked polymers [5–8].

Thus, special research and development of optimal carbonaceous structures become one of the most important stages in materials design for the successful realization of the hybrid hydrogen storage conception. In this work, the attempt of the structure investigation of 3D-nano/microstructures with the carbon nanotubes deposited on the surface of carbon fiber is undertaken.

8.2 Experimental Procedure

The research was conducted using variable tilt angle pyrolytic deposition reactor (Fig. 8.1). It was shown experimentally that the convection and gravitation processes, which are dependent on a reactor position, have a considerable impact on nanostructures formation.

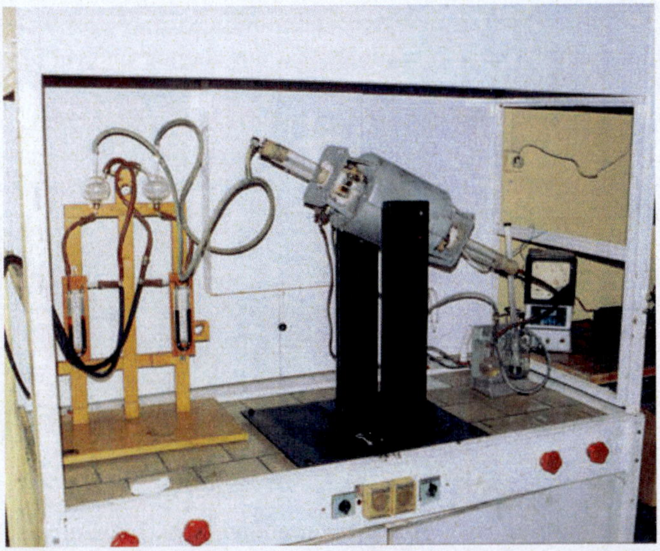

Fig. 8.1 Pyrolytic deposition reactor with variable tilt angle for the synthesis of carbon Pomytkin. A.P.

Employing the reactor, carbon nanotube (CNT) synthesis was carried out in the temperature range from 350°C to 800°C on the surface of carbon fibre filaments with cross-sectional dimensions from 4 to 10 μm, preliminary impregnated with non-metal catalysts. Gas mixtures of acetylene and helium were used as precursors.

Advantage was taken of new materials produced by original (developed by authors) technology with the use of non-metallic catalysts [9–12]. It is common knowledge that metals of iron group (Fe, Ni, Co and their mixtures) are most often used as an catalysts for synthesis of carbon nanostructures. After synthesis the catalyst which content amounts to as much as 30 wt.% is dissolved in mineral acids by boiling.

In our previous papers on the use of fullerene solutions for nanostructures synthesis it was mentioned that at the arc evaporation of nickel in the fullerene solution in toluene the net from carbon nanotubes was formed on the surface of nickel particles [9]. At the evaporation of $LaNi_5$ alloy in the fullerene solution in toluene (compound – 2,355) the nanotubes were identified on the surface of metallic particles [10].

In consequence of these research works we made a conclusion that fullerene molecules in the reaction zone can serve not only as a source of carbon pair (arc discharge) but as the catalyst of synthesis of carbon nanostructures (pyrolysis of hydrocarbons).

The majority of hydrocarbons as a sources of carbon atoms at the pyrolytic synthesis of nanostructures are sublimated and decompose in the temperature range from 200°C to 500°C. For this reason the thermostable catalyst is required over this interval. The carbon fullerene molecules that are coming into sublimation at the temperature higher 600°C can go to work as such catalysts.

The carbon nanostructures prepared by pyrolysis of hydrocarbons on the non-metal catalysts don't require the harmful and power-intensive stage of chemical treatment and they contain on their surface the molecules of inorganic acids in adsorbed form and another associated impurities.

In the present work the carbon fibres were impregnated by fullerene solution in toluene that catalyzed the process of carbon nanotubes growth at the fibres surface.

Morphology and microstructure of manufactured samples were studied by SEM/EDX methods using Philips XL 30 SFEG machine.

8.3 Results and Discussion

The formation of carbon nanostructures (nanotubes and nanofibers of various diameters from 15 nm to 0.5 μm) were determined by SEM on the filaments surface of treated carbon fibre. The obtained experimental results showed correlation between operating modes and characteristics of nanostructures.

Several different types of formations were detected: dense networks of "snarls" (nanotubes of 10–15 nm diameter), hollow nanotubes (usually, of 20–150 nm diameter), "coalescent" nanotubes (100–200 nm) and nanofiber with greater dimensions (up to 0.5 μm), tortuous in shape and rolled like a cigar or a flower bud (Figs. 8.2 and 8.3).

Depending on the nature of initial carbon fiber and operating modes the pyrolytic carbon structures were formed in interfilament spacing occasionally. These structures were of various shapes: from plane needle-shaped (jagged) to three-dimensional smooth pseudo-triangular with negative curvature in cross-section (Fig. 8.4).

Crumbling of deposits from the surface of treated fibers was different as well as the structure of crumbled particles, constituting the "snarls" of nanotubes, pyrolytic formations and nanotubes, wreathed around them (Fig. 8.5).

Under the optimum conditions of carbon fiber filaments treatment ($T = 630°C$, $\gamma = 50^{o}$), which were determined experimentally, the carbon nanostructures, which formed on the surface of carbon fibre, have the sufficient adhesion to the substrate (Fig. 8.6a). More likely, the mechanism of carbon nanotube growth on the specially treated carbon fibre surface is connected with the formation of different carbon atoms cycles: pentagons, hexagons and heptagons (Fig. 8.6b).

The carbon 3D-nano/microstructures, prepared by the manufacturing method mentioned above, can be used in hybrid hydrogen storage as the medium (matrix), which is active for physisorption of molecular hydrogen and containing the dispersion of metal or metal compound nanoparticles inside of itself.

Fig. 8.2 The cluster formation of carbon nanostructures on the surface of treated carbon fiber filaments

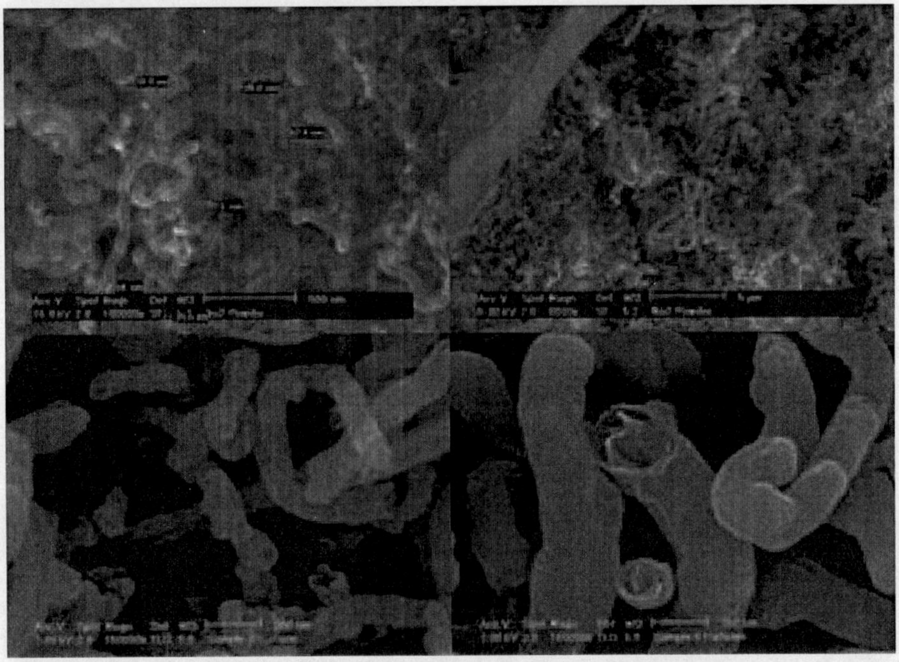

Fig. 8.3 Microstructures of carbon nanotubes deposited on the surface of treated carbon fiber

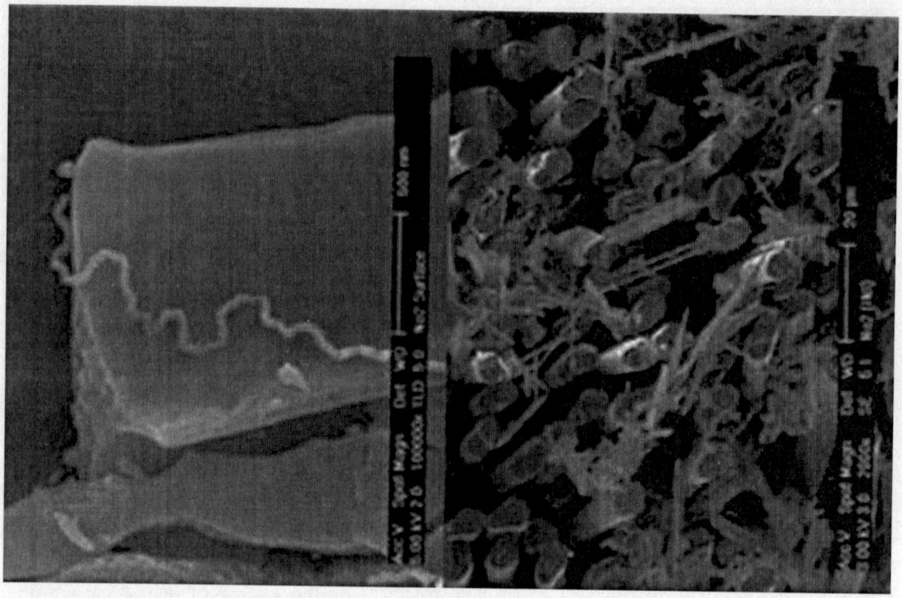

Fig. 8.4 Pyrocarbon deposited structures at the interfilament spacing

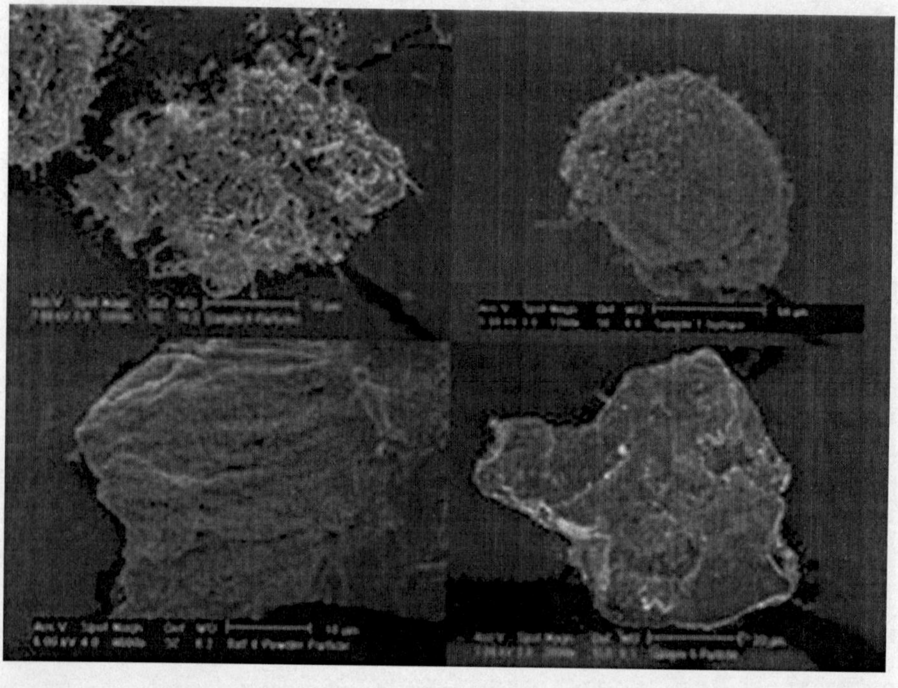

Fig. 8.5 Powder-like particles crumbled from the surface of treated carbon fibers

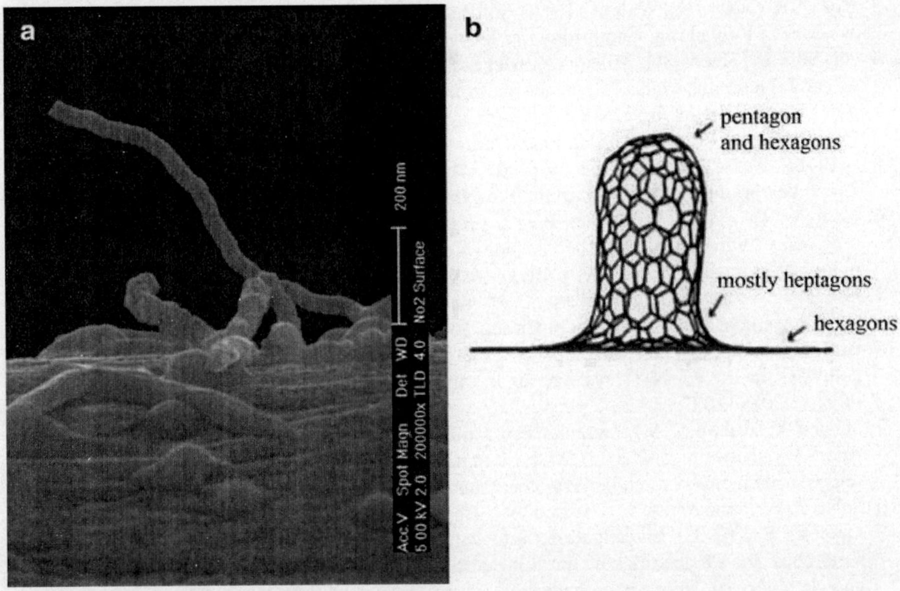

Fig. 8.6 Growth of carbon nanotubes on the treated carbon fibre filament surface. (**a**) SEM image; (**b**) proposed nanostructure growth mechanism

8.4 Conclusions

The manufacturing method, developed experimentally on the basis of the special treatment of commercially available carbon fibers, can be used for the production of special 3D-nano/microstructures. Deposited from gas phase carbon nanostructures have good adhesion to the carbon fibre filaments surface. The character of carbon 3D-nano/microstructures is varied depending on the synthesis conditions. The carbon fibers with the deposited nanostructure clusters have extended surface area. The mechanism of carbon nanotube growth on the specially treated carbon fibre surface is proposed. The carbon 3D-nano/microstructures can be used in hybrid hydrogen storage as the medium (matrix), which is active for physisorption of molecular hydrogen.

References

1. Ross DK (2006) Hydrogen storage: the major technological barrier to the development of hydrogen fuel cell cars. Vacuum 80:1084–1089
2. Hydrogen, Fuel Cells and Infrastructure Technologies Program. Multi-Year Research, Development and demonstration plan. Planned program activities for 2005–2015. U.S. Department of Energy. 3.3 Hydrogen Storage, pp 1–9

3. Ross DK (2008) In: Walker G (ed) Solid state hydrogen storage: materials and chemistry. Woodhead Publishing, Cambridge, pp 135–170
4. Shabalin IL, Keens SG, Mileeva ZhA et al (2009) Hydrogen materials science and chemistry of carbon nanomaterials. The use of neutron scattering techniques for investigation of hydrogen storage systems. In: Schur DV, Zaginaichenko SYu, Veziroglu TN, Skorokhod VV (eds) Proceedings of the 11th International Conference (ICHMS'2009). AHEU, Kyiv, pp 44–45
5. Seifi M, Ross DK, Giannasi A (2007) Raman characterization of single-walled carbon nanotubes produced by the catalytic pyrolysis of methane. Carbon 45:1871–1879
6. Georgiev PA, Ross DK, Albers P et al (2006) The rotational and translational dynamics of molecular hydrogen physisorbed in activated carbon: a direct probe of microporosity and hydrogen storage performance. Carbon 44:2724–2738
7. Lee J-Y, Wood CD, Bradshaw D et al (2006) Hydrogen adsorption in microporous hypercrosslinked polymers. Chem Comm 26:2670–2672
8. Bull DJ, Weidner EW, Shabalin IL et al (2010) Pressure-dependant deuterium reaction pathways in the Li–N–D system: an in situ deutron diffraction study. Phys Chem Chem Phys 12:2089–2097
9. Schur DV, Dubovoy AG, Lysenko EA, Golovchenko TN, Zaginaichenko SYu, Savenko AF, Adeev VM, Kaverina SN (2004) Synthesis of nanotubes in the liquid phase. In: Hydrogen materials science and chemistry of carbon nanomaterials. NATO Science Series, vol II/172, pp 147–151
10. Schur DV, Dubovoy AG, Savenko AF, Bogolepov VA, Koval AYu, Zaginaichenko SYu, Lysenko EA (2003) Investigations into catalytic activity of $LaNi_5$ in synthesis of carbon nanotubes. In: Proceedings of the 8th international conference "hydrogen materials science and chemistry of carbon nanomaterials", Sudak, Crimea, 14–20 Sep 2003, pp 410–413
11. Bogolepov VA, Schur DV, Adeev VM, Golovchenko TN, Voronaya TV, Kotko AV, Lysenko EA (2009) Synthesis of carbon nanotubes on the surface of carbon fibers. In: Proceedings of the 11th international conference "hydrogen materials science and chemistry of carbon nanomaterials", Yalta, Crimea, 25–31 Aug 2009, pp 406–409
12. Mileeva ZhA, Bogolepov VA, Schur DV, Zaginaichenko SYu, Begenev VA, Shabalin IL, Ross DK (2009) Hybrid 3D-nano/microstructures obtaining on the basis of pretreated carbon fibers. In: Proceedings of the 11th international conference "hydrogen materials science and chemistry of carbon nanomaterials", Yalta, Crimea, 25–31 Aug 2009, pp 746–749

Chapter 9
Cyclic Hydrocarbon Decomposition to Carbon Nanoparticles via Spark Discharge

M. Konstantinova and N. Koprinarov

Abstract A method is proposed for decomposition of cyclic hydrocarbons using spark energy for this purpose. The electrical discharge is between electrodes separated at 0.5 mm one from another. An 8,000 V transformer is employed as spark source operating by 50 Hz. Hydrocarbons are mixed with water in the proportion 1/1 and the process is carried out in an ultrasonic tank with the aim to reduce deposition on the electrode surfaces. Graphenes, nanohorns, carbon bands, and carbon cones of different spatial angle and carbon nanobeams have been obtained. The catalyst (Fe in the considered case) influence has been studied through adding ferrocene to the hydrocarbon-water mixture (proportion 1/1/0.2 wt%) during synthesis. The presence of Fe stimulates particle growth but does not change particle type. The synthesized nanohorns and spheres show tendency to order in parallel one to another due to the high electrical field between the spark discharges. The carbon beams are in agglomerates with a direction almost radial to the agglomerate centre.

Keywords Spark discharge · Hydrocarbon decomposition · Carbon particles · Carbon nanocones

9.1 Introduction

The number of new-found carbon structures is permanently growing, recently. This is stimulated by the researchers' interest in the structure peculiarities and properties and the vast potential for practical applications. Once a new particle type has been described, various methods for its synthesis appear. These are mainly based on starting substance decomposition through out-fed energy and creation of those

M. Konstantinova (✉) and N. Koprinarov
Central Laboratory on Solar Energy and New Energy Sources
Bulgarian Academy of Science, Sofia, Bulgaria
e-mail: koprin@phys.bas.bg

S.Yu. Zaginaichenko et al. (eds.), *Carbon Nanomaterials in Clean Energy Hydrogen Systems - II*, NATO Science for Peace and Security Series C: Environmental Security 2, DOI 10.1007/978-94-007-0899-0_9,

conditions that would be the most favorable for the given particle type to arise. That happens with or without catalyst presence. The needed for decomposition energy could be supplied through different kinds of discharge; when it is liberated in chemical processes; received through direct heating or by high-energy irradiation. The processes can be implemented in vacuum; in active or inert gas; as well as in liquid ambient. The processes going on are different in their specificity and strongly depend on the conditions they are realized in. That allows variation of the synthesis conditions and, therefore, a change of the obtained structure types and properties. In the end, however, researchers trying to change the conditions under which the processes of carbon structure synthesis are carried out, came to a level, at which methods practically become close in ideas and way of realization. To a certain extent that concerns the method for carbon nanostructure synthesis we are offering. A similarity of the proposed method might be looked for among the ones known so far, by different signs, for instance, the type of starting materials used, the kind of atmosphere synthesis proceeds in, the energy source kind, the type of synthesized particles and so on, but the obtained results are different. The results discussed in this work could be best match up with the results obtained for arc discharge synthesis in water with the support of gas injection [1], synthesis of single-walled carbon nanotubes and nanohorns by arc in liquid nitrogen [2–4] or in water [5–7] and pulse arc synthesis [8], which are the closest in their root. Such a comparison gives the opportunity to assess the method's specific character and its advantages.

9.2 Experiment

An 8,000 V transformer directly fed from the grid (50 Hz) served as feeding source for the experiments made. Discharges were performed in a vessel full of xylene and water at a volume ratio 1/1. The metal electrodes at a distance of 0.5 mm were dipped in the liquid to avoid air access and liquid ignition. The close location is needed to supply a strong electrostatic field sufficient to cause electric breakdown in the liquid mixture. In order to remove the products deposited on the electrodes during electric breakdown the liquid is constantly subjected to ultrasonic impact. A series of experiments have been done to examine the influence of Fe as catalyst of carbon nanostructure growth by adding ferrocene in the xylene-water mixture in a proportion of 0.2/1/1 wt%. The obtained products were separated after precipitation and drying. Microscopic examinations were carried out with a JEM 2100 – 200 kV TEM and JEOL 100B with no additional material processing.

9.3 Results and Discussion

Synthesis of nanostructures by spark discharge in liquids has a number of peculiarities which reflect on the type of the particles and formations originated. These peculiarities give the method advantages over the other methods of synthesis in liquid.

The first feature is the strong current impulse due to the breakdown in a liquid. Between the electrodes arises a cord of high temperature plasma wreathed by liquid vapor. The pressure in this zone increases and the zone tries to grow in volume. The liquid around is inert and has a very low contractibility. That hampers the expansion and provokes the second peculiarity; namely, synthesis remains in the plasma zone and under a higher than normal pressure. The products synthesized can not overcome the plasma-liquid boundary easily and they remain in the breakdown zone, practically. That is another peculiarity. Ionized atoms get acceleration along the field orientation and they are moving in parallel. At the same time, the unionized atoms and a part of the not accelerated by the field ions turn out to be affected by the strong temperature gradient aroused between the plasma and the surrounding liquid. This gradient accelerates them radially to a plasma cord. After discharge cessation and till the next spark is lighted regeneration of the liquid starts and the obtained products get wet. Thus they penetrate into the liquid but because of the short time to the next impulse they are not able to move substantially away from the place they have been produced. This is of particular importance for the process further repetition. The liquid between the electrodes regenerates during the voltage polarity exchange and its resistance increases. The electric field grows up proportionally to voltage growth till the conditions for the following discharge appear. Due to that the electric field gradient between the electrodes reaches very high values before the breakdown. All this engenders the next peculiarity. The strong electric field tries to put the particles synthesized on its own orientation. Thus conditions are created for the synthesized structures to be arranged in parallel and oriented along the lines of force.

Analyzing the obtained results and the possible processes taking place during the different synthesis stages we came to the conclusion that unattached carbon nanotubes, nanohorns and nanocones originate when around there are no synthesized particles from previous spark discharges. If that is not the case synthesized particles are willing to agglomerate with the neighboring particles. The conditions for nanotube growth (Fig. 9.1), moreover, parallelly ordered should be the most favorable along

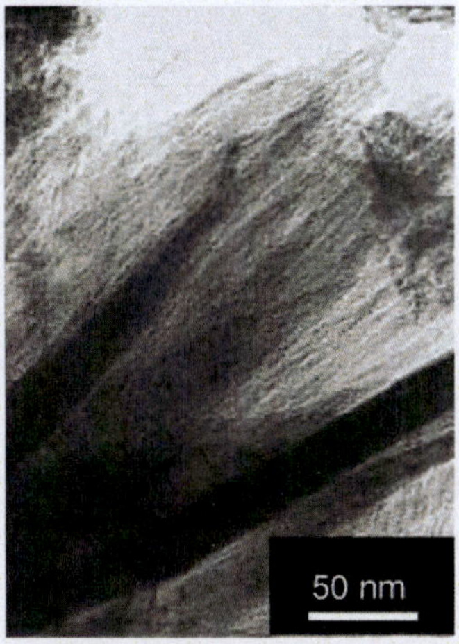

Fig. 9.1 Carbon nanohorns

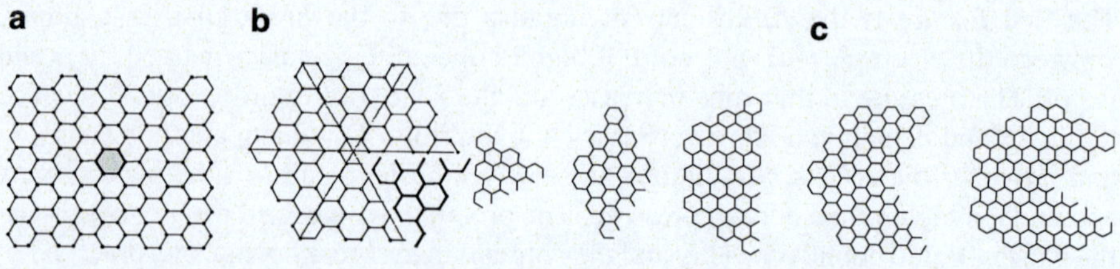

Fig. 9.2 (**a**) Carbon honeycomb network, (**b**) the network divided in six equal parts, (**c**) the networks building cones with angles 19.2°, 38.9°, 60.0°, 83.6°, and 112.9°

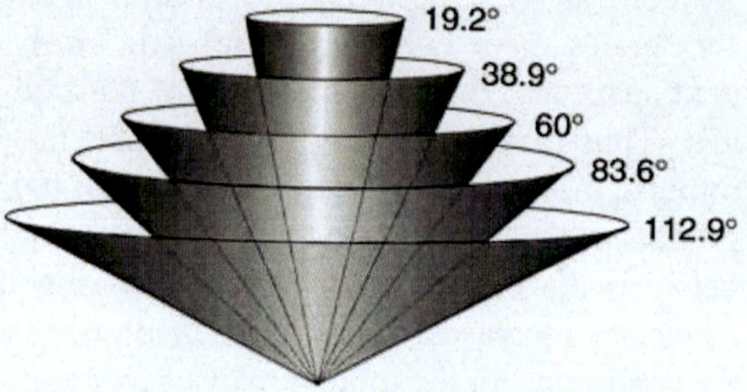

Fig. 9.3 Carbon nanocones with angles 19.2°, 38.9°, 60.0°, 83.6°, and 112.9°

field direction where carbon ions are moving in parallel. This resembles nanotube synthesis under arc discharge in gaseous ambient.

Single wall nanohorns (SWNHs), produced by carbon laser ablation are structures with conical tips with an average cone angle of 20°. They are grouped in spherical aggregates (diameter 80 nm) and with the conical tips of individual tubules protruding out of the surface of the aggregate like horns. The tubules have a typical diameter of about 2 nm with lengths in the range 30 ± 50 nm [9], unlike the ones produced by us are arranged in parallel one to another (Fig. 9.1) and for some of them the length of the conical part approaches 30–40 nm. The reason is that atoms produced by the spark method are not in gas at a comparatively low pressure and have no possibility to build packs (like those after each laser impulse) and to form conical structures there. The ordered longer cones and their going over into nanotubes of bigger diameters could be explained as a result of the atom directed movement under field influence.

Conditions for cone origin are created by atom moving radially to the plasma cord. The cone network arises when in the perfect graphite lattice (Fig. 9.3a) a ring (n-gon) of number of sides $n < 6$ appears. That can be illustrated and explained in the following way: if a carbon net built around a six atom ring is divided into six equal 60° sectors (Fig. 9.2b) and if the creation of the whole network in Fig. 9.2a starts with bonding between atoms from the marked ring, one could say that each sector will grow independently from the others when its first atom is included in this ring. A missing atom in the marked ring will constitute a local defect. The corresponding sector will not be created and, as a result, will be absent in the growing network. Hence, a local defect in the marked ring will transform the whole network despite of the fact that all the other rings have not been altered. The rings of n atoms (n-gon), for $n = 1$ up to infinity, and the changes in the network due to an *n*-gonal defect, are widely discussed and explained in [10]. The sectors grown around the *n*-gons ($n \neq 6$) cannot be flatter than are those in the honeycomb and bend. The calculated by different methods carbon network transition shows that the surfaces in Fig. 9.3c go over in cone-like (a Gaussian positively curved surfaces, with the *n*-gon on the top and rotational axis, crossing perpendicular their *n*-gon surface in its center). From a purely geometric point of view the *n*-gons for $n = 1$, 2, 3, 4 and 5 can be replaced in the cone network by five atomic rings, correspondingly 5, 4, 3, 2 and 1 in number. On a suitable location they lead to an entirely analogous shape of the provoked by them conic surface. Since the possibility of obtaining five atoms carbon rings is quite close to the one to obtain six atom ones and much greater than the one of obtaining other kinds of n-gons it could be assumed with a great extend of reliability that cone origin happens because of corresponding number of five atom rings building in. The apex angles α of the cones can be calculated from:

$$\sin\left(\frac{\alpha}{2}\right) = \frac{2\pi - p\pi/3}{2\pi},$$

where *p* is the pentagon number. For $p = 1, 2, 3, 4$ and 5 the angle α equals to 112.9°, 83.6°, 60.0°, 38.9° and 19.2°, correspondingly (Fig. 9.3).

The synthesis of a great amount of five and six atom rings during spark discharge is a premise for the simultaneous appearance of all kinds of cones in the material synthesized (Fig. 9.4). Several examples of such cones are shown in the figure, together with their angles, though still a number of such cones can be identified on the same photo.

The nanocones produced by the described here method have a height of several tens of nanometers and resemble the ones described in [11]. They are several hundred times smaller than the cones synthesized by accident [12] under pyrolysis of heavy oil in a cycle known as Kvaerner's carbon-black and hydrogen process [13].

We identify the oriented atom moving in the field direction as a reason for the origin of carbon bands, as that shown in Fig. 9.5. They are single-layered and consist of graphene of well arranged atoms. The graphene breakage at an angle of 60° in the zone pointed with an arrow proves the atom perfect arrangement in a honeycomb carbon lattice.

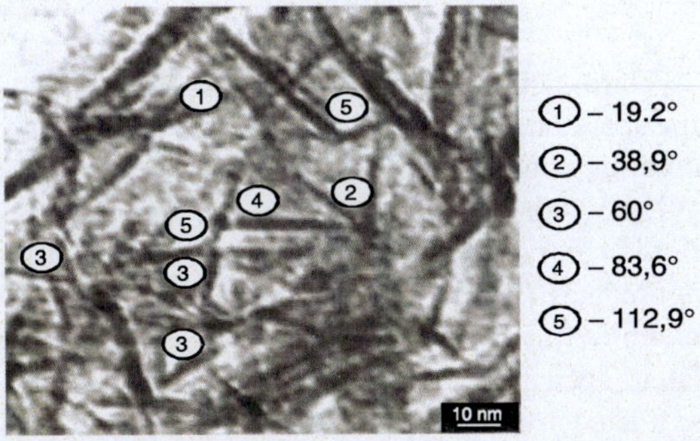

Fig. 9.4 Nanocones into the obtained material

Fig. 9.5 Carbon band

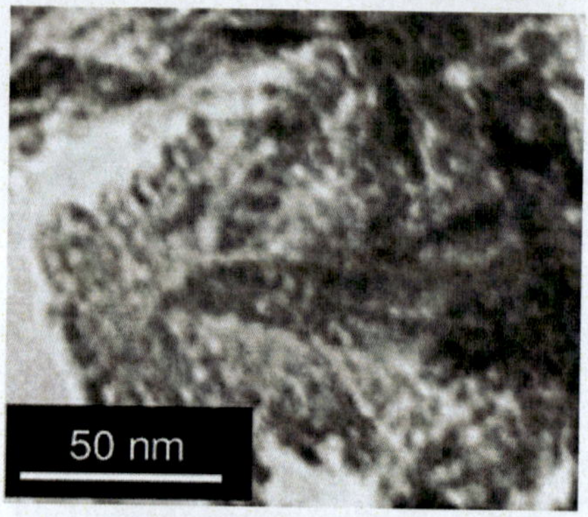

Fig. 9.6 Agglomerate of carbon spheres

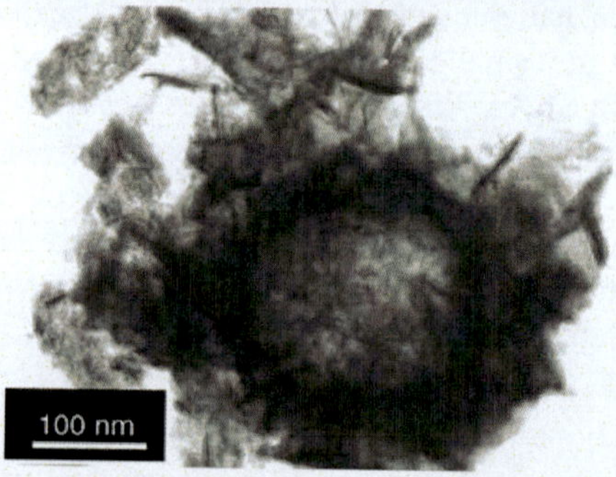

Fig. 9.7 Particles in ring-like aggregates

During the spark flaming the liquid is pressed due to the plasma expending. After the spark burns out the liquid shrinks and presses the atoms not bounded so far, thus creating conditions of their agglomeration in hollow spheric formations (Fig. 9.6). When synthesized particle density becomes high after a long spark number the liquid contraction often gathers the produced particles in big and compact ring-like aggregates (Fig. 9.7). A lot of the discussed here particles and formation can be identified in them. Particle conglomeration causes problems for their separation later. Their microscopic observation is hampered, also, if they are not previously extracted from the produced material. For instance, nanocones are covered and

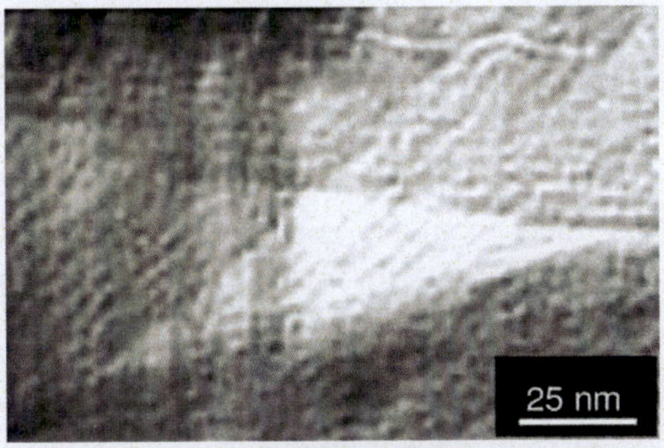

Fig. 9.8 Graphenes one on another with different orientation

Fig. 9.9 Clusters obtained as result of the spark synthesis method

filled in by amorphous carbon and different in size structures. That impedes cone identification and exact determination of their dimensions.

Carbon particles are in a strong electrostatic field between the spark lightings. They start orienting along the field direction, as well as striving to arrange one after another under field influence. That helps building groups of parallel nanotubes, nanohorns and spheres. This effect explains the structure arrangements observed in Figs. 9.1 and 9.6.

The already built up particles or carbon formations can serve nuclei from which similar or new types of structures could be built during the time following spark discharges. The effect of such a mechanism could explain the different orientation of the graphenes shown in Fig. 9.8. In our opinion, when building up the overlying grapheme started the grapheme that serves a base, has not been oriented precisely along the field but it reclined at an angle close to 60º. The overlying grows in the field direction and for that reason both are turned at 60º. In the material synthesized a number of clusters of the type shown in Fig. 9.9 are available, also. The beam orientation out of the center is a clear tendency. Measuring the angles between them shows something unexpected. The angles have no a rate frequency of 60º, as follows, if for some reasons the carbon network doubles or has a common beginning. Their angles are corresponding to those of the conic structures. The possible explanation of this phenomenon is the assumption, that when atoms are radially scattered due to the temperature gradient, cones with their tips directed to the agglomerate center are formed and they divide the flux in parts. As a result nanobeams grow around the cones. That is proved by the fact, that almost all of the cone tips are oriented towards the agglomerate center.

Adding ferrocene to the water-xylene mixture has shown that using iron as catalyst leads to an increase of synthesized particle quantity but no appearance of new kinds of structures and formations has been registered

Because of some unique properties possessed by the spark method produce particles they inspire purely scientific interest in their origin and how they grow besides the opportunities they give for practical application. The particles described above except the cones and the nanohorns with enlarged conic parts have been produces in many ways and undergone very detailed observations concerning their structure and possible applications. For example, nanotubes can find application as electronic devices [14, 15] computing devices [16, 17], big palette of sensors [18–20], actuators [21–23], super capacitors [24, 25], field effect emitters [26]; graphenes are expected to find application in creation of nanodevices [27–29] and the large size cones can serve as sorbent material for solid-phase extraction [30], as field emission sources [31], as tunnel diode tips [32] and as nanosize nozzles for liquid injection, provided a part of the edges has been removed. For that reason this type of particles will not be concerned here.

Recently the search for efficient and reliable methods of hydrogen storage turned to be one of the most important tasks relating the mankind energy problems. It is a task that appeared as a consequence of the technical progress. The creation of mobile machines brought to the necessity of perpetual feed from an energy source, directly connected and moving simultaneously with them. That bases the main requirements for the energy source: to be light, highly efficient and of small volume. The assessments made so far indicate that hydrogen meets to a great extend the requirements for weight and caloricity and the carbon nanostructures are a prospective material for its storage. The first studies on hydrogen storage in carbon structures have been made with carbon nanotubes [33, 34]. The tip which practically is a cone occurred to be the place of most consistent filling up. This conclusion directed researchers to nanohornes [35]. Hydrogen isotherms at 20 K show that the average density of confined hydrogen inside single wall nanohorns (SWNHs) is more densely packed than that in the free liquid and approaches the solid hydrogen density [36]. This phenomenon is attributed to strong quantum effects [37] at low temperatures but it is expected that it will be still effective if hydrogen is in volumes commensurable to hydrogen molecule size and at temperatures near about room temperature [38]. Studies by TPD, XPS and UPS measurements performed in [39] show that hydrogen accumulation proceeds through physisorption, and hydrogen cluster formation can take place in narrow spaces, too [40]. But clusters can be formed if enough space is available inside the carbon particle [35]. The area adjacent to the closed top end inside nanohorns could bind hydrogen molecule much stronger compared to a graphene sheet. However, due to the limited space inside, the adsorption of hydrogen could not contribute a high percentage for hydrogen storage. For the cones the availability of large inner space is not a problem since their diameters are of big values.

One of the disadvantages of hydrogen storage in nanohorns is, that they themselves have been synthesized gathered in agglomerates of low density. For that a necessity occurs to use huge volumes in order to store big amounts of hydrogen [41]. The nanohorns produced by the spark method (see Fig. 9.1) are of an order, which substantially increases density of the obtained material and, hence, an increase of the stored amount of hydrogen in a unit volume should be expected.

The expectations carbon cones to be used for hydrogen storage are based on experiments carried out on a material consisting cones produced under pyrolysis of heavy oil at the Institute for Energy Technology [42]. It has been calculated, also, that when carbon is situated along cone edges [43], the interaction potential due to the dipole moment of a cone and the quadrupole moment of a H_2 molecule is attractive when averaged over all space angles.

9.4 Conclusions

The adduced spark discharge experiments in mix of water and liquid hydrocarbons show that synthesis process interruption and regeneration of liquid around the electrodes create periodically repeated nontraditional conditions for carbon structure synthesis and arrangement. These conditions are favourable for the synthesized structure arrangement. Time and again structures of the same type and with an overall beginning and growth in different direction have been observed. The angles between the growth directions are multiple to 60º and at angles belonging to the carbon nanocones. It is logical to assume that in the first case structures are concerned starting from the same carbon network and preserving its orientation. In the second case the structures have grown drawing atoms from one and the same source but being forced to separate by the carbon cone appearing between them. The reason for the cone to appear is building in of one or several pentagons in the carbon network formed. As this is a common event in carbon structure creation the registered availability of many and of different angles cones should be admitted normal.

The proposed method is a simple one and does not require application of sophisticated apparatuses. The simultaneous synthesis of different structures may be considered a disadvantage because one method is good enough when is capable of producing a single-type product at a high yield.

Acknowledgment This work has been supported by the Bulgarian Science Foundation (contract no DOO2-241/18.12.2008) which is gratefully acknowledged.

References

1. Sano N (2005) Separated syntheses of Gd-hybridized single-wall carbon nanohorns, single-wall nanotubes and multi-wall nanostructures by arc discharge in water with support of gas injection, Letters to the Editor. Carbon 43:447–453

2. Sano N, Nakano J, Kanki T (2004) Synthesis of single-walled carbon nanotubes with nanohorns by arc in liquid nitrogen, Letters to the Editor. Carbon 42:667–691
3. Ishigami M, Cumings J, Zettl A, Chen S (2000) A simple method for the continuous production of carbon nanotubes. J Chem Phys Lett 319:457–459
4. Wang H, Chhowalla M, Sano N, Jia S, Amaratunga GAJ (2004) Large-scale synthesis of single-walled carbon nanohorns by submerged arc. Nanotechnol 15:546–550
5. Sano N (2004) Low-cost synthesis of single-walled carbon nanohorns using the arc in water method with gas injection. J Phys D Appl Phys 37(8):L17–L20
6. Lange H, Sioda M, Huczko A, Zhu YQ, Kroto HW, Walton DRM (2003) Nanocarbon production by arc discharge in water. Carbon 41:1617–1623
7. Hsin YL, Hwang KC, Chen F-R, Kai J-J (2001) Production and in-situ metal filling of carbon nanotubes in water. Adv Mater 13(11):830–833
8. Yamaguchi T, Bandow S, Iijima S (2004) Synthesis of carbon nanohorn particles by simple pulsed arc discharge ignited between pre-heated carbon rods. Chem Phys Lett 389:181–185
9. Adelene NJ, Yudasaka M, Bandow S, Kokai F, Takahashi K, Iijima S (2000) Adsorption and catalytic properties of single-wall carbon nanohorns. Chem Phys Lett 328:381–386
10. Ihara S, Itoh S, Akagi K, Tamura R, Tsukada M (1996) Structure of polygonal defects in graphitic carbon sheets. Phys Rev B 54(20):14713–14719
11. Ge M, Sattler K (1994) Observation of fullerene cones. Chem Phys Lett 220:192–196
12. Krishnan A, Dujardin E, Treacy NMJ, Hugdahl J, Lynum S, Ebbesen TW (1997) Graphitic cones and the nucleation of curved carbon surfaces. Nature 388:451–454
13. Kvaerner's patent No. PCT/NO98/00093 for production of micro domain particles by use of a plasma process
14. Burke PJ (2004) AC performance of nanoelectronics: towards a ballistic THz nanotube transistor. Solid State Electron 48:1981–1986
15. Bachtold A, Hadley P, Nakanishi T, Dekker C (2001) Logic circuits with carbon nanotube transistors. Science 294(5545):1317–1320
16. Mateiu R, Davis ZJ, Madsen DN, Molhave K, Boggild P, Rassmusen A-M, Brorson M, Jacobsen CJH, Boisen A (2004) An approach to a multi-walled carbon nanotube based mass sensor. Microelectron Eng 73–74:670–674
17. Guo L, Wang R, Xu H, Liang J (2005) Why can the carbon nanotube tips increase resolution and quality of image in biological systems? Physica E 27:240–244
18. Liu L, Zhang Y (2004) Multi-wall carbon nanotube as a new infrared detected material. Sensors Actuators A 11:6394–6397
19. Ionescu R, Espinosa EH, Sotter E, Llobet E, Vilanova X, Correig X, Felten A, Bittencourt C, Van Lier G, Charlier J-C, Pireaux JJ (2006) Oxygen functionalisation of MWNT and their use as gas sensitive thick-film layers. Sensors Actuators B 113:36–46
20. Sayago I, Terrado E, Lafuente E, Horrillo MC, Maser WK, Benito AM, Navarro R, Urriolabeitia EP, Martinez MT, Gutierrez J (2005) Hydrogen sensors based on carbon nanotubes thin films. Synth Met 148:15–19
21. Baughman RH, Cui Ch, Zakhidov AA, Iqbal Z, Barisci JN, Spinks GM, Wallace GG, Mazzoldi A, De Rossi D, Rinzler AG, Jaschinski O, Roth S, Kertesz M (1999) Carbon nanotube actuators. Science 284(5418):1340–1344
22. Portet C, Taberna PL, Simon P, Flahaut E (2005) Influence of carbon nanotubes addition on carbon-carbon supercapacitor performances in organic electrolyte. J Power Sources 139:371–378
23. Qin X, Durbach S, Wu GT (2004) Electrochemical characterization on RuO_2 .x H_2O/carbon nanotubes composite electrodes for high energy density supercapacitors, Letters to the Editor. Carbon 42:423–460
24. Ye J-S, Liu X, Cui HF, Zhang W-D, Sheu F-S, Lim TM (2005) Electrochemical oxidation of multi-walled carbon nanotubes and its application to electrochemical double layer capacitors. Electrochem Commun 7:7249–7255
25. Chen Q-L, Xue K-H, Shen W, Tao FF, Yin S-Y, Xu W (2004) Fabrication and electrochemical properties of carbon nanotube array electrode for supercapacitors. Electrochim Acta 49:4157–4161

26. Fursey GN, Novikov DV, Dyuzhev GA, Kotcheryzhenkov AV, Vassiliev PO (2003) The field emission from carbon nanotubes. Appl Surf Sci 215:135–140
27. Wang X, Zhi L, Mullen K (2008) Transparent, conductive graphene electrodes for dye-sensitized solar cells. Nano Lett 8(1):323–327
28. Geim AK, Novoselov KS (2007) The rise of grapheme. Nature 6:183–191
29. Nomura K, MacDonald AH (2006) Quantum Hall ferromagnetism in graphene. Phys Rev Lett 96:256602.1–256602.4
30. Jimenez-Soto JM, Cardenas S, Valcarcel M (2009) Evaluation of carbon nanocones/disks as sorbent material for solid-phase extraction. J Chromatogr A 1216:5626–5633
31. Qu CQ, Qiao L, Wanga C, Yu SS, Jiang Q, Zheng WT (2010) Density functional theory study of the electronic and field emission properties of nitrogen- and boron-doped carbon nanocones. Phys Lett A 374:782–787
32. Shenderova OA, Lawson BL, Areshkin D, Brenner DW (2001) Predicted structure and electronic properties of individual carbon nanocones and nanostructures assembled from nanocones. Nanotechnol 12:191–197
33. Harris PJF, Tsang SC, Claridge JB, Green MLM (1994) High-resolution electron microscopy studies of a microporous carbon produced by arc-evaporation. J Chem Soc Faraday Trans 90:2799–2802
34. Iijima S, Yudasaka M, Yamada R, Bandow S, Suenage K, Kokai F, Takahashi K (1999) Nano-aggregates of single-walled graphitic carbon nano-horns. Chem Phys Lett 309:165–170
35. Murata K, Kaneko K, Kanoh H, Kasuya D, Takahashi K, Kokai F, Yudasaka M, Iijima S (2002) Adsorption mechanism of supercritical hydrogen in internal and interstitial nanospaces of single-wall carbon nanohorn assembly. J Phys Chem B 106:11132–11138
36. Tanaka H, Kanoh H, El-Merraoui M, Steele WA, Yudasaka M, Iijima S, Kaneko K (2004) Quantum effects on hydrogen adsorption in internal nanospaces of single-wall carbon nanohorns. J Phys Chem B 108:17457–17465
37. Anil Kumar AV, Jobic H, Bhatia SK (2006) Quantum effects on adsorption and diffusion of hydrogen and deuterium in microporous materials. J Phys Chem B 110:16666–16671
38. Wang Q, Johnson JK, Broughton JQ (1997) Path integral grand canonical Monte Carlo. J Chem Phys 107:5108–5118
39. Tverdal X, Yu M, Raaen S, Helgesen G, Knudsen KD (2008) Hydrogen adsorption on carbon nanocone material studied by thermal desorption and photoemission. Appl Surf Sci 255:1906–1910
40. Dillon AC, Jones KM, Bekkedahl TA, Kiang CH, Bethune DS, Heben MJ (1997) Storage of hydrogen in single-walled carbon nanotubes. Nature 386:377–379
41. Bekyarova E, Kaneko K, Kasuya D, Murata K, Yudasaka M, Iijima S (2002) Oxidation and porosity evaluation of budlike single-wall carbon nanohorn aggregates. Langmuir 18:4138–4141
42. Maeland et al. (2001) US patent No. 6,290,753, B1, Hydrogen storage in carbon material
43. Heiberg-Andersen H, Skjeltorp AT, Sattler K (2008) Carbon nanocones: a variety of non-crystalline graphite. J Non Cryst Solid 354:5247–5249

Chapter 10
Encapsulated Ferromagnetic Nanoparticles in Carbon Shells

Al.D. Zolotarenko, An.D. Zolotarenko, V.A. Lavrenko, S.Yu. Zaginaichenko, N.A. Shvachko, O.V. Milto, V.B. Molodkin, A.E. Perekos, V.M. Nadutov, and Yu.A. Tarasenko

Abstract This paper discusses the method of arc discharge in gaseous phase (ADG) and it has been used for production of iron, nickel and their mechanical mixtures based on Me-C composites. The product has been synthesized using are plasma -chemical installation with vertical location of the reactor which has the mobile cathode. The mechanical mixture of graphite and metals powders has been added into the anode along its axis. Investigations of micro- and nanostructures of produced composites have been carried out using the transmission electron microscope (TEM). The Me-C nanocomposites (Me = Fe, Ni, Fe + Ni) with the structure simillar to the nucleus – cover structure have been produced. It has been noticed their response to the magnetic field action. Me-C nanocomposite material has been taken out of soot by magnetic separation of parietal soot suspension in hydrocarbons. The initial fullerite crystals, which are contained in soot before extraction process, have been shown. The self-descriptiveness and the high speed performance of qualitative and quantitative analysis of the method for soluble nanomaterials have been pointed out.

Al.D. Zolotarenko (✉), An.D. Zolotarenko, V.A. Lavrenko, S.Yu. Zaginaichenko, N.A. Shvachko, and O.V. Milto
Institute for Problems of Materials Science of NAS of Ukraine, Krzhyzhanovsky str. 3, 03142 Kiev, Ukraine
e-mail: o.d.zolotarenko@gmail.com

V.B. Molodkin, A.E. Perekos, and V.M. Nadutov
Institute for Metal Physics of NAS of Ukraine, 36 Acad. Vernadsky Boulevard, UA-03680 Kiev-142, Ukraine
e-mail: perekos@ukr.net

Yu.A. Tarasenko
O. Chuiko Institute of Surface Chemistry of NAS of Ukraine, 17 General Naumov str, Kiev 03164, Ukraine
e-mail: nikar@kartel.kiev.ua

S.Yu. Zaginaichenko et al. (eds.), *Carbon Nanomaterials in Clean Energy Hydrogen Systems - II*, NATO Science for Peace and Security Series C: Environmental Security 2, DOI 10.1007/978-94-007-0899-0_10, © Springer Science+Business Media B.V. 2011

Keywords Nanoferromagnetics · Ni · Fe · FeNi · Magnetic susceptibility · Curie point · Arc discharge

10.1 Introduction

While the pyrolytic method is based on thermocatalytic destruction of precursors to obtain free carbon atoms or molecules starting from the processes of hydrocarbon dehydrogenation or CO disproportionation, the arc synthesis processes are based on the reaction of graphite or hydrocarbons thermodestruction in arc discharge to achieve the same object [1–10]. At present the researchers have no clear idea of the question in which state the carbon is on arc heating. The supporters of fullerene formation from graphene fragments believe that graphite is not evaporated but sputtered by the arc in the form of graphenes because the arc energy is not sufficient to transform carbon into a vaporous state.

Thermodynamic stability region of gaseous carbon is in the near-plasma temperature range, ~8,000 K [2]. For this reason, the carbon atom transfer into an excited state is possible only at the temperatures that exceed this range. According to Fig. 10.1, the interelectrode arc exhibits the temperature zones (8,000–12,000 K) that allow the transfer of carbon atoms from graphite electrodes into a free atomic state. Besides plasma conditions, these temperatures can be reached by graphite heating with a laser beam. As the laser method is low-output, power-consuming and, therefore, expensive and inaccessible, we have used the arc method of graphite evaporation.

According to our previous experimental and theoretical research [3–5], one of the variants of such a process can be our scheme suggested in Fig. 10.2.

Discussing the peculiarities of the synthesis, we shall proceed that the arc synthesis allows the generation of carbon vapor (i.e. atomic carbon). The processes of carbon vapor generation (Fig. 10.2) in the arc synthesis of carbon nanostructures in gas [3, 4] and in liquid [5] are identical, but the processes of carbon vapor transfer into a solid state in different medium demonstrate many peculiarities and specific physical effects. Some of them will be discussed below.

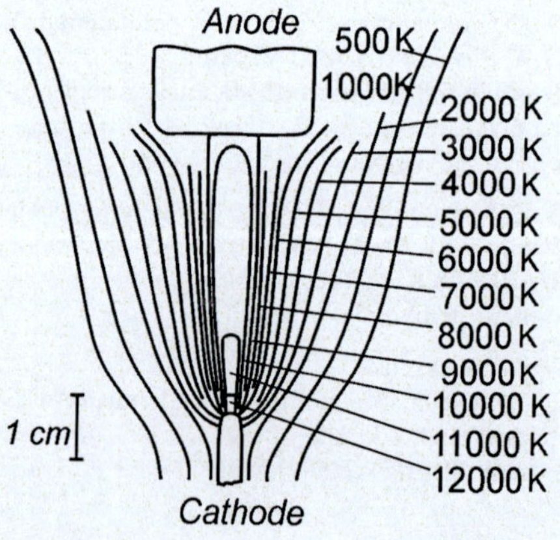

Fig. 10.1 Temperature zone distribution (K) along the axis of an electric arc between the graphite electrodes at the current of 200 A

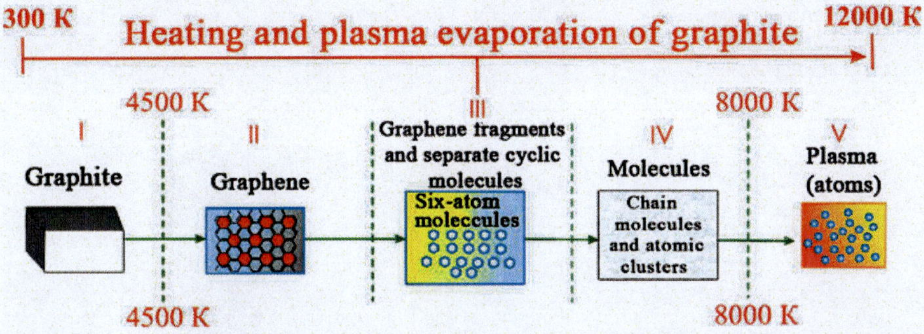

Fig. 10.2 Levels of substance organization that enable one to consider the mechanism of the carbon vapor formation by changing the temperature from room temperature to 12,000 K

10.2 Equipment and Methods of Product Synthesis and Analysis Experimental Setup for Me-C Nanocomposite Synthesis

The electric arc method is based on the electric arc discharge generated between two graphite electrodes in helium or argon inert atmosphere. Krätschmer [7, 8] developed this method for the fullerene synthesis. After its modernization that involved the increase in temperature of plasma formed between two rods the method allows production not only of fullerenes, but also different carbon nanostructures. Carbon nanotubes first attracted a serious interest when fullerene-like materials produced by arc method [8] were studied.

In our case, the electric arc vacuum plasmochemical setup was used to synthesize Me-C nanocomposites for the electric arc carbon evaporation.

Experiments on producing all CNMs were carried out in the stainless steel reactor with inner diameter of 150 mm equipped with a water jacket. The temperature of thermostat water-cooling jacket was kept constant (25–30°C).

A high-quality graphite of the MPG-7 grade has been used in the work. An anode graphite rod had a cross section of $9 \times 9\ \text{mm}^2$ and a length of 800 mm. The graphite was evaporated in vacuum, under the helium pressure of 0.02–0.09 MPa, at the electrode voltage of 22–30 V and the current of 250–300 A.

Me-C nanocomposites have been synthesized by co-evaporation. For metal and graphite co-evaporation, their mechanical mixture was pressed into the electrode cavity that was drilled along its axis (Fig. 10.3).

10.3 Methods of Analysis of Me-C Nanocomposites Composition and Structure

Carbon nanomaterials (CNM) have been studied by the method of oxidative thermogravimetry on the Q-1500D derivatograph. The samples have been heated linearly at the rate of $5°\text{C}\cdot\text{min}^{-1}$ in the air in the temperature range from 25 to 990°C.

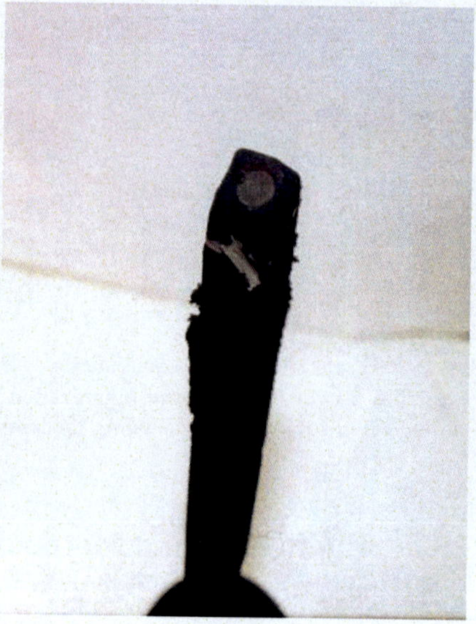

Fig. 10.3 Anode containing the metal and graphite mixture along the electrode axis

The carbon content in the studied CNM samples has been analyzed by the burning method at 1,400°C, and the metals content in the same samples has been evaluated by the atomic absorption spectroscopy.

X-ray diffraction patterns from the carbon products have been taken by the ADP-1 and DRON-2 diffractometers. Exposure of the produced milled deposit cages has been carried out on the X-ray apparatus DRON-3 M using filtered CuKα radiation followed by the identification of the X-ray diffraction patterns.

The carbon product structures have been studied on scanning (JSM-T20) and transmission (JEM 100 CXII) electron microscopes. The presence of different nanostructures in studied samples from prepared CNMs has been determined by the method of stepwise thermofractional oxidation (oxidative carbon extraction in CO_2 form in the purified oxygen flow using coulometric measurements of carbon oxide (IV).

To evaluate the elemental composition of the synthesis products, CNMs have been studied by the atomic emission spectral analysis on the ICXA-733 apparatus. The analysis is based on the study of spectra of free atoms and ions in a gas phase in the wavelength range of 150–800 nm.

10.4 Findings of Investigations

In the Me-C composites evaporation, the metal-carbon nanostructures, in which metal nanostructures are encapsulated into the carbon matrix (Fig. 10.4), are condensed at the reactor walls.

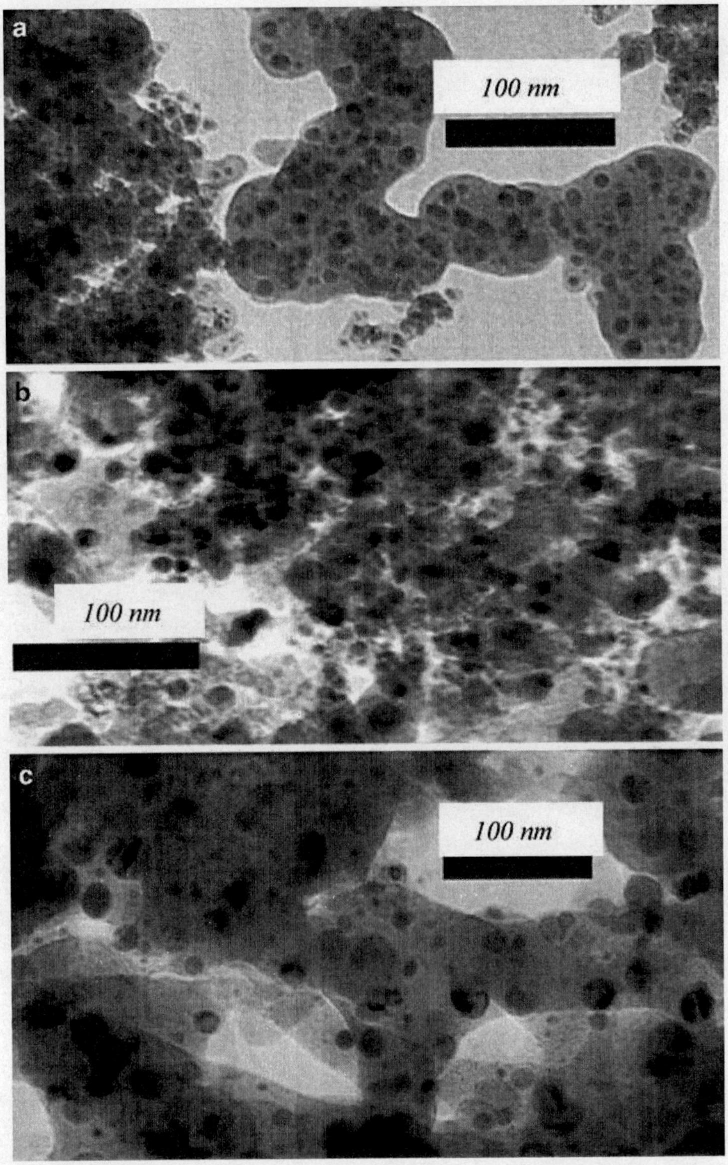

Fig. 10.4 Ferromagnetic metal nanoparticles covered with a carbon film: (**a**) – Ni; (**b**) – Fe; (**c**) – (Fe + Ni)

The soluble component of the product has been extracted and separated by the chromatographic method, the solutions have been analyzed on the SF-2000 spectrophotometer to detect fullerenes, endofullerenes and their derivatives. Extraction of the soluble component from the product, which was formed by arc evaporation of the mixture of graphite and ferromagnetic metal powders, left the component that contained metal nanoparticles of ferromagnetic metals (2–20 nm)

The soluble component of the product has been extracted and separated by the chromatographic method, the solutions have been analyzed on the SF-2000 spectrophotometer to detect fullerenes, endofullerenes and their derivatives.

Extraction of the soluble component from the product, which was formed by are evaporation of the mixture of graphite and ferromagnetic metal powders, left the component that contained metal nanoparticles of ferromagnetic metals (2–20 nm) encapsulated into the carbon matrix. The TEM-investigations have shown that Me-C composites consist of metallic particles substantially covered by carbon (Fig. 10.4 (a)–Ni; (b)–Fe). As it is observed, metallic particles of 1–30 nm in diameter are completely enclosed in multi-layer carbon capsules. These metallic particles have the quasi-spherical morphology.

The X-ray investigations have shown that nickel nanoparticles covered by carbon shell basically are the monocrystalline metal (of FCC lattice). The iron particles form the Fe_3C carbide.

It should be pointed out that on exposure to external magnetic field action on the suspension of produced Me-C composites their particles move along the strength lines of the field. The process of these particles separation has been based on this property of composite materials. The lest mobile phase contained composite with encapsulated particles of 5–10 nm in diameter.

The X-ray structural analysis, Raman scattering and atomic-force microscopy have shown that the prepared samples contain impurities of different carbon modifications and metal carbides. The investigation of the field and temperature dependences of saturation magnetization have demonstrated the decrease in its value at the 300 K as compared with the pure Ni (Fig. 10.5).

The products containing ferromagnetics begin to oxidize at lower temperatures (206–209°C) and this process occurs in more broader temperature range. A decrease in the products mass equals 61–72,3%. Under oxidation of the product containing Fe the shoulder on DTG curve is noticed at temperature about 386°C, and large asymmetric peak at 668°C (T_{max} = 668°C). DTG curve of oxidation for

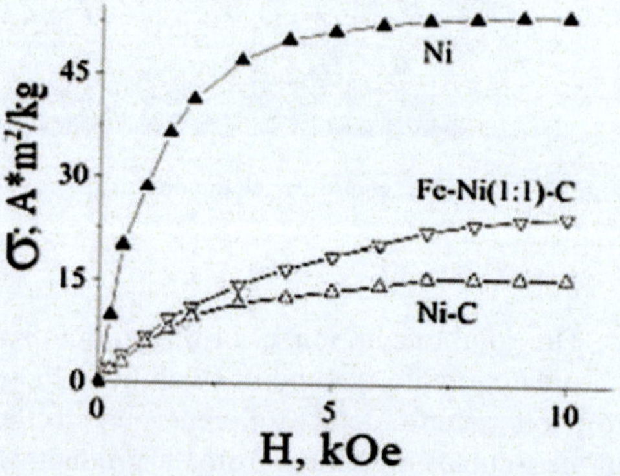

Fig. 10.5 Field dependence of specific magnetization

the product with Ni contains as low temperature shoulder (346°C) and so the second shoulder ($T_{2\ shoulder} = 530°C$). $T_{1\ max}$ of large asymmetric peak equals 683°C, as in the case of the product with Fe oxidation. The features of high temperature behavior for the product containing Fe-Ni are the shoulder occurrence at 505°C only shift of large asymmetric peak to higher temperature range (690°C), appearance of one little peak more at 776°C.

The results of X-ray structural studies of the Me-C composites, produced after magnetic separation, are given in Table 10.1.

The temperature studies into magnetic properties of the Me-C composites have shown that the metal particles encapsulated into the carbon matrix remain magnetic with the Curie temperature of 400°C for Ni-C and 585°C for FeNi-C (Fig. 10.6).

The magneto-force microscopy of the compressed samples of the Me-C composites has revealed the periodicity in magnetic structure on their surface that is not related to their surface relief. The period is 8.5 μm (Fig. 10.7).

The studies have been carried out for the particles of 200–600 nm in size, which retain their magnetic properties.

Table 10.1 Results of X-ray structural analysis of the Me-C composite samples before and after oxidation

Material	Phase composition before oxidation	Phase composition after oxidation
1. C(soot)	Graphite with rhombohedral distortion	—
2. C + Fe	Soot (amorphous phase, halo at $2\theta \approx 12$–220), graphite traces, (line of mean intensity with d = 3.36), α-Fe (bcc) and γ-Fe (fcc)	Master phase – Fe_2O_3, amorphous phase, (halo at $2\theta \approx 14$–240)
3. C + Ni	Soot, Ni(C)- solid solution (interplanar spacing is increased), NiC traces	Master phase – NiO, amorphous phase, (halo at $2\theta \approx 16$–250)
4. C + Fe + Ni	$Fe_{0.64}Ni_{0.36}$ (cubic system), soot, graphite traces with rhombohedral distortion	Master phase – $Ni_{1.43}Fe_{1.704}$ (fcc), NiO, amorphous phase, (halo at $2\theta \approx 12$–240)

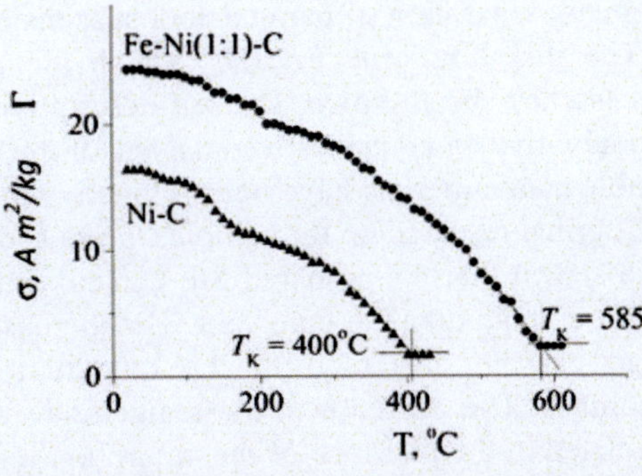

Fig. 10.6 Temperature dependence of magnetic properties

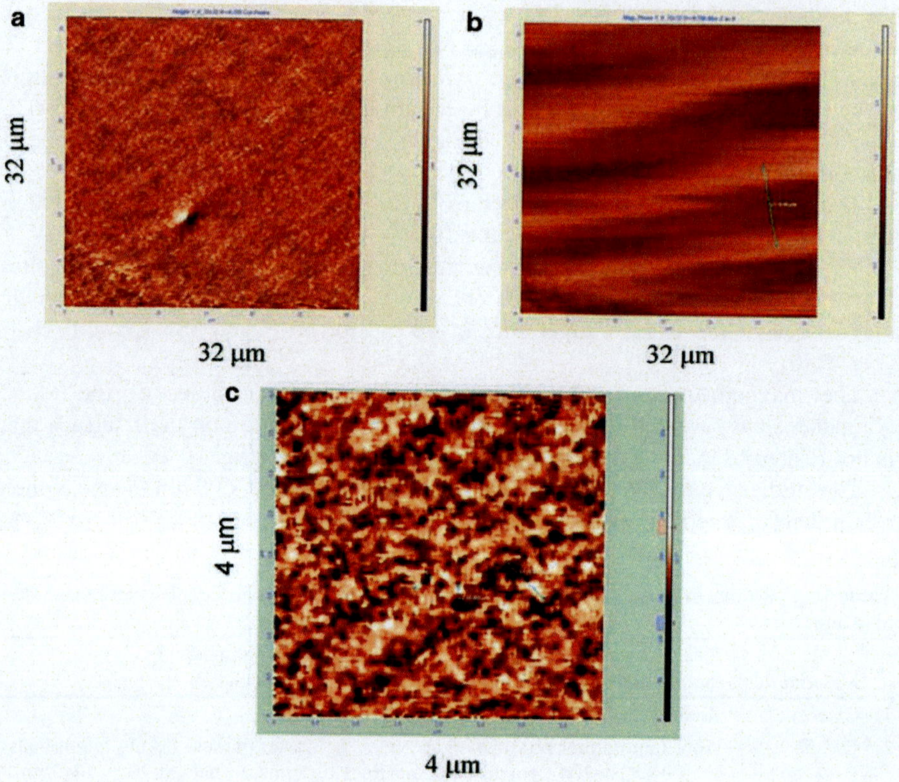

Fig. 10.7 The results of atomic-force (**a**) and magneto-force (**b**, **c**) microscopy of the Ni-C composite surface

10.5 Conclusions

The Me-C nanocomposites (Me = Fe, Ni, Fe + Ni) with the structure simillar to the nucleus – cover structure have been produced. It has been noticed their response to magnetic field action. The Me-C nanocomposite has been taken out of soot by magnetic separation of parietal soot suspension in hydrocarbons.

The initial fullerite crystals, which are contained in soot before extraction process, have been shown. The self-descriptiveness and the high speed performance of qualitative and quantitative analysis of the method developed by the authors for soluble nanomaterials have been pointed out.

Thermal research on the samples, produced by Me and graphite co-evaporation, have shown that the prepared Me-C nanocomposites oxidize over the wider temperature range than the pure soot. The increase in the upper oxidation temperature range can be attributed to the oxidation of ferromagnetics and multi-wall nanotubes. The decrease in the temperature range for the onset of oxidation can be related to the presence of the amorphous carbon component in the composite.

References

1. Schur DV, Zaginaichenko SYu, Lysenko EA (2008) The forming peculiarities of C_{60} molecule. In: Carbon nanomaterials in clean energy hydrogen systems, NATO Science Series. Springer, Netherlands, pp 53–65
2. Schur DV, Zaginaichenko SYu, Matysina ZA (2008) The special features of formation of carbon nanostructures, their classification and site on the state diagram of carbon. In: Carbon nanomaterials in clean energy hydrogen systems, NATO Science Series. Springer, Netherlands, pp 67–83
3. Zolotarenko AD, Zolotarenko AlD, Zolotarenko AnD, Schur DV, Pomytkin AP, et. al (2009) I. About processes of carbon nanostructures formation on cathode under electric arc discharge conditions. In: Proceedings of 11th international conference on hydrogen materials science and chemistry of carbon nanomaterials, Yalta, 25–31 Aug 2009, pp 392–394
4. Zolotarenko AD, Zolotarenko, AnD, Zolotarenko AlD, Schur DV, Zaginaichenko SYu, et.al. (2009) II. About the processes of carbon nanostructures formation in the gaseous phase and on the reactor walls under electric arc discharge condition. In: Proceedings of 11th international conference on hydrogen materials science and chemistry of carbon nanomaterials, Yalta, 25–31 Aug 2009, pp 398–399
5. Zolotarenko AnD, Zolotarenko AlD, Schur DV, Zaginaichenko SYu, et. al. (2009) III. On processes of carbon nanostructures formation in liquid phase. In: Proceedings of 11th International conference on hydrogen materials science and chemistry of carbon nanomaterials, Yalta, 25–31 Aug 2009, pp 402–403
6. Iijima S (1991) Helical microtubules of graphitic carbon. Nature 354:56–58
7. Kratschmer W, Huffman D (1992) Fullerites: new form of crystalline carbon. Carbon 30:1143–1147
8. Kratschmer W, Lamp LD, Fotiropoulos K, Huffman DR (1990) Solid C_{60}: a new form of carbon. Nature 347:354–358
9. Golovko EI, Bogolepov VA, Schur DV (2005) The use of thermogravimetric analysis for certification of nanostructural materials. Nanosyst Nanomater Nanotechnol 3(3):633–643, in Russian
10. Eletskii AV (2002) Carbon nanotubes and their emissivity. Phys Usp 172(4):401–439, in Russian

Chapter 11
The Peculiarities of Nanostructures Formation in Liquid Phase

An.D. Zolotarenko, Al.D. Zolotarenko, E. Rudakova, S.Yu. Zaginaichenko, A.G. Dubovoy, D.V. Schur, V.A. Lavrenko, A.P. Pomytkin, A.E. Perekos, V.P. Zalutskiy, M.M. Divizinyuk, E.V. Azarenko, and Yu.A. Tarasenko

Abstract The processes occurring on the electrodes and in the liquid phase during the arc discharge in the liquid phase (ADLP) have been considered in the present work and we explain the mechanism of carbon nanostructures (CNS) formation proposing the model based on the analysis of existing regularities in behaviour of charged particles under extreme temperature and pressure gradients.

The CNS synthesis by ADLP method has been performed in dielectric liquids: hydrocarbons, liquid gases (N_2, Ar, He, etc.), deionized water and others. Suspension containing clusters of synthesized nanostructures has been formed by the synthesis.

The efficiency of this method is sharply increased by using arc discharge in the liquid phase where powder reagent layer is used as anode. To increase the frequency of electrodes clamping and moving apart, an electromagnetic vibrator has been used in this method and it brings and takes away the cathode from the powder reagent at a specified frequency. For ADLP, nanostructures form simultaneously at several points on the conducting particle surface as a result of microscopic acts of arc discharge.

An.D. Zolotarenko (✉), Al.D. Zolotarenko, E. Rudakova, S.Yu. Zaginaichenko, A.G. Dubovoy, D.V. Schur, V.A. Lavrenko, and A.P. Pomytkin
Institute for Problems of Materials Science of NAS of Ukraine, Krzhyzhanovsky str. 3, Kiev 03142, Ukraine
e-mail: a.d.zolotarenko@gmail.com

A.E. Perekos and V.P. Zalutskiy
Institute for Metal Physics of NAS of Ukraine, 36 Acad., Vernadsky Boulevard, UA-03680, Kiev-142, Ukraine
e-mail: perekos@ukr.net

M.M. Divizinyuk and E.V. Azarenko
Sevastopol National University of Nuclear Energy and Industry, Kurchatov str. 7, Sevastopol 99033, Ukraine
e-mail: shumira_07@mail.ru

Yu.A. Tarasenko
O. Chuiko Institute of Surface Chemistry of NAS of Ukraine, 17 General Naumov street, Kiev 03164, Ukraine
e-mail: nikar@kartel.kiev.ua

S.Yu. Zaginaichenko et al. (eds.), *Carbon Nanomaterials in Clean Energy Hydrogen Systems - II*, NATO Science for Peace and Security Series C: Environmental Security 2, DOI 10.1007/978-94-007-0899-0_11,

These nanostructures are generated from the liquid phase and anode vapors and represent the product exhibiting rather interesting physical and chemical properties.

Based on the analysis of the observations performed in the course of carbon nanostructures synthesis, the model of nanostructures formation by arc discharge in the liquid phase has been proposed in this paper. Presence and absence of deposit on the cathode have been explained.

Keywords Me-C nanocomposite · Structure · Phase composition · Arc discharge · Liquid phase · Specific magnetization

11.1 Introduction

Materials which physical properties can be controlled by varying them within wide limits occupy a particular place among modern metallic materials. These materials are used extensively in different fields of engineering and industry and define to a large extent the pace of scientific and technical progress. Magnetic metals and alloys can be realistically assigned to such materials [1, 2].

Magnetic properties of metals and alloys depend on many technological and physical factors: conditions of their production, chemical composition, structural state, number and distribution of different stable and metastable phases in a material, etc. which in turn are determined to a large measure by different types of external actions on a material during its production (mechanical, thermal, magnetic, thermo-mechanical, thermomagnetic, ultrasonic, etc. treatments) [3].

Synthesis of superfine, ultradispersed and nanodispersed magnetic materials provides unique possibilities for physicists and technologists. Magnetic properties of materials with the superfine or ultradispersed structure can be changed over very a wide limits by varying their dispersion, phase state, surface state and other factors [4, 5]. This paper realizes one of such possibilities by the example of superfine iron and nickel powders produced by the method of electric arc dispersion in dielectric liquid media (DDLM).

After the discovery of fullerenes and carbon nanotubes methods of their synthesis has been constantly investigated and improved. In parallel with the arc method in the gaseous phase and the pyrolytic method of synthesis of carbon nanostructures, since 2000 we have investigated and developed the method of arc synthesis in the liquid phase (ASLP). For the last decade, this method is used in increasing frequency to produce different nanostructures as the method alternative to the arc discharge in the gaseous phase (ADGP).

In the eighties we began our work on producing ultradispersed metal powders by the electroerosion method [6–8] and continue it today. Besides carbon nanostructures produced by evaporation of carbon electrodes in the liquid phase, there appears a possibility to produce metal-carbon composites by sublimation of metal in the carbon-containing liquid. In this case the metal nanoparticles form along with carbon nanostructures on their surface.

The main positive features [5, 8] of the method used are as follows:

1. high temperature in the arc zone, >4,000 K;
2. high cooling rate of evaporated products, >10^9 K/s;
3. high degree of dispersion. The particles range in size from 1 to 100 nm;
4. high nucleation rate at a low growth rate of a particle.

This method does not require using of unhealthy gases, vacuum equipment or expensive lasers. The proposed method provides a possibility of producing a wide range of materials by varying the conditions for synthesis and it presents a way of modifying the chemical composition of electrodes and a medium, in which the synthesis is carried out [9]. At present time different research groups over the world are engaged in such studies [10–19]. The liquid phase may be of different chemical compositions that affect the structure and composition of the produced nanoobjects being studied (Fig. 11.1).

In the present work we have considered the processes occurring on the electrodes and in the liquid phase during the arc discharge in the liquid phase process and explained the mechanism of carbon nanostructures (CNS) formation proposing the model based on the analysis of existing regularities in behaviour of charged particles under extreme temperature and pressure gradients.

The Me-C composites have been produced by the arc discharge method in toluene using a powder metal anode (ADIP); their structure, phase composition and magnetic properties have been studied.

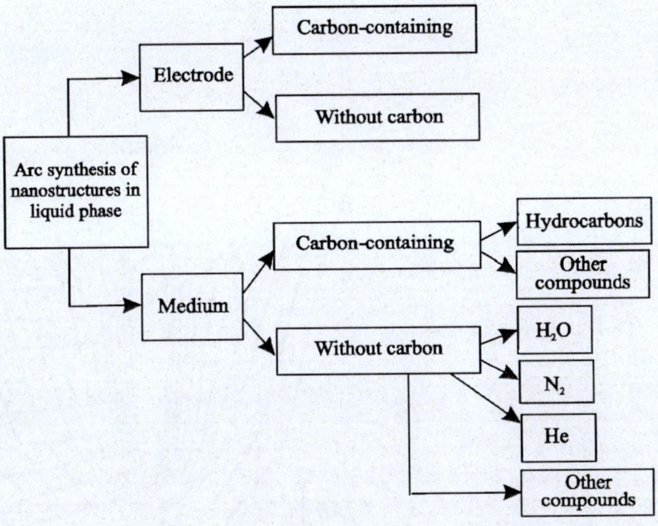

Fig. 11.1 Diagram for possible combinations of medium and electrode materials in synthesis of nanostructures by the arc method in the liquid phase

11.2 Experimental

Synthesis of nanoparticles in the liquid phase has been carried out in the installation specially designed for these studies (Fig. 11.2). This installation allows the metal and graphite electrodes to be evaporated by an electric arc in the liquid medium in the temperature range from 4 to 340 K. In the neighbourhood of the cathode the arc temperature may be as much as $1{,}2\cdot10^4$ K at currents of 200–300A (Fig. 11.3).

The electronic control block is simple to operate and provides a possibility of varying and measuring voltage and electric current. These changes make it possible

Fig. 11.2 The installation for synthesis of nanocarbon structures and Me-carbon composites in the liquid phase

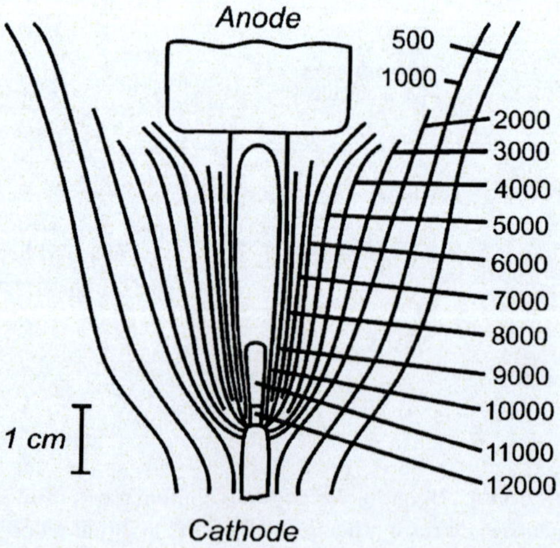

Fig. 11.3 Temperature distribution (in K) in different regions of the electric arc between the carbon electrodes at strength of current equal to 200A [19]

to affect the conditions of the plasmochemical process in the reactor and the product morphology and yield.

All the chemical reagents used in the synthesis have been subjected to the prior purification and rectification. Type MPG-7 graphite has been used. The graphite rods have been pre-annealed in vacuum. The metallic rods have been remelted repeatedly in an arc furnace in the spectro-pure argon medium.

Crystalline structures and phase compositions of powders have been determined using X-ray diffractometer DRON-3.0 in Cok_{α} irradiation; magnetic properties have been measured on a ballistic magnetometer; dimensions of the coherent-scattering region (CSR) have been calculated by X-ray lines broadening with Selyakov-Sherrer formula.

11.3 Results and Discussion

Table 11.1 represents the results of studies on initial iron and nickel powders and the product prepared by arc discharge in the liquid phase using a powder anode (ADIP) in toluene. Also, this table gives the data on Fe(B-5-2) and Ni(B-2) powders before and after thermo-magnetic measurements. Figures 11.4–11.9 and Table 11.2 illustrate the results of measurements of magnetic properties: specific saturation magnetization, σ_s, coercive force, H, and residual induction, I_R; in this case Figs. 11.4–11.7 demonstrate field dependences of specific magnetization and Figs. 11.8–11.9 show temperature dependences.

Diffraction patterns of initial iron and nickel powders show only lines of *bcc* iron and *fcc* nickel, respectively. After ADIP treatment in toluene, the phase composition of synthesized powders has changed. The diffraction pattern of Fe (B-5-2) powder demonstrates two crystalline phases, α-Fe (~24%) and Fe_3C (~76%). In addition to the lines for pure nickel, the diffraction pattern of Ni(B-2) powder has the most intensive line for carbon solid solution in nickel. In magnetic

Table 11.1 Phase composition and dimension of CSR for iron and nickel powders

Sample	Phase composition	Content, %	D, nm
Fe powder, initial state	α- Fe	100	270
	α- Fe	24	–
Fe powder (B-5-2)	Fe_3C	76	24
Ni powder, initial state	Ni	100	150
	Ni	96	
Ni powder (B-2)	Ni-C	4	150
	α-Fe		
	Fe_3C	73	110
	Fe_3O_4	11	40
Fe powder (B-5-2), after heating	FeO	16	100
Ni powder (B-2), after heating	Ni	100	130

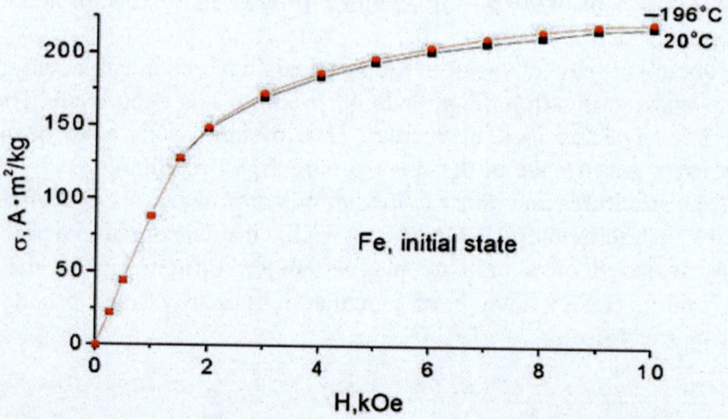

Fig. 11.4 Field dependence of specific magnetization of initial iron powder at temperatures T = −196°C and 20°C

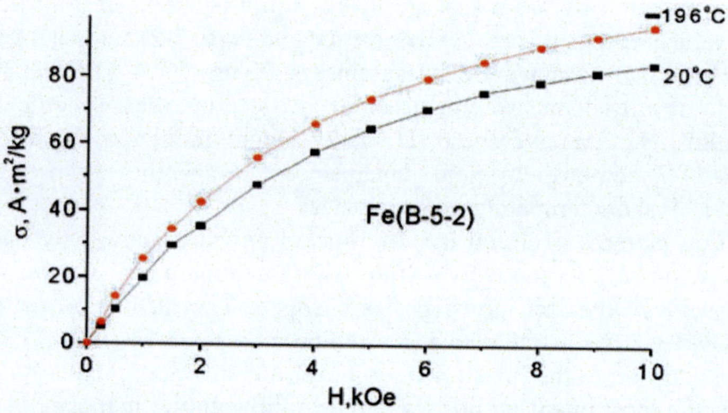

Fig. 11.5 Field dependence of specific magnetization of Fe-C nanocomposite at temperatures T = −196°C and 20°C

thermograms, Curie temperatures 400 and 760°C for iron powders (Fig. 11.8) and 225 and 360 C for nickel powder (Fig. 11.9) correspond to these phases. For Me-C composites, specific saturation magnetization of nickel powder is almost unchanged (Figs. 11.6 and 11.7, Table 11.2); this is related to a low amount of Ni-C crystalline phase that is formed in powder by ADIP treatment. Specific saturation magnetization of Fe-C composites changes more significantly (Figs. 11.4 and 11.5, Table 11.2). This is conditioned by the considerable change in the phase composition of iron powders after ADIP treatment. In the course of synthesis, initial α-Fe transforms into carbide Fe_3C almost completely (Table 11.1). For the products, changes in coercive force, H, and residual induction, I_R, are attributable to the corresponding changes in the phase composition and dispersivity of the powders.

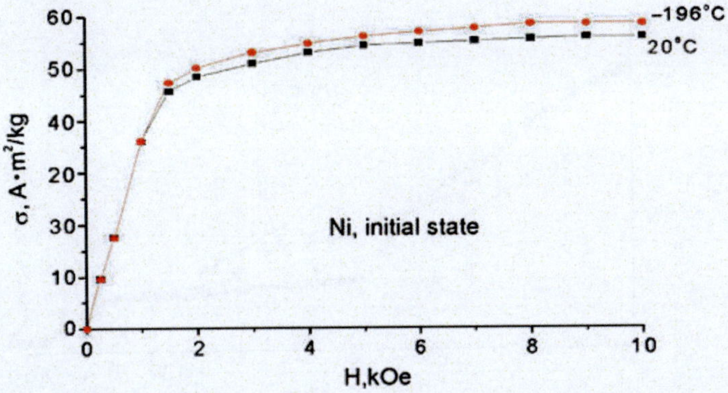

Fig. 11.6 Field dependence of specific magnetization of initial nickel powder at temperatures T = −196°C and 20°C

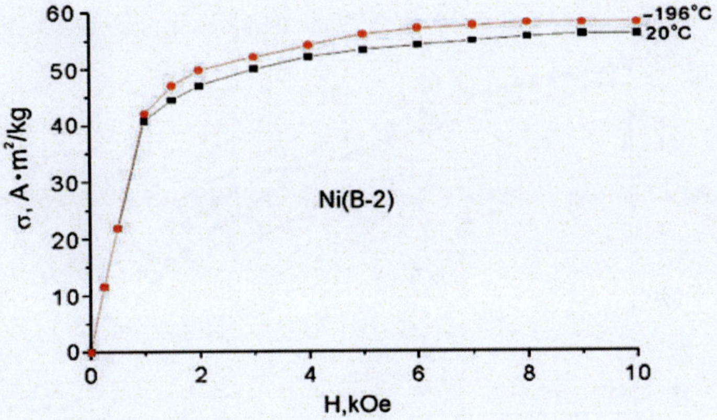

Fig. 11.7 Field dependence of specific magnetization of Ni-C nanocomposite at temperatures T = −196°C and 20°C

The results of the change in the powders phase composition after heating in measuring the temperature dependence of specific saturation magnetization are also of interest. As can be seen from Figs. 11.8, 11.9 and Table 11.1, the phase compositions of iron and nickel in the synthesized powders are changed significantly on heating. After heating, the line of Ni-C solid solution has disappeared from the diffraction pattern of Ni(B-2) powder. After heating in the diffraction pattern of Fe(B-5-2) powder intensities of the lines of Fe_3C carbide have reduced, intensities of lines of α-Fe increased and the lines of crystalline phases FeO and Fe_3O_4 appeared. After heating and on subsequent cooling the knee at ~225°C (Ni-C solid solution) has disappeared from the magnetic thermogram of Ni(B-2) powder. On further cooling, in addition to the knees at ~760°C (α-Fe) and ~400°C

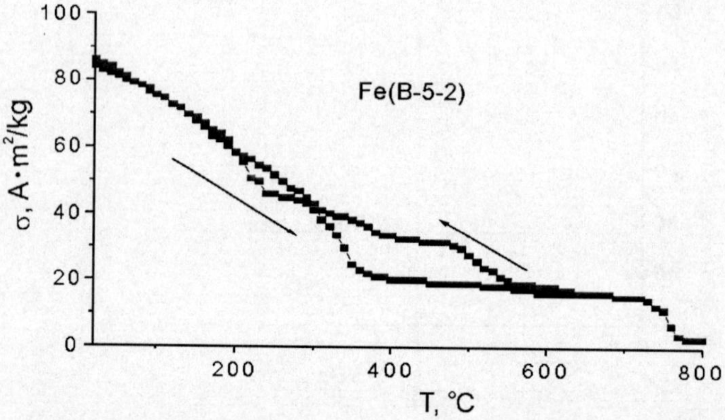

Fig. 11.8 Temperature dependence of specific magnetization of Fe-C nanocomposite before and after heating to 800°C

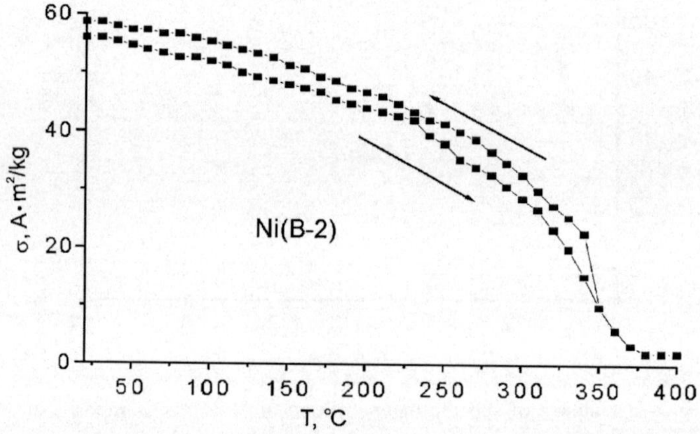

Fig. 11.9 Temperature dependence of specific magnetization of Ni-C nanocomposite before and after heating to 400°C

Table 11.2 Magnetic properties of iron and nickel powders

	σ_s, A·m²/kg			
Material	20°C	−196°C	H, Oe	I_R, Gs
Ni, initial	56	58.5	50	1,434
Fe, initial	219	222	6	102.2
Ni(B-2)	56	58	50	860
Fe(B-5-2)	84.3	95.8	20	108.7

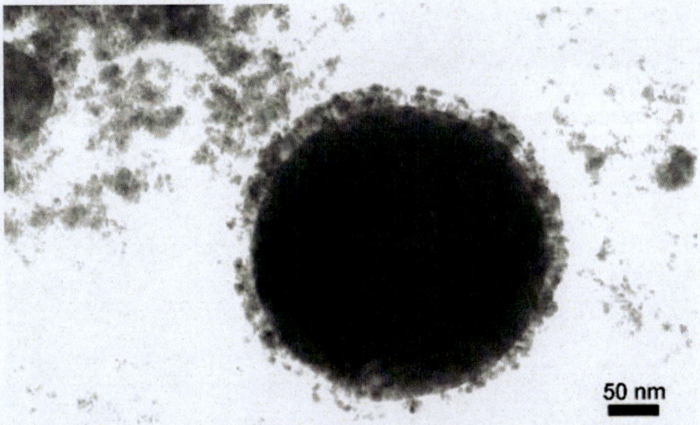

Fig. 11.10 TEM photograph of Ni microgranule (250 nm) in nanocarbon shell

(Fe_3C), the knee at ~600°C corresponding to the Curie temperature for Fe_3O_4 oxide has also appeared in the thermogram of Fe(B-5-2) powder.

On the basis of TEM observations, one can note that the synthesized nanocomposites of both iron and nickel contain particles 1–400 nm in diameter. The most part of particles have a diameter ranging from 10 to 20 nm (Figs. 11.10 and 11.11).

The shell on the large nickel particles (Fig. 11.10) indicates that this particle is formed from the melt. Interaction of melted Ni and carbon vapor gives rise to Ni_3C carbide that decomposes during the alloy crystallization with liberation of carbon. The carbon forms graphite-like nanostructures on the particle surface.

Large iron particles (~150 nm) do not have such prominent boundaries, although all particles of 10–30 nm fractions are enclosed in the carbon shells.

In addition, iron nanoparticles formed by the arc magnetic field exhibit residual magnetization. This causes the nanoparticles to agglomerate in spherical clusters up to 1 μm in diameter (Fig. 11.12).

11.4 Model of Process

The CNS synthesis by the ADLP method has been performed in dielectric liquids: hydrocarbons, liquid gases (N_2, Ar, He, etc.), deionized water and others. Suspension containing clusters of synthesized nanostructures has been formed by the synthesis.

Discharge in liquid is initiated by moving apart electrodes that were initially clamped. The high-temperature arc column that appears between the electrodes converts both the anode material and the liquid phase surrounding this anode into the vapor phase. In the case that electrode spacing does not exceed 1 mm, the deposit similar to that formed in ADGP has been generated on the cathode. Deposit

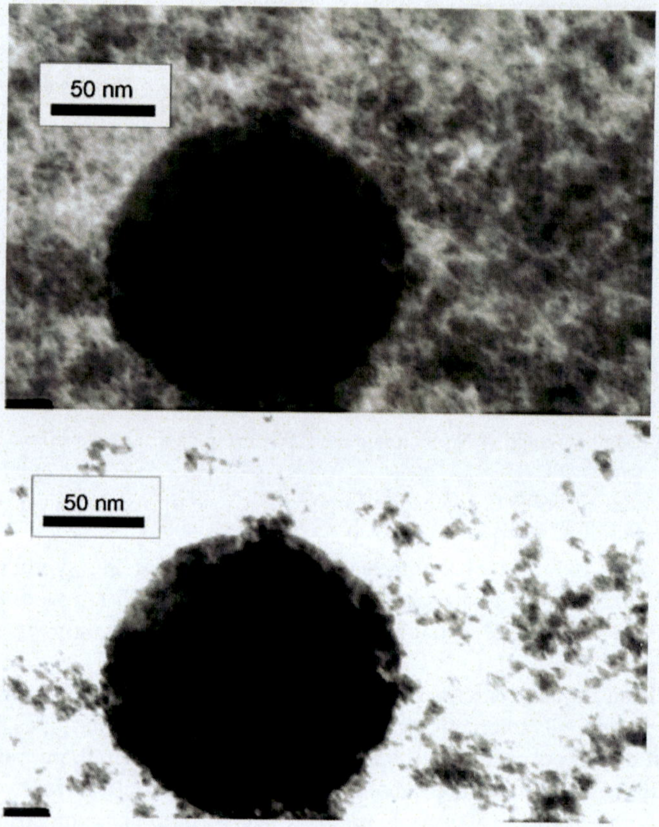

Fig. 11.11 TEM photographs of Ni microgranule (100 nm) in nanocarbon shell

on the cathode is not formed as the electrodes moving apart is more than 1 mm. In this situation the whole resulting product is either in suspension in the liquid phase or on the bottom as sediment.

The efficiency of this method is sharply increased by using the arc discharge in the liquid phase where a layer of powder reagent is used as an anode (Fig. 11.13). In this case in displacement of electrodes, each conducting particle being among the similar ones is, on the one hand, an anode and, on the other hand, a cathode. To increase the frequency of electrodes moving together and apart, an electromagnetic vibrator has been used in this method and it brings and takes away the cathode from the powder reagent at a specified frequency. A large amount of nanoproduct has been formed as a result of great number of electric discharges.

On the basis of our phenomenological model of processes occurring in the interelectrode space in the liquid phase, the following variants of the process course can be assumed.

1. During ADLP, when the electrode separation is less than 1 mm, liquid phase transformed into a vapor state (Fig. 11.14), thus providing conditions similar to

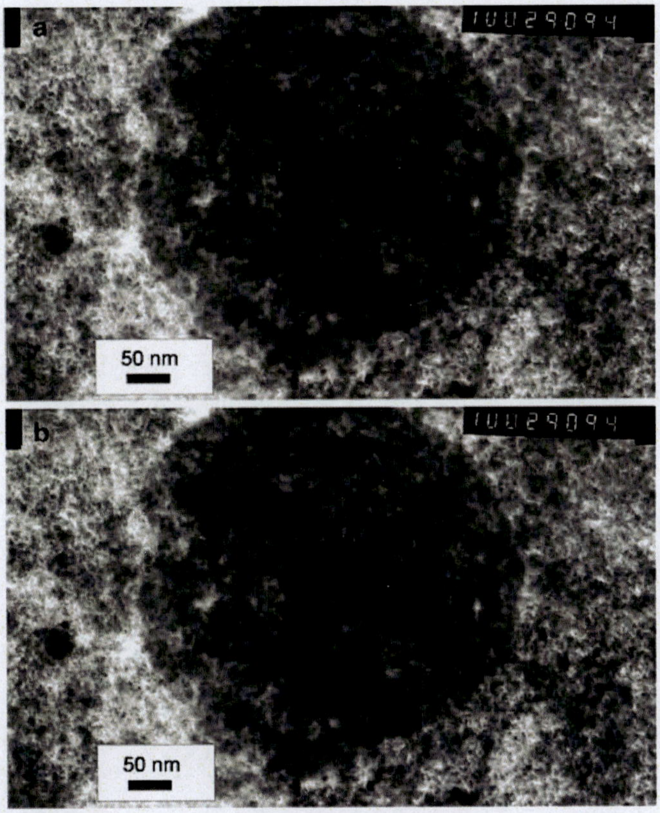

Fig. 11.12 Spherical clusters consisting from ferromagnetic nanoparticles

Fig. 11.13 Schematic diagram of operation of the arc discharge unit in the liquid phase with a dispersed anode

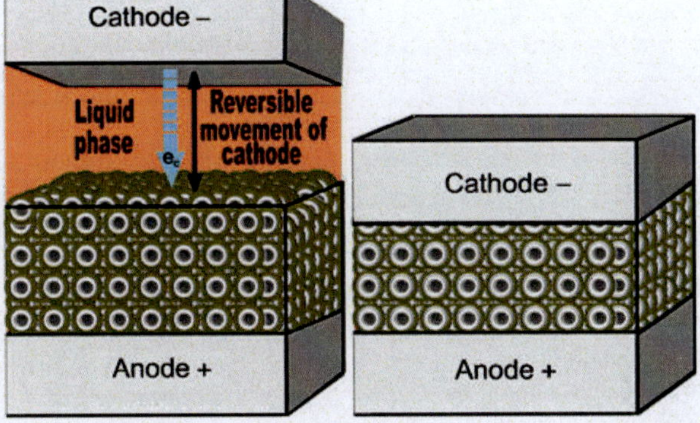

those in ADGP. As this takes place, carbon vapor, carbon nanostructures and fragments of graphene sheets interact with each other under the action of electromagnetic forces and move in different directions.

At the interface (g-*l*) the vapor phase condenses due to the temperature gradient. The charged particles, moving from the anode to the cathode, form the deposit. A minor amounts of these particles, by virtue of collision with an electron stream, in concert with neutral particles are ejected from the arc zone and quenched in crossing the interface.

Near the quenching zone, the particles comprising the gaseous phase agglomerate through the saturation of nonsaturated bonds, create different nanoforms and assemble in clusters.

2. When electrodes are moved apart at the distance exceeding a vapor bubble diameter, the deposit formation will stop on the cathode (Fig. 11.14b). This can be derived from the fact that the particles, forming the deposit and having plasma temperatures, are bound now to overcome the layer of liquid to reach the cathode. Approaching to the interface (g-*l*), these particles undergo the quenching process. In the quenching zone, the particles begin to agglomerate, form clusters and lose completely their reactivity.

Breaking away from the anode surface, the bubbles go into the volume of liquid phase (Fig. 11.14c). All structures contained in a bubble and formed as a result of the anode evaporation remain the volume-enclosed of the bubble.

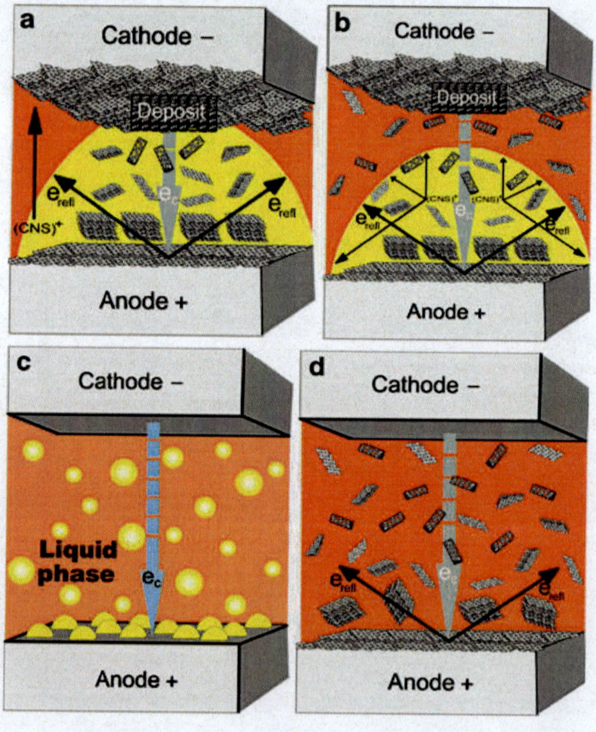

Fig. 11.14 The mechanism of formation of carbon nanostructures and their composites in the liquid phase

Vapor, getting to the zone of lower temperatures, condenses, the bubbles are shuted and their content turn into the liquid phase (Fig. 11.14d). In the liquid nanoparticles can assemble in clusters and precipitate or can be in suspension.

In the case of co-evaporation of metal and graphite or metal in the hydrocarbons medium, the metallic nanoparticles encapsulated in the carbon matrix or other composites can be produced by condensation of vapours mixture in the shutting bubbles.

3. In the case that powder reagent layer is used as an anode, the nanostructures form simultaneously at several points on the surface of current-conducting particle as a result of microscopic acts of arc discharge, similarly as shown in Fig. 11.14a. These nanostructures are generated from the vapours of liquid phase and anode and represent the product exhibitied rather interesting physical and chemical properties.

11.5 Conclusions

In the present work Me-C nanocomposites have been produced by the ADIP method. Their structure, phase composition and magnetic properties have been studied.

The performed studies have shown that Me-C composites have a significantly changed phase composition. The α-Fe powder transforms into Fe_3C carbide almost completely and solid carbon solution in nickel (Ni-C) forms in the Ni powder.

Heating the synthesized nanocomposites also leads to the change in their phase composition: after heating, a crystalline phase Ni-C disappears in the Ni(B-2) powder, and oxides (FeO and Fe_3O_4) appear in the powder Fe(B-5-2) as well as the ratio of crystalline phases (α-Fe and Fe_3C) changes.

Based on the analysis of the observations performed in the course of carbon nanostructures synthesis, the model of nanostructures formation by arc discharge in the liquid phase has been proposed in this paper.

The presence and absence of deposit on the cathode have been explained in consequence of the experiments performance.

Acknowledgment The work has been done within the framework of STCU project 4919.

References

1. Vonsovskiy SV (1971) Magnetism. Nauka, Moscow, p 1032 (in Russian)
2. Vonsovskiy SV (ed) (1961) Magnetic properties of metals and alloys. Gostekhizdat, Moscow, p 560 (in Russian)
3. Gusev AI, Rempel AA (2004) Nanocrystalline materials. Cambridge International Science, Cambridge, p 149

4. Chuistov KV, Shpak AP, Perekos AE et al (2003) Small-size metallic particles: production conditions, atomic and electron structure, magnetic properties and practical implementation. Usp Fiz Met 4(4):235–245
5. Chuistov KV, Perekos AE, Zalutskiy VP et al (1997) The effect of production conditions on structural state, phase composition and fineness of iron and iron-based powders made by electric-spark erosion. Met Phys Adv Technol 16(8):865–875
6. Dubovoy AG, Perekos AE, Chuistov KV (1985) Structure and magnetic properties of small amorphous particles of metallic Fe-15 at.% B alloy. Phys Met 6(5):1085–1088
7. Dubovoy AG, Zalutskiy VP, Ignat'ev IYu (1990) Structure, magnetic parameters and thermal stability for small amorphous particles and amorphous strips of Fe-15 at.% B. Phys Met 8 (4):804–807
8. Chuistov KV, Perekos AE (1998) Structure and properties of small-size metallic particles. 1. Phase-structure state and magnetic characteristics (Review). Met Phys Adv Technol 17 (1):57–84
9. Schur DV, Dubovoy AG, Zaginaichenko SYu, Adejev VM, Kotko AV, Bogolepov VA, Savenko AF, Zolotarenko AD (2007) Production of carbon nanostructures by arc synthesis in the liquid phase. Carbon 45(6):1322–1329
10. Loiseau A, Demoncy N, Stephan O et al (2000) Filling carbon nanotubes using an ARC discharge, Science and application of nanotubes. Kluwer Academic Publishers, New York, p 398
11. Schur DV, Dubovoy AG, Lysenko EA et al (2003) Synthesis of nanotubes in the liquid phase. In: Extended abstracts of 8th international conference on hydrogen materials science and chemistry of carbon nanomaterials (ICHMS'2003), Sudak (Crimea, Ukraine), p 399–402
12. Schur DV, Dubovoy AG, Zaginaichenko SYu, Savenko AF (2004) Method for synthesis of carbon nanotubes in the liquid phase.In: Extended abstracts of international conference on carbon, providence (Rhode Island, USA). American Carbon Society p 196–198
13. Antisari MV, Marazzi R, Krsmanovic R (2003) Synthesis of multiwall carbon nanotubes by electric arc discharge in liquid environments. Carbon 41(12):2393–2401
14. Biro LP, Horvath ZE, Szalmas L et al (2003) Continuous carbon nanotube production in underwater AC electric arc. Chem Phys Lett 372(3–4):399–402
15. Sano N, Nakano J, Kanki T (2004) Synthesis of single-walled carbon nanotubes with nanohorns by arc in liquid nitrogen. Carbon 42(3):686–688
16. Qui J, Li Y, Wang Yu et al (2004) Synthesis of carbon-encapsulated nickel nanocrystals by arc-discharge of coal-based carbons in water. Fuel 83(4–5):615–617
17. Bera D, Kuiry SC, McCutchen M et al (2004) In-situ synthesis of palladium nanoparticles-filled carbon nanotubes using arc discharge in solution. Chem Phys Lett 386(4–6):364–368
18. Montoro LA, Lobrano Renata CZ, Rosolen JM (2005) Synthesis of single-walled and multi-walled carbon nanotubes by arc-water method. Carbon 43(1):200–203
19. Ishlinsky AYu (1989) Polytechnic dictionary. Soviet Encyclopedia, Moscow, p 611 (in Russian)

Chapter 12
The Temperature Dependence of Chemical Shifts of Individual Peaks in the ^{13}C NMR Spectrum of the Fullerite C_{60}, Doped with Molecular Oxygen

O.V. Val'ba, E.M. Anokhin, A.V. Maksimychev, A. Michtchenko, D.V. Schur, and Yu.M. Shulga

Abstract The methods of solid-state high-resolution NMR have been applied to investigate samples of fullerite C_{60}, doped with molecular oxygen. It was found that in the case of well-crystallized $(O_2)_{0.44}C_{60}$ sample the temperature dependence of the relative chemical shift of the satellite peaks reveals jump nearly the orientational phase transition temperature. At temperatures above the phase transition, the shift due to the redistribution of electron density was ~ 0.1 ppm, which corresponds to the transfer of 0.05 electron from O_2 molecule to the fullerene molecule.

Keywords Fullerite $C_{60}(O_2)_x$ · Phase transition · Oxygen content · Electron density · NMR spectrum · Chemical shift · Temperature

12.1 Introduction

Fullerene C_{60} doped on the octahedral pores with paramagnetic molecules, it is extremely interesting object in terms of study characteristics of the dipole-dipole interaction between the magnetic moment of the dopant and the spin of the

O.V. Val'ba, E.M. Anokhin, and A.V. Maksimychev
Moscow Institute of Physics and Technology (State University), 141700 Dolgoprudny, Moscow District, Russia

A. Michtchenko
Department of Electronic Engineering, Instituto Politecnico Nacional, SEPI-ESIME-IPN, Mexico C.P. 07738, D.F., Mexico

D.V. Schur
Institute for Problems of Materials Science of NAS of Ukraine, Krzhyzhanovsky str., 3, Kiev 03142, Ukraine

Yu.M. Shulga (✉)
Institute of Problems of Chemical Physics of Russian Academy of Sciences, 142432 Chernogolovka, Moscow District, Russia
e-mail: shulga@icp.ac.ru

S.Yu. Zaginaichenko et al. (eds.), *Carbon Nanomaterials in Clean Energy Hydrogen Systems - II*, NATO Science for Peace and Security Series C: Environmental Security 2, DOI 10.1007/978-94-007-0899-0_12, © Springer Science+Business Media B.V. 2011

nucleus ^{13}C C_{60} molecule. It is known that only a narrow singlet with chemical shift of 143.7 ppm is observed in the spectrum of the ^{13}C NMR of pure C_{60} at room temperature [1]. This is due to the fact that molecules of C_{60} in the crystal are in a state of almost free rotation with three degrees of freedom. The average period of rotation at room temperature is 0.9–1.2 * 10^{-11} s [1]. This value is only 3–4 times greater than for the free rotation of the fullerene in the solution. ^{13}C NMR spectrum of C_{60}-doped molecular oxygen (S = 1) or NO (S = 1/2), consists of seven equally spaced peaks of different intensity [2–6]. The peak with a minimum value of shift is located at 143.7 ppm, which coincides with the singlet in the spectrum of unmodified C_{60}. The distances between the nearest peaks at approximately 0.7 ppm in the case of $(O_2)_xC_{60}$ and 0.4 ppm in the case of $(NO)_xC_{60}$.

Assink [2] suggested that the six satellite peaks, which are located in the weaker field compared with the peak at 143.7 ppm, arise from the interaction between spin magnetic moment of O_2 molecules and nucleus magnetic moment of ^{13}C atom of C_{60}. Each fullerene molecules in the lattice is surrounded by six octahedral pores. Depending on the number of pores filled by oxygen, there is a different chemical shift of the ^{13}C nuclei of the molecule C_{60}. If to designate intensity of peak n $(0 \leq n \leq 6)$ as I_n than the share of molecules C_{60} surrounded by i molecules of oxygen (f_i), will be expressed by a parity $f_i = I_i/\Sigma I_n$ (summation it is conducted on all n from 0 to 6). Then concentration of oxygen in the sample or value x in the formula $(O_2)\ _xC_{60}$ will be defined by expression $x = \Sigma n f_n/6$ [2, 6, 7]. In [8] showed that the oxygen content in the fresh sample $(O_2)_xC_{60}$ determined from the NMR spectrum by the Assink's method, the same as such, as determined by elemental analysis.

Temperature dependence of the individual peaks in the NMR spectrum of C_{60}-doped paramagnetic molecules was investigated earlier [4]. However, if the temperature dependence of chemical shift in the case of $(NO)_xC_{60}$ seen a jump at the orientational phase transition (opt), in the case of $(O_2)_xC_{60}$ such a jump was absent. The question arises, this difference is due to the different nature of NO and O_2, or still deciding proved inadequate samples of $(O_2)_xC_{60}$ and small value of x? We have developed methods of synthesis [9] allows to obtain well crystallized samples of $(O_2)_xC_{60}$ with high value of x. The study of such samples, in our opinion, can definitely answer the question.

12.2 Experimental

Samples of fullerite intercalated by oxygen were synthesized by the low temperature method described elsewhere [9]. The sample is a polycrystalline powder of black colour with metallic lustier.

The NMR spectra of high-resolution ^{13}C nuclei were recorded using an NMR spectrometer Varian Unity Inova 500 M WB with an operating frequency for protons of 500 MHz. Measurements of the chemical shift was performed in a wide temperature range (from −70°C to 100°C in increments of 10°C). In the temperature range from −20°C to −30°C until the step size is reduced to 3°C. To obtain high resolution spectra used the method of sample rotation at the magic angle with a frequency of 8 kHz. Fullerite was placed in a ceramic rotor with an outer diameter of 3.2 mm, sample volume was 22 ml. For the NMR signal was used 90° pulse, the delay between pulses was 25 s. ^{13}C-NMR spectra with signal to noise ratio of at least 10 for the natural ^{13}C isotope content were obtained by averaging 64 individual spectra. The chemical shift was measured relative to the shift of adamantane (37.7 ppm relative to tetramethylsilane), which, for simplicity, in the beginning of each series of experiments was tuned to 38 ppm. Further, chemical shifts were recalculated relative to the shift of tetramethylsilane.

12.3 Results and Discussion

NMR spectrum of the sample $(O_2)_xC_{60}$ obtained by us is shown in Fig.12.1. The maximum intensity has a peak with n = 4, but not with n = 0, as in the case of the sample $(O_2)_{0,20}C_{60}$, that was investigated in [3]. The calculation of the value of x from the measured integral intensities of individual peaks of In the above expression allowed us to determine that in the sample composition corresponds to the formula $(O_2)_{0,44}C_{60}$.

Note that the number of satellite peaks in the spectrum indicates that the spin of O_2 molecules rapidly changes its orientation in space. In the case of slow shifts (when the coup comparable with the time of the experiment) was observed in the NMR spectrum would be 12 satellite lines of the interaction of ^{13}C nuclei with spin of O_2, whose projection on the axis Z: +1 or −1.

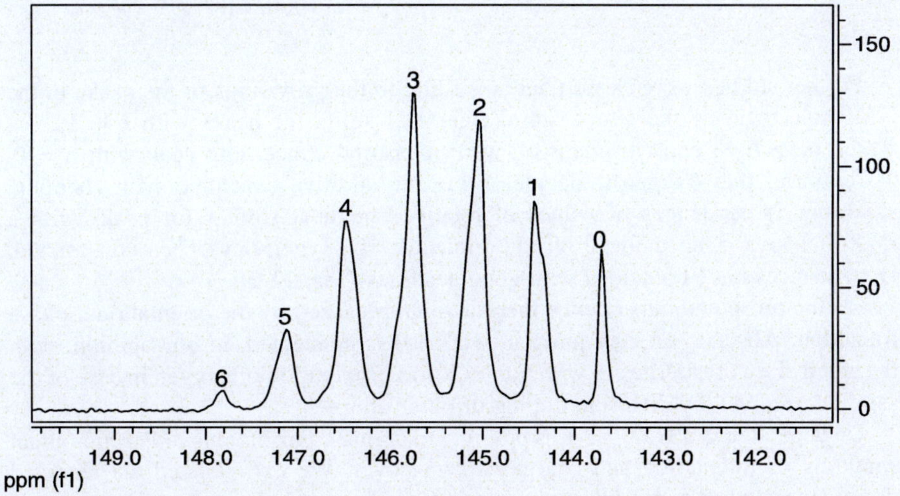

Fig. 12.1 ^{13}C-NMR spectrum of C_{60} intercalated with molecular oxygen

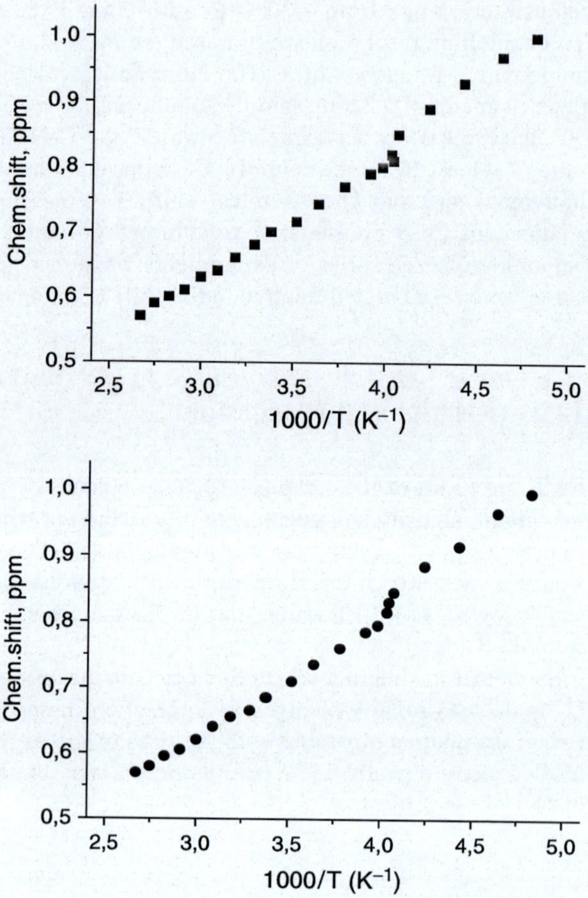

Fig. 12.2 The dependence of the relative chemical shifts σ/n on temperature for peaks with n = 1(*top*) and n = 2 (*bottom*)

Further studies were carried out according to the provisions of the peaks in the NMR spectrum of the temperature. Chemical shifts for peaks with n = 1,2,3,4 which have high enough intensity, were measured concerning peak with n = 0. To confirm, that oxygen molecules bring the additive contribution in chemical shielding, dependences of values of relative chemical shift σ for peaks with a miscellaneous n, normalized on one molecule of oxygen have been constructed. Dependences have coincided with good accuracy (Fig. 12.2).

On the presented dependence there is a jump caused by the orientational phase transition. Absence of such jump in work [4] is connected, in our opinion, with the small sizes crystallites or with non-uniform distribution of oxygen in case of the samples received by diffusion doping of pure fullerites.

On dependence σ from $1/T$ it is possible to divide contributions to Fermi-contact shielding of interaction and charge carrying over. We extrapolated the received dependence two straight lines $\sigma/n = A + B/T$. Extrapolation for a case n = 1 is shown on Fig. 12.3. The coefficients of extrapolation for peaks with different values of n are presented in Table 12.1.

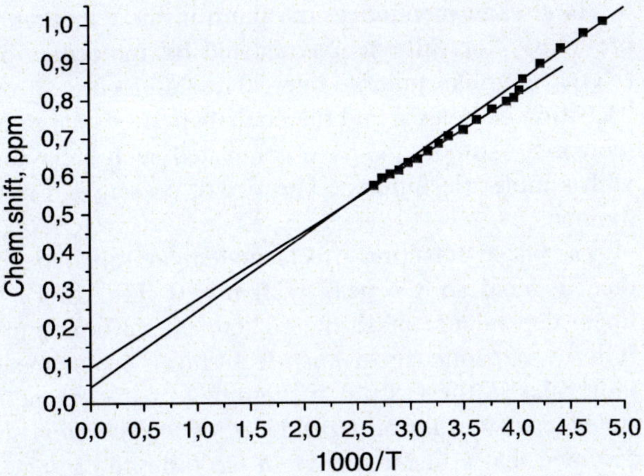

Fig. 12.3 Extrapolation of the temperature dependence for the peak with n = 1 in the high (*1*) and low temperature (*2*) phase

Table 12.1 Parameters of extrapolation lines

	$T > T_{opt}$		$T < T_{opt}$	
q	A, ppm	B, ppm*K^{-1}	A, ppm	B, ppm*K^{-1}
1	0.095 ± 0.008	178 ± 2	0.043 ± 0.036	199 ± 8
2	0.093 ± 0.007	177 ± 2	0.034 ± 0.033	200 ± 7
3	0.096 ± 0.006	176 ± 2	0.035 ± 0.027	200 ± 6
4	0.098 ± 0.004	176 ± 1	0.013 ± 0.012	205 ± 3

At temperatures above the phase transition, the shift due to the redistribution of electron density was ~ 0.1 ppm, which corresponds to the transfer of 0.05 e of O_2 molecules to the fullerene molecule, since the transfer of one electron per molecule of C_{60} corresponds to a shift equal to ~ 2 ppm [4, 11]. The authors of [4] for the charge transfer amount received 0.065 e per molecule of C_{60}. Experiments to study the dielectric constant yield for charge transfer value 0.04 e [10]. Thus, our results confirm that the system $(O_2)_xC_{60}$ holds the charge transfer, but the magnitude of the charge being carried is small.

On the line inclination it is possible to estimate a constant of hyperfine interaction A_{eff} in accordance with the known formula:

$$\sigma_{contact} = n\left(\frac{A_{eff}}{\hbar}\right)\frac{g\beta S(S+1)}{3kT\,\gamma_I}.$$

It turned out that in our case A_{eff} ~ 0.013 MHz. For comparison, in the case of fullerites intercalated with alkali metals A_{eff} ~ 100–1,000 MHz [11].

At temperatures below temperature of phase transition the constant of hyperfine interaction increases approximately by 14%. Interaction Fermi-contact increase is caused by reorganization of a crystal lattice – the distance between molecules C_{60} decreases and O_2, changes in distribution of electronic density of a crystal also are possible. In it specifies also change in factor A (Table. 12.1). In low temperature phase the contribution caused by carrying over of a charge, practically is absent.

As already mentioned, the jump in the temperature dependence was observed previously for fullerite intercalated by molecules of NO [4]. However, for the $(NO)_xC_{60}$ unlike fullerite intercalated molecules of oxygen, the slope of the phase transition decreases, and the contribution of charge transfer in chemical shielding increases. Authors [4] have connected such behavior with strong interaction NO with a molecule fullerene because of presence at molecule NO permanent dipole moment.

The phase transition affects the temperature dependence of the absolute value of the chemical shift δ peak with n = 0 (Fig. 12.4). In a high-temperature phase linear dependence of absolute chemical shift on temperature is observed, for pure fullerite dependence of chemical shift also is linear [12]. It is known that for molecules in the gas phase averaging of the magnetic shielding of the rotational states leads to a linear dependence of the chemical shift on temperature [13]. It is assumed that a linear change in the fullerite can also be caused by averaging the magnetic shielding of the rotational states of the molecule C_{60} [12].

More interesting is the temperature dependence of the absolute chemical shift δ peak with n = 0 in the low-temperature phase. Dependencies δ = δ (T) for fullerite x ~ 0.22 and x ~ 0.007 have the same character: the temperature is lowered to T_{fz} chemical shift reaches a certain size and with a further decrease of temperature does not change (Figs. 12.4 and 12.5). Perhaps this behavior is due to the fact that the low-temperature phase of the fullerene molecule did not rotate freely, and by turning between equivalent states. Parameters of the chemical shielding tensor orientation for the equivalent states are the same, and do not depend on the frequency hopping of the molecule of C_{60} between these states, and thus does not depend on the temperature. It should also be noted that the intercalation of molecules of oxygen leads to "spreading" of the phase transition; the chemical shift reaches a constant value at lower temperatures. Such conduct, we believe, due to the uneven distribution of oxygen molecules on fullerite's octapores, which leads to non-uniform expansion of the lattice of fullerite.

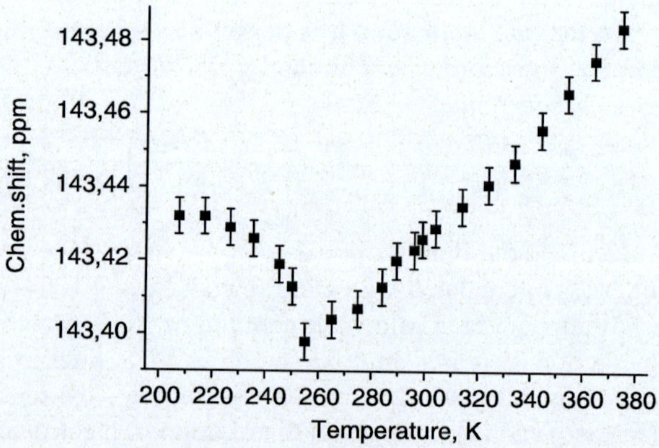

Fig. 12.4 Dependence of chemical shift on temperature for peak with n = 0

Fig. 12.5 Dependence of chemical shift on temperature for peak with n = 0 in the case of fullerite with low oxygen content

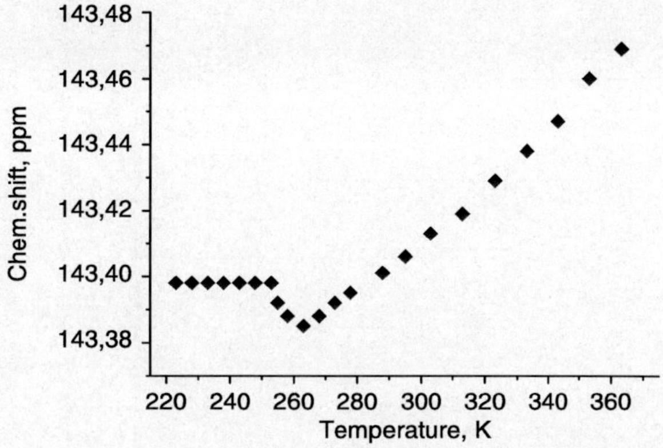

Acknowledgement This work was supported by the Russian Foundation for Basic Research (project No 09-03-00597-a).

References

1. Yannoni CS, Johnson RD, Meijer G et al (1991) Carbon-13 NMR study of the C_{60} cluster in the solid state: molecular motion and carbon chemical shift anisotropy. J Phys Chem 95:9–10
2. Assink RA, Shirber JE, Loy DA et al (1992) Intercalation of molecular species into the interstitial sites of fullerene. J Mater Res 7:2136–2143
3. Gu M, Tang TB, Hu C, Feng D (1998) Order-disorder transition in solid C_{60} charged with O_2 and with N_2: a study with dielectric and ^{13}C NMR spectroscopies. Phys Rev B 58:659–663
4. Gu M, Wang S, Wu J et al (2005) NMR evidence for the charge transfer from interstitial NO or O_2 to molecule C_{60} in solid C_{60}. Chem Phys Lett 34:167–170
5. Belahmer Z, Bernier P, Firlej L et al (1993) Intercalation of O_2 in solid C_{60} and molecular-rotation hindrance. Phys Rev B 47:980–983
6. Bernier P, Lukyanchuk I, Belahmer Z et al (1996) High-resolution ^{13}C NMR study of oxygen intercalation in C_{60}. Phys Rev 53:7535–7538
7. Renker B, Schober H, Fernandez-Diaz MT et al (2000) Structure and dynamics of C_{60} intercalation compounds: N_2C_{60} and O_2C_{60}. Phys Rev B 61:13960–13968
8. Shulga YM, Martynenko VM, Anokhin EM et al (2010) Structure of C_{60} intercalated with molecular oxygen. Chem Phys 4:543–547 (in Russian)
9. Shulga YM, Martynenko VM, Shestakov AF et al (2006) Doping fullerite with molecular oxygen at low temperature and pressure. Izv AN Ser Khim 55(4):687–696 (in Russian)
10. Pevsner B, Hebard AF (1997) Role of molecular oxygen and other impurities in the electrical transport and dielectric properties of C_{60} films. Phys Rev B 55:16439–16449
11. Pennington CH, Stenger VA (1996) Nuclear magnetic resonance of C_{60} and fulleride superconductors. Rev Mod Phys 68:856–910
12. Tarasov VP, Muravlev YV, Izotov DE (2001) ^{13}C NMR of fullerite C_{60} at temperatures 295–1000 K. Dokl Phys Chem 381:271–274 (in Russian)
13. Jameson CJ (1991) Gas-phase NMR spectroscopy. Chem Rev 91:1375–1395

Chapter 13
Synthesis of Carbon Nanotubes on Zirconium Alloys Surface

V.A. Bogolepov, D.V. Schur, A.F. Savenko, V.M. Adeev, S.Yu. Zaginaichenko, K.A. Meleshevich, A.P. Pomytkin, M.M. Diviziniuk, and E.V. Azarenko

Abstract The radiation-stimulated structural phase transformations can significantly affect the processes influencing on the structural materials operation under radiation and high temperatures conditions. These processes lower reliability and service life of shells and covers thus reducing the times of safe operation of nuclear reactors.

Therefore the search for the way and technologies directed to the increase in reliability and durability of fuel element (tvels) shells and thermal converter covers is a special issue of the present day.

The purpose of the present paper is to give the description of the performed series of experiments on carbidization of the zirconium substrate surface. Such carbidization has been conducted through the carbon nanostructures dissolution in the matrix volume under various conditions of heat treatment. Metallic and nonmetallic catalysts have been used to synthesize carbon nanostructures on the zirconium alloys surfaces. The surface of test specimens of thermal fuel elements (tvels) has been subjected to grinding and also to the electrolytic polishing in acid-water solution for taking the strain hardening. Carbon nanostructures have been synthesized on the substrate surfaces by the method of acetylene pyrolysis. Nickel nanoparticles deposited on the substrate by cladding have been used as a metallic catalyst. Synthesis of carbon nanostructures on such catalysts allows the formation of uniform layer of carbon nanotubes on the sample surface.

The microhardness has been measured for the synthesized specimens with nanotubes on the zirconium substrate cladded with nickel and compared with initial samples. The study has revealed that the microhardness of treated specimens increases, resulting in the hardening of tvel covers by carbidization and in doing

V.A. Bogolepov, D.V. Schur (✉), A.F. Savenko, V.M. Adeev, S.Yu. Zaginaichenko, K.A. Meleshevich, A.P. Pomytkin, M.M. Diviziniuk, and E.V. Azarenko
Institute for Problems of Materials Science of NAS of Ukraine, Krzhyzhanovsky str. 3, 03142 Kiev, Ukraine
e-mail: shurzag@ipms.kiev.ua

S.Yu. Zaginaichenko et al. (eds.), *Carbon Nanomaterials in Clean Energy Hydrogen Systems - II*, NATO Science for Peace and Security Series C: Environmental Security 2, DOI 10.1007/978-94-007-0899-0_13,

so authors have found a way to the rise of reliability and durability of tvel shells and thermal converter covers hardenable by heat-treatment.

Keywords Tvel · Nanostructure · Carbidization · Zr-alloy surface · Microhardness

13.1 Introduction

In solving materials science problems related to the design and operation of reactors of different types it is necessary to consider peculiarities of structural materials operation under radiation and higher temperatures conditions, i.e. at maximal radiation-enhanced diffusion of interstitial impurities in metals.

The above processes lower reliability and service life of shells and cases thus reducing hours of safe operation of nuclear reactors. Therefore the search for the way and technologies directed to the increase in reliability and durability of fuel element shells and thermal converter cases is a special issue of the day. As found in searching, it is such radiation-stimulated structural phase transformations that can significantly affect the processes influencing on the structural materials operation under radiation conditions[1–10].

Authors have carried out the hardening of the zirconium substrate surface by its carbidization. Carbidization was conducted through carbon nanostructures (CNS) dissolution in the matrix volume under different conditions of thermal treatment.

The present work considers the results obtained in refining the technology of CNS synthesis on the zirconium alloys surfaces.

13.2 Results and Discussion

Metallic and nonmetallic catalysts were used to synthesize CNS on the zirconium alloys surfaces.

A special emphasis was placed on the preparation of the surfaces of samples under study.

Subjects of inquiry in this work were:

- experimental batches of reactor alloys based on alloy KTC-110 (99 Zr – 1 Nb, wt. %), and alloys prepared on the basis of binary alloys and multi-component systems, as follows:
- zirconium iodine;
- Zr-Nb system;

- Zr-C system;
- Nb-C system;
- alloys alloyed with carbon, i.e. Zr – Nb – C system;
- alloys of Zr-Nb system with a fullerene film;
- alloys prepared and smelted on the basis of Zr-Nb system with synthesized CNS on the surface.

Before experiment, the surfaces of fuel element samples were subjected to grinding as well as to electrolytic polishing in acid-water solution to relief strain hardening.

Microhardness was measured on apparatus PMT-3 with a Vickers pyramid at the end and side sections of a fuel element case, and on the external and internal surfaces of a tube at a load of 40 g. Microhardness values were calculated by the formula:

$$H_{\mu} = \frac{18544 \cdot P}{d^2_{av.}} \left[kg/mm^2\right],$$

where P is load, kg; d_{cp} is an average value of an indentation diagonal, mm.

To evaluate the average value and the interval of scatter of readings, the frequency dependence of microhardness was plotted for each measurement (at most 20 indentations). The results are given in Table 13.1:

The studies have shown that the interval of scatter of microhardness average value is practically unchanged both for the sections of a fuel element (end, side) and for the tube surfaces (external, internal).

The values of microhardness on the surfaces and the interval of scatter are somewhat higher than those in the tube sections. Obviously, this is due to structural heterogeneity of the fuel element case and thermal mechanical treatment at the technological stages of the tube formation.

CNS on the substrate surfaces were synthesized by the acetylene pyrolysis method. Nickel nanoparticles deposited on the substrate by cladding were used as a metallic catalyst (Fig. 13.1).

CNS synthesis on the prepared catalysts proceeded rather intensively, but CNS were distributed along the surface irregularly (Fig. 13.2). Large bundles of CNS formed at the positions of big nickel clusters. Organic substances were used as metal-free catalysts. CNS synthesis on such catalysts allows the formation of a uniform CNT layer on the sample surface (Fig. 13.3).

Table 13.1 The results of microhardness investigation

	Section of fuel element case		Surface of fuel element case	
	End (1)	Side (2)	External (3)	Internal (4)
Average value and interval of scatter of microhardness H_{μ}, kg/mm^2	258[225–275]	265[235–280]	280[250–320]	285[260–315]

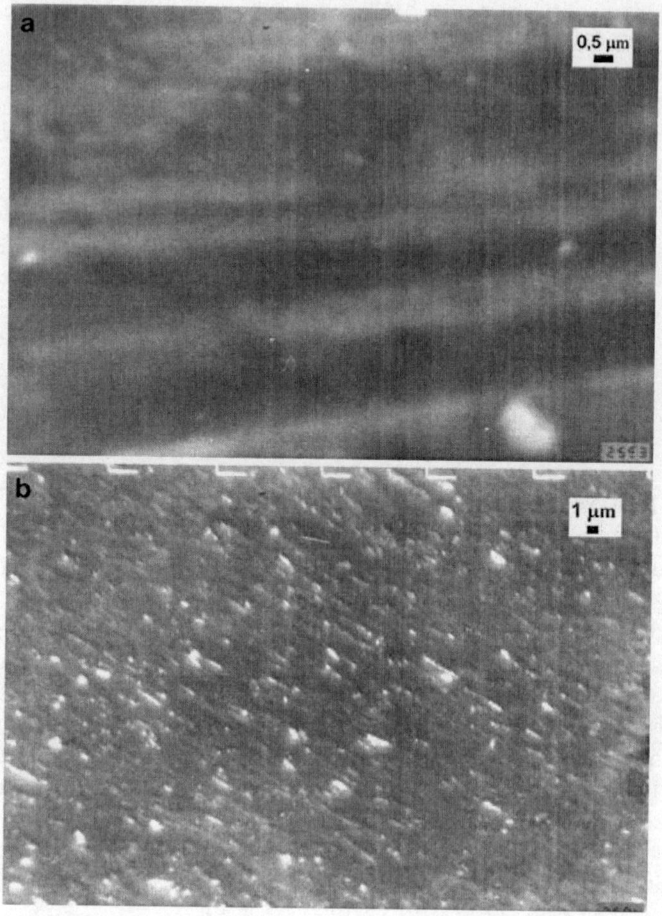

Fig. 13.1 Zirconium substrate clad with nickel: (**a**) initial, (**b**) after cladding with nickel

Fig. 13.2 Nanotubes synthesized on the zirconium substrate clad with nickel

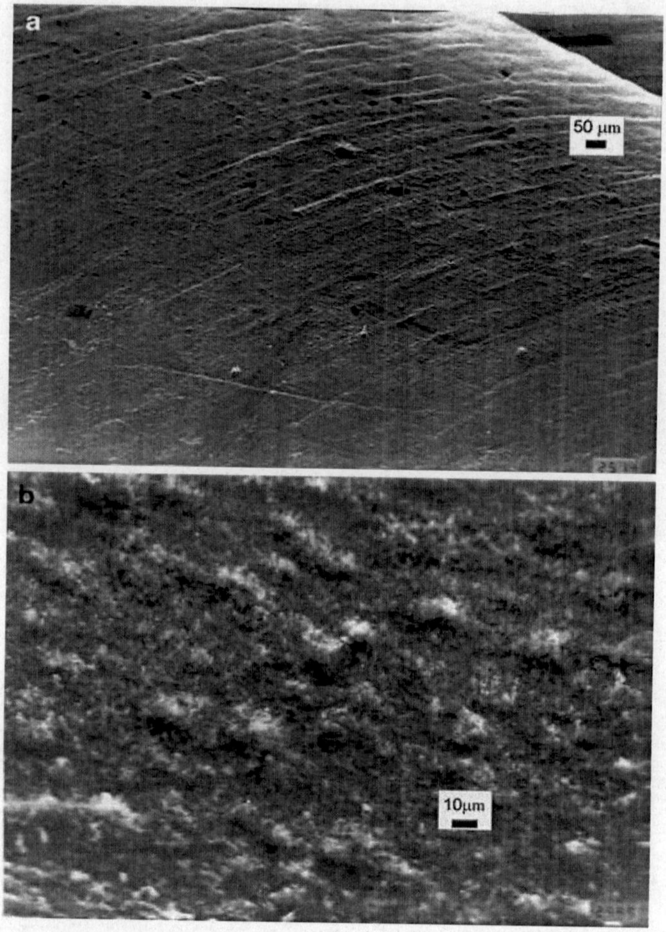

Fig. 13.3 Carbon nanotubes (CNT) synthesized by the pyrolytic method on the zirconium substrate using a non-metallic catalyst: (**a**) initial with a clad catalyst, (**b**) carbon nanostructure

13.3 Conclusions

The technology of metal-free catalytic synthesis of carbon nanotubes on the zirconium substrate has been refined.

Acknowledgment The work has been done under financial support of STCU grant 4012.

References

1. Iseki M, Uchida K, Kirihara T (1977) Electron radiation damage of electrode steel in a high voltage electron microscope. In: Proceedings of the 5th international conference on high voltage electron microscopy. Kyoto pp 585–588
2. Boulanger L (1977) Precipitation ordounee de carbures dans des aciers toadies. J Nucl Mater 64(1/2):176–182
3. Stanley JT, Hadricson LE (1979) Ferrit formation in neutron-irradiated austeritic stainless steel. J Nucl Mater 80(1):69–78
4. Williams TN, Titchmarsh JM (1979) The occurrence of silicon-rich phase of the M6C type in neutron-irradiated FV548 steel. J Nucl Mater 87(2–3):398–400
5. Maradhi M (1978) Untersuchungen an neutronenbestralten kupferdotieren Eisenpoben mit Hilfe der Neutronenkleinwinkelstreuung. Ges Kernenergieverwert Schiffhau schi ffahrt E40:159
6. Chambered A, Laugier J, Penisson JM (1979) Electron irradiation effects on iron-nikel invar alloys. J Magn Magn Mater 10(2/3):139–144
7. Bian Z, Wang RJ, Wang WH, Zhang T, Inoue A (2004) Carbon-nanotube-reinforced Zr-based bulk metallic glasscomposites and their properties. Adv Funct Mater 14(1):59–63
8. Dubovtsev IA, Maslov VI et al (1980) Effect of ionic irradiation on Fe-Ni and Fe-Si alloys. Izv Sev Kavk Nauch Cent Vyssh Shkoly Estestv nauki 2:33–37
9. Huang L, Lau SP, Zhang YB, Tay BK, Fu YQ (2004) The synthesis of carbon nanotubes and zirconium carbide composite films on a glass substrate. Nanotechnology 15(5):663
10. Lee WJ, Smyrl WH (2005) Zirconium oxide nanotubes synthesis via direct electrochemical anodization. Electrochem Solid State Lett 8(3):B7–B9

Chapter 14
Small Size Particles of Different Metal Alloys with Protective Shell for Hydrogen Storage

G.N. Churilov, G.A. Glushenko, A.S. Fedorov, Z.I. Popov, A.M. Zhizhaev, A.V. Cherepahin, I.V. Osipova, Ye.V. Tomashevich, and S.N. Vereshchagin

Abstract The work presents both theoretical and experimental studies of Mg–C, Ni–C, Mg–Ni–C composites. The composites were produced in carbon-helium plasma flow. The composites were hydrogenised directly in synthesis process. Out of three composites under study only Mg–Ni–C contains MgH_2 hydride. Hydride content is 69.99 at.%, the remaining magnesium is in oxidized state – 30.06 at.%. Photographs of Mg–Ni–C composite particles dehydrogenation were made with a scanning microscope. Ab initio theoretical studies established that diffusion rate of hydrogen atoms in magnesium hydride with Ni impurities is increased substantially in the vicinity of Ni atoms. It can be used for the magnesium hydrogenation process acceleration. Also it was defined that nickel prefers to form many layers covering on magnesium surface by island growth mechanism.

Keywords Nanoparticles · Metal alloys · Hydrogen sorption · DFT calculations

Subscript

HF	high-frequency
XPS	X-ray photoelectron spectroscopy
DFT	density-functional theory
PBC	periodical boundary conditions

G.N. Churilov (✉), G.A. Glushenko, A.S. Fedorov, Z.I. Popov,
A.M. Zhizhaev, A.V. Cherepahin, and I.V. Osipova
L.V. Kirensky Institute of Physics SB RAS, Akademgorodok, 50, bld. 38,
660036 Krasnoyarsk, Russia
e-mail: churilov@iph.krasn.ru

Ye.V. Tomashevich and S.N. Vereshchagin
Institute of Chemistry and Chemical Technology SB RAS,
Marks st., 42, 660049 Krasnoyarsk, Russia

S.Yu. Zaginaichenko et al. (eds.), *Carbon Nanomaterials in Clean Energy Hydrogen Systems - II*, NATO Science for Peace and Security Series C: Environmental Security 2,
DOI 10.1007/978-94-007-0899-0_14,

14.1 Introduction

Further technological development of civilization depends nowadays on assimilation of alternative sources of energy. Ecological catastrophes (e.g. the recent one in Mexican Gulf), pollution of cities, depletion of oil reserves draw the attention of society to hydrogen energy. In this aspect urgent and significant is again the issue of safe storage and transportation of hydrogen. Presently hydrogen content in sorbent is considered economically sound at the level of 6.5 wt.%, so far it was not achieved.

Numerous studies indicate that this problem cannot be solved by physical sorption alone. Feasibility of chemical sorption and combination of both these mechanisms is the general avenue of attack on the problem.

However, there is a nanotechnology field producing and studying various nanostructures. From our viewpoint to solve the problem of an efficient "hydrogen storage" most promising are the nanoparticles with nucleus-shell structure and nanocomposites, i.e. a mixture of a catalyst metal with a hydride forming metal. For the first case the shell defines not only stability of nucleus within it, but also the sorption and desorption rates. For the second case the catalyst metal can accelerate sorption and desorption processes thousandfold.

Indeed, imagine an Mg particle covered with Pd layer, an H_2 molecule on the surface of Mg will dissociate into atoms and through the Pd layer these atoms will penetrate into Mg nucleus to form its hydride. Our earlier quantum-chemical calculations showed the first Pd layer on Mg surface to form with benefit 3.38 eV per atom, while the second – 2.14 eV per atom. In this manner Pd film forms on the Mg surface layer-by-layer and not in the form of antennas or projections [1].

14.2 Ab Initio Modeling of MgH_2 and Mg-Ni Systems

For better understanding of Ni impurities influence to magnesium hydrogenation process we carried out ab initio calculations of this process. Assuming that enthalpy $H = E_{bind} + p\Omega$ under ambient conditions is close to it under temperature $T = 0$ K and pressure $P = 0$ conditions, zero temperature and pressure were used for all calculations. At that the formation enthalpy would be equal to the total binding energy E_{bind}, which was defined during this energy minimum search. The calculations of all systems binding energies were performed using density-functional theory (DFT) [2] within generalized-gradient approximation (GGA) using the functional of Perdew, Burke, and Ernzerhof (PBE) for the exchange-correlation energy. At that periodical boundary conditions (PBC) were used. The calculations were implemented in the fully selfconsistent ab initio package VASP package [3]). There only valence electrons were considered using plane wave basis functions, at that their interactions with atom nuclei and core electrons were treated by ultrasoft scalar relativistic Vanderbilt-kind pseudopotential formalism. The Vanderbilt pseudopotentials allow one to reduce significantly a maximal kinetic energy Ecutoff of basis functions without loss of accuracy. This lead to decreasing of the necessary number of plane waves and to essential reduction of the calculation time. The geometry optimization in all calculations has been carried out until the forces acting on all atoms become lower than 0.05 eV/Å. To improve the convergence of the binding energy the Gaussian broadening for occupation number of all electronic levels was used with a smearing width of 0.1 eV. Because of large size of supercells for the integration over Brillouin-zone (BZ) the G-point was used only in all calculations.

First of all, the Ni impurity in the pure Mg was modeled. The calculations details were similar with calculations in our previous work [4]. The nickel binding energy $E_{bind,Ni-Mg}$, equal 1.275 eV was calculated for isolated Ni atom in supercell consisted of 53 Mg atoms and contained 3 × 3 × 3 unit cells. The $E_{bind,Ni-Mg}$ was defined as

$$E_{bind,Ni-Mg} = E_{bind,Mg53Ni} - 53E_{bind,Mg} - E_{bind,Ni}, \tag{14.1}$$

where binding energies per atom $E_{bind,Mg}$ and $E_{bind,Ni}$ were determined from calculations of single Mg and Ni crystals. The positive value $E_{bind,Ni-Mg}$ means that Ni atoms can be solved inside bulk magnesium in small concentrations only that is confirmed by experiments. At that contribution of the configurationally entropy $-TS = T * \log(\frac{\Theta}{1-\Theta})$($\Theta$ – concentration of Ni atoms in Mg matrix) into Gibbs free energy $G = E_{bind} + P\Omega - TS$ become essential and lead to profitable penetration of Ni atoms into bulk Mg.

After that the process of hydrogen atom migration with help of H vacancy inside MgH_2 was investigated. To find a transition state that is a potential barrier height $E_{barrier}$ for hydrogen atom movement from one potential minimum to nearest one the nudged elastic band method (NEB) [5] has been used. Then ratios k of hydrogen atom hopping between minima were calculated within the transition state theory by well-known formula

$$k = Ae^{-\frac{E_{barrier}}{kT}}, \tag{14.2}$$

where T is the temperature, $E_{barrier}$ is the potential barrier height value for the atom hopping and factor A is defined with help of products of all vibration frequencies in the potential minima and saddle points, (see Slater, 1956,1959; Vineyard, 1957). The potential barrier for H migration inside pure MgH_2 was calculated as 1.05 eV. After substitution of any Mg atom into Ni atom the potential barriers for H migration in the vicinity of Ni atom were calculated, see Fig. 14.1.

One can see the barrier value for H jump between energy minima in the second (0.896 eV) and especially in the first neighbour atom shell (0.396 eV) from Ni atom are substantially smaller that this for pure MgH_2 case. At that the binding energy of H atom in first neighbour atom shell is on ~ 1 eV lower then that for second shell. It means that at first stage H atoms concentrated in the close vicinity of Ni impurities and H jump rate in the Ni second neighbour atom shell would be increased. Assuming (14.2) it is easy calculate increasing of H jump rate due to the Ni impurities inside MgH_2 at T = 300 K to be equal ~ 350 times.

Also the structure of Ni covering at Mg (0001) surface was modeled. We calculate binding energy of one and two layer clusters contained 16 Ni atoms both at magnesium surface in the periodical slab geometry. Assuming formulas

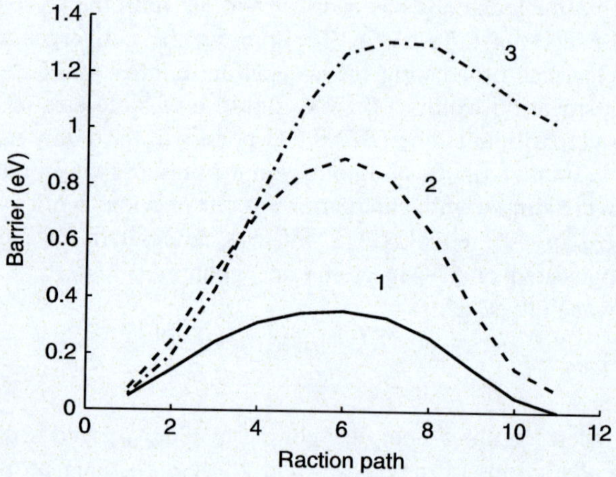

Fig. 14.1 Dependence of the potential barriers for H jump inside MgH_2 in the vicinity of Ni atom. *1* – barrier for H jump in the first neighbour atom shell; *2* – barrier for the second shell; *3* – barrier for H jump from the first to the second shell

similar with (14.1) the binding energy per Ni atom $E_{bind,Nicover}$ in the both coverings were calculated. Values $E_{bind,Nicover}$ for one and two Ni layer covering (+1.425 and +1.248 eV correspondently) were defined. It means that nickel prefer to form many layers covering on magnesium surface by island growth mechanism. Whereas the Ni covering thickness is increased the Ni binding energy $E_{bind,Nicover}$ would aspire to bulk Ni binding energy.

14.3 Experimental Results and Discussion

Figure 14.2 shows a diagram of a setup we designed to synthesize metallocarbon composite nanoparticles. Synthesis was carried out in the flow of helium in HF arc plasma. The arc was burning in the analytical gap formed by two coaxial electrodes, graphite bushing and graphite rod. Metals in powdered form were added with the helium flow during the synthesis process, at that, hydrogen was added into the chamber.

X-ray phase analysis of samples was performed with DRON-4 automated powder diffractometer, X-ray tube radiation Cu K_a. XPZ studies were carried out with SPECS (SPECS Gmbh, Germany) ultra-high vacuum photoelectron spectrometer, X-ray tube radiation Al K_a. Electron microscopy of samples was carried out with SEM Hitachi S-5500 high-resolution microscope. Thermogravimetric analysis was performed with NETZSCH STA 449 C – QMS from 40°C to 1,000°C, 10°C/min, in Ar flow 40 ml/min, oxygen impurities about 0.01 vol.%.

In addition to amorphous halo corresponding to fullerenes and amorphous carbon X-ray powder diffraction of carbon condensate, produced by addition of Mg and H_2, shows reflections of graphite and Mg. Mg reflections are broadened which is due to small size of the particles. XPS made possible to identify composition of carbon composite with Mg (at.%): C – 70.1, O – 20.9, Mg – 9. Carbon binding energy, Fig. 14.3a, corresponds to hybridization sp^2 (63.98%) and sp^3

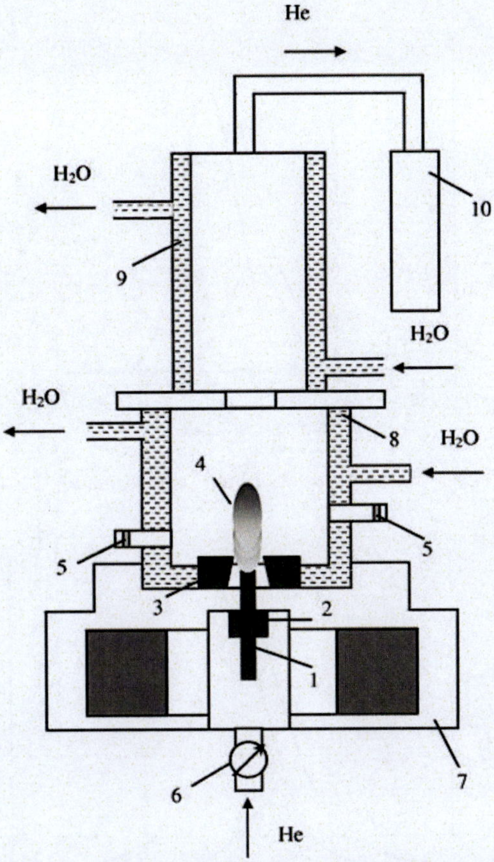

Fig. 14.2 Draft diagram of plasma-chemical setup for nanotubes and nanoparticles synthesis at atmospheric pressure: *1* – inner graphite electrode; *2* – graphite contacts; *3* – outer electrode; *4* – carbon-helium plasma; *5* – quartz windows; *6* – flow meter; *7* – reducing transformer; *8* – lower chamber; *9* – higher chamber;10 – liquid nitrogen trap

(24.44%), and to C-O (3.52%) and C = O (1.06%) bonds. Slight carbon content is due to Mg. Mg binding energy (Fig. 14.3b) corresponds to the free state of Mg (62.08%) and Mg-O state (37.92%). Differential thermal analysis showed that the loss of mass early in the heating process is associated with the loss of adsorbed water, amorphous carbon burning and graphite burning. All methods of analysis unambiguously showed no MgH_2 in the sample.

The carbon condensate produced by addition of Ni and H_2 was synthesized with use of Fe-containing graphite rods. X-ray diffraction image of carbon condensate shows presence of amorphous halo (fullerenes and amorphous carbon) and reflections corresponding to graphite, Ni and Fe with Ni phases (Fe_3Ni_2, Ni_3Fe). XPS allowed to identify the following composition of carbon composite with Ni (at.%): C – 72.3,O – 10.5, Ni – 10.3, Fe – 6.9. Spectrum in the area of C1s line given in Fig. 14.4a shows that carbon is in hybridization sp^2 (77.43%) and sp^3 (15.52%), and in C–O state (7.05%). Figure 14.4b shows intensity of Ni binding energy (94.89%), in whose cubical lattice several atoms are replaced by Fe and Ni in oxidized state (5.11%). Similar to the case of carbon condensate produced by addition of Mg, no metal hydride was found.

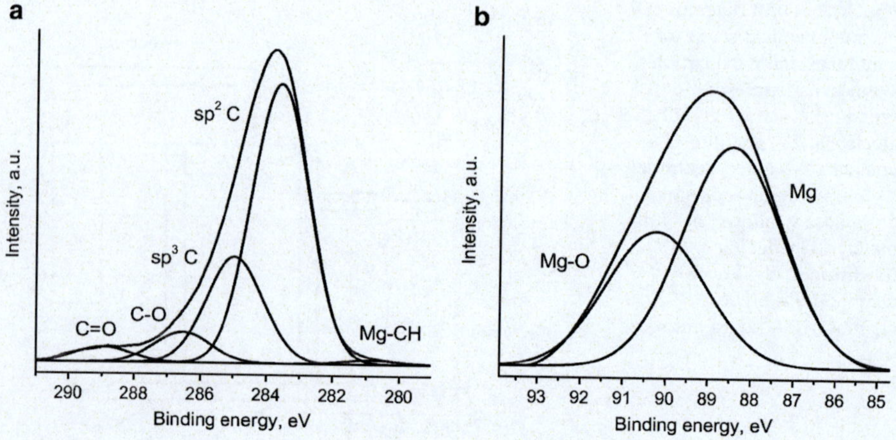

Fig. 14.3 XPS spectra of Mg-C composite: (**a**) C1s (**b**) Mg2s

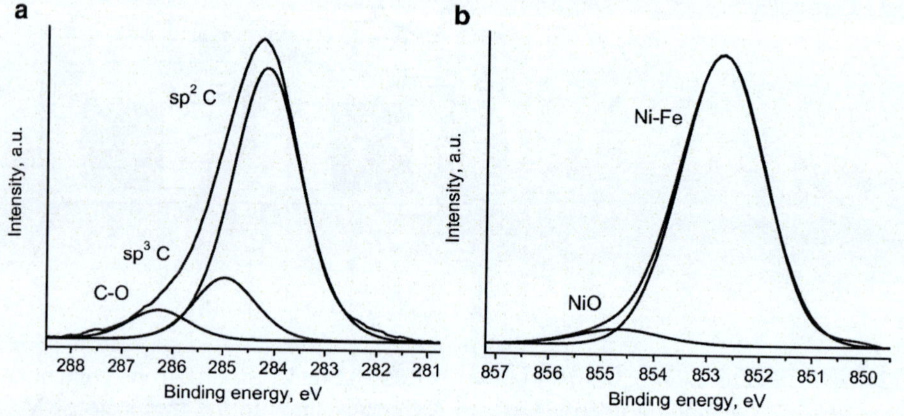

Fig. 14.4 XPS spectra of Ni-C composite: (**a**) C1s (**b**) Ni2p

X-ray powder diffraction of carbon condensate produced by addition of Ni and Mg allowed to identify graphite, Mg and Ni in whose cubical lattice some atoms were replaced with Fe. Mg-Ni-C composite was also studied by XPS. Decomposition of carbon line (Fig. 14.5a) made possible to isolate components corresponding to carbon-hydrogen bonds (62.76 at.%) and oxygen, double (3.75 at.%) and single bonds (7.62 at.%), carbon in the sample is present in hybridization sp^2 (25.87 at.%). Most of oxygen is bound with carbon and hydrogen (81.72 at.%), and in combination with Mg and Ni (18.28 at.%), Fig. 14.5b. Ni in the carbon condensate is present in free state (45.41 at.%), and in bound state with Fe (43.79 at.%) and O (10.8 at.%). In Fig. 14.5c XPS unambiguously showed most of Mg to form hydride – 69.99 at. %, the remaining part – oxide – 30.06 at.%, Fig. 14.5d.

This sample was further analyzed by thermogravimetry, Fig. 14.6. Loss of mass early in the heating process corresponds to the loss of adsorbed water – 1.4 wt.%. From 120°C to 640°C the loss of mass is due to burning of amorphous carbon,

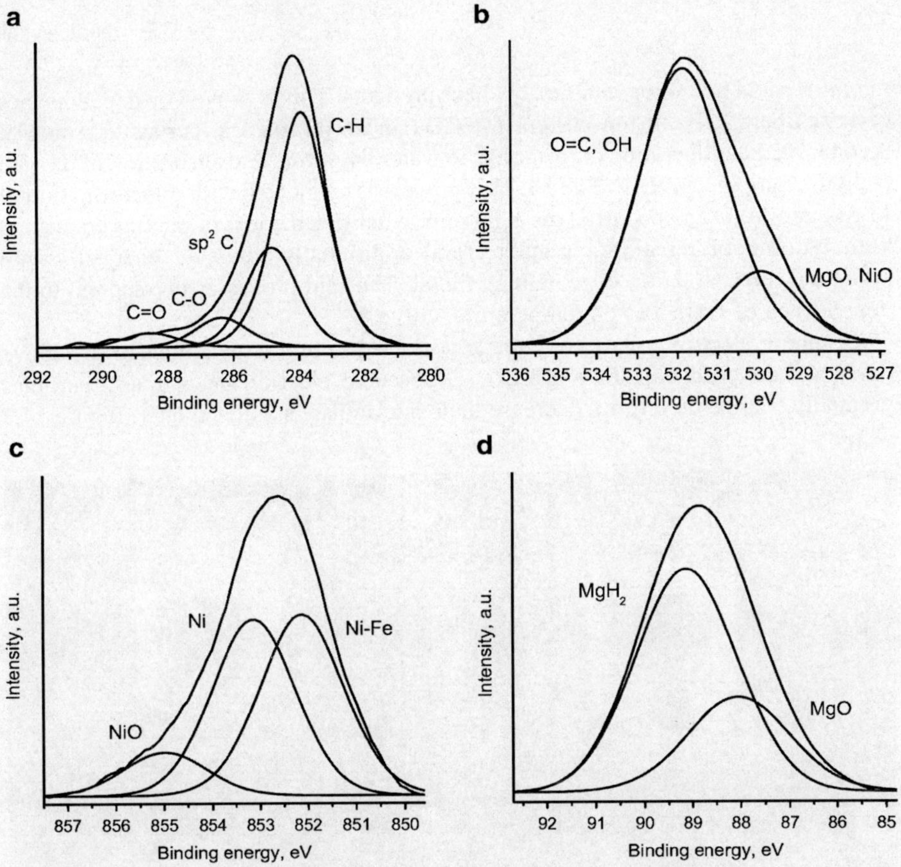

Fig. 14.5 XPS spectra of Mg-Ni-C composite: **(a)** C1s **(b)** O1s **(c)** Ni2p **(d)** Mg2s

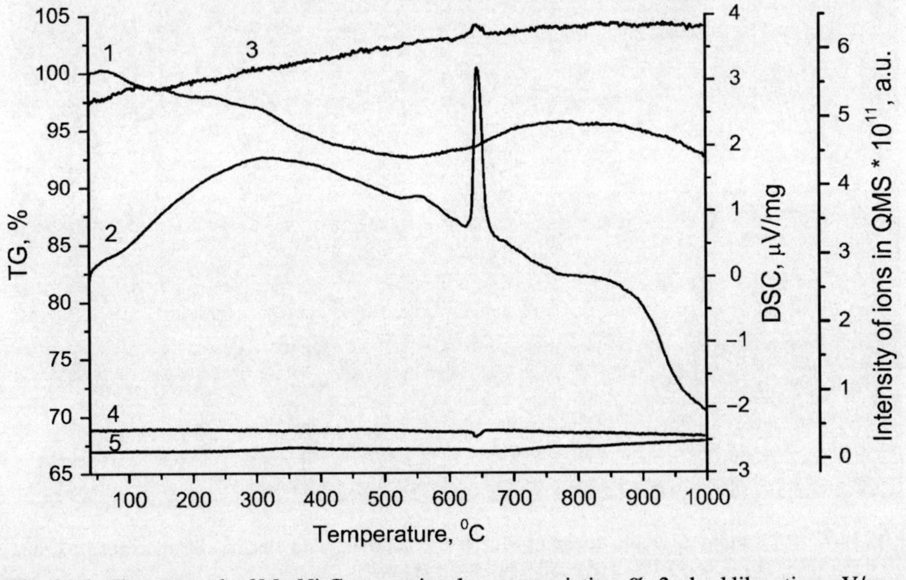

Fig. 14.6 Thermograph of Mg-Ni-C composite: *1* – mass variation, %; *2* – heal liberation, µV/mg; *3*, *4*, *5* – intensity of m/z = 18 (H_2O), 32 (O_2), 44 (CO_2), a.u.

nanotubes and hydrogen retained by these products. This is proved by dependencies of heat liberation and ion current for CO_2 and H_2O masses, curves 2, 5 and 3. At 644.3°C heat liberation corresponds to water liberation and adsorption of oxygen from the sample, curves 2, 3 and 4. This process is connected with transformation of MgH_2 into MgO and occurred during 4 min. After that the loss of mass is connected with burning of remaining graphite, and additionally liberated energy – with structural reconstruction of remaining metal. The said processes correspond to the dependence of mass loss on temperature, curve 1.

Scanning electron microscopy made possible to observe dynamics of the above described dehydrogenization process. Exposed to electron beam many particles gradually change their form, decrease their size and are stabilized later. Figure 14.7

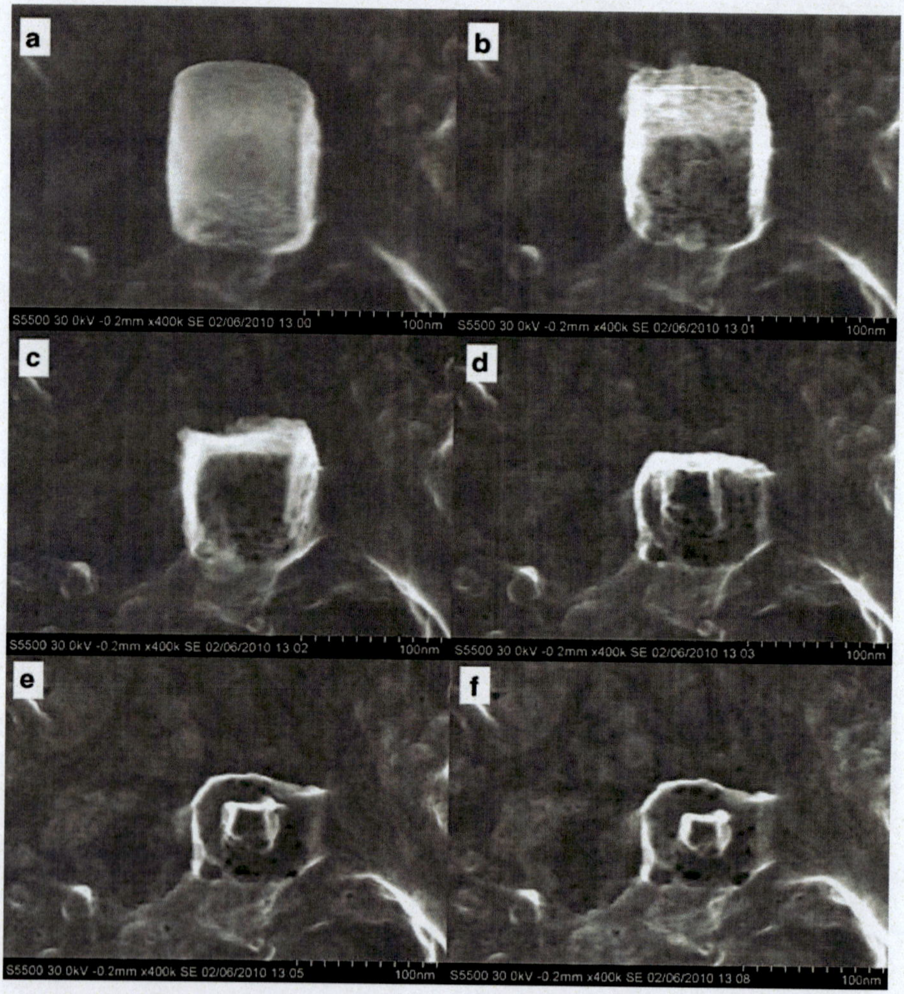

Fig. 14.7 Electron microscopic image of Mg-Ni-C composite with electron beam exposure time, 30 kV: (**a**) 0 s (**b**) 60 s (**c**) 120 s (**d**) 180 s (**e**) 300 s (**f**) 480 s

shows a series of successive photographs of Mg-Ni-C composite particle demonstrating the dehydrogenization process.

14.4 Conclusions

The DFT calculations showed that the diffusion rate of hydrogen atoms in magnesium hydride with Ni impurities is increased substantially (~ 350 times at $T = 300$ K) in the vicinity of Ni atoms. Also it was defined that nickel prefers to form many layers covering on magnesium surface by island growth mechanism. Experiments showed that Mg–Ni–C composite hydrogenates in plasma during their formation. Dehydrogenization process takes place after amorphous carbon is removed at 644.3°C during 4 min.

Acknowledgements Financial support from the Russian Foundation for Basic Research under grant No. 09-03-00383 is gratefully acknowledged. The authors appreciate Institute of Computational Modeling SB RAS for opportunity to use their cluster computer for all calculations.

References

1. Churilov GN (2006) Synthesis and characterization of novel nanomaterials for hydrogen adsorption. In: Proceedings of the 38th ISTC Japan workshop on advanced technologies in Russia, pp 173–193
2. Kohn W, Sham LJ (1965) Self-consistent equations including exchange and correlation effects. Phys Rev 140(4A):1133–1138
3. Kresse G, Furthmüller J (1994) Efficient iterative schemes for ab initio total-energy calculations using a plane-wave basis set. Phys Rev B 54:11169–11186
4. Fedorov AS, Serzhantova MV, Kuzubov AA (2008) Analysis of hydrogen adsorption in the bulk and on the surface of magnesium nanoparticles. J Exp Theor Phys 107(1):126–132
5. Henkelman G, Uberuaga BP, Jónsson H (2000) A climbing image nudged elastic band method for finding saddle points and minimum energy paths. J Chem Phys 113(22):9901–9904

Chapter 15
CVD-Synthesis Peculiarities of Carbon Nanomaterials from Ethylene with Gaseous Additions

A.A. Volodin and B.P. Tarasov

Abstract A catalytic pyrolysis of ethylene at 500–700°C in the presence of vapors of H_2O, C_2H_5OH, $(CH_3)_2SO_4$, $(CH_3O)_3P$ and $(CH_3O)_3B$ is carried out. On the basis of the obtained results a number of conclusions about the processes in the gas phase is made and also the relationship between the conditions of catalytic pyrolysis of ethylene and structure of the resulting carbon nanofibers is revealed. It is found that the introduction of gaseous additives significantly affects on the formation, growth and structure of carbon nanofibers.

Keywords Synthesis · Pyrolysis · Ethylene · Gaseous addition · Carbon nanofibers

15.1 Introduction

During the last years a set of carbonaceous compounds used for synthesis of carbon nanostructures (CNS) has noticeably grown. Feeding various additives in the reaction zone may lead to a change in both the yield of target product and structure of carbon filaments. For example, if sulfur is fed in the reaction zone double walled nanotubes may be mainly produced [1], a little amount (250 ppm) of phosphor gives carbon nanofibers (CNF) with the length of 6 cm [2], addition of CH_2Cl_2 to acetylene leads to the formation of bamboo-like nanotubes [3]. Helix-like nanotubes were synthesized by the pyrolysis of diethyl ether over zinc catalyst [4]. Feeding various additives changes not only the structure and the yield of carbon nanostructures but also their properties. For example, the paper [5] reported synthesis of nanotubes with big amount of SO_3H-groups, which reveal both the electron and proton conductivities. Thus the mere little feeding various additives in the reaction zone may influence the yield, structure and the properties of carbon

A.A. Volodin (✉) and B.P. Tarasov
Institute of Problems of Chemical Physics of RAS, Akademician Semenov av. 1,
Chernogolovka, Moscow Region 142432, Russian Federation
e-mail: alexvol@icp.ac.ru

S.Yu. Zaginaichenko et al. (eds.), *Carbon Nanomaterials in Clean Energy Hydrogen Systems - II*, NATO Science for Peace and Security Series C: Environmental Security 2,
DOI 10.1007/978-94-007-0899-0_15, © Springer Science+Business Media B.V. 2011

nanostructures. In the present work we studied the influence of gaseous oxygen containing additives to ethylene on the yield and the structure of the formed carbon nanofibers.

15.2 Experimental

MgO was used as a support for the catalysts. The catalysts were deposited on the surface of the support by precipitation of Ni from its aqueous solution $Ni(NO_3)_2$ followed by air drying at 150°C for half an hour and subsequent reduction to the metal just in the ethylene pyrolysis. Elemental analysis evidenced the contents of the deposited catalyst was 5% Ni/MgO.

The catalytic pyrolysis of ethylene was performed in a horizontal flow gas quartz reactor at atmosphere pressure and in the temperature range of 500–700°C. During the pyrolysis a gas mixture of the ratio C_2H_4:H_2:Ar = 1.5:3:1 was used. Gaseous additives were fed in the reaction zone by passing an argon flow through a bubbler filled with a liquid substance chosen from the set H_2O, C_2H_5OH, $(CH_3)_2SO_4$, $(CH_3O)_3P$ or $(CH_3O)_3B$. Products of the synthesis were purified from MgO and Ni by ultrasound treatment in the concentrated hydrochloric acid at 70°C for one hour and a half.

15.3 Results and Discussion

Mass spectra of the initial gas mixture of the ratio C_2H_4:H_2:Ar = 1.5:3:1 revealed the peaks m/z = 24, 25, 26, 27, 28, 29 with the maximum at m/z = 28 ($[C_2H_4]^+$), as well as the peak at m/z = 2 ($[H_2]^+$). Gas reaction products have the peaks at m/z = 12, 13, 14, 15, 16, 17 with the maximum at m/z = 16 ($[CH_4]^+$). The peaks at $[Ar]^+$ and $[Ar]^{2+}$ corresponding to m/z = 40 and m/z = 20 are also present.

The data obtained allow one to make an assumption that the following reactions take place in the reaction zone:

$$H_2C = CH_2 \rightarrow 2C + 2H_2 \tag{15.1}$$

$$H_2C = CH_2 \rightarrow C + CH_4 \tag{15.2}$$

$$H_2C = CH_2 + 2H_2 \rightarrow 2CH_4 \tag{15.3}$$

$$H_2C = CH_2 + H_2 \rightarrow H_3C - CH_3 \tag{15.4}$$

When ethanol vapors are fed in the reaction zone additional groups of peaks occur with the maxima at m/z = 31 and 45, that is peculiar to the spectra of ethanol

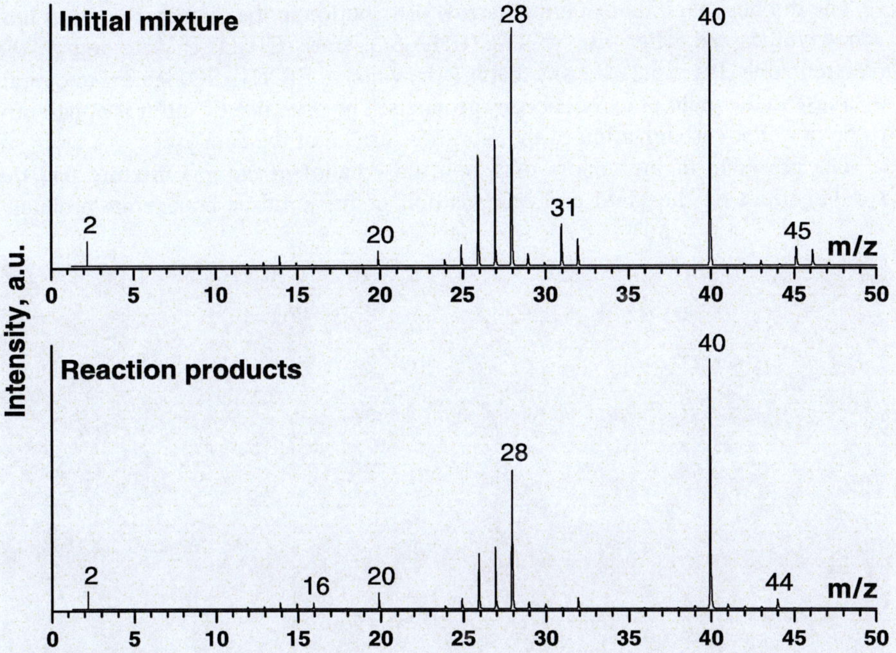

Fig. 15.1 Mass spectra of the initial gas mixture at with C_2H_5OH vapors and the pyrolysis products

(Fig. 15.1). The mass spectra of the pyrolysis products had the peaks with the maximum at m/z = 16 corresponding to $[CH_4]^+$. In addition to that the peak at m/z = 28 occurred, which might correspond to the non-reacted $[C_2H_4]^+$ and the formed $[CO]^+$, as well as the peak at m/z = 44 corresponding to $[CO_2]^+$. The following additional reactions are feasible during the pyrolysis:

$$C_2H_5OH \rightarrow CH_4 + CO + H_2 \tag{15.5}$$

$$2CO \rightarrow C + CO2 \tag{15.6}$$

When $(CH_3O)_3P$ vapors are fed at 20°C the mass spectra of the products have only main peaks at m/z = 40 ($[Ar]^+$), m/z = 28 ($[C_2H_4]^+$), and m/z = 2 ($[H_2]^+$), while at 50°C – additional peak being present at m/z = 79, which corresponds to the $[CH_3OPOH]^+$ ion. Traces of $(CH_3O)_3P$ were not found in the reaction products.

Transmission electron microscopy (TEM) data evidence that the mere little amounts of gaseous additives fed in the reaction zone leads to significant change of the structure of the formed CNF. For example, during the synthesis without additives we produced nanofibers with the diameters varied within the large (10–40 nm) range.

The thinnest nanofibers having narrow distribution in the diameters (10–15 nm) were synthesized when the vapors $(CH_3)_2SO_4$ and $(CH_3O)_3P$ were fed in the reaction zone (Fig. 15.2). At this feeding the vapors of $(CH_3)_2SO_4$ leaded to overall decrease in the yield of carbonaceous products. The presence of sulfur is apparently to favor quick carbonization of the working surface of the catalyst.

The presence of the vapors of water and ethanol in the gas mixture had the similar effect on the yield and composition of the solid carbonaceous products.

Fig. 15.2 TEM image of the pyrolysis products at feeding $(CH_3O)_3P$ vapors

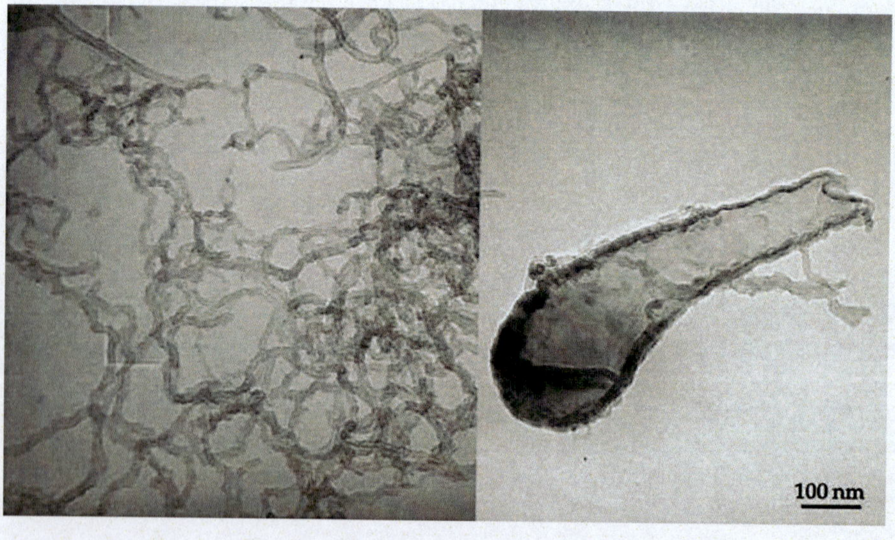

Fig. 15.3 TEM image of the pyrolysis products at feeding $(CH_3O)_3B$ vapors

The synthesized fibers had highly defected structure of the walls. The diameter of these fibers varied within the range of 10–20 nm, and the channel diameter was within the range of 3–5 nm. Sometimes the inner channel was absent. Feeding the vapors of $(CH_3O)_3B$ in the system leaded to the formation of thicker fibers with diameters of 15–30 nm and inner channel diameters of 5–15 nm (Fig. 15.3). Along with the nanofibers very big structures were formed with the outer diameter up to 250 nm the inner channel – up to 150 nm.

The data of elemental analysis evidenced that the contents of sulfur in the carbonaceous products of the pyrolysis conducted at the feeding of $(CH_3)_2SO_4$ was 5.3 mass. %, and the hydrogen content was the largest in the fibers synthesized when the vapor additives (1.5%) of water and ethanol were fed. In the other cases the hydrogen contents was less than 1%.

15.4 Conclusion

Feeding oxygen containing additives to the process of catalytic pyrolysis of ethylene allows one to vary the structure of CNF that is of importance for organizing direct synthesis of various types of carbon nanostructures.

Acknowledgement The work was performed under the financial support of Russian Foundation for Basic Research (Grant No 08-03-01117).

References

1. Zhou Z, Ci L, Chen X et al (2003) Controllable growth of double wall carbon nanotubes in a floating catalytic system. Carbon 41:337–342
2. Benissad-Aissani F, Ait-Amar H, Schouler MC, Gadelle P (2004) The role of phosphorus in the growth of vapour-grown carbon fibres obtained by catalytic decomposition of hydrocarbons. Carbon 42(11):2163–2168
3. Brichka YaS, Prikhod'ko GP et al (2004) Synthesis of carbon nanotubes from a chlorine-containing precursor and their properties. Carbon 42(12–13):2581–2587
4. Luo T, Liu J, Chen L et al (2005) Synthesis of helically coiled carbon nanotubes by reducing ethyl ether with metallic zinc. Carbon 43(4):755–759
5. Peng F, Zhang L, Wang H et al (2005) Sulfonated carbon nanotubes as a strong protonic acid catalyst. Carbon 43(11):2397–2429

Chapter 16
Carbon Nanotubes Filled Composite Materials

Yu. Sementsov, G. Prikhod'ko, M. Kartel, M. Tsebrenko, T. Aleksyeyeva, and N. Ulyanchychi

Abstract It was shown that insertion of carbon nanotubes (CNT) in different matrixes, such as polymers, hydroxyapatite (HAP), elastomers and liquid Selenium, leads to significant changes of their parameters. The influence of filler appears on strength characteristics of obtained composite materials. Such changes were due to CNT net-working in initial matrix. Also it was shown that not only volume characteristics of filled composites but surface properties are changing and this explains the better biocompatibility of nanocomposites, which is observed in vivo experiments.

Keywords Carbon nanotubes · Nanocomposites · Physical properties nanomaterials

16.1 Introduction

In the field of materials science there are developed new objects based on well-known matrixes by filling them with nanoparticles. Last decade was devoted to elaboration of new nanocomposites with multiwall carbon nanotubes (CNT) or

Yu. Sementsov, G. Prikhod'ko, and M. Kartel (✉)
O. Chuiko Institute of Surface Chemistry of NAS of Ukraine,
17 General Naumov Str., 03164 Kiev, Ukraine
e-mail: Nikar@kartel.kiev.ua

M. Tsebrenko
Kiev National University of Technologies and Design,
2 Nemirovich-Danchenko Str., 01011 Kiev
e-mail: mfibers@i.com.ua

T. Aleksyeyeva
G. Kurdyumov Institute of Metal Physics of NAS of Ukraine,
36 Vernadsky av., 03142 Kiev, Ukraine

N. Ulyanchychi
I. Frantsevich Institute for Problems of Materials Science, NAS of Ukraine,
3 Krzhizhanovsky Str.,, 03142 Kiev, Ukraine

S.Yu. Zaginaichenko et al. (eds.), *Carbon Nanomaterials in Clean Energy Hydrogen Systems - II*, NATO Science for Peace and Security Series C: Environmental Security 2,
DOI 10.1007/978-94-007-0899-0_16,

carbon fibers as additives. The goal of this study is to investigate the influence of nanoparticles addition on different matrixes.

16.2 Materials and Methods

CNT was obtained by the CCVD (catalytic chemical vapor deposition) method by pyrolysis of ethylene, propylene or propane – butane on complex metal oxides catalysts [1–3]. The average diameter of CNT was 10–20 nm, specific surface area determined by argon desorption was 200–400 m^2/g, and bulk density was within 20–40 g/dm^3. According to TEM (Fig. 16.1), X-ray diffraction, Raman spectroscopy data, the noticeable amount of amorphous carbon presence was not detected.

Unfortunately the industrial CNT production by CVD method resulted in receiving them as agglomerates of entangled tubes with dimensions of 20–500 μm [4] (Fig. 16.2). At the same time composites filled with nanotubes and with extraordinary properties can be obtained under the condition of uniform distribution ones in the polymer matrix [5, 6]. This leads to the necessity of finding an effective method to disperse CNT agglomerates [7].

Different methods were used for multiwall CNT dispersion: treating different CNT concentrations (3 and 5 wt.%) in ethyl alcohol and water solutions in a homogenizer, based on the cavitation effect and ultrasonic processing in glucose aqueous solutions by UZDN–2 T. Analysis of agglomerates status was carried out by laser correlation spectroscopy. Distribution of particles' size was determined by laser correlation spectrometer « ZetaSizer-3 » (Malvern Instrument, UK) with the adaptor 7032 and Helium-Neon laser PH-111, 25 mW, $\lambda = 633$ nm. Studied suspension, 1 ml, was put in a cylindrical optical glass cell with 10 mm in diameter and placed in a thermostatically controlled cell of laser correlation spectrometer. Registration and statistical

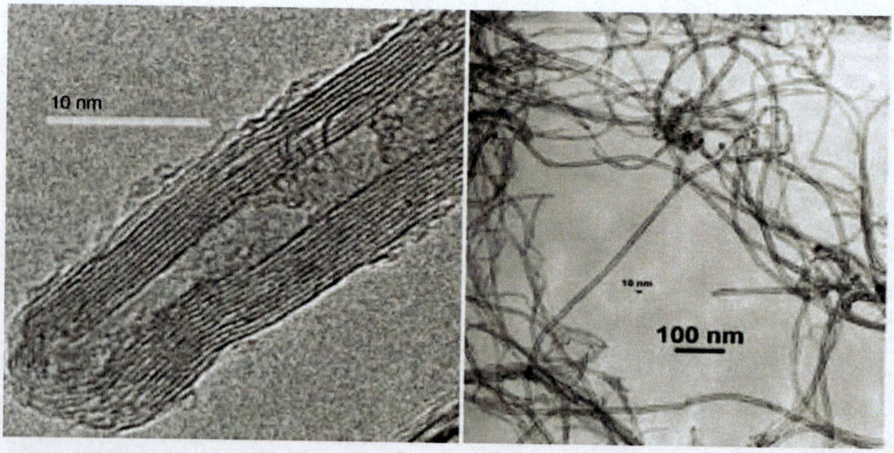

Fig. 16.1 TEM images of multi-walled CNT

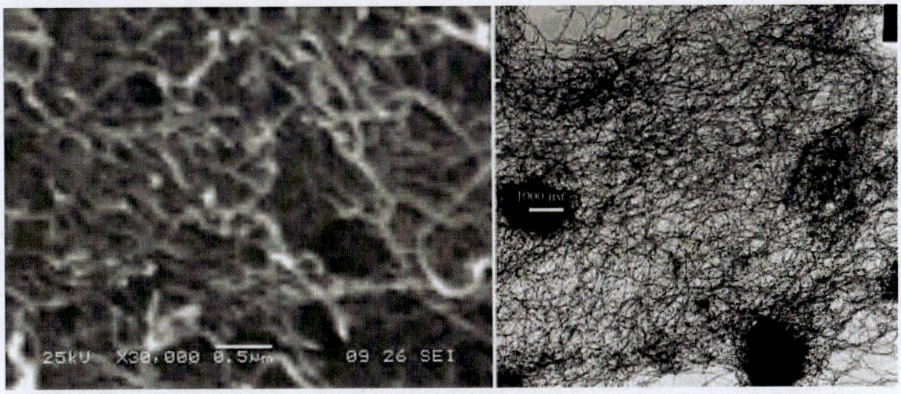

Fig. 16.2 SEM and TEM images of multi-walled CNT agglomerates

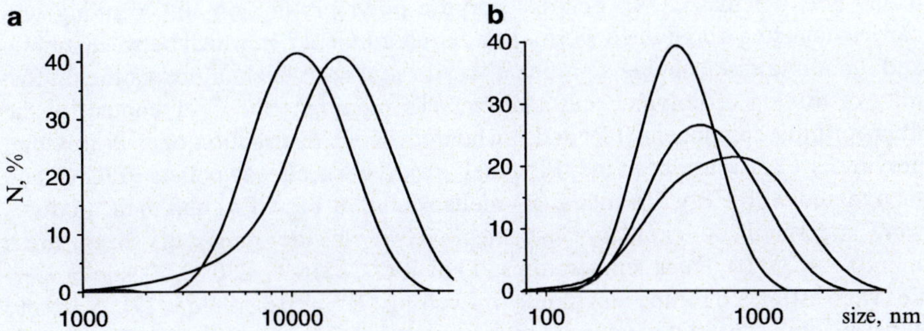

Fig. 16.3 Size distribution of CNT agglomerates: (**a**) 0.067 wt. % CNT in 3% glucose solution after ultrasonic treatment; (**b**) system 0.05 wt. % CNT – water treated by cavitation methods

processing of the laser radiation was performed for 300–400 s. The resulting autocorrelation function (ACF) was treated using the PCS-Size mode [8]. Fig. 16.3a, b represented particle size distribution for different systems.

Data analysis suggests the following conclusions:

- CNT agglomerates disintegration by ultrasonic method in water did not significantly change agglomerates size in comparison with baseline (20–500 μm), but extended ones as particles with 0.5–1.0 μm in diameter and 5–100 μm in length; the same system in glucose yields of spherical particles with 1–100 μm in size (Fig. 16.1a);
- Homogeneity of system CNT-water possessing by cavitation method is much higher and depends on CNT concentration. So at 0.05 wt.% CNT concentration particles dimensions are 0.2–1.0 μm in range (Fig. 16.3b).

Also it should be noted that the processing system 0.2 wt.% CNT-water gives two sizes of particles: 0.01–0.1 μm and 1.0–5.0 mm.

16.2.1 Carbon Nanotubes Filled Composite Materials

16.2.1.1 Rheological Properties of Molten Mixtures Polypropylene/Copolyamides/CNT

As native matrix was used melt blends polypropylene, PP (grade 21060, TC 05-1756-78, Lysychansk Chemical Plant, Ukraine)/copolyamides, CPA copolymer ε-caprolactam and AG salt (salt hexamethylenediamine and adipic acid) of 50:50 (brand PA-6/66, OST 6-05-438-78, Sverdlovsk Chemical Plant, Russia) in proportion 30/70 wt.% respectively with the addition of CNT. Characteristics of the initial PP and CPA are listed in Table 16.1.

Different concentrations of CNT: 0.05, 0.1, 0.5, 1.0 and 5.0 wt. % were selected to modify PP microfibers properties. It is known that the excess surface energy leads to the agglomeration of nanoparticles and their aggregation. Therefore an important task in creating nanocomposites is necessity to ensure uniform distribution of filler in the polymer matrix [4]. For this purpose polymers mixing and blending was carried out using a combined worm-disk extruder LGP 25, in which between mobile and immobile disks having significant tensile stresses, which improves the uniformity of mixing of polymer and additives. To maximize the CNT content in the fiber-forming component (PP) and for homogeneous distribution ones in mixture, previously CNT introduced into PP's melt, receiving such way pellets PP/CNT and then mixed with CPA. Viscosity (η) melts source of PP, CPA, and their mixtures were measured by capillary viscometry CF-2 in the range of shear stress $\tau = (0.1\text{–}5.7)*10^4$ Pa, at temperatures (T) of 190°C, 210°C, 220°C. Elastic properties were estimated for the magnitude of swelling "B"; as described in [7]. Warranty experimental error in the determination of η and "B" was ±(2 ÷ 5)%. Flow mode "η" was determined by the value of the slope of the tangent to the abscissa at a given point of the flow curve. The ability of the melt to the longitudinal deformation was estimated by the maximum plated hood (F_{max}) with accuracy of ±7%.

16.2.1.2 CNT Network-Forming Properties in Hydroxyapatite Based Ceramic

The different concentrations of CNT homogeneous suspension in water, alcohol or 3% glucose solution (from 0.01 up to 0.2 wt. %) were injected into HAP at precipitation stage. Received samples undergo the thermal processing from 150 up to 900°C. Specific surface of obtained nanocomposites were determined by argon desorption.

Table 16.1 Characteristics of the initial polymers

Polymer	T_m, °C	η^a, Pa·c	n^a	B^a
PP	169	300	1.8	2.1
CPA	170	1,230	1.2	1.4

$^a T = 190°C$; $\tau = 5.7 \cdot 10^4$ Pa

16.2.1.3 System Liquid Selenium – CNT

Quartz ampoules with equal volumes selenium and CNT (about 1 wt. %) pumped out, sealed, then heated over the temperature of selenium melt, mix and cooled.

16.2.1.4 Biofunctional Composites Polymer/CNT

Four types of nanocomposites based on polymeric matrices – polytetrafluoroethylene (PTFE), polypropylene (PP), and two types of rubbers – were studied. The composites based on PTFE were obtained by methods of mixing PTFE powder (F4-PN20) with CNT in the presence of liquids and following coagulation of PTFE aqueous dispersion with CNT. The dried powdered mixtures were melded by hot-pressing. The samples of composites with different CNT contents (0.05, 0.1, 1.0, 2.0, 5, 10, 15, and 20 wt. %) were tested for non-axial compression by using 2167-P50 recording device with automatic deformation diagrams recording. Thus, conventional yield compression strength ($\sigma_{0.2}$) and compression elasticity modulus (Ec) were determined.

Nanocomposites on the base of PP (brand 21060) with content of CNT within 0.05–5.0 wt. % range were obtained by stirring a mixture of molten PP and CNT in an extruder at 50 rpm. Primary samples were got as granules and processed further by hot-pressing. Structure parameters of newly obtained systems polymer-CNT (level of crystallinity) determined be Rh-structure analysis on diffractometer DRON 3 M.

16.2.1.5 Surface Study with AFM As Biosensor

The surface of initial PTFE and PTFE–15% CNT nanocomposite were studied using a NanoScope IIIa atomic-force microscope (Veeco corp). The data obtained were processed with the help of GWIDDION software. Samples' surfaces were also studied by AFM tip loaded with antibodies, immunoglobulin type G (IgG) obtained from the animal's blood serum.

16.3 Results and Discussion

16.3.1 Rheological Properties

Filled polymers are often viewed as a concentrated suspension. For such systems should take into account the possibility of interaction between only particles and with the additives in dispersion medium, that may influence on flow nature. Thus, the study of dynamic viscosity of polyethylene filled with powder of barium sulfate and calcium carbonate with different size particles, showed that with increasing

Table 16.2 Effect nanoparticles additives on rheological properties of PP melt

Concentration CNT, %wt	η^a, Pa·s		n^a	B^a	F_{max}, %[a]
	$\eta_{exp.}$	η_E			
0	300	–	1.8	2.1	18,000
0.05	305	300	1.8	1.6	22,000
0.10	315	301	1.8	1.6	27,000
0.50	350	302	1.9	1.6	29,000
1.00	450	305	1.9	1.5	21,000
5.00	480	323	1.9	1.4	15,000

[a]$\tau = 5{,}7 \cdot 10^4$ Pa; $T = 190°C$

content of additive dynamic viscosity increases, and for suspensions, which contain finely dispersed filler, it increases abruptly, while for compositions with relatively large size particles the critical concentration is not achieved even with the large modifier amount [8].

Investigation of the rheological properties of molten polypropylene filled with CNT, gives an increasing viscosity of the suspension, as linear function of additive concentration (from 0.05 to 5.0 wt. %) (Table 16.2).

This is agreement with the conclusion that nanoparticle additives determine stiffener thixotropic effect, which leads to an increase in the viscosity of polymer melts [9]. For compositions with low concentrations of CNT (0.05 ÷ 0.10%wt.) "*n*" increases insignificant and within the error coincides with the effective viscosity (ηE), calculated for the Einstein equation for dilute suspensions:

$$\eta_E = \eta_0(1 + 2.5F),$$

where η_0 – viscosity of the medium, F – volumetric concentration of suspended particles.

The nature of the flow source and the modified polypropylene melts almost unchanged from the content of CNT at all investigated temperatures and obeys a power law. As it was expected, the elasticity of the melts' compositions decreases with increasing filler concentration, so evidenced by the decrease in the swelling extrudates (Table 16.3).

The effective viscosity experimental data of binary and ternary systems melts in 5–6 times lower fillers significances. Fixed patterns are explained due to change of rheological processes that occur during the flow of mixtures' melts, namely: the deformation of droplets at dispersed phase in the jet and the orientation of the latter in the direction flow. With the introduction of 0.05 ÷ 0.1 wt. % CNT into molten PP/CPA is a tendency to increase effective viscosity of compositions. With further increasing of fillers' concentration η of ternary mixture also rise, but remains significantly lower for additive value. This can be explained by the fact that the mixture's melt viscosity is the result of several opposite factors. Solid additives, CNT, structure the melt and increase its viscosity and, on the other hand, distribute homogeneously due to the formation of liquid jets of polymer dispersed phase (PP) in CPA matrix. Thus, we can conclude beyond what the impact of fiber formation as

Table 16.3 Effect of additives on the rheological properties of system CNT/molten mixture of PP/CPA

Ingredients tracks PP/CPA/CNT, %wt	η^a, Pa·s		n^a	B^a	F_{max}, %[a]
	η exp.	$\eta_{ад}$			
100/0/0	300	–	1.8	2.1	18,000
0/100/0	1,230	–	1.2	1.4	95,600
30/70/0	150	951	1.8	5.7	10,500
30/70/0.05	160	954	1.7	6.8	7,800
30/70/0.10	170	961	1.7	7.2	7,600
30/70/0.50	190	966	1.7	7.0	7,300
30/70/1.00	210	996	1.6	6.7	6,900

[a]$\tau = 5{,}7 \cdot 10^4$ Pa; $T = 190°C$

Table 16.4 Effect of CNT supplementation on the activation energy of binder melt flow PP/CPA

Concentration CNT, wt. %	E, kJ/mol, $\tau \cdot 10^4$ Pa		
	$\tau = 5.69$	$\tau = 3.47$	$\tau = 1.61$
0	48.3	52.1	52.4
0.05	50.4	53.0	55.0
0.10	50.4	53.0	56.0
0.50	50.0	52.0	53.1
1.00	50.0	52.0	53.1

evidenced due to a sharp fall in viscosity nanosized fillers mixture in comparison with the η of initial PP and CPA. Modified mixture of PP/CPA, as original one, is non-Newtonian liquids. The degree of deviation from Newtonian mode, judging by the size "*n*", practically does not depend on the amount of additives (Table 16.3). Effect of fillers on elastic properties of melts system PP/CPA/CNT can be seen from changing the value of swelling extrudate *B*, which is an indirect elasticity parameter. The data from Table 16.4 shows that value *B* increases in 1.2–1.3 times for all CNT concentrations. The reason for increasing swelling is that both components of the mixture are characterized by high elasticity and store elastic energy during the flow in the inlet zone. These anisotropic structures are new relaxing elements and cause of elasticity growth.

An important technological characteristic of polymer melts and their mixtures is ability to processing in fiber and film, which is determined by the maximum plated hood (F_{max}). As can be seen from Table 16.3, melts bi-and three-component mixtures have less ability to longitudinal deformation in comparison with the original components.

Mixtures' spinnability reduction is the result of several factors – incompatibility between the components and the weak interaction between the PP and co-polyamides on the border dividing the phases, a sharp drop in shear viscosity and increasing heterogeneity of ternary mixtures. At the same time it must be emphasized that the values of maximum plated exhaust are in the range, which allows processing composition PP/ CPA/CNT in fibers and films.

Table 16.5 Thickness of fibrous-connective capsule (10^{-6} m) around nanocomposites with different CNT concentration

Sample	Thickness of fibrous-connective capsule (10^{-6} m)
PTFE	253
PTFE + 15 wt. % CNT	25
PP + 0.05 wt. % CNT	156
PP + 0.1 wt. % CNT	52
PP + 3.00 wt. %. CNT	135
PP + 5.00 wt .% CNT	206

To determine the mechanism of the CNT influence on the patterns of flow of melts of PP/CPA it was investigated the temperature dependence of viscosity at different shear stresses and the calculated activation energy of viscous flow (*E*) by Frenkel-Eiring equation:

$$\eta = A_0 \cdot e^{E/RT},$$

where A_0 – ratio, which depends on the molecular nature of the liquid; *R* – gas constant, which equals 8.3 kJ/mol; *T* – absolute temperature.

It was established that the temperature dependence of viscosity in coordinates: $lg\eta = f(1/T)$ at different shear stresses is expressed by straight lines whose slope remains almost unchanged for the melts of mixtures with different filler content. Data in Table 16.5 the values of activation energy of viscous flow of composition E is increased in the presence of CNT, which indicates a change in the kinetic part of the flow under the influence of additives.

The activation energy of binder flow naturally increases with decreasing shear stress (Table 16.4).Thus, studies have shown that the presence of CNT in molten polypropylene and initial blends polypropylene/co-polyamides significantly affect the regularity of their flow.

16.3.1.1 Biofunctional Composites PP/CNT and PTFE/CNT

According to obtained data the characteristics of the melting temperature were slightly sensitive to the presence of filler and its concentration. At the same time, the beginning of crystallization temperature and other temperature characteristics were significantly influenced by CNT and changed in non-monotone mode with increasing CNT concentration in the polymer matrix. Thus, the lowest temperature for melting to begin in the system PP/CNT was observed at 0.05% mass CNT. This concentration meets the maximum temperature process range, which decreases both at increasing and decreasing CNT concentration. For the crystallization process, there is monotonous, but non-linear temperature dependence of characteristic with increasing concentration of CNT in the studied range. The data obtained agree well with the results [10] for the system PP + 0.8% single-walled carbon nanotubes (SWNT) when compared with PP + 1.0% CNT: the crystallization temperature of PP + CNT is higher as compared with pure PP.

Dependence of crystallinity level of systems PTFE/CNT, PTFE/SiO_2 and PP/CNT on concentration of nanosized filler SiO_2 and CNT is represented at Fig. 16.4. Changes in elastic properties of these systems are well correlated with changes of structure parameters at low filler concentration that are shown in Fig. 16.5.

It was suggested in works [11, 12] to determine the compatibility of artificial materials with living body by the thickness of fibrous-connective capsule (pocket) that formed around foreign material. Testing of samples was carried out by implantation of the preformed ones into the muscle "pouch" on the back of experimental animals. In order to decrease the number of animals in study and to receive the unbiased results by comparison of the body reaction to different material types the samples were implanted on one animal in different parts of the back. The two

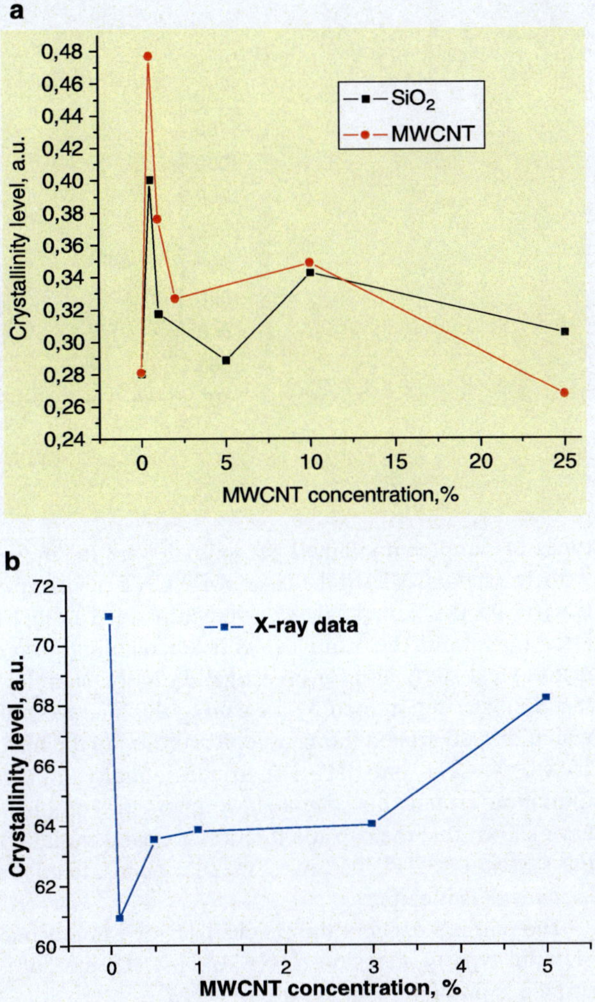

Fig. 16.4 Dependence of crystallinity level of PTFE-CNT (**a**) and PP – CNT (**b**) nanocomposite on SiO_2 and CNT concentration

Fig. 16.5 Dependence of mechanical characteristics of PTFE/CNT (**a**) and PP/CNT (**b**) nanocomposite on SiO_2 and CNT concentration

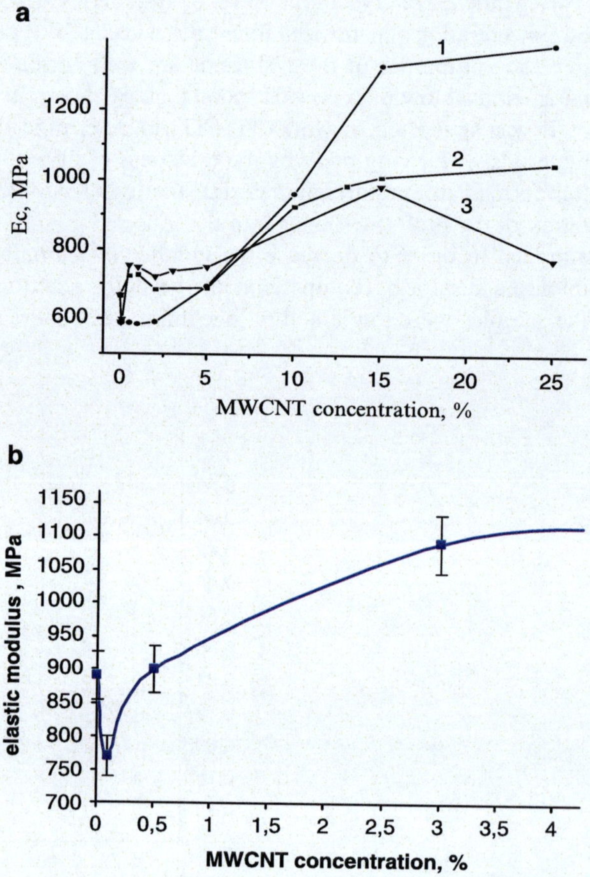

types of samples implanted per animal were (a) initial PTFE and (b) PTFE filled with 15 wt. % CNT. In the case of PP/CNT the samples with different contents of CNT (0.05, 0.1, 3, and 5 wt. %) were implanted along both back's sides. Four weeks after operation, the samples with surrounding tissue were excised for further histological study and for investigation of the samples' surface. The level of body reaction was determined by measuring the thickness of fibrous-connective capsule which formed around the sample. According to the histological data the insertion of 15 wt. % CNT into PTFE matrix essentially influences the fibrous-connective formation around the sample. Morphometry of histological slides (Table 16.5) have shown that the capsule thickness around initial PTFE was appreciably larger. For system PP-CNT the lowest mean for capsule thickness was for 0.5 wt. % CNT in matrix (Table 16.5).

The surface triggers the mechanism of foreign body recognition by immunoglobulin type G adhesion (IgG) [13, 14]. To measure the adherent IgG forces to surface is possible with the use of AFM.

In all studies standard silicon nitride AFM tip was used (DNP-20, Veeco corp.) with nominal cantilever flexibility of 0.06 N/m. The force of interaction of modified AFM tip with initial matrix surface was higher than with nanocomposite. It means that the initial PTFE matrix is recognized by living organism as foreign material which triggers bodies' protective reaction.

The interaction of PTFE/CNT system and modified AFM tip is lower and it signifies that body's reaction would be lower. The obtained data agree with histological results which unambiguously show that PTFE/CNT composites cause the least fibrous-connective capsule formation around them. In other words, presence of CNT in matrix improves their biocompatibility.

16.3.1.2 CNT Network-Forming Properties in Hydroxyapatite Based Ceramic

The presence of CNT as filler resulted in increasing of samples strength parameters. At the same time, the strength characteristics were significantly influenced by CNT and changed in non-monotone mode with increasing CNT concentration in the HAP ceramics. Thus, the lowest compression strength HAP/CNT was observed at 0.42% wt CNT and the highest one was at 0.082%wt CNT. Samples' specific surface also was depended on changing CNT concentration (Table 16.6). The data obtained well agree with the results [15–19], for the system HAP +0.8% single-walled carbon nanotubes (SWNT).

CNT insertion into HAP ceramics resulted in not linear increasing of specific area. At low CNT concentration (below percolation threshold) the structure peculiarities were observed due to the presence of disordered phase. This phase probably appears around nanopores, and its existence correlates with other characteristics of the system (Fig. 16.6). At the same time the composites elements analyse shows the lack of CNT carbon after heating up to 600°C.

16.3.1.3 System Liquid Selenium – CNT

As a result there was obtained solid amorphous structure of system liquid Selenium-CNT in comparison with polycrystalline metal one in the case of alone selenium. This fact allow us to speculate that CNT dissolve in molecular Selenium.

Table 16.6 Specific HAP + CNT surface after thermal treatment at 900°C

Sample	CNT, %wt	Specific surface m^2/g
Initial HAP	–	7.5
HAP + CNT with SiO_2	0.41	14.3
HAP + CNT	0.42	19.6
HAP + CNT	0.12	15.3
HAP + CNT	0.08	22.3

Fig. 16.6 SEM images (**a**) System HAP/CNT; (**b**) HAP

16.4 Conclusions

1. Carbon nanotubes (CNT) are prospective fillers for polymer materials due to their unique structure and outstanding combination of strength, electrical and thermal properties of CNT. Reinforcement of polymers including elastomers with infinite CNT net results in change of both three-dimensional material parameters and surface properties.
2. Filling of polymers with CNT changes their surface properties and improves biocompatibility. This was demonstrated by experiments in vivo in PTFE/CNT and PP-CNT systems.

3. At low CNT concentration (below percolation threshold) the structure peculiarities were observed due to the presence of disordered phase. This phase probably appears around nanopores, and its existence correlates with other characteristics of the system.
4. CNT influences on generation of HAP nanostructure and HAP/CNT composites are prospective materials for manufacturing coatings for medical endoprosthesis.

References

1. Sementsov YuI, Melezhek OV, Prikhod'ko GP et al (2007) Synthesis, structure, physicochemical properties of nanocarbon materials. In: Shpak AP, Gorbyk PP (eds) Physical chemistry on nanomaterials and supramolecular structures, vol 2. Naukova dumka, Kyiv
2. Melezhyk AV, Sementsov YuI, Yanchenko VV (2005) Synthesis of thin carbon nanotubes on co-precipitated metaloxide catalysts. Russ J Appl Chem 78(6):938–946
3. Yanchenko VV, Sementsov YuI, Melezhyk AV (2004) Method of obtaining of catalysts for CVD of carbon nanotubes. Ukrainian Patent Application 20041008154, Int. Cl.7 C01B11/00, D01F9/12, 8 Oct 2004
4. Rakov EG (2007) Fibers with carbon nanotubes. Market Light Ind 48:51–57
5. Sokolov YA, Shubanov SM, Kandyrin LB, Kalugin EV (2009) Polymer nanocomposites. Struct Properties Plast 3:18–23
6. Malysheva TL (2005) Wonders of technology and the era of « smart » textiles. Market 471(21): 194–201
7. Zhang H, Harwood W, Ross G (2006) Antistatic polymer monofilament, method for making an antistatic polymer monofilament for production of spiral fabrics and spiral fabrics formed with such monofilaments. US Patent N7094467, DCA D 01 F 6/00, 2006
8. Tsebrenko MV (1991) Ultrathin synthetic fibers. Chemistry, Moscow 214 (in Russian)
9. Tsebrenko MV, Rozanov NM, Kuvaev EP, Sapyanenko AA, Dzyubenko LS, Gorbik PP (2007) Patterns for polypropylene microfibers containing filler in nano state. Chem Fibers 5:16–21
10. Utracki LA, Bakerdjiane Z, Kamal MR (1975) A method for the measurement of the true die swell of polymer melts. J Appl Polym Sci 19(2):481–501
11. Khan ChD (1979) Rheology in processing of polymers. Chemistry, Moscow 367 (in Russian)
12. Lacerda L, Bianco A, Prato M, Kostarelos K (2006) Carbon nanotubes as nanomedicines: from toxicology to pharmacology. Adv Drug Deliv Rev 58:1460–1470
13. Castner DG, Ratner BD (2002) Biomedical surface science: foundations to frontiers. Surf Sci 500:28–60
14. Lazarenko ON, Aleksyeyeva TA (2009) Method of individual testing implant on biocompatibility of recipient organism. Patent of Ukraine 87387, 10 July 2009
15. Tirrell M, Kokkoli E (2002) Biesalski M the role of surface science in bioengineered materials. Surf Sci 500:61–83
16. Aleksyeyeva TA, Sementsov YuI, Gun'ko GS et al (2009) Deaglomerawbz mnogostennyh CNT v etylovom spirte i yego vodnyh rastvorah/tesisy dokladov Vseukrainskoy konferencii c mezhdunarodnym uchastiyem. Prikladnay phisicheskayay himiaya I nanohimiya 10–14 Oct 2009, Sudak, Crimea, pp 162–163
17. Masa-aki T, Takamasa O, Mamoru O, Akira O, Toshiyuki H (2006) Mechanical properties of carbon nanotubes/hydroxyapatite composites prepared by Spark plasma sintering. AIP Conf Proc 832:430–432
18. White A, Best S, Kinloch I (2005) Hydroxyapatite–carbon nanotube composites for biomedical applications. Rev Appl Ceram Technol 4(1):1–13
19. Kealley C, Elcombe M, van Riessen A (2008) Microstrain in hydroxyapatite carbon nanotube composites. J Synchrotron Rad 15:86–90

Chapter 17
Analysis of the Interrelation of the Thermal Stability of Hydrides of the Intermetallic Compounds of Composition AB_2 with the Nature of Their Chemical Bonds Character Me–H

V.D. Dobrovolsky, O.G. Ershova, Yu.M. Solonin, and I.Yu. Zavaliy

Abstract Charge state of Zr atoms in ZrV_2 intermetallic compound and its hydride ZrV_2H_4 has been studied using the X-ray absorption spectroscopy (XAS) method. Thermal stability of the hydride has been investigated employing the thermodesorption spectroscopy (TDS) method. It has been established the positive charge on Zr ions in ZrV_2H_4 hydride (i.e., the "transfer" of electronic charge from Zr atoms during formation of the hydride), and this fact indicates the presence of ionic component of metal-hydrogen bonds in the hydride. The conclusion about the existence of correlation between observed increasing thermal stability of ZrV_2H_4 hydride and ionic component of its metal-hydrogen bonds has been made.

Keywords X-ray absorption spectroscopy · Thermodesorption spectroscopy · Hydrogen sorption · Metal-hydrogen bonds · Thermal stability

17.1 Introduction

As a result of a number of experiment studied reported in Ref. [1], the following statement was made: metal-hydrides possessing equilibrium pressure at room temperature close to atmospheric one and revealing low thermal stability with dissociation at sufficiently low temperatures, as a rule demonstrate practically complete absence of ionic component of metal-hydrogen bonds, which have predominantly covalent or metallic-covalent nature. Correlation between the charge transfer and thermal stability of hydrides of intermetallic AB_2 compounds, in particular the intermetallic ZrV_2 compound, has not been studied yet employing

V.D. Dobrovolsky (✉), O.G. Ershova, and Yu.M. Solonin,
Institute for Problems of Materials Science NAS of Ukraine,
Krzhyzhanovsky Str. 3, 03142 Kiev, Ukraine
e-mail: dobersh@ipms.kiev.ua

I.Yu. Zavaliy
Physicomechanical Institute NAS of Ukraine, Naukova Str. 5, 79601 Lviv, Ukraine
e-mail: zavaliy@ipm.lviv.ua

S.Yu. Zaginaichenko et al. (eds.), *Carbon Nanomaterials in Clean Energy Hydrogen Systems - II*, NATO Science for Peace and Security Series C: Environmental Security 2,
DOI 10.1007/978-94-007-0899-0_17,

the methods of X-ray absorption and thermodesorption spectroscopy (TDS). The information about the charge transfer from hydrogen atoms to Zr atoms in ZrV_2H_4 hydride [2], which was made adopting data of X-ray emission spectroscopy, causes doubt and it is necessary to testify this information employing a direct method of investigating the charge transfer in solids, namely the method of X-ray absorption spectroscopy (XAS).

In the present work, on the basis of XAS and TDS studies of ZrV_2H_4 hydride, we aim in clarifying the following question: is there a correlation between thermal stability of AB_2 intermetallic hydrides and the degree of ionicity of their metal-hydrogen bonds? Such a correlation, as it was established in Ref. [1], is characteristic of a number of metallic hydrides, partly to intermetallic AB and AB_2 hydrides.

17.2 Experiment and Results

The ZrV_2alloy was obtained by arched melting: zirconium (99,99%) and vanadium (99,999%) were used. The as-prepared alloy was homogenized at temperature 800°C for 4 days, and at temperature 1,100°C for 2 days. Hydride ZrV_2H_4 was obtained using a Siverts-type apparatus by direct hydrogenation from the gas phase of ZrV_2 alloy as a result of three cycles of hydrogenation-dehydrogenation. Before the hydrogenation, ZrV_2 alloy was thermally activated at temperature 225°C. After the thermal activation of ZrV_2 alloy, its hydrogenation was achieved by supplying hydrogen into a reactor at pressure 5 bar and temperature 225°C with subsequent cooling the reactor to room temperature, maintaining constant pressure of hydrogen in the reactor and measuring in this case the volume of hydrogen absorbed by the sample taking into account displacements of piston in a measuring cylinder (i.e., using the isobar-volumetric method of hydrogenation).

X-ray spectroscopy studies of the nature of chemical Zr-H bonds in ZrV_2H_4 hydride and of ionic component of the metal-hydrogen bonds were carried using an X-ray absorption spectrometer by measuring the zirconium K absorption spectra in metallic Zr, intermetallic compound ZrV_2, hydride ZrV_2H_4, and reference compound ZrO_2. The mentioned K absorption spectra of zirconium were obtained using Cochouis' focusing method and employing the method of "variable field of absorption" [3]. As a dispersion element, a quartz crystal-analyzer with the (1,010) reflection plane was used. Radius of curvature of the crystal-analyzer was 890 mm. All spectra were obtained in the second order of reflection. The value of charge transfer from Zr atoms to H atoms in ZrV_2H_2 hydride was determined by measuring the shift of the Zr K absorption edge with respect to the energy position of the edge in ZrV_2 (Fig. 17.1).

The value of charge transfer (and, respectively, effective charge on zirconium atoms) in the hydride was evaluated taking into account the energy shift of the zirconium K absorption edge in the reference ZrO_2 compound with respect to the position of the edge in metallic Zr (Fig. 17.2).

Values of shifts of zirconium XAS K edges in ZrO_2, ZrV_2, and ZrV_2H_4 compounds compared to the position of the K absorption edge in pure metallic zirconium are listed in Table 17.1. As it is obvious from data of Table 17.1, there is no

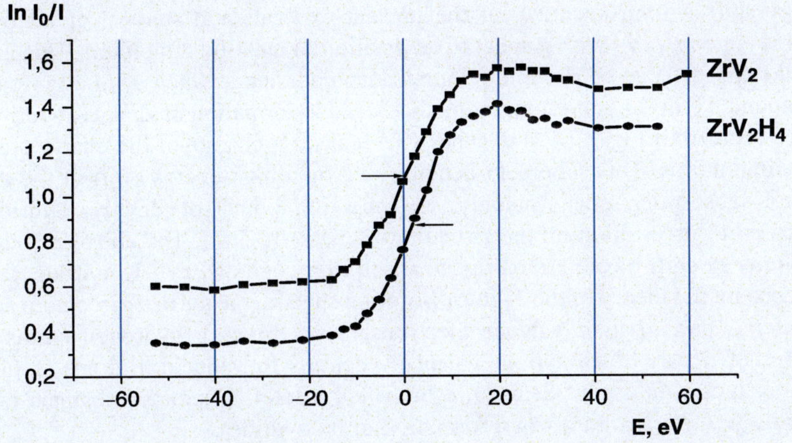

Fig. 17.1 Zirconium K absorption edges of intermetallic compound ZrV_2 and its hydride ZrV_2H_4

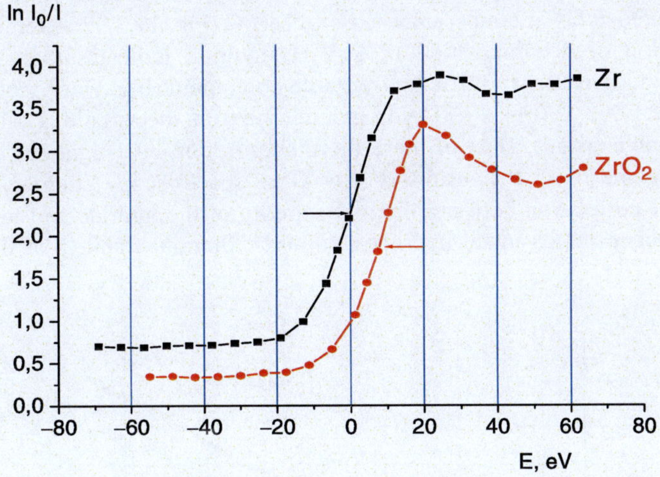

Fig. 17.2 Zirconium K absorption edges of metallic zirconium and ZrO_2

Table 17.1 Shift of the zirconium K absorption edge in ZrV_2, its hydride, and ZrO_2 oxide with respect to its position in metallic zirconium

Compound	Shift of zirconium K edge, eV
Zr	0
ZrV_2	0 ± 0.3
ZrV_2H_4	1.9 ± 0.3
ZrO_2	7.5 ± 0.3

energy shift, within accuracy of the present experimental studies, of zirconium XAS K edge in ZrV_2 with respect to its position in metallic zirconium. This means that the charge state of zirconium atoms does not change when synthesizing ZrV_2 compound. Hydriding this compound that leads to formation of ZrV_2H_4 hydride, as can be seen from Fig. 17.1 and Table 17.1, causes a change of the charge state of zirconium atoms. This statement is confirmed by a high-energy shift of the zirconium K absorption edge (namely, its inflection point) of ZrV_2H_4 hydride as compared to its position on the zirconium K edge of ZrV_2. The above mentioned **high-energy** shift of the zirconium K absorption edge of ZrV_2H_4 hydride, taking into account the theory of the X-ray absorption spectra, means that zirconium atoms **transfer** a part of their valence electrons to neighboring hydrogen atoms (the transfer of a part of zirconium valence electrons to vanadium atoms is highly unlikely taking into consideration established in Ref. [2] minor changes of the charge states of vanadium when formation of its hydride).

The thermal desorption of hydrogen from ZrV_2H_2 hydride was investigated by isobar- volumetric method using the Siverts-type apparatus as mentioned above. Heating the hydride sample was accomplished in hydrogen medium at normal pressure with speed 5 deg/min. During the heating, the hydride starts to release hydrogen intensively at temperature close to 240°C (Fig. 17.3). The temperature of the beginning of decomposition of ZrV_2H_4 hydride indicates that at ambient conditions the hydride is a sufficiently stable compound (it is worth noting that at temperature 1,050°C (Fig. 17.3) hydrogen release from the sample is still observed with noticeable speed). Data of these thermodesorption investigations of ZrV_2H_4 hydride, together with the results of the X-ray spectroscopy studies, testify the existence of correlation between thermal stability of the hydride and the value of ionic component of its metal-hydrogen bonds, if thermal stability of the hydride

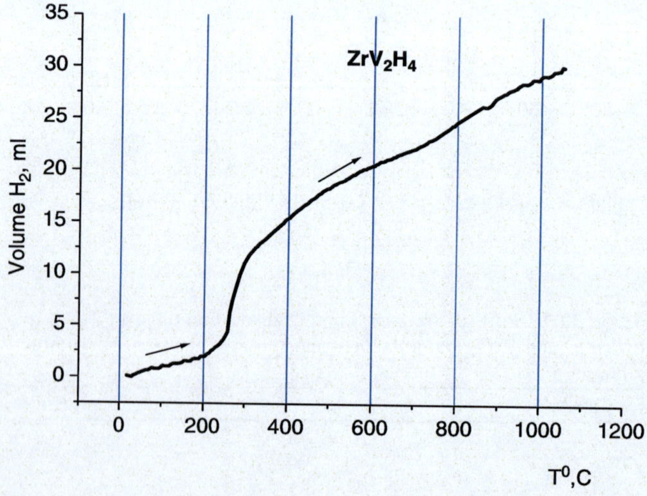

Fig. 17.3 Curve of the thermal desorption of hydrogen from hydride ZrV_2H_4

is established taking into consideration the temperature of the beginning of its intensive decomposition, and ionic component of the metal-hydrogen bonds to evaluate by measuring shifts of the zirconium K absorption edge in ZrV_2H_4 hydride as compared to its position in Zr and ZrV_2. The high-energy shift of Zr K edge in the hydride was found to be noticeable (1.9 eV) and it testifies that the charge transfer occurs from zirconium atoms to hydrogen atoms (but not in the opposite direction, as it was stated by authors [2]).

17.3 Conclusions

The XAS Zr K edges of metallic zirconium, ZrV_2, ZrV_2H_4, as well as of reference ZrO_2 compound have been studied. A change of the charge state of zirconium atoms when going from ZrV_2 to ZrV_2H_4 (i.e., a transfer of electron charge from zirconium atoms towards hydrogen atoms during formation of the hydride) has been established. Hydrogen-sorption properties and thermal stability of ZrV_2H_4 have been studied using the TDS method.

On the basis of the present X-ray spectroscopy and termodesorption investigations, it was possible to make a conclusion about the existence of correlation between the thermal stability of the hydride of the intermetallic AB_2 compound and the presence of ionic component of its metal-hydrogen bonds.

References

1. Dobrovolsky VD (2006) The correlation between ionicity of metal-hydrogen bonds in hydrides and their thermal firmness, hydrogen materials science and chemistry of carbon nanomaterials, NATO Science Series. II. Mathematics, Physics and Chemistry. Springer, Berlin, pp 407–414
2. Porutsky SG, Zhurakovsky EA (1990) Soft x-ray emission studies of metal-hydrogen bonding states in zirconium-based Laves phase hydrides – II: p-states and conclusion. J Less Common Met 166:283–292
3. Dobrovolsky VD (1968) Registration of x-ray absorption spectra by "variable field absorption method". In: Nemoshkalenko VV (ed) Electronic structure of transition metals and their alloys. IPM NANU, Kiev, pp 296–299, in Russian

Chapter 18
Molecular-Kinetic Theory of Phase Transitions in Crystals of Fluorofullerenes $C_{60}F_{48} \rightarrow C_{60}F_{36}$ and Their Heat Capacity

S.Yu. Zaginaichenko, D.V. Schur, M.M. Diviziniuk, and Z.A. Matysina

Abstract The theory of phase transition of order-disorder type on the molecular-kinetic grounds has been developed for the mixture of fluorofullerenes $C_{60}F_{48}$, $C_{60}F_{36}$: transition from the ordered body-centered tetragonal (bct) structure to the disordered face-centered cubic (fcc) structure. The free energies of these phases have been found, their dependence on temperature, composition of material, the degree of ordering and energetic constants has been determined. The temperature of transition between phases has been calculated. The constitution diagram has been constructed and it defines the temperature and concentration areas of bct, fcc phases formation and also the region of realization of both bct and fcc phases. The configuration heat capacity of bct phase and its temperature dependence $C_p(T)$ has been defined. The peak-shaped increase of heat capacity in the neighbourhood of the temperature of phase transition has been estimated and this is in agreement with experimental data.

Keywords Fluorofullerene · Average energies method · Phase transformation · Constitution diagram · Configuration heat capacity

18.1 Introduction

The intensive research studies of fullerenes and features of their formation of last years led to the findings of thousands new compounds possessing the unique physicochemical properties. The fluorofullerenes are of great interest for scientific groups. Among the inorganic derivatives of fullerenes they have the high

---

S.Yu. Zaginaichenko (✉), D.V. Schur, and M.M. Diviziniuk,
Institute for Problems of Materials Science of NAS of Ukraine,
Krzhyzhanovsky Str. 3, 03142 Kiev, Ukraine
e-mail: shurzag@materials.kiev.ua

Z.A. Matysina
Dnepropetrovsk National University, 72 Gagarin Str., 49000 Dnepropetrovsk, Ukraine

S.Yu. Zaginaichenko et al. (eds.), *Carbon Nanomaterials in Clean Energy Hydrogen Systems - II*, NATO Science for Peace and Security Series C: Environmental Security 2,
DOI 10.1007/978-94-007-0899-0_18,

thermodynamic stability in comparison, for example, with chlorides, bromides, oxides and manifest much wide variability. The attachment of fluorine atoms to C_{60} molecules affect much their physical and chemical properties. This brings up the prospects of production of new materials with unusual properties and the awakening of interest to the fluoride derivatives of fullerenes.

The solid-phase fluorination of C_{60} crystals by the molecular flow of fluorine F_2 brings primarily into existence the molecules $C_{60}F_{36}$ which possess enhanced stability [1–7]. The atomic bond C-F in them is sufficiently strong. The next fluorine atoms have the less bonding force with molecules C_{60}. Depending on temperature, pressure and duration of interaction of fluorine molecules flow with fullerite the addition of 12 supplementary fluorine atoms gives the formation of fluorofullerene $C_{60}F_{48}$ [3, 7–10]. The last 12 fluorine atoms migrate readily over the carbon frame, redistribute in condensate and even break away [11–13], resulting in rearrangement of structure of fluorofullerene molecules, formation of structures of different type, production of distinct their isomers and consequently change of physical and chemical properties. The exothermicity of fluorination reaction and temperature increase provides the intensity of regrouping of fluorine atoms.

The extensive experimental and theoretical investigation of physicochemical and thermodynamic properties of fluorofullerenes begins and in the most of research papers just more stable fluorofullerenes $\Phi_1 = C_{60}F_{48}$, $\Phi_2 = C_{60}F_{36}$ are being examined. The laboratories perform research on their crystalline and electronic structure, degree of stability, mass composition, the availability of isomers of different level symmetry, phase transitions in condensate with the external pressure and temperature variations, carry out the estimation of phase transition temperature, measure the lattice constants, determine their bulk modulus, conduct the plotting of constitution diagram with indication of temperature and pressure regions of realization of fullerite, fullerene polymers, fluorofullerenes and amorphous phase, evaluate and calculate the heat of fluorofullerenes formation, the enthalpy of their forming, thermal coefficient of expansion, heat capacity [12, 14–39].

Experimental investigation of heat capacity C_p of fluorofullerene Φ_1 establish the anomaly in its temperature dependence [22, 30, 39], there is the peak in the curve $C_p(T)$ in the temperature region $T_0 \sim 330K$ (Fig. 18.1). The further determination of crystalline and molecular structure of condensate demonstrates that at this temperature the phase transition of order-disorder type takes place. The fluorofullerene structure is changed from the ordered body-centered tetragonal (bct) to the disordered face-centered cubic (fcc) [21, 23]. The anomalous change of lattice parameter [23] is observed at this temperature, it seems reasonable to say that it is caused by realization of phase transition.

Theoretical investigation of fluorofullerenes, the development of statistical theory of transition between phases at the temperature T_0, the construction of constitution diagram of the system being studied, the elucidation of temperature dependence of fluorofullerite heat capacity, the explanation and justification of appearance of possible anomaly in this dependence at the temperature T_0 are of our interest.

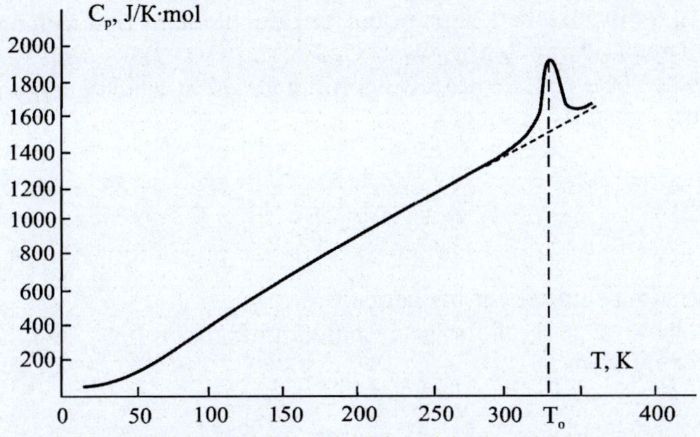

Fig. 18.1 Experimental plot of temperature dependence of heat capacity of $C_{60}H_{48}$ fluorofullerenes [22, 30, 39]. T_o is the temperature of structural phase transition bct → fcc. The *dotted line* shows the dependence $C_o(T)$ in the absence of phase transition

To solve this problem we calculate the free energy of condensate. In theoretical calculations the mixture of fluorofullerenes Φ_1, Φ_2 is considered and according to experimental data this system has ordered bct crystalline lattice at temperatures below T_0 and disordered fcc lattice at temperatures higher than T_0. The method of average energies [40], the approximation of account of interaction energies of nearest fluorofullerene pairs [40, 41] and the model of symmetric spherically rigid spheres are used in calculations.

18.2 Calculation of Free Energies

The computation of free energies of bct and fcc phases is performed by the known formula

$$F_i = E_i - kTlnG_i, \tag{18.1}$$

where E_i is the configuration internal energy equal to the sum of energies of pair interaction of fluorofullerenes, G_i is the thermodynamic probability determined by the amount of discernible distributions of fluorofullerenes over all their positions, k is Boltzmanns constant, T is absolute temperature. The bct phase is named as first, the fcc phase – as second.

At first we calculate the free energy F_1 of bct phase.

The configuration energy is equal to

$$E_1 = -N_{11}\upsilon_{11} - N_{22}\,\upsilon_{22} - N_{12}\,\upsilon_{12}, \tag{18.2}$$

where N_{ij}, υ_{ij} are the numbers of pairs and energies of interaction with opposite sign of fluorofullerenes $\Phi_1 = C_{60}F_{48}$, $\Phi_2 = C_{60}F_{36}$ (i, j = 1, 2).

The distance between the nearest fluorofullerenes, at which energy υ_{ij} is determined for bct structure, is equal to

$$r_1 = \sqrt{2a_1^2 + a_2^2/2}, \tag{18.3}$$

where a_1, a_2 are parameters of bct lattice.

The numbers of pairs of the nearest fluorofullerenes $\Phi_1\Phi_1$, $\Phi_2\Phi_2$, $\Phi_1\Phi_2$ are defined by expressions

$$\begin{gathered} N_{11} = \tfrac{1}{2} zNP_1^{(1)}P_1^{(2)}, N_{22} = \tfrac{1}{2} zNP_2^{(1)}P_2^{(2)}, \\ N_{12} = \tfrac{1}{2} zN\left(P_1^{(1)}P_2^{(2)} + P_1^{(2)}P_2^{(1)}\right), \end{gathered} \tag{18.4}$$

where z = 8 is the coordination number, N is the number of all sites (fluorofullerenes), $P_i^{(\alpha)}$ are the a priori probabilities of substitution of bct lattice sites of $\alpha = 1$, 2 type with fluorofullerenes of $i = \Phi_i$ $(i = 1, 2)$ kind. For the studied structure they are equal to

$$P_1^{(1)} = c_1 + \frac{1}{2}\eta,\ P_1^{(2)} = c_1 - \frac{1}{2}\eta,\ P_2^{(1)} = c_2 - \frac{1}{2}\eta,\quad P_2^{(2)} = c_2 + \frac{1}{2}\eta, \tag{18.5}$$

η is the order parameter in fluorofullerenes distribution over the sites of crystal bct lattice

$$\eta = 2\left(P_1^{(1)} - c_1\right). \tag{18.6}$$

In view of relations (18.5) we rewrite the numbers N_{ij} as

$$\begin{gathered} N_{11} = \frac{1}{2} zN\left(c_1^2 - \frac{1}{4}\eta^2\right),\ N_{22} = \frac{1}{2} zN\left(c_2^2 - \frac{1}{4}\eta^2\right), \\ N_{12} = zN\left(c_1c_2 + \frac{1}{4}\eta^2\right). \end{gathered} \tag{18.7}$$

Then the configuration energy E_1 is determined by formula

$$E_1 = E_{01} - \frac{1}{8} zN\omega_1\eta^2, \tag{18.8}$$

where augend for E_{O1} energy

$$E_{O1} = -\frac{1}{2}zN\left(c_1^2\upsilon_{11} + c_2^2\upsilon_{22} + 2c_1c_2\upsilon_{12}\right) = -\frac{1}{2}zN(c_1\upsilon_{11} + c_2\upsilon_{22} + c_1c_2\,\omega_1) \tag{18.9}$$

is independent of energy order and

$$\omega_1 = 2\upsilon_{12} - \upsilon_{11} - \upsilon_{22} \tag{18.10}$$

is the ordering energy of first phase. The thermodynamic probability G_1 is calculated according to the rules of combinatorics

$$G_1 = \frac{N_1!}{N_1^{(1)}!N_2^{(1)}!} \cdot \frac{N_2!}{N_1^{(2)}!N_2^{(2)}!}, \tag{18.11}$$

where N_1, N_2 are the numbers of lattice sites of $\alpha = 1, 2$ type and $N_i^{(\alpha)}$ are the numbers of fluorofullerenes of *i* kind on the sites of α type

$$N_i^{(\alpha)} = N_i P_i^{(\alpha)}, \quad i = 1, 2; \quad \alpha = 1, 2. \tag{18.12}$$

Taking into consideration formulae (18.5) and using Stirling formula $ln X! = X(ln X - 1)$ for large X numbers, we find the thermodynamic probability

$$lnG_1 = -\frac{1}{2}N\left[\left(c_1 + \frac{1}{2}\eta\right)ln\left(c_1 + \frac{1}{2}\eta\right) + \left(c_1 - \frac{1}{2}\eta\right)ln\left(c_1 - \frac{1}{2}\eta\right) + \left(c_2 - \frac{1}{2}\eta\right)ln\left(c_2 - \frac{1}{2}\eta\right) + \left(c_2 + \frac{1}{2}\eta\right)ln\left(c_2 + \frac{1}{2}\eta\right)\right]. \tag{18.13}$$

Substituting configuration energy E_1 (18.8) and thermodynamic probability G_1 (18.13) in Eq. 18.1, we define the free energy of bct phase calculated for one site of crystal lattice as follows

$$f_1 = F_1/N = e_1 - \omega_1\eta^2 + \frac{1}{2}kT\Delta_1, \tag{18.14}$$

where the following designations are used

$$\Delta_1 = \left(c_1 + \frac{1}{2}\eta\right)ln\left(c_1 + \frac{1}{2}\eta\right) + \left(c_1 - \frac{1}{2}\eta\right)ln\left(c_1 - \frac{1}{2}\eta\right) + \left(c_2 - \frac{1}{2}\eta\right)ln\left(c_2 - \frac{1}{2}\eta\right) + \left(c_2 + \frac{1}{2}\eta\right)ln\left(c_2 + \frac{1}{2}\eta\right), \tag{18.15}$$

$$e_1 = E_{O1}/N = -4(c_1\upsilon_{11} + c_2\upsilon_{22} + c_1c_2\,\omega_1). \tag{18.16}$$

The formulae (18.14)–(18.16) determine the dependence of free energy of bct phase on temperature, order parameter, phase composition (concentrations c_1, c_2) and energetic parametres $\upsilon_{11}, \upsilon_{22}, \omega_1$.

The calculation of free energy F_2 of the second fcc phase is similar to the foregoing

$$f_2 = F_2/N = e_2 + kT\Delta_2, \tag{18.17}$$

where

$$\Delta_2 = c_1 lnc_1 + c_2 lnc_2, \tag{18.18}$$

$$e_2 = E_2/N = -6\left(c_1^2\upsilon'_{11} + c_2^2\upsilon'_{22} + 2c_1c_2\upsilon'_{12}\right), \tag{18.19}$$

and $z_2 = 12$ is taken into account for fcc phase. The υ'_{ij} energies of interaction of the nearest fluorofullerenes in fcc phase are determined at the distances

$$r_2 = a\big/\sqrt{2}, \tag{18.20}$$

a is the parameter of fcc lattice.

According to formulae (18.17)–(18.19), the free energy of second fcc phase depends on temperature, concentrations c_1, c_2 and energetic parameters $\upsilon'_{11}, \upsilon'_{22}, \upsilon'_{12}$.

18.3 Temperature of Phase Transition. Equations of Thermodynamic Equilibrium. Constitution Diagram

The phase transition bct → fcc occurs at the temperature $T = T_o$ when free energies of both phases are equal

$$f_1 = f_2. \tag{18.21}$$

We equate the two expressions (18.14), (18.17) and find the temperature T_o of phase transition

$$kT_o = \left(e_1 - e_2 - \omega_1\eta^2\right)\Big/\left(\frac{1}{2}\Delta_1 - \Delta_2\right). \tag{18.22}$$

The equilibrium value of order parameter η in (18.22) is estimated from the condition of thermodynamic equilibrium, i.e. by equation

$$\partial f_1/\partial\eta = 0, \tag{18.23}$$

and this gives the relation

$$kTln\frac{\left(c_1+\frac{1}{2}\eta\right)\left(c_2+\frac{1}{2}\eta\right)}{\left(c_1-\frac{1}{2}\eta\right)\left(c_2-\frac{1}{2}\eta\right)}=8\ \omega_1\eta. \tag{18.24}$$

Considering $\eta \rightarrow 0$ in (18.24) and straightforward rearranging gives the temperature of ordering of the first bct phase

$$kT_C = 8c_1c_2\ \omega_1. \tag{18.25}$$

The equilibrium value of order parameter depends on temperature and concentrations c_1, c_2. The maximum order ($\eta = 1$) is realized for the infinitely low temperatures and concentrations corresponding to the stoichiometric composition $c_1 = c_2 = 0{,}5$.

For bct phase of stoichiometric composition the Eq. 18.24 in terms of (18.25) takes the form

$$ln\frac{1+\eta}{1-\eta}=2\frac{T_C}{T}\eta. \tag{18.26}$$

For the numerical assessment of temperatures of phase transitions T_o of bct $\rightarrow$ fcc type and T_C of order-disorder type the evaluation of energetic parametres e_1, e_2, ω_1 should be made. These parametres are estimated approximately using the experimental data for the temperature T_o of phase transition bct $\rightarrow$ fcc ($T_o \approx 330$K or $kT_o = 0{,}028$ eV). This estimation shows that

$$e_1=-0,01\ \text{eV},\quad e_2=-0,005\ \text{eV},\quad \omega_1=0,01\ \text{eV}. \tag{18.27}$$

As an example we evaluate this temperature T_o for the maximum value of order parameter defined by equalities

$$\eta_m=\begin{cases}2c_1\ \text{at}\ c_1\leqslant 0,5,\\ 2c_2\ \text{at}\ c_1\geqslant 0,5.\end{cases} \tag{18.28}$$

It is evident that $\eta_m = 1$ for the stoichiometric composition of the system. In this case $kT_C = 0{,}02$ eV, i.e. the temperatures T_o and T_C are close to each other.

The value Δ_1 (18.15) for order parameter η_m (18.28) takes the following form

$$\Delta_1=\begin{cases}2c_1ln2c_1+(c_2-c_1)ln(c_2-c_1)\ \text{at}\ c_1\leqslant 0,5\\ 2c_2ln2c_2+(c_1-c_2)ln(c_1-c_2)\ \text{at}\ c_1\geqslant 0,5.\end{cases} \tag{18.29}$$

Furthermore, the calculation of free energies f_1, f_2 is carried out by formulae (18.14), (18.17) using the energetic parametres (18.27), magnitude η_m (18.28), values of Δ_1 (18.29), Δ_2 (18.18).

Figure 18.2 gives the plots of concentration dependences of free energies constructed for different temperatures. These graphical representations of free energies make it possible to estimate the temperature T_o of phase transition bct $\rightarrow$ fcc and to construct the constitution diagram of the system under study. The concentration and temperature regions of realization of one- and twophase condensates are determined in the method of common tangents to the curves $f_1(c_1)$, $f_2(c_1)$.

The constitution diagram of the system being studied is presented in Fig. 18.3. It is constructed by the use of method of common tangents to the curves $f_1(c_1)$, $f_2(c_1)$ and by the intersection points of these curves. It is seen from this figure that at low temperatures the ordered bct phase is bound to be realized from the mixture of fluorofullerenes Φ_1, Φ_2 in accordance with experimental data. The phase transition into disordered state of fluorofullerenes Φ_1, Φ_2 mixture with fcc lattice takes place with increase in temperature that also corresponds to experimental data. As the temperature increases, the concentration range of fcc phase formation broaden and at the sufficiently high temperatures the fcc phase is realized over all concentration range. The existence of bct phase is retained at the rather high temperatures, but for the low or high concentrations of fluorofullerenes Φ_1, Φ_2. In the region near stoichiometric composition the fcc phase is realized over the wide temperature range beginning with $kT_o > 0.01$ eV. The two-phase regions of bct and fcc phases formation appear at $kT_o \approx 0.01$ eV, which at first broaden with increase in temperature, thereafter converge and disappear at the high temperatures.

The experimental check of the emerged regularities of realization of concentration and temperature ranges of ordered and disordered fcc phases is of interest for physicochemical engineers.

18.4 Configuration Heat Capacity

The heat capacity in dependence on temperature can be determined from formula

$$C_V = \frac{\partial E_1}{\partial T} = -2N\,\omega_1 \eta \frac{\partial \eta}{\partial T}, \tag{18.30}$$

in which the order parameter η and its temperature derivative $\partial\eta/\partial T$ should be determined from the equation of thermodynamic equilibrium (18.24). The configuration heat capacity of disordered fcc phase is equal to zero, because the E_2 energy is temperature independent.

The calculation of configuration heat capacity with account of formula (18.25) gives the result

$$\frac{C_V}{kN} = ln^2 \frac{\left(c_1 + \frac{1}{2}\eta\right)\left(c_2 + \frac{1}{2}\eta\right)}{\left(c_1 - \frac{1}{2}\eta\right)\left(c_2 - \frac{1}{2}\eta\right)} \Bigg/ 4\left[\frac{c_1}{c_1^2 - \frac{1}{4}\eta^2} + \frac{c_2}{c_2^2 - \frac{1}{4}\eta^2} - \frac{1}{\eta} ln \frac{\left(c_1 + \frac{1}{2}\eta\right)\left(c_2 + \frac{1}{2}\eta\right)}{\left(c_1 - \frac{1}{2}\eta\right)\left(c_2 - \frac{1}{2}\eta\right)}\right]. \tag{18.31}$$

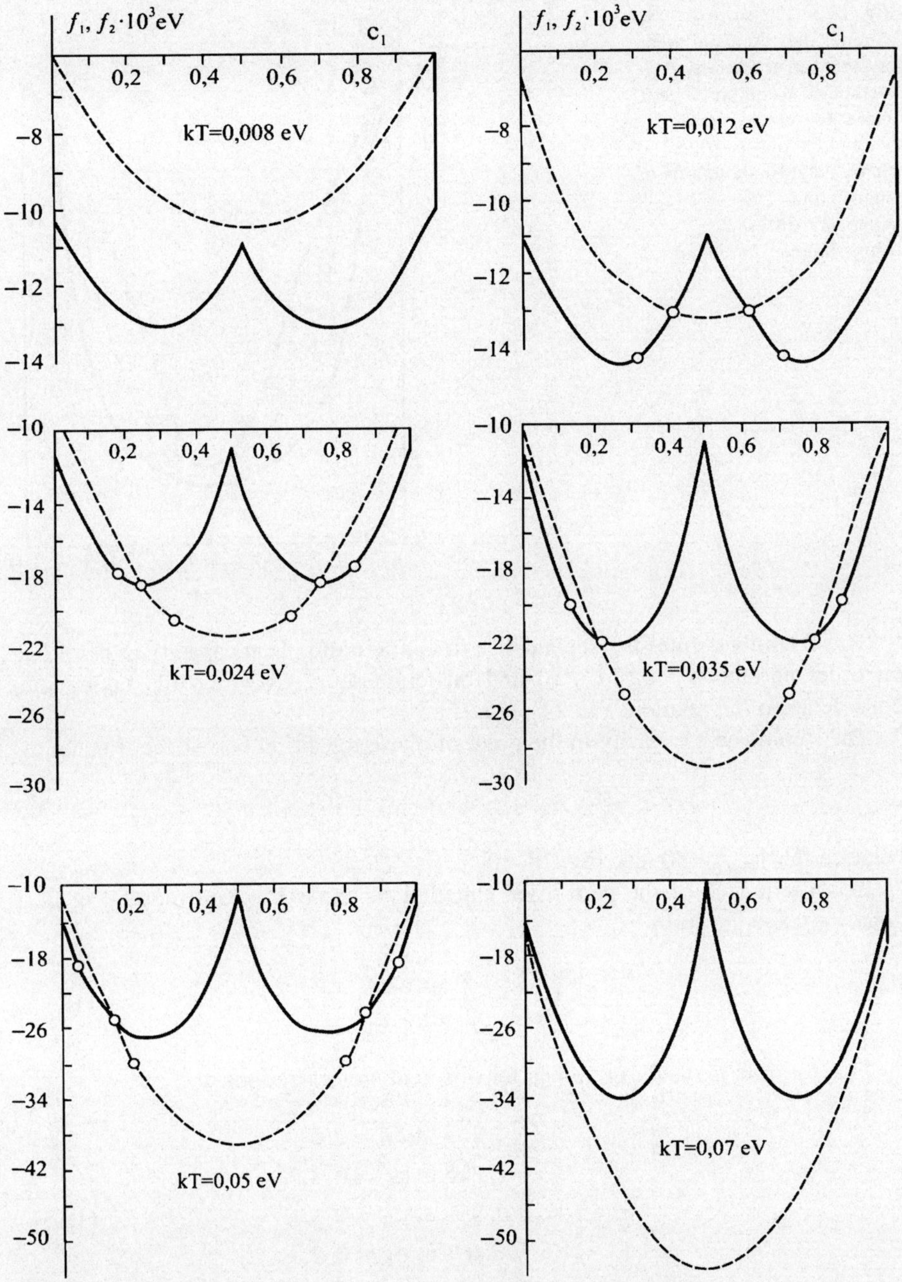

Fig. 18.2 The design plots of concentration dependence of free energies of ordered bct phase (*unbroked curves*) and disordered fcc phase (*dotted curves*) of flurofullerenes constructed for different temperatures. The intersection points of $f_1(c_1)$, $f_2(c_1)$ functions and points of common tangent to them are marked off by the open circles on the curves

Fig. 18.3 The state diagram of molecular fluorofullerene crystals constructed by method of common tangent curves to free energies f_1, f_2 of bct and fcc phases (*full curves*) and by the points of intersection of $f_1(c_1), f_2(c_2)$ curves for different temperatures (*dotted curve*)

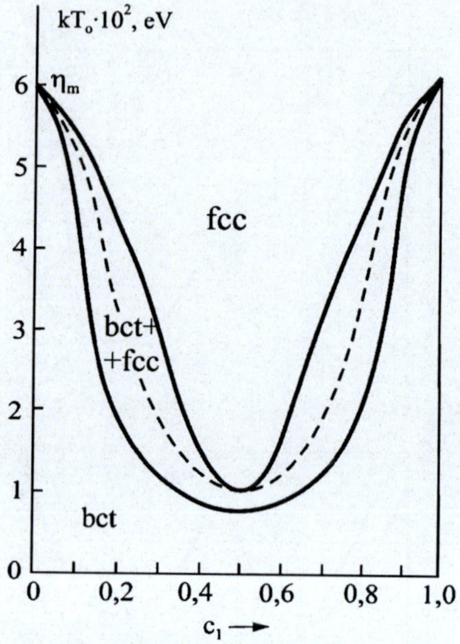

This formula defines the dependence of configuration heat capacity of bct phase on order parameter $C_V = C_V(\eta)$ and on temperature $C_V = C_V$ (T) taking into consideration the relations (18.24), (18.25).

The step of heat capacity in the point of phase transition bct → fcc is equal to

$$\Delta C = (C_V)_{T_C-0} - (C_V)_{T_C+0} = (C_V)_{T_C-0}, \qquad (18.32)$$

because $(C_V)_{T_C+0} = 0$ (see Fig. 18.4a).

In consequence of the performed calculations the expression for step of heat capacity takes the form

$$\Delta \frac{C}{kN} = \frac{3}{2} \cdot \frac{c_1 c_2}{c_1^3 + c_2^3}. \qquad (18.33)$$

This formula gives $\Delta \frac{C}{kN}$ values for different concentrations c_1

$$\Delta \frac{C}{kN} = \begin{cases} 1,5 \text{ at } c_1 = 0,5, \\ 1,286 \text{ at } c_1 = 0,4, \\ 0,851 \text{ at } c_1 = 0,3, \\ 0,462 \text{ at } c_1 = 0,2, \\ 0,185 \text{ at } c_1 = 0,1, \end{cases} \qquad (18.34)$$

i.e. the heat capacity step decreases with deviation of composition of bct phase from stoichiometry.

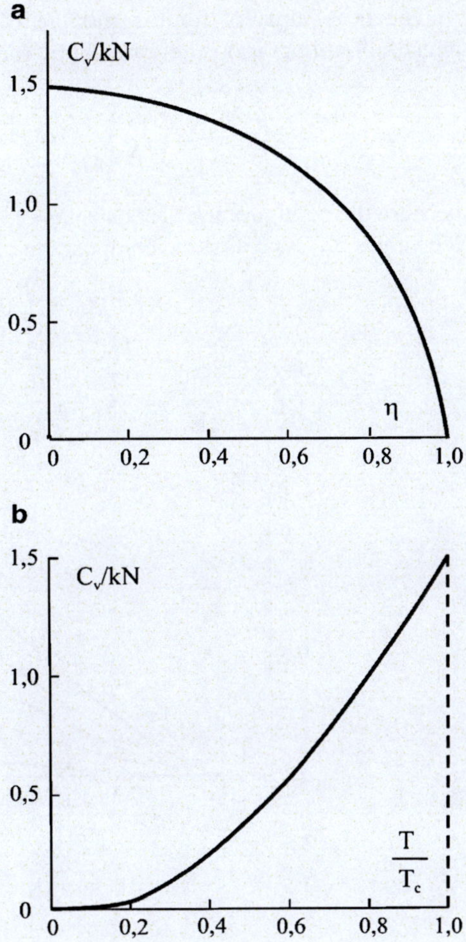

Fig. 18.4 The plots of configuration heat capacity of bct phase of stoichiometric composition as a function of order parameter (**a**) and temperature (**b**)

In the case of bct phase of stoichiometric composition the formula (18.31) is simplified

$$\frac{C_V}{kN} = \eta(1-\eta^2)ln^2\frac{1+\eta}{1-\eta}\Big/2\left[2\,\eta-(1-\eta^2)ln\frac{1+\eta}{1-\eta}\right]. \qquad (18.35)$$

In Fig. 18.4 the plots of configuration heat capacity dependence on order parameter (*a*) and on temperature (b) are shown for bct phase of stoichiometric composition. From these curves one can see that heat capacity falls with a decrease in order parameter and increases with a rise in temperature.

The heat capacity for the mixture of two phases (two-phase region at the constitution diagram) is determined by formula

$$\frac{C_v}{kN} = X_1\left(\frac{C_v}{kN}\right)_{\text{bct}} + X_2\left(\frac{C_v}{kN}\right)_{\text{fcc}} = X_1\left(\frac{C_v}{kN}\right)_{\text{bct}}, \quad (18.36)$$

because the configuration heat capacity of fcc phase is equal to zero. In this formula the values X_1, X_2 define the composition of bct and fcc phases mixture, in this case

$$X_1 + X_2 = 1, 0 \leqslant X_1, X_2 \leqslant 1. \quad (18.37)$$

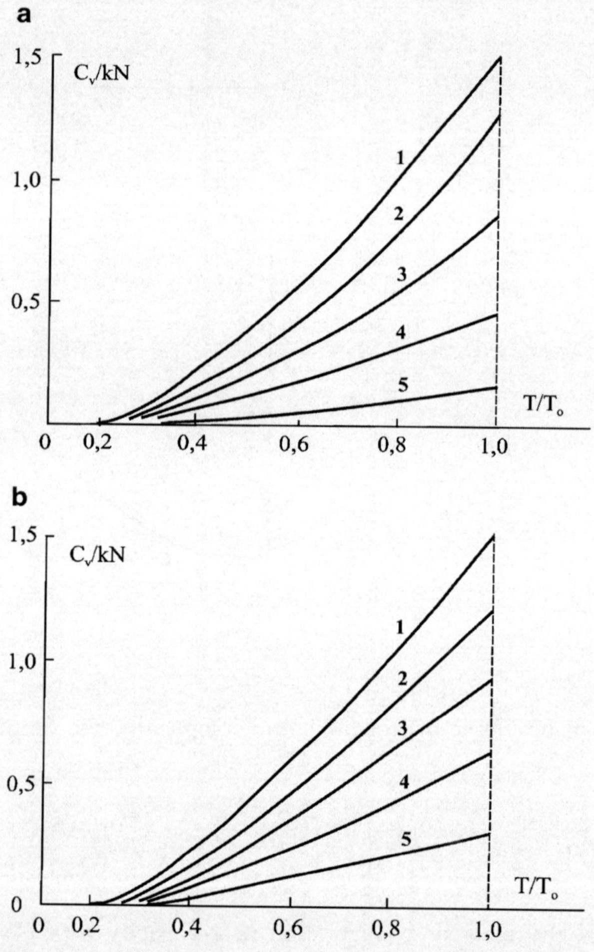

Fig. 18.5 The design plots of temperature dependence of configuration heat capacity: (**a**) by formula (31) for fct phase and different concentrations c_1 = 0,5; 0,4; 0,3; 0,2; 0,1 (curves 1–5); (**b**) by formulae (31), (36) for two-phase mixture (bct and fcc), when the content of first phase is X_1 = 1; 0,8; 0,6; 0,4; 0,2 (curves 1–5)

The temperature dependences of configuration heat capacity for different values of concentration c_1 (*a*) and of composition X_1 of the first bct phase (b) are given in Fig. 18.5. The character of dependences $C_V(T_1)$ is identical for both cases (*a*) and (b). The heat capacity and its step at $T = T_C$ show a decrease with decreasing value c_1 or X_1. The theory gives results presented in Fig. 18.5 in qualitative agreement with experimental data (Fig. 18.1) for the heat capacity dependence on temperature for the ordered bct phase. The decrease of heat capacity C(T) in a gradual manner at $T > T_C$ on the experimental plot can be caused by the presence of short-range ordering, which is not taken into account in our calculations.

18.5 Conclusions

The developed statistical theory makes possible to justify and to provide an explanation for the phase transition of order–disorder type in the mixture of fluorofullerenes $C_{60}F_{48}$, $C_{60}F_{36}$ which is revealed experimentally at the temperature of 330 K when transformation of ordered bct phase into disordered fcc phase is realized.

The calculation of free energies of both phases has been carried out, the evaluation of phase transition temperature in dependence on composition of phases and order parameter has been performed.

The construction of constitution diagram has been made for determination of temperature and concentration regions of constituent phases in the formation of pure bct and fcc phases as well as their mixture. It follows from this diagram that the ordered bct phase has been realized at low temperatures, the temperature rise has stimulated the formation of disordered fcc phase. Both theoretical conclusions are in agreement with experimental data.

The calculation of configuration heat capacity of bct phase has shown its peak-shaped increase in the neighbourhood of phase transition temperature that also corresponds to experimental observations. The evaluation of heat capacity step in the point of phase transition has been carried out and the value of this step decreases with deviation of fluorofullerenes mixture composition from stoichiometry.

References

1. Selig H, Lifshitz C, Peres T et al (1991) Fluorinated fullerenes. J Am Chem Soc 113:5475–5476
2. Gakh AA, Tuinmann AA, Adcock JL et al (1994) Selective synthesis and structure determination of $C_{60}F_{48}$. J Am Chem Soc 116:819–820
3. Clare BW, Kepert DL (1999) The structures of $C_{60}F_{36}$ and new possible structures for $C_{60}H_{36}$. J Mol Struct Theochem 466(1–3):177–186
4. Meier MS (2000) Fullerenes. In: Kadish KM, Ruoff RS (eds) Chemistry, physics and technology. Wiley, New York, 129 p

5. Bagryantsev VF, Zapolskii AS, Boltalina OV et al (2000) Reaction of fullerenes with molecular fluorine. J Neorg Khim 45(7):1121–1127 (in Russian)
6. Avent AG, Taylor R (2002) Fluorine takes a hike: remarkable room-temperature rearrangement of the C1 isomer of C60F36 into the C3 isomer via a 1, 3-fluorine shift. Chem Commun 22:2726–2727
7. Sidorov LN, Yurovskaya MA, Borschevskii A et al (2005) Fullerenes. Ekzamen, Moscow, 688 p (in Russian)
8. Boltalina OV, Sidorov LN, Bagryantsev VF et al (1996) Formation of $C_{60}F_{48}$ and fluorides of higher fullerenes. J Am Chem Soc 2:2275–2278
9. Privalov VI, Boltalina OV, Galeva NA, Taylor R (1998) Structure of crystalline $C_{60}F_{48}$ by NMR data with rotation under angle. Dokl RAN Ser Khim 360(4):499–502 (in Russian)
10. Troyanov SI, Troshin PA, Boltalina OV et al (2001) Two isomers of $C_{60}F_{48}$: an indente fullerene. Ange Chem Int Ed 40(12):2285–2287
11. Papina TS, Kolesov VP, Lukyanova VA et al (1999) The standard molar enthalpy of formation of fluorofullerene $C_{60}F_{48}$. J Chem Thermodyn 31(10):1321–1328
12. Papina TS, Kolesov VP, Lukyanova VA et al (2000) Enthalpy of formation and C-F bond enthalpy of fluorofullerene $C_{60}F_{36}$. J Phys Chem B 104(23):5403–5405
13. Gakh AA, Tuinmann AA (2001) 'Fluorine dance' on the fullerene surface. Tetrahed Lett 42 (41):7137–7139
14. Tuinman A, Gakh A, Adcock J, Compton R (1993) Hyperfluorination of buckminsterfullerene. J Am Chem Soc 115:5885–5886
15. Kniaz K, Fischer JE, Selig H et al (1993) Fluorinated fullerenes: synthesis, structure and properties. J Am Chem Soc 115(4):6060–6064
16. Fowler PW, Sandall JPB, Taylor R (1997) Structural parallels in hydrogenated and fluorinated [60] – and [70] – fullerenes. J Am Chem Soc 2:419–423
17. Boltalina OV, Galeva NA, Markov VYu et al (1997) A mass spectrometric study of $C_{60}F_{48}$. Mendeleev Commun 5:169–212
18. Clare BW, Kepert DL (1997) An analysis of the 94 possible isomers of $C_{60}F_{48}$ containing a three-fold axis. Theochem J Mol Struct 389(1–2):97–103
19. Mitsumoto R, Araki T, Ito E et al (1998) Electronic structures and chemical bonding of fluorinated fullerenes studied by NEXAFS, UPS, and Vacuum-UV adsorption spectroscopies. J Phys Chem A 102(3):552–560
20. Taylor R (1998) Progress in fullerenes fluorination. Russ Chem Bull 47(5):823–832
21. Kawasaki S, Aketa T, Touhara H et al (1999) Crystal structures of the fluorinated fullerenes $C_{60}F_{36}$ and $C_{60}F_{48}$. J Phys Chem B 103(8):1223–1225
22. Druzhinina AI, Galeva NA, Varushchenko RM et al (1999) The low temperature heat capacities of fluorofullerenes. J Chem Thermodyn 31(11):1469–1482
23. Kawasaki S, Okino F, Touhara H (2000) Crystal structures and phase transformations of the fluorinated fullerenes. Mol Cryst Liq Cryst 340:629–633
24. Boltalina OV, Galeva NA (2000) Direct fluorination of fullerenes. Russ Chem Rev 69 (7):661–674
25. Gakh AA, Tuinman AA (2001) The structure of $C_{60}F_{36}$. Tetrahed Lett 42(41):7133–7135
26. Slanina Z, Uhlik F, Boltalina OV, Kolesov VP (2002) Isomeric $C_{60}F_{36}$ (g) species: computed structures and heats of formation. Solid State Phys 44(3):534–535
27. Kawasaki S, Yao A, Okino F et al (2002) High pressure phases of fullerenes, hydrofullerenes and fluorofullerenes. Mol Cryst Liq Cryst Sci Technol 386:106–114
28. Yao A, Matsuoka Yu, Komiyama S et al (2002) Structural properties of fluorinated fullerenes at high pressures and high temperatures. Solid State Sci 4(11–12):1443–1447
29. Avent AG, Clare BW, Hitchcock PB et al (2002) $C_{60}F_{36}$: there is a third isomer and it has C_1 symmetry. Chem Commun 20:2370–2371
30. Druzhinina AI, Varuschenko RM, Boltalina OV, Sidorov LN (2002) Proceedings of international conference on chemical thermodynamics, Saint-Petersburg, Russia, Paper I-P 28, 2002 (in Russian)

31. Gakh AA, Romanovich AYu, Bax A (2003) Thermodynamic rearrangement synthesis and NMR structures of C_1, C_3 and T isomers of $C_{60}F_{36}$. J Am Chem Soc 125(26):7902–7906
32. Rau JV, Cesaro SN, Boltalina OV et al (2004) Raman and infrared spectroscopic study of $C_{60}F_{18}$, $C_{60}F_{36}$ and $C_{60}F_{48}$. Vib Spectrosc 34(1):137–147
33. Boltalina OV, Strauss SH (2004) Fluorofullerenes. In: Scharz JA, Contescu CI, Putyera K (eds) Dekker encyclopedia of nanoscience and nanotechnology, vol 2., pp 1175–1190
34. Papoular RJ, Allouchi H, Dzyabchenko AV et al (2006) High-resolution x-ray powder diffraction structure determination of $C_{60}F_{48}$. Fullerenes Nanotubes Carbon Nanostruct 14 (2–3):279–285
35. Popov A, Senyavin V, Boltalina OV et al (2006) Infrared, Raman and DFT spectroscopic studies of $C_{60}F_{36}$ and $C_{60}F_{48}$. J Phys Chem A 110:8645–8652
36. Bulusheva LG, Okotrub AV, Shnitov VV et al (2009) Electronic structure of $C_{60}F_{36}$ studied by quantum-chemical modeling of experimental photoemission and x-ray absorption spectra. J Chem Phys 130:014704
37. Sheka EF (2009) Step-wise computational synthesis of fullerene C_{60} derivatives. 1. Fluorinated fullerenes $C_{60}F_{2k}$. In: Lecture notes in computer science, pp 1–32
38. Mikoushkin VM, Shnitov VV, Bryzgalov VV et al (2009) Core electron level structure in $C_{60}F_{18}$ and $C_{60}F_{36}$ fluorinated fullerenes. Tech Phys Lett 35(3):256–259
39. Boltalina OV, Sidorov LV Buckminsterfullerene, higher fullerenes and their endo and fluorinated derivatives, Russian Chemical Reviews (accepted for publication)
40. Matysina ZA, Zaginaichenko SYu, Schur DV (2005) Orders of various type in crystals and phase transitions in carbon materials. Nauka i obrazovanie, Dnepropetrovsk, 524 p (in Russian)
41. Schur DV, Matysina ZA, Zaginaichenko SYu (2007) Carbon nanomaterials and phase transformations in them. Nauka i obrazovanie, Dnepropetrovsk, 680 p (in Russian)

Chapter 19
The Designed Metal-Hydride Torches and Hydrogen Accumulators for Various Purposes

D.V. Schur, A.F. Savenko, V.A. Bogolepov, S.Yu. Zaginaichenko, L.O. Teslenko, and T.N. Veziroglu

Abstract The hydrogen storage in metal hydrides is the urgent problem of hydrogen power engineering and the demand for metal hydrides as capacitive, safe and convenient in service sources of hydrogen has stimulated the study of hydrogen capacity of multicomponent alloys. In recent years much attention has been given by scientists to the investigation of hydrogen-sorption and desorption properties of different materials including nanocarbon structures and composites on their base, the study of peculiarities of the reversible hydrogen interaction with hydride forming metals and alloys, the development of high-pure hydrogen storage and transportation in solids. This paper deals with the designed hydrogen metal-hydride torches with piezoelectric firing of flame, two models of accumulators/compressors of great capacity on hydrogen, and three modifications of laboratory hydrogen accumulators used in operation of fuel cells. We show the construction of all torches and accumulators, their technical operating characteristics, the special features and advantages of devices developed and produced in our department and their extremely effective applications in conditions of high ecological requirements.

Keywords Metal hydride · Hydrogen storage · Oxygen · Compressor · Application · Torch

---

D.V. Schur (✉), A.F. Savenko, V.A. Bogolepov, S.Yu. Zaginaichenko, and L.O. Teslenko
Institute for Problems of Materials Science of NAS of Ukraine,
3 Krzhyzhanovsky Str., 03142 Kiev, Ukraine
e-mail: shurzag@materials.kiev.ua

T.N. Veziroglu
Clean Energy Research Institute, University of Miami, EB 219 McArthur Engineering Building, 33124 Coral Gables, FL, USA
e-mail: veziroglu@iahe.or

S.Yu. Zaginaichenko et al. (eds.), *Carbon Nanomaterials in Clean Energy Hydrogen Systems - II*, NATO Science for Peace and Security Series C: Environmental Security 2,
DOI 10.1007/978-94-007-0899-0_19,

19.1 Introduction

The world's research society has tried to use the hydrogen-sorption properties of different metals and alloys to solve a great variety of scientific and technical problems for over 50 years [1–18]. The properties of many chemical elements and their hydride alloys and compounds have been studied for this period. Metal hydrides have already found a wide application owing to the peculiarities of thermodynamics of hydrogen sorption and desorption processes that depend on the chemical composition of a solid subjected to the process of hydrogen pickup.

In recent years much attention has been given to the development of hydrogen absorbing alloys due to their high hydrogen capacity. Under normal conditions the amount of hydrogen stored in a vessel filled with metal hydride can be larger than that in the same vessel filled with liquid hydrogen. The method of hydrogen storage in solids advantageously differs from that in gas-cylinders and cryogenic. This method is safe and requires lower service costs. Therefore the world-known companies have put in order the serial production of various modifications of metal hydride accumulators of hydrogen of different construction.

The application of metal hydride technologies allows the manufacture of compact, safe and technologically flexible hydrogen treatment units. Also, peculiarities of the reversible hydrogen interaction with hydride forming metals and alloys makes it possible to purify hydrogen from gas admixtures in the MH units. The possibility to control the output hydrogen pressure by controlling heat influence on the MH sorbent allows the realization of controlled hydrogen supply to a consumer under the preset pressures. Storage purification, compression/controlled supply can be combined in a single multi-functional unit. This feature makes such applications extremely effective.

At present time it has found applications both as the energy carrier and as the power source. If the change of electric current by moved hydrogen in pipe-lines is a rather complicated process, its transferring in containers of different construction has been put into operation firmly.

19.2 Hydrogen Accumulators and Compressors

We have developed a series of laboratory MH sources of high-pure hydrogen with hydrogen output under controlled increased pressure (up to 200 bar). The sources use MH placed in a high pressure container equipped with an internal heat exchanger. The hydrogen accumulators have different operating characteristics that can be changed in dependence on requirements of user. The development work has been done for the production of the series of metal hydride storage accumulators for diversified applications (Fig. 19.1).

The metal-hydrogen accumulators have been designed for the laboratory use. Figures 19.1–19.4 show three modifications of the desk hydrogen accumulators of

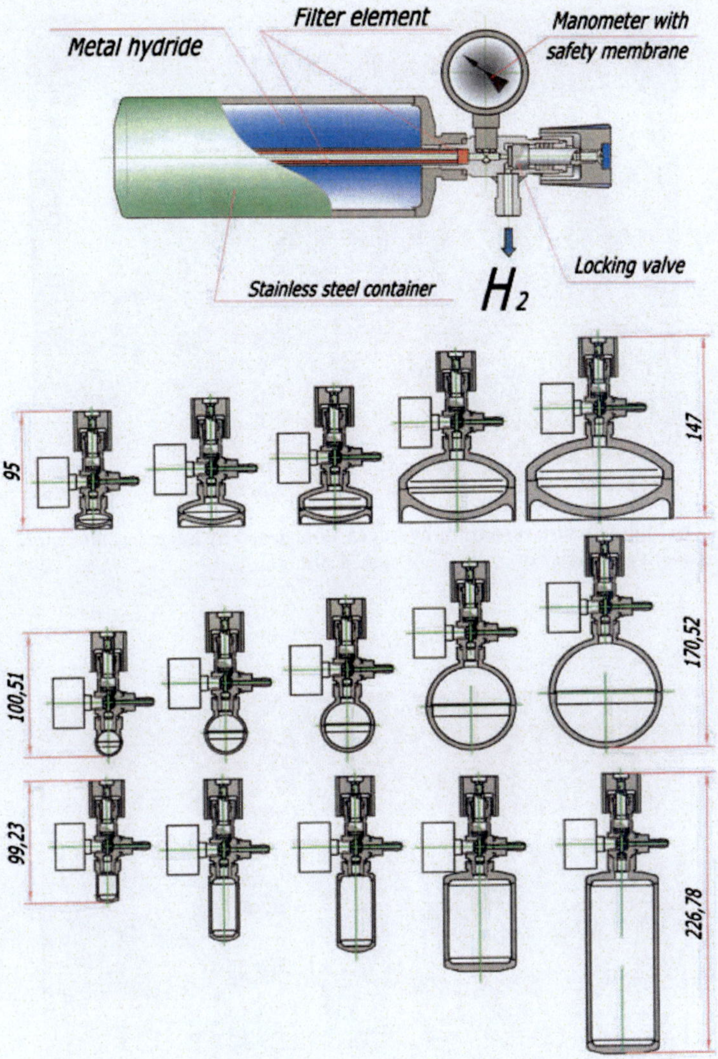

Fig. 19.1 The series of metal hydride storage elements for various applications

different capacities ("Alsav", "Viachbog", "Dmisch"), which have been designed for operation under laboratory conditions completed with the laboratory fuel cells to perform laboratory training on hydrogen energy.

Intermetallic compounds used in these accumulators were selected on the basis of consumers' individual requirements concerning temperature and hydrogen pressure. Alloys of AB_5 and AB types with different additives are typically used.

The vessels in which the metal hydrides are placed have been designed for pressures from 15 to 20 MPa with a double margin of safety. Hydrogen from the

Fig. 19.2 The high-pressure vessels for hydrogen storage and hydrogen accumulators of "Alsav" modification with a capacity of 3, 10, 15, 75 and 170 l

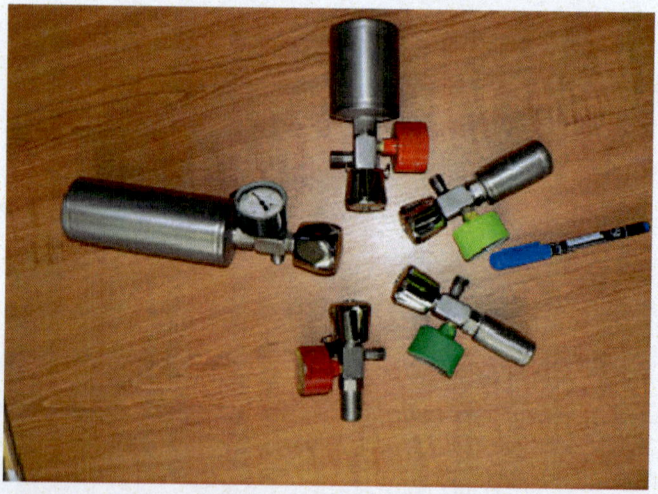

Fig. 19.3 The high-pressure vessels for hydrogen storage and hydrogen accumulators of "Dmisch" modification with a capacity of 3, 10, 15, 75 and 170 l

storage units can be provided at the room temperature under a pressure from 0.1 to 3 MPa, and on heating to 100°C from 4 to 16 MPa . Hydrogen can be produced at 25 MPa on heating to 300°C. The internal (Fig. 19.3) and external (Fig. 19.4) heat exchangers are used in the accumulators of this modification (Fig.19.5).

Fig. 19.4 The hydrogen accumulators "Dmisch" supplied with thermostated heat exchangers energized from the constant-current sources of 12 V

Fig. 19.5 The high-pressure vessels for hydrogen storage and hydrogen accumulators of "Viach-bog" modification with a capacity of 3, 10, 15 and 75 l

The capacity of hydrogen accumulators can be variable from litres to several thousand of litres. Each accumulator is equipped with a manometer which is simultaneously used as a safety valve.

We also have worked out two models of accumulators of great capacity (2,000–7,000 l) on hydrogen. The model *"VEZAYF"* is designed to be operated for hydrogen delivery at low pressure (up to 0,5 MPa), but has the high hydrogen capacity. The model *"SVETZAG"* has three times smaller capacity on hydrogen, but owing to the presence of the temperature control system it permits to conduct the hydrogen under controlled pressure up to 20.0 MPa.

19.2.1 *"VEZAYF"*

The laboratory metal-hydride storage of high-pure hydrogen is designed for operation in the laboratory setups (Fig. 19.6) in the cases when the high hydrogen demands of low pressure have been in existence.

The RE(Ni,Fe,Al)5 hydrogen storage alloy made on the basis of the commercial cerium ligature (Ce/83%/La Pr Nd Fe Al), lanthanum and nickel (technical purity grade both) was used in the unit. The composition of the alloy was selected to provide hydrogen equilibrium pressure over the MH of ~0.5 MPa at room temperature.

Specifications "*VEZAYF*":
Overall dimensions of "*VA – 7000*":
Length 480 mm
Height 440 mm
Width 300 mm
Number of modules six pieces
Mass of a module 8.8 kg
Mass of metal hydride in a module 7 kg
Hydrogen capacity of a module 1,200 l
General hydrogen capacity of a storage element 7,000 l
Lump of a storage element 55 kg
Lump of metal hydride 42 kg
Working pressure 0.3–0.5 MP (T = 30°C)

19.2.2 *"SVETZAG"*

The laboratory metal hydride storage/compressor of high-pure hydrogen is designed for operation in the laboratory setups.

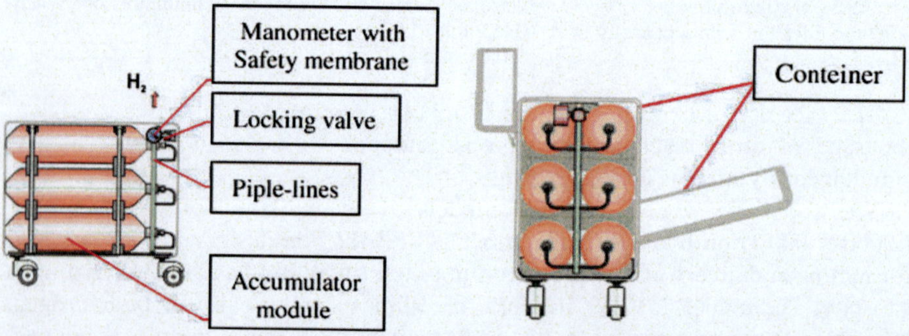

Fig. 19.6 Schematic view of the laboratory metal hydride storage of high-pure hydrogen "*VA – 7000*"

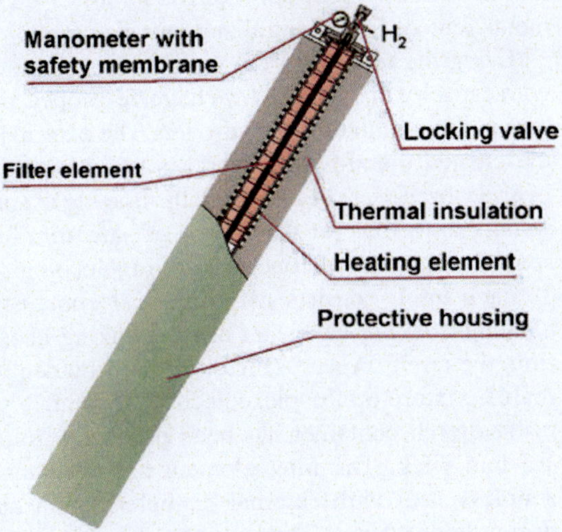

Fig. 19.7 Schematic view of the laboratory metal hydride storage/compressor of high-pure hydrogen "*Svetzag – 2000*"

The performed strength calculations (GOST 14249–89, margin of safety of 1.5 and correction for strength reduction by welding of 0.8) have shown that the allowed working pressure in the MH container can be as high as 400 bar at 300°C (Fig. 19.7).

The developed metal hydride unit for hydrogen storage and compression is characterized by high compactness and relatively low reheat temperature of MH with the delivery of sufficiently high hydrogen pressure and good dynamic performance.

Specifications "*Svetzag*":
Overall dimensions of "*Svetzag - 2000*":
  Length. . ..1,410 mm
  Flange diameter . . ..180 mm
Hydrogen capacity 2,000 l
Diameter of a working reactor 70 mm
Working pressure 25 MPa (300°C)
Mass of a storage element 25.5 kg
Working pressure 0.3 ÷ 0.5 MPa (30°C)
Mass of metal hydride 11 kg
Chemical composition of metal hydride $AB_5 + X$ (X = Fe,Al,Mg . . .) alloys

19.3 Hydrogen Torches

In our work we show the combination in one construction of container for hydrogen transportation and gaseous torch representing the instrument – demand of hydrogen. The hydrogen metal-hydride torch is the self-contained, compact device and it works without additional sources of energy. This torch is usable in settlements remote from centralized energy supply.

Using the totality of our experience on the creation of hydrogen accumulators with capacity of 1 ÷ 100 l we have developed and produced the torches with metal-hydride accumulators of hydrogen. The external appearance and schematic view of the created metal-hydride torches are presented in Figs. 19.8–19.10. The metal-hydride torches are dedicated to the brazing of small-sized parts by high-temperature solders as well as for the cutting of part from foil and in the realization of another specialized works in conditions of high ecological requirements.

Each torch consists of cylindrical container filled with metal hydride, filter element, locking valve, jet orifice, mixing chamber, nozzle, manometer. In addition, the torch "Alsav" (Fig. 19.9) is provided with the device of firing of inflammable mixture on the piezoelectric element.

The torch container has been produced from stainless steel with cylinder wall 1.5 mm thick. The filter element has been made of pipe frame fabricated from stainless steel with external diameter 8 mm and 5 μm filter micron-insert from porous fluoroplastic. The locking valve, having the centreline channel for hydrogen emission up to the jet orifice, provides the necessary flow of hydrogen in pursuance of specific works.

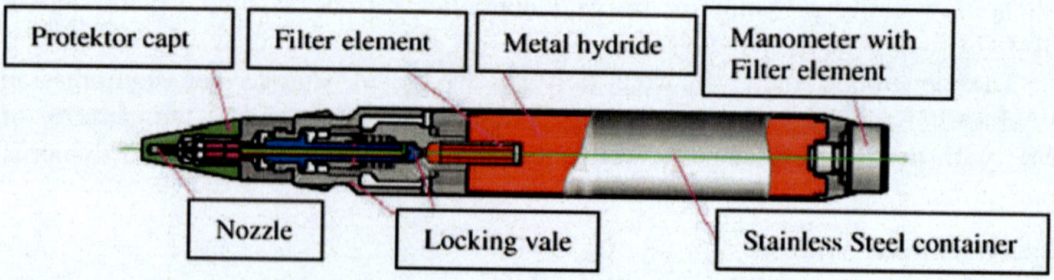

Fig. 19.8 Schematic view of the metal-hydride torch "Viachbog-30"

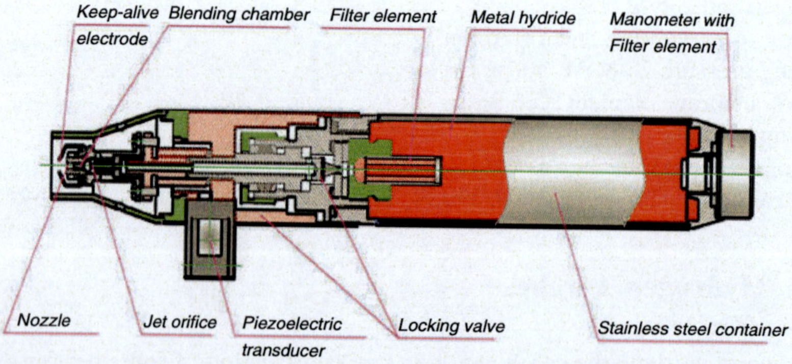

Fig. 19.9 Schematic sketch of the hydrogen self-contained metal-hydride torch "Alsav" with piezoelectric firing of flame

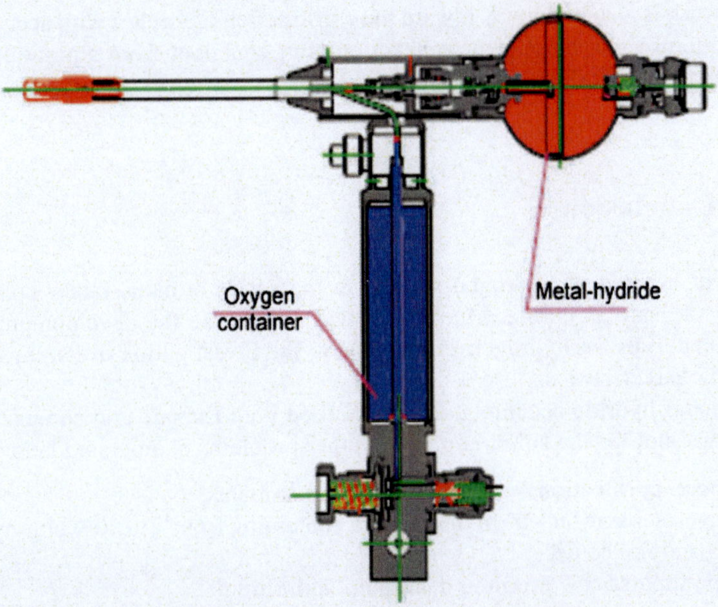

Fig. 19.10 Schematic representation of the oxydric metal-hydride torch

The main technical characteristics of these torches are:

- inner volume of container – 60 cm^3;
- mass of metal hydride – 0.18 kg;
- hydrogen capacity – 30 l;
- total mass – 0.45 kg;
- working pressure at room temperature – 0.2 ÷ 0.5 MPa;
- maximum working pressure – 1 MPa;
- length – 245 mm;
- diameter – 29 mm.

To increase the temperature of gaseous flame the oxyhydrogen torch has been also developed and manufactured with metal-hydride accumulator of hydrogen, cylinder of high pressure for oxygen and piezoelectric firing of flame.

- The main technical characteristics of this oxyhydrogen torch (Fig. 19.10) are:
- capacity of metal-hydride accumulator – 50 l;
- capacity of cylinder for oxygen – 15 l;
- mass – 2.8 kg.

The special features and advantages of the created metal-hydride torches are:

- the high reliability in the work, convenience in operation and maintenance;
- the ease of process of brazing and cutting;
- the possibility of their use in the most severe conditions of surrounding medium;
- the lack of environment pollution.

The work is also underway toward the construction of torches with accumulators of hydrogen of non-permanent application with container from aluminium alloys based on alkali, alkali-earth and another metals.

19.4 Conclusions

At present method of hydrogen storage in the solids remains rather convenient, efficient and safe as before. Much work was devoted to the development of new hydride materials for hydrogen storage and to the investigation of its physical and chemical characteristics.

The metal-hydride accumulators can be used both for safe and compact hydrogen storage and for the solution of some other problems enumerated below:

- hydrogen purification – scavenging to ppm amounts;
- hydrogen separation – from the mixture containing from 1 to 90% non-hydrogen atoms and molecules;
- isotope separation - protium, deuterium and tritium;
- hydrogen compression – hydrogen is sorbed at low temperature and desorbed at higher temperature and creates high pressure;
- heat accumulation – the process is based on the use of heat effect (20–60 kJ/mol) of the hydrogenation/dehydrogenation reaction for heat
- absorption and evolution.

The designed torches, accumulators and compressors are feasible for large-scale industrial applications and for use in the wide areas of man's activities. These devices are needed to fill the gaps in instrumental provision those are bridged by another less suitable devices or are not closed at all at present.

References

1. Ven Mal HH, Bushow KHJ, Miedema AR (1974) Hydrogen absorption in $LaNi_5$ and related compounds: experimental observation and their explanation. J Less Common Met 35(1):65–76
2. Semenenko KN, Malyshev VP, Petrova LA, Bumasheva VV, Sarynin VK (1977) The interaction of $LaNi_5$ with hydrogen. Izv Akad Nauk SSSR Neorg Mater 13(11):2009–2013
3. Shinar J, Shaltiel D, Davidov D (1978) Hydrogen sorptoin properties of $La_{1-X}Ca_XNi_5$ and La $(Ni_{1-X}Cu_X)_5$ systems. J Less Common Met 60:209–219
4. Lartigue C, Percheron A, Acherd JC (1980) Thermodynamic and structural properties of $LaNi_{5-X}Mn_X$ compounds and their related hydrides. J Less Common Met 75(1):23–29
5. Patrikeev YuB, Levenskii YuV, Badovskii VV, Filyand YuM (1984) Thermodynamics and hydrogen diffusion in $LaCo_5H_X$ alloys. Izv Akad Nauk SSSR Neorg Mater 20(9):1503–1506
6. Percheron-Guegan A, Lartigue C, Achard JC (1985) Correlation between the structural properties, the stability and the hydrogen content of substituted lantanum-nickel ($LaNi_5$) compounds. J Less Common Met 114(2):287–309

7. Colinet C, Pasturel A (1987) Enthalpies of formation and hydrogenation of $La(Ni_{1-X}Co_X)_5$ compounds. J Less Common Met 134:109–22
8. Majer G, Kaess U, Bowman RC Jr (1998) Nuclear magnetic resonance studies of hydrogen diffusion in $LaNi_5H_6$ and $LaNi_{4,8}Sn_{0,2}H_{5,8}$. Phys Rev B 57(21):13599–13603
9. Matysina ZA, Zaginaichenko SYu, Schur DV, Pishuk VK (1997) Hydrogen in lanthan-nickel alloys-accumulators. In: 5th International conference hydrogen materials science and chemistry of metal hydrides, 2–8 Sept 1997. Katsiveli, Crimea, Ukraine, pp 62–63
10. Matysina ZA (1997) Isotherms of hydrogen solubility in hydrogen-storage lanthanum-nickel alloys. Phys Met Metallogr 84(5):495–500
11. Schur DV, Zaginaichenko SYu, Matysina ZA, Pishuk VK (2002) Hydrogen in lanthanum-nickel storage alloys. J Alloys Comp 330–332(1):70–75
12. Kadir K, Sakai T, Uehara I (1997) Synthesis and structure determination of a new series of hydrogen storage alloys: RMg_2Ni_9 (R = La, Ce, Pr, Nd, Sm and Gd). J Alloys Comp 257:115–121
13. Koimo T, Yoshida H, Kawashima F, Inada T, Sakai T, Yamamota M, Kanda M (2000) Hydrogen storage properties of new ternary system alloys: La_5MgNi_9, $La_5Mg_2Ni_{23}$, La_3MgNi_{14}. J Alloys Comp 311:L5–L7
14. Chen J, Kuriyama N, Takashita HT, Tanada H, Sakai T, Haruta M (2000) Hydrogen storage alloys with $PuNi_3$ type structure as metal hydride electrodes. Electrochem Solid State Lett 3 (6):249–252
15. Liao B, Lei YQ, Chen LX, Lu GL, Pan HG, Wang QD (2004) Effect of the La/Mg ratio on the structure and electrochemical properties of $La_XMg_{3-X}Ni_9$ (x = 1.6–2.2) hydrogen storage electrode alloys for nickel-metal hydride batteries. J Power Sources 129:358–367
16. Chen J, Takashita HT, Tanaka H, Kuriyama N, Sakai T, Uehamra I, Haruta M (2004) Hydriding properties of $LaNi_3$ and $CaNi_3$ and their substitutes with $PuNi_3$–type structure. J Alloys Comp 302:304–313
17. Matysina ZA, Chumak VA (2001) Deformational hysteresis and elastic compliance of crystals with H4 structure near Curie point. Ukr Fiz Zhurn 46(9):957–959
18. Matysina ZA, Zaginaichenko SY, Schur DV (2005) Orders of different type in crystals and phase transformations in carbon materials. Nauka i obrazovanie, Dnepropetrovsk, p 522, in Russian

Chapter 20
Electric Field Gradients at Hydrogen and Metal Sites in Light Metal Hydrides

V.P. Tarasov, D.E. Izotov, and Yu.M. Shul'ga

Abstract The results of measuring the quadrupole coupling constants and asymmetry parameters at deuterium and metal sites in amorphous BeD_2, crystalline α-MgD_2, and crystalline α-AlD_3 by solid state NMR, as well as the results of ab initio Hartree–Fock calculations of the EFG tensors at hydrogen and metal positions for some (Be_nH_m), (Mg_nH_m), and (Al_nH_m) clusters, are reported. It have been found that the magnitude of the EFG at the hydrogen sites is more than an order of magnitude larger than at the metal sites. The sign, amplitude, and asymmetry parameter at hydrogen depend on the M–H–M angle.

Keywords Be · Mg · Al deuterides · Quadrupole coupling constant · Electric field gradient · Hartree–Fock calculation

20.1 Introduction

Great interest in simple p-metal hydride systems has been stimulated by both the fundamental aspects and various applications. Such kinds of metal hydrides, e.g., beryllium, magnesium, and aluminum hydrides, which all have high weight percentage of hydrogen, are promising candidates for hydrogen storage. If they are subjected to high pressures, they become metallic and may be candidates for high-temperature superconductivity [1]. Solid state hydrides of light p-metals with

---

V.P. Tarasov (✉)
Institute of General and Inorganic Chemistry, Russian Academy of Sciences, Leninskii pr. 31, 119991 Moscow, Russia
e-mail: tarasov@igic.ras.ru

D.E. Izotov,
Chemical Department, University of Pacific, Stockton, CA, USA

Yu.M. Shul'ga
Institute of Problems of Chemical Physics, Russian Academy of Sciences, 142432 Chernogolovka, Russia

S.Yu. Zaginaichenko et al. (eds.), *Carbon Nanomaterials in Clean Energy Hydrogen Systems - II*, NATO Science for Peace and Security Series C: Environmental Security 2, DOI 10.1007/978-94-007-0899-0_20, © Springer Science+Business Media B.V. 2011

high contents of the hydrogen isotopes (H, D, T) are an engaging compounds/materials in the fields of nuclear fusion, hydrogen storage, and solid propellants. For example, beryllium dideuteride, BeD_2, is considered a material for target shell (ablator) for thermonuclear reactions [2].

Binary hydrides of light metals – BeH_2, MgH_2, and AlH_3 – have a three-dimensional polymeric structure in which chemical bonding is through bridging hydrogen atoms. The metal–hydrogen bond in these hydrides is characterized by electron density transfer from the metal to hydrogen. Several theoretical investigations on charge density distribution in MgH_2 [3–5], BeH_2 [6, 7] and AlH_3 [8–10] have been reported. The bonding nature and ionic state of MgH_2 have been studied experimentally by using X-ray synchrotron radiation [9]. The specific features of the charge distribution at the atoms of these hydrides manifest themselves in the values of the electric field gradients (EFGs) at the hydrogen and metal positions. Here, we report the results of measuring the quadrupole coupling constants (χ_Q) and EFG tensor asymmetry parameters (η) at the deuterium and metal sites in amorphous BeD_2, crystalline α-MgD_2, and crystalline α-AlD_3 by solid state NMR. The results of ab initio Hartree–Fock calculations of EFG tensors at hydrogen and metal sites for some (Be_nH_m), (Mg_nH_m), and (Al_nH_m) clusters are also presented.

20.2 Experimental

Amorphous beryllium dideuteride BeD_2 was produced as described in [11]. The purity of the sample was 98% and the protium/deuterium ratio was about 0.01 as determined by mass spectroscopy. Aluminum trideuteride AlD_3 was synthesized as described in [12]. The purity of the sample was 98.6% and the protium/deuterium ratio was 0.045 (Fig. 20.1).

The magnesium dideuteride, MgD_2, was prepared by deuterium activation treatment of the metal magnesium powder (purity 99%). The sample was homogenized in deuterium at 600 K and pressure 1 MPa. The X-ray diffraction pattern shows that the material is composed of two main phases, tetragonal α-MgD_2 (90%) and hexagonal Mg metal (10%). The protium/deuterium ratio was about 0.9. Mass spectra of gases produced from the samples during the heating in vacuum were recorded using an MI 1201B mass spectrometer at electron ionization energy of 70 eV. IR spectra were taken on an FTIR spectrometer (Bruker IFS113s) at room temperature (Figs. 20.1 and 20.2). ^{2}H, ^{9}Be, ^{27}Al NMR spectra were recorded in a field of 7.04 T using a solenoid RF-coil 10 mm in diameter. ^{1}H, ^{25}Mg NMR spectra were taken in a field of 14.1 T using a solenoid RF-coil 5 mm in diameter (Fig. 20.3). A spin system was excited using a single-pulse sequence with a pulse width of 4 μs (^{2}H, ^{9}Be) and 1 μs (^{25}Mg, ^{27}Al) following by FT of the FID or a two-pulse sequence $(\theta_x$–τ–$\theta_y)$-echo with $\theta = 2$–3 μs, $\tau = 14$–17 μs following by FT of the echo. The same samples of AlD_3 and BeD_2 were earlier studied by NMR [13, 14].

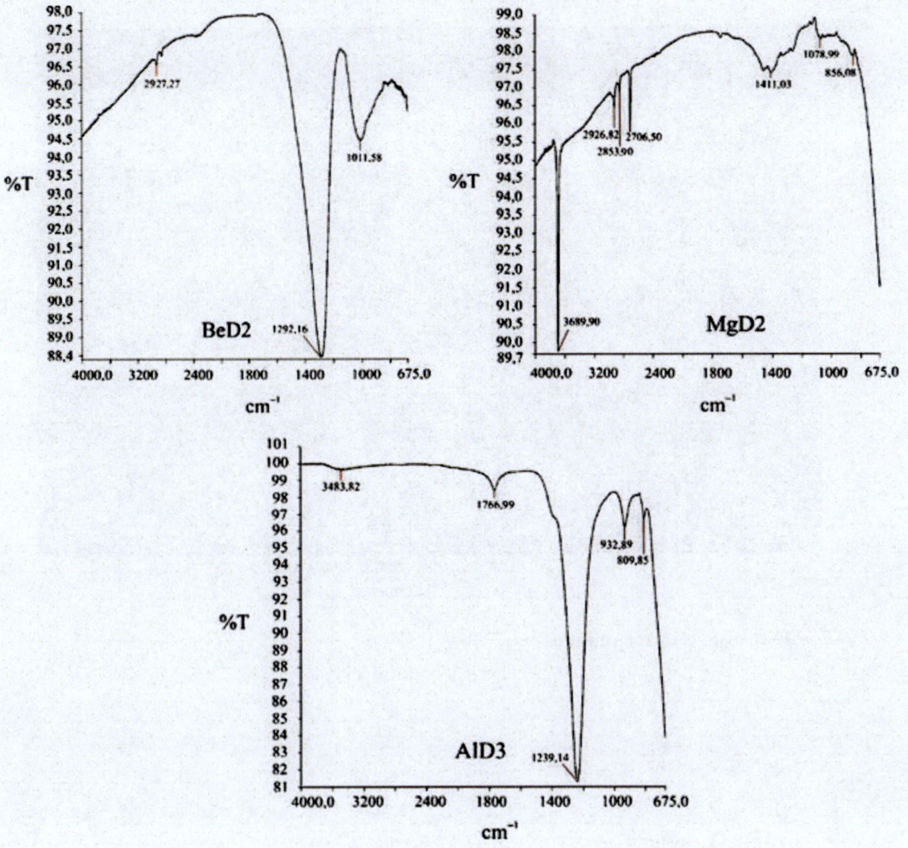

Fig. 20.1 IR spectra of the Be, Mg, and Al deuterides

20.3 Computations

We carried out *ab initio* Hartree–Fock calculations of the EFG tensors at the metal and hydrogen sites for some Be_nH_m, Mg_nH_m, and Al_nH_m clusters. Calculations were performed with the GAMESS program package [15]. The 6-311G** basis set with polarization functions on metal and hydrogen atoms was used.

20.4 Theory

The nuclear quadrupole coupling constant (QCC) χ_Q is the energy of interaction of the electric quadrupole moment (Q) of the atomic nucleus and the EFG at the site of the nucleus. Components of the QCC tensor are related to those of the EFG q_{ij} by the equation

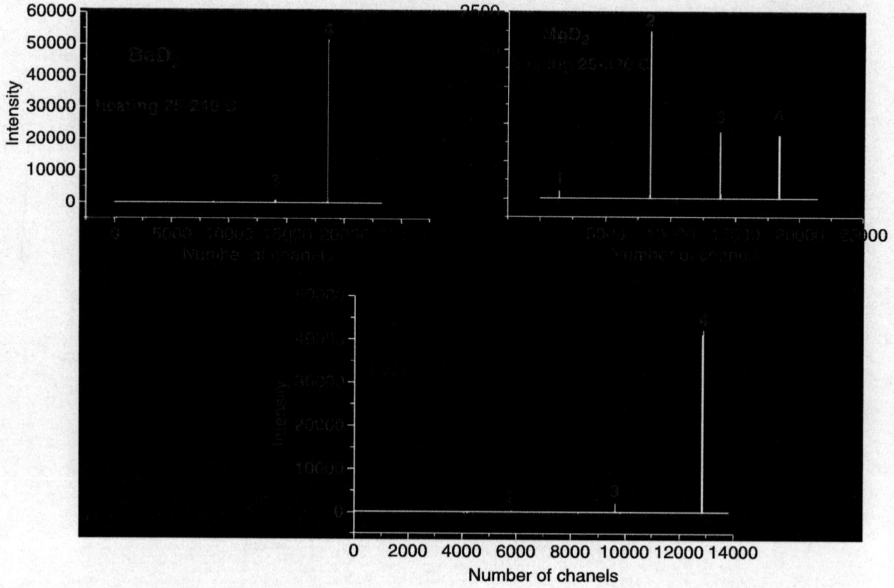

Fig. 20.2 Mass spectra of the samples

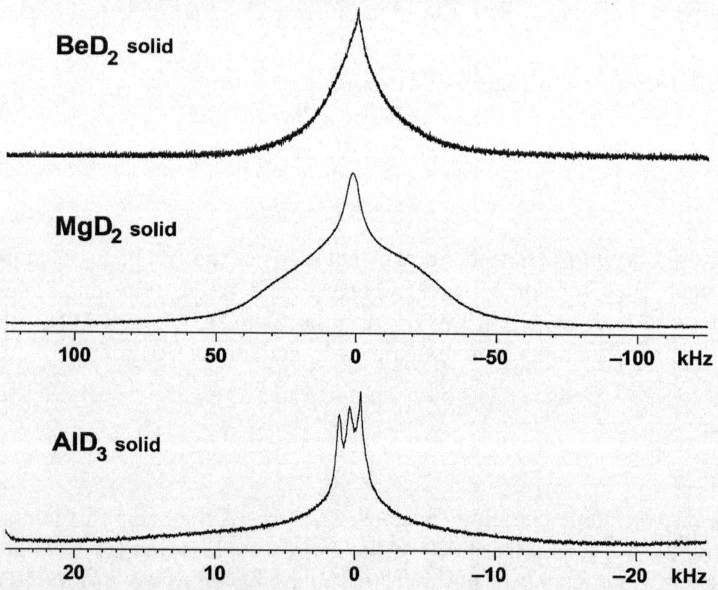

Fig. 20.3 ^{1}H NMR of residual protium in the samples

$$\chi_{ij} = (eQ/h)q_{ij}, \tag{20.1}$$

where subscripts i, j = x,y,z are principal axes of the tensor. The asymmetry parameter is

$$\eta = \left|(q_{yy} - q_{xx})/q_{zz}\right|$$

measures the departure of the EFG from cylindrical symmetry.

The EFG q_{zz} at a given nucleus is the sum of the electronic (q^{el}) and nuclear (q^{nucl}) contributions. In atomic units q_{zz} is expressed as

$$q_{zz} = (1 - \gamma_\infty)\sum\nolimits_n \xi_n(3\cos^2\theta - 1)R^{-3} - \left\{\varphi * \left|\sum\nolimits_i (3z_i^2 - r_i^2)r_i^{-5}\right|\varphi\right\}, \tag{20.2}$$

where γ_∞ is the Sternheimer antishielding factor (for deuterium $\gamma_\infty = 0$), ξ_n is the nuclear charge, R_n is the interatomic distance, θ_n is the angle between R_n and the EFG direction, i.e., the z axis, r_i is the distance between the nucleus and electron, z_i is the z coordinate of the ith electron radius vector, and φ is the wave function of the ground state. In the first term q^{nucl}, the summation is over nuclei, in the second term q^{el}, the summation is over electrons. The electronic contribution is negative and slightly smaller in magnitude than the nuclear contribution. The linear relation between the electronic and nuclear parts of the deuterium EFG has been found by Huber [16]:

$$q^{el} \approx -0.87q^{nucl} \tag{20.3}$$

20.5 Results

BeD_2. Neutron diffraction study showed that the structural block of amorphous $(BeD_2)_n$ is a tetrahedron with a Be–D distance of 1.43 Å and an average angle ~110° [11] or 135° [17] between tetrahedra. According to [18], crystalline beryllium dihydride $(BeH_2)_n$ has a body-centered orthorhombic lattice with space group I_{bam} and unit cell parameters a = 9.082 Å, b = 4.160 Å, c = 7.707 Å, Z = 12. The beryllium and hydrogen atoms occupy positions of two types: four Be(1) atoms are located in the 4a positions (in Wyckoff notation), and eight Be(2) are in 8j; sixteen H(1) are in the 16 k, and eight H(2) are in 8j positions. The Be(1)–H(1) and Be(2)–H(2) distances are 1.38 and 1.44 Å, respectively (Fig. 20.4).

The ^{9}Be NMR spectrum of amorphous $(BeD_2)_n$ shows a single featureless line with a width of 3.4 ± 0.1 kHz at –8 ± 2 ppm (Fig. 20.5a). In the temperature range 130–400 K, the ^{9}Be NMR line shape and chemical shift remain unaltered, which points to the "rigidity" of the lattice on the NMR scale. The presence of the featureless line is evidence that the shifts for the Be(1) and Be(2) positions are close to each other and the second-order quadrupole splitting for ^{9}Be is smaller than

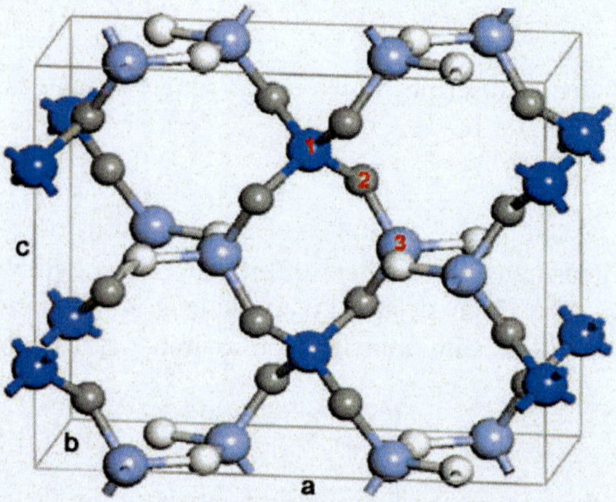

Fig. 20.4 Unit cell for body-centered orthorhombic BeH_2. Both Be and H have two kinds of atomic occupations

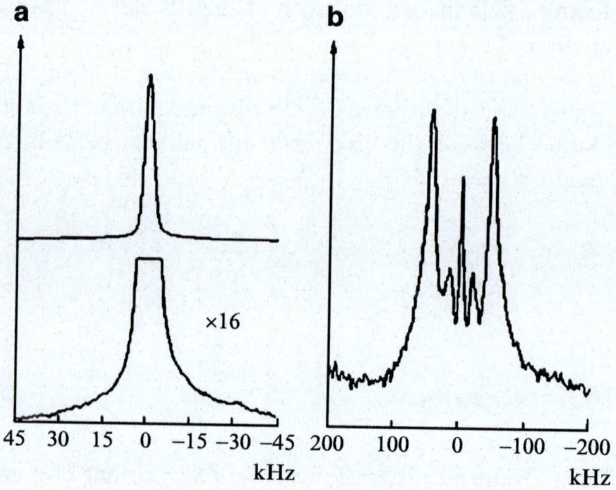

Fig. 20.5 NMR spectra of amorphous beryllium dideuteride $(BeD_2)_n$. **(a)** ^{9}Be NMR (42.16 MHz) at 130 K. The lower spectrum is magnified 16 times. **(b)** ^{2}H NMR (46.07 MHz) at 291 K

the linewidth. This result means that the BeD_4 clusters in amorphous $(BeD_2)_n$ have a virtually regular tetrahedral structure.

The ^{2}H NMR line of an amorphous $(BeD_2)_n$ powder at 130, 295, and 400 K is composite and consists of a narrow central signal and two pairs of lines caused by quadrupole splitting (Pake doublets) (Fig. 20.5b). Since the ^{2}H NMR spectral pattern weakly depends on temperature (i.e., the lattice is rigid), we interpret this spectrum as a superposition of two Pake doublets that arise from two types of deuterium positions, D(1) and D(2). The presence of the central signal with a width of less than 0.5 kHz can be caused by the existence of mobile D_2 or HD molecules that appear at the the grain surface upon partial decomposition of beryllium dideuteride [14]. Inasmuch as the partial decomposition $BeD_2 = Be + 1/2D_2$ leads to the formation of metallic Be, the ^{9}Be NMR spectrum should show a weak signal that arise from metallic beryllium. Indeed, the ^{9}Be NMR spectrum magnified 16 times shows broad weak wings at ~ ±14 kHz on both sides of the basic signal (Fig. 20.5a), which we assign to the metallic beryllium phase [19]. The assignment is based on comparison of this spectrum with the ^{9}Be NMR spectrum of metallic beryllium, which is a typical powder pattern caused by first-order quadrupole interaction with the (±1/2 ↔ ±3/2) satellites at ±14 kHz and the central transition (±1/2) at –10(2) ppm [19] (Fig. 20.5).

Taking into account the structural data on the populations of the deuterium states (2:1), we assigned the stronger doublet to D(1) in the 16 k positions and the weaker doublet, to D(2) in the 8j positions. The narrow central signal was assigned to high-mobility D_2 molecules.

The ^{2}H NMR spectrum of a powder on the frequency scale is characterized by three singularities:

$$\begin{aligned} \nu_{xx} &= \pm(1-\eta)\,0.375\,e^2qQ/h;\ \nu_{yy} = \pm(1+\eta)\ e^2qQ/h; \\ \nu_{zz} &= \pm 3/4\ e^2qQ/h \end{aligned} \tag{20.4}$$

From consideration of the singularities of the experimental lineshape, we found that, for D(1) position, $\eta = 0.18$ and $\chi_Q = 150$ kHz, while, for D(2) position, $\eta = 0{,}25$ and $\chi_Q = 65$ kHz. It is worth noting that such a significant difference in the QCC and η between D(1) and D(2) positions in amorphous $(BeD_2)_n$ is unexpected. No experimental ^{2}H QCC values for deuterium bonded to beryllium are available, and the calculated ^{2}H QCC for the BeD_2 molecule is 69.5 kHz [20].

MgD_2. Magnesium hydride is tetragonal (space group $P4_2/mnm$, rutile type). The structure of α-MgD_2 is built of magnesium-centered hydrogen octahedra that share edges along one direction and corners in the other two directions. The octahedra in α-MgD_2 are slightly distorted (distances Mg–D = 1.94 and 1.95 Å; angles D–Mg–D = 90° and 180°). The cell parameters are $a = 4.5180$ Å and $c = 3.021$ Å for MgH_2 and $a = 4.5010$ Å and $c = 3.01$ Å for MgD_2 [21]. Figure 20.6 shows the crystal structure of α-MgH_2.

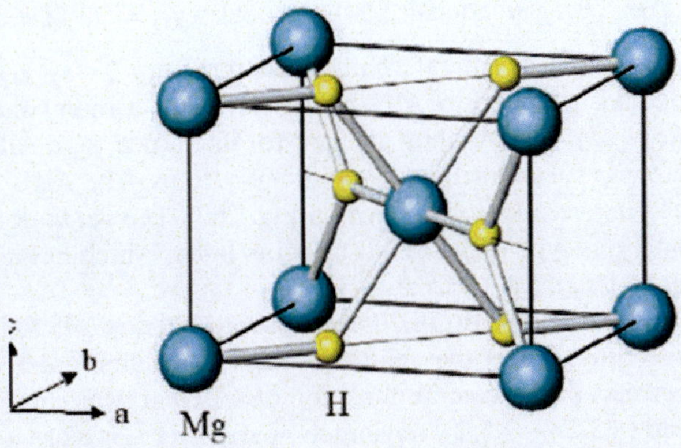

Fig. 20.6 Arrangement of H and Mg atoms in the tetragonal cell of the rutile type α-MgH_2. The large circles are Mg atoms, and small circles represent H atoms

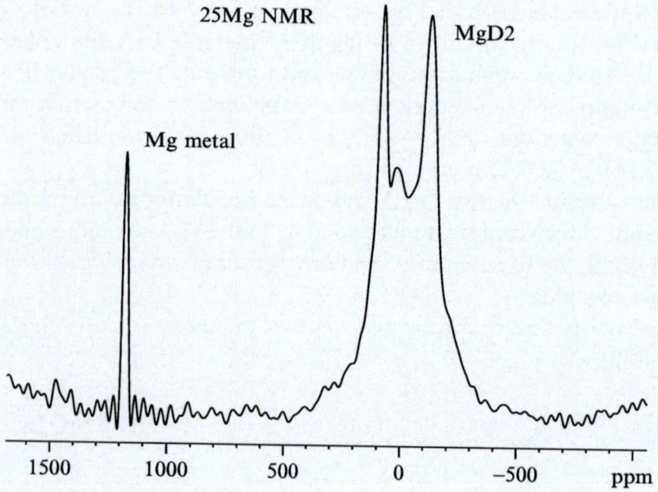

Fig. 20.7 ^{25}Mg NMR (36.4 MHz) in the MgD_2 sample at room temperature

The static ^{25}Mg NMR (36.4 MHz) spectrum of crystalline α-MgD_2 at room temperature is shown in Fig. 20.7. The spectrum shows two resonances. One of them is due to the Mg metal with a Knight shift of 1,179 ppm and frequency separation $\nu_Q = 47$ kHz between the inner ($\pm1/2 \leftrightarrow \pm 3/2$) satellites. The values of the Knight shift and quadrupole frequency are well consistent with those reported earlier for metallic magnesium [22]. The other resonance at –27 ppm is due to MgD_2 and shows a second-order quadrupole interaction. The lineshape analysis the gives QCC $\chi_Q(^{25}\text{Mg}) = 2.9$ MHz and $\eta = 0.1$.

The static ^{2}H NMR (46.06 MHz) spectrum of the MgD_2 sample is shown in Fig. 20.8. The spread around the central frequency should be very large, ranging between +40 and −40 kHz. The high content of protium in this MgD_2 sample is due to a strong dipole broadening of the deuterium resonance. This spectrum shows three singularities ("horns") on either side of the central frequency ν_0. The distances between the horns are as follows: $\Delta\nu_{xx} = 11$ kHz; $\Delta\nu_{yy} = 43$ kHz and $\Delta\nu_{zz} = 55$ kHz. Using relations (20.4), we obtain $\chi_Q = 73$ kHz and $\eta = 0.58$.

AlD_3. The structural characterization from X-ray and neutron diffraction study demonstrates that α-AlH_3 and α-AlD_3 has a rhombohedral lattice of space group *R3c* [23]. The building element for the lattice is an AlH_6 octahedron where the Al atom is surrounded by six hydrogen atoms. The AlH_6 octahedra are connected by sharing vertices as is shown in Fig. 20.9. The network of these octahedra produces only one type of Al–H–Al bridging bond, which has a bond angle of 141.2° and a bond length of 1.715 Å.

The ^{27}Al NMR (78,205 MHz) of $(AlD_3)_n$ at 295 K is shown in Fig. 20.10. This spectrum shows the splittings due to of a first-order quadrupole interaction and consists of the central transition ($\pm1/2$) and two pairs of satellites ($\pm1/2 \leftrightarrow \pm3/2$) and ($\pm3/2 \leftrightarrow \pm5/2$). Asymmetry parameter η is close to zero since the Al atoms are

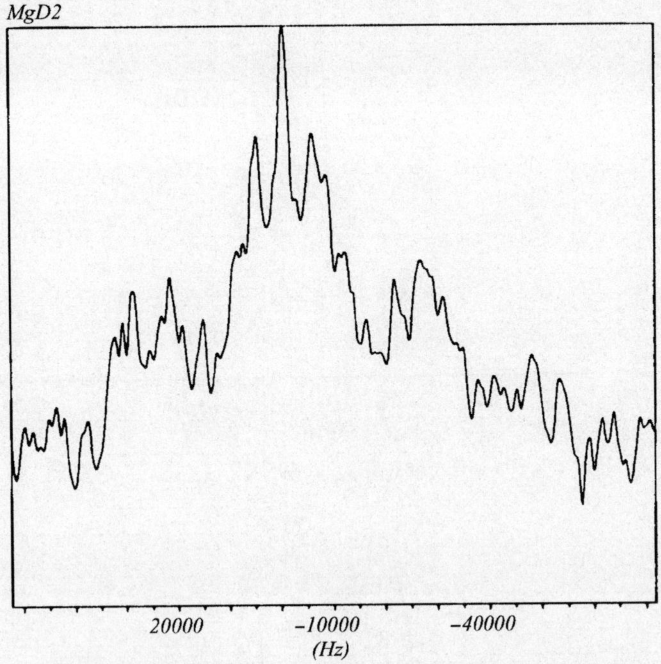

Fig. 20.8 ^{2}H NMR (46.04 MHz) spectrum in MgD_2 sample at room temperature

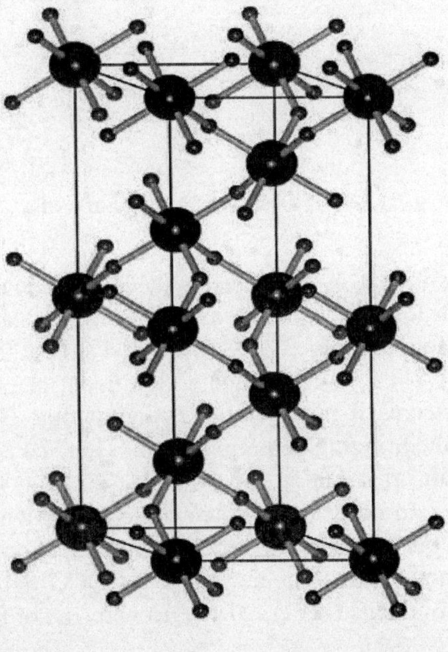

Fig. 20.9 The structure of α-AlH_3 with the trigonal space group *R*-3*c*. The large and small spheres correspond to Al atoms and H atoms, respectively. This structure consists of AlH_6 octahedra linked by Al–H–Al bridges

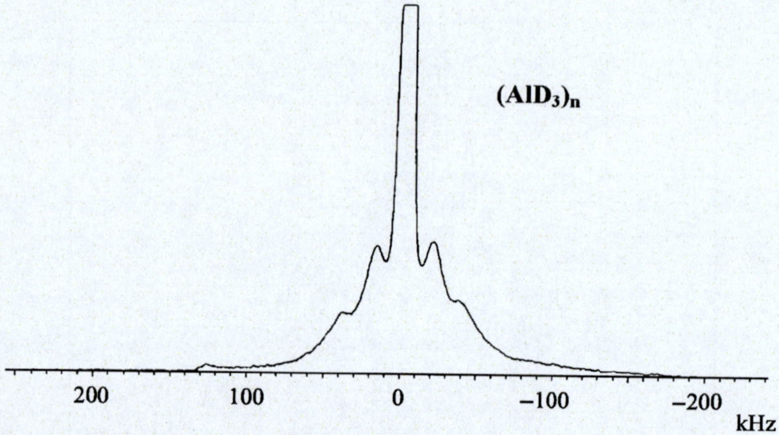

Fig. 20.10 ^{27}Al NMR (78.205 MHz) of $(AlD_3)_n$ at 295 K

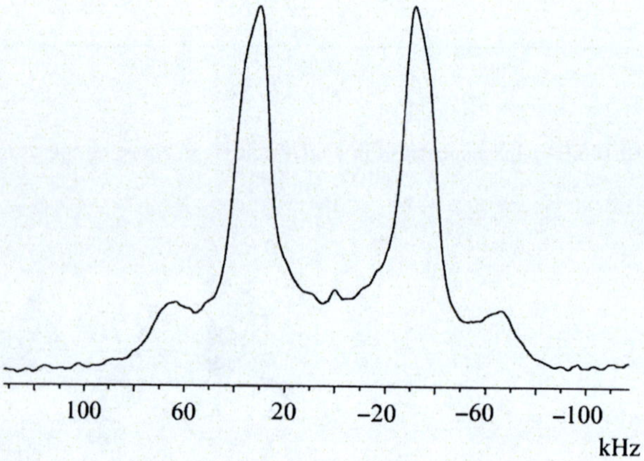

Fig. 20.11 ^{2}H NMR (46.06 MHz) of $(AlD_3)_n$ at 295 K

situated in positions with C_3 symmetry. The value of the minimal quadrupole frequency ν_Q as measured from the satellite structure is equal to 39.6 $\pm$ 2.0 kHz and, therefore, QCC $\chi_Q = 264$ kHz at 295 K. At 120 K, $\nu_Q = 34.7 \pm 2.0$ kHz and $\chi_Q = 231$ kHz. Such a significant change in QCC χ_Q from 264 to 231 kHz with decreasing temperature is anomalous. The doublet contour of the ^{2}H NMR signal of the bridging deuterium atoms (Fig. 20.11) is characterized by $\chi_Q(^{2}H) = 89 \pm 2$ kHz and $\eta = 0.08$ at 295 K (Table 20.1).

In order to compare these values with the calculated ones, it is convenient to convert them into atomic units. The conversion factors of EFG values from atomic units into frequency units of QCC depend on the values of nuclear quadrupole moments Q (Fig. 20.11). Hence, 1au of EFG =672 kHz for ^{2}H, 12.43 MHz for ^{9}Be,

Table 20.1 Experimental values of the QCCs (kHz) and η at the deuterium and metal sites

Compound	Deuterium site	Metal site
BeD_2 amorphous	D(1): 150 ± 2	Be(1): 0
	$\eta = 0.18$	
	D(2): 65 ± 2	Be(2): 0
	$\eta = 0.25$	
MgD_2 crystalline	73 ± 5	2,900 ± 50
	$\eta = 0.58$	$\eta = 0.1$
AlD_3 crystalline	89 ± 2	264 ± 2
	$\eta = 0.08$	$\eta = 0$

Table 20.2 Eigenvalues of the EFG ($\times 10^{-3}$ au) at the metal and hydrogen sites in some clusters

Atom	Clusters	$q_{\alpha\alpha}$
Be(1)	$Be_{11}H_{20}$	−7.03
Be(2)	$Be_{10}H_{25}$	−8.25
H(1)	Be_7H_{14}	−124.5
H(2)	$Be_{11}H_{22}$	−122.2
Mg	$Mg_{11}H_{22}$	56
H	$Mg_{12}H_{25}$	−72
Al	$Al_{13}H_{36}$	2.7
H	Al_9H_{19}	−64

46.85 MHz for ^{25}Mg, and 35.06 MHz for ^{27}Al. Then, we obtain the following EFGs in atomic units: q(D1) = 0.223 and q(D2) = 0.097 for BeD_2; q(D) = 0.109 and q(Mg) = 0.062 for MgD_2; q(D) = 0.132 and q(Al) = 0.008 for AlD_3.

20.6 Calculations of EFGs

To elucidate the nature of the EFG at the hydrogen and metal sites we carried out ab initio Hartree–Fock calculation of the EFG tensor for some (Be_nH_m), Mg_nH_n, and Al_nH_m clusters. Clusters were chosen to obey the following criteria. The positions of the atoms in the cluster should coincide with their positions in the crystal lattice. The nucleus at which the EFG is calculated is located in the center of the cluster. The site symmetry of the atom is the same as that of the corresponding nucleus in the crystal. The calculated principal components of the EFG tensor at hydrogen and metal sites are given in Table 20.2.

It follows from these calculations that EFGs at both sites are positive for beryllium hydride. The EFG values at the H(1) and H(2) sites are an order of magnitude higher than the EFG values at the Be(1) and Be(2) sites, which is presumably associated with the tetrahedral environment of the beryllium atoms. For the Be(1) position, the calculated QCC is χ_Q = 180 kHz, η = 0.07, and for the Be(2) position, χ_Q = 135 kHz, η = 0.43. It should be noted that the calculated principal components of the EFG at ^{9}Be in the clusters under consideration can serve only as the upper estimate for the EFG tensor in the crystal. Amorphization is considered to be the continuous disruptions of topological order and is accompanied by distribution of structural parameters (distances and angles). Indeed, the EFG tensor at the Be sites turns out to be very sensitive to the distances to neighboring atoms. The calculation, for example, for BeH_4^{2-} (symmetry D_2) with the Be atom shifted by 0.05 Å (~4% of the Be–H bond length) along the c axis gives $q_{\alpha\alpha} = 0.0059$; $q_{\beta\beta} = 0.0091$, and $q_{\gamma\gamma} = -0.0150$ au, which differs considerably from the EFG of the initial cluster: −0.0026; −0.0028, and 0.0054 au, respectively. As distinct from the Be atoms, displacement of the hydrogen atom only slightly changes the values of the EFG tensor at the hydrogen atoms.

The theoretical ^{2}H NMR spectra reconstructed using the calculated values of the EFG tensors at the H(1) and H(2) sites are slightly different. The calculations show that the z axis of the EFG tensor at hydrogen is roughly aligned with the Be–Be direction and x axis is virtually coincident with the bisector of the Be–H–Be angle. Figure 20.12 shows the angular dependences of the principal components of the EFG tensor and asymmetry parameter η calculated for the $Be_2H_7^{3-}$ cluster. Indeed, with a change in the Be–H–Be angle, the q_{zz} component of the EFG tensor changes its sign at ~105°. This result clearly shows that the sign of the EFG at hydrogen is determined by the angle at the bridging hydrogen atom and accounts for the negative EFG in B_2H_6 molecule (B–H–B angle = 85°) [24]. As distinct from the negative sign of the EFG at bridging hydrogen atom in the diborane molecule B_2H_6, the calculated EFG at hydrogen in crystalline $(BeH_2)_n$ turns out to be positive since the BeH(1)Be and BeH(2)Be angles in it are 127º and 130°, respectively [18]. The average angle between tetrahedra in amorphous BeD_2 is 110° [11] or 135° [17]. Therefore, the corresponding EFGs at the hydrogen atoms are positive.

The ionicity of the metal–hydrogen bonds is mainly characterized by charge transfer from the metal Be 2s, Mg 3s, and Al 3s to H 1s atomic orbitals, while their

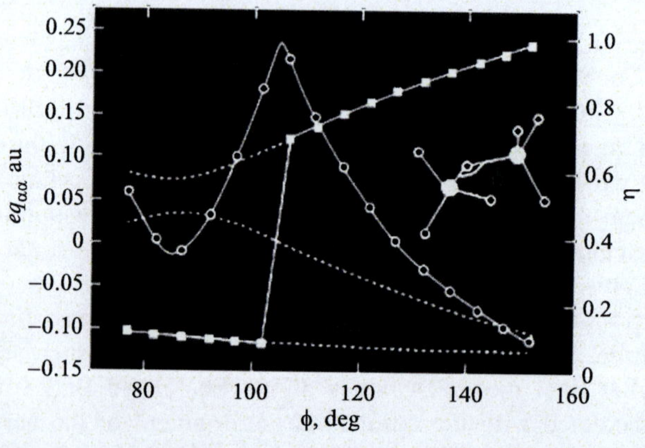

Fig. 20.12 Eigenvalues of the EFG tensor at the central hydrogen atom in $Be_2H_7^{3-}$ (symmetry *m*) at various BeHBe angles. The values of the EFG $q_{zz} = \max|q_{ii}|$ and the asymmetry parameter η are shown with black squares and open circles, respectively. The cluster symmetry and bond lengths remain unaltered on varying the angle

Table 20.3 Estimation of bond ionicity and the s contribution to metal orbital

Compounds	Ionicity of bond (%)	s contribution
BeD_2	33	0.86
	71	0.68
MgD_2	67	0.33
AlD_3	60	0.98

covalence is dominated by the hybridization of H 1 s and metal Be 2p, Mg 3p, or Al 3p states. The EFGs at hydrogen can be interpreted in terms of the ionic character (**i**) of the metal–hydrogen bond:

$$i = 1 - \{q_{M-D}{}^{el}/q_{HD}{}^{el}\},$$

where $q_{M-D}{}^{el}$ is an electronic contribution to EFG at hydrogen for metal-D bond, and $q_{HD}{}^{el}$ is electronic contribution to EFG for HD molecule. The QCC $^2H = 225$ kHz ($q = 0.335$ au) for HD molecule [24]. The q^{el} terms are obtained from relations (2) and (3). Now, using the data for i and q_{met} it is possible to give a crude estimation of the s contribution to a metal hybrid orbital. According to the Towns–Delay treatment, the EFGs q_{met} is given by

$$q_{met} \approx (1 - s)(1 - i)q_{at},$$

where q_{at} is the EFG at the atom. The q_{at} are calculated from the spin–orbital splitting in atoms and have the following values in atomic units: $q_{at}(Be) = 0.139$ au, $q_{at}(Mg) = 0.616$ au, $q_{at}(Al) = 1.016$ au. The i and s values are given in Table 20.3.

20.7 Conclusions

The results of measuring the quadrupole coupling constants and asymmetry parameters at deuterium and metal sites in amorphous BeD_2, crystalline α-MgD_2, and crystalline α-AlD_3 by solid state NMR, as well as the results of ab initio Hartree—Fock calculations of the EFG tensors at hydrogen and metal positions for some (Be_nH_m), (Mg_nH_m), and (Al_nH_m) clusters, are reported. It has been found that the EFG at the hydrogen sites is more than an order of magnitude larger than at metal sites. The sign, amplitude, and asymmetry parameter at hydrogen depend on the M–H–M angle. The experimental and calculated EFGs at hydrogen positions depend on the metal to which the deuteron is bonded. The EFGs at hydrogen decrease with increasing the metal–hydrogen bond length.

Acknowledgments This work was supported by the Russian Foundation for Basic Research, project no.10-03-00055a

References

1. Islam AKMA, Ali MM, Ali MS (2010) AlH_3 between 65–110 GPa: implications of electronic band and phonon structures. Physika C 470:403–406
2. Bel'kov SA, Dolgoleva GV, Kochemasov GG, Mitrofanov EI (2002) Application of beryllium deuteride as shell matter of laser X-ray targets. Quantum Electron 32(1):27 (in Russian)
3. Pfrommer B, Elsasser C, Fahnle M (1994) Possibility of Li–Mg and Al–Mg hydrides being metallic. Phys Rev B 50:5089
4. Yu R, Lam PK (1988) Electronic and structural properties of MgH_2. Phys Rev B 37:8730
5. Baraille I, Pouchan C, Causa M, Pisani C (1994) An *ab initio* Hartree–Fock study of electronic and structural properties of MgH_2. Chem Phys 179:39
6. Vajeeston P (2004) Theoretical modeling of hydrides. Dissertation Phd, N. 309, Department of Physics, University of Oslo, Norway
7. Wang B-T, Zhang P, Shi H et al (2010) Mechanical and chemical bonding properties of ground state BeH_2. Eur Phys J B 74:303–308
8. Kato H, Yamaguchi K, Yonezawa T, Fukui K (1965) The electronic structure of some hydrides, halides, and alkyl compounds of boron and aluminum. I. Monomers and ions. Bull Chem Soc Jpn 38:2144
9. Ke X, Kuwabara A, Tanaka I (2005) Cubic and orthorhombic structures of aluminum hydride AlH_3 predicted by a first-principal study. Phys Rev B 71:184107-1-8
10. Noritake T, Towata S, Aoki M et al (2003) Charge density measurement in MgH_2 by synchrotron X-ray diffraction. J Alloy Comp 356–357:84–86
11. Senin MD, Akhachinskii VV, Markushkin YuE et al (1993) Preparation, structure, and properties of beryllium hydryde. Neorg Mater 29(12):1582
12. Brower FM, Matzek NE, Reigler PF et al (1976) Preparation and properties aluminum hydride. J Am Chem Soc 98(9):2450–2453
13. Tarasov VP, Muravlev YuB (2004) ^{2}H and ^{27}Al nuclear magnetic resonance of aluminum trihydride and trideuderide. Khim Fiz 23(4):3–15 (in Russian)
14. Tarasov VP, Muravlev YuB, Izotov DE (2005) Electric field gradients in beryllium hydride. Dokl Phys Chem 404:190–194
15. Schmidt M, Baldridge K, Boatz J et al (1993) General atomic and molecular electronic structure system. J Comput Chem 14:1347–1363
16. Huber H (1985) Deuterium quadrupole couplings. A theoretical investigation. J Chem Phys 83:4591
17. Sampath S, Lantzky KM, Benmore CJ et al (2003) Structural quantum isotope effects in amorphous beryllium hydride. J Chem Phys 119:12499
18. Smith GS, Johnson QC, Smith DK et al (1988) The crystal and molecular structure of beryllium hydride. Solid State Commun 67(5):491–494
19. Tarasov VP, Muravlev YuB, Kirakosyan GA (2008) Angular dependence of the Knight shift, electric field gradient, and spin–lattice relaxation time of ^{9}Be NMR in beryllium metal. Phys Solid State 50(6):1009–1013
20. Mokarram M, Rangle JL (1973) On the relationship between deuteron quadrupole coupling constants and force constants in diatomic hydrides. J Chem Phys 59:2770–2771
21. Bortz M, Bertheville B, Bottner G, Yvon K (1999) Structure of the high pressure phase γ-MgH_2 by neutron powder diffraction. J Alloy Comp 287(1–2):L4–L6
22. Bastow TJ (1991) Temperature dependence of the nuclear quadrupole coupling and relaxation time of ^{25}Mg in Mg metal. J Phys Condens Matter 3(6):753
23. Turley JW, Rinn HW (1969) The crystal structure of aluminum hydride. Inorg Chem 8:18–22
24. Snyder LC (1978) Deuteron quadrupole coupling in molecules. J Chem Phys 68(1):291–294

Chapter 21
Polymer Membranes for Fuel Cells: Achievements and Problems

S.S. Ivanchev and S.V. Myakin

Abstract The current state in the field of synthesis, structural modification and implementation of polymer membranes for fuel cells is analyzed. Synthetic methods, physicochemical characteristics and specific features of their composition, microphase separation and the structure of water ionic channels are considered for different types of proton-conducting materials together with the approaches to enhancement of their technical performances. The main focus is addressed to Nafion and other similar perfluorinated proton conducting polymer membranes as currently leading materials in respect of both R&D progress and commercial implementation in fuel cells of different purpose. The recent advances in submarine and automotive applications of Nafion based fuel cells are discussed.

Keywords Fuel cells classification · Polymer electrolyte copolymerization · Hydrogen exploration · Fluorinated proton conducting membrane · Nafion membrane

21.1 Introduction. Classification of Fuel Cells

Energy generation and supply is the technical background of the modern society. The progress in the relating branches determines the level of economic and cultural development of the mankind facing a potential threat of exhausting the conventional energy reserves (particularly oil and gas) and necessity to search for new efficient and environment friendly sources.

In view of these problems the study and implementation of the concept based on hydrogen exploration is considered as a promising and important alternative approach to the development of efficient and sustainable energy generating systems [1].

S.S. Ivanchev (✉) and S.V. Myakin
St-Petersburg Department of the Boreskov Institute of Catalysis,
Siberian Branch of the Russian Academy of Sciences,
14, prospect Dobrolubova, St-Petersburg 197198, Russia
e-mail: ivanchev@SM2270.spb.edu

S.Yu. Zaginaichenko et al. (eds.), *Carbon Nanomaterials in Clean Energy Hydrogen Systems - II*, NATO Science for Peace and Security Series C: Environmental Security 2,
DOI 10.1007/978-94-007-0899-0_21, © Springer Science+Business Media B.V. 2011

A significant impact into the promotion of studies in this field was contributed by concluding the agreement on cooperation between the European Community and USA on the strategy for the development and implementation of hydrogen energetic and fuel cells concept in 2003 [2].

The present review presents the continuation and extension of our recent publication [3] relating to a brief analysis of the main steps in the development and achievements in the field of polymer membrane based fuel cells for the recent 5 years.

21.2 Main Types of Fuel Cells and Prospects for Their Enhancement

Fuel cell (FC) is a power generation device based on a direct conversion of chemical reaction energy into electric power provided by a permanent supply of reagents to electrodes and removal of the reaction products.

Fuel cells and power supply systems on their basis provide the following advantages over heat generators:

- significantly (by about one order) reduced hazardous emissions;
- high efficiency (up to 90%) of fuel conversion into electric power;
- noiseless operation;
- possibility for using various types of fuel.

The most commonly used fuel for FC is hydrogen obtained from various sources in combination with oxygen or air as an oxidizer. Fuel cells are classified according to different approach with the most widely applied one based on the type and structure of applied electrolytes. The main performances of fuel cells involving different electrolytes are summarized in Table 21.1.

The functioning of all the considered FC is provided by oxygen or air as oxidizers. In alkaline FC air should be thoroughly cleaned from CO_2 to prevent electrolyte carbonization. In systems based on polymer electrolytes, pure oxygen is used in order to enhance the FC efficiency. It is difficult to prevent degradation of catalyst layers on electrodes and disintegration of a porous matrix used for the H_3PO_4 immobilization, which is a drawback of phosphoric acid-based FC. The problem of retention of the carbonate melt within a porous matrix and insufficient stability of construction materials at working temperatures in the range of 700–1,000°C limits the use of molten carbonate FC.

In contrast, for membrane FC an increase in the working temperature is desirable to reduce sorption of CO contained in technical hydrogen thus preventing poisoning of the platinum catalyst.

Membrane systems of the Nafion type are inefficient at elevated temperatures that is probably determined by drastically reduced water retention of the membrane under these conditions rather than by the thermal instability of the polymeric electrolyte. The efficient functioning of FC relating to all the considered types is determined by the following common factors:

- specific design that can subject to perfection;
- catalyst efficiency;
- specific features of electrolyte behavior.

Table 21.1 Main types and some characteristics of fuel cells

FC type	Electrolyte	Working temperature, K	Fuel	Service life, h	Manufacturers, power	Problems
Alkali	KOH ~30%	353–370	Pure H_2	Up to 10,000	United Technology Co., Ural Electrochemical Plant in cooperation in with S.P. Korolev RSC "Energiya", 27 and 100 kW	– Sensitivity to poisoning impurities in H_2, – Requires Pt
Phosphoric acid	H_3PO_4 98%	463–490	Technical purity H_2	Up to 50,000	United Technology Co., Toshiba, IFC 12 kW – 11 MW	– Requires Pt; cathode corrosion; – Sensitive to poisoning impurities
Molten carbonate	$Li_2CO_3 + K_2CO_3$	893–923	H_2 + CO, CH_4 etc.	Up to 20,000	USA, Japan, Germany, Netherlands, Russia (Institute of High Temperature Chemistry of the Russian Acad. Sci.)	– Cathode corrosion – Electrolyte migration – Changes in electrode humidity
Solid oxide	$ZrO_2 + Y_2O_3$	1,073–1,273	H_2 + CO, CH_4 etc.	Up to 60,000	USA, Japan, Germany, Italy, France, Russia [Institute of High Temperature Chemistry of the Russian Acad. Sci., Russian Federal Nuclear Centre – E I Zababakhin All-Russian Research Institute of Technical Physics (Snezhinsk), Special Boiler Design Plant 'Kotlostroenie' (St Petersburg)], ~100 kW	– Interaction between solid layers – Technological problems
Solid polymer	Polymer membranes (Nafion, etc.)	343–363	H_2, CH_3OH	Up to 20,000–30,000	Du Pont, Ballard; highly membrane 30 000 Power Systems, (Nafion, etc.) up to 500 kWi	– Requires Pt – High sensitivity to poisoning impurities and membrane wetting

The electrolyte is one of the key elements of a FC providing ion transport between the electrodes and preventing fuel transfer from the anodic region to the cathodic one and diffusion of the oxidant in the opposite direction thus eliminating their direct interaction. Oxide-based solid electrolytes and polymer electrolytes are preferable compared to liquid counterparts due to a wider potential in respect of FC design and operation.

Presently the most significant advances are achieved in the development of FC based on polymer electrolytes in the form of thin membranes. The progress in this area is determined by extended application of polymer membrane based FC mainly as compact portative power sources. The application of solid oxide FC is likely limited to high power systems in which the heat loss to the environment can be minimized.

The first prominent success in the field of polymer electrolyte membranes fuel cells (PEMFC) relates to the development of FC containing a sulfonated polystyrene membrane used in the "Gemini" satellite [4]. It was the first step of intensive studies of PEMFC and particularly polymer membranes required for their effective functioning.

The analysis of exploration conditions of the first PEMFC allowed the formulation of the following main requirements to the relating polymers [1, 3, 5]:

- high ionic (proton) conductivity;
- stability in redox media, particularly at elevated temperatures;
- inertness relating to the catalyst supported on the considered material;
- low permeability to the fuel (hydrogen, methanol, etc.), its components and oxidants (oxygen, etc.);
- mechanical stability, because the polymer electrolyte in the form of a thin membrane should retain a sufficient mechanical strength during a long-term (up to 30,000 h) exploration in a hydrated state;
- water retention ability under the exploration conditions because the proton transfer in such membranes proceeds in the hydrated state;
- stability of operating performances during long-term (several tens thousand hours) functioning;
- relatively low cost.

The compliance to the above requirements is largely determined by the chemical structure of the elementary unit of the polymeric base of the membrane. Particularly, the proton conductivity is defined by the presence of ionic (mainly acidic – commonly sulfonic, phosphate or carboxyl) groups. Sulfonic acid groups are preferable over phosphate and carboxyl ones due to higher dissociation ability. The structure of an organic fragment bound to the sulfonic acid group is also important: the proton conductivity decreases in the series fluorinated > aromatic > aliphatic group. In addition, the proton-conducting properties depend on the ability of the corresponding group to undergo decomposition (e.g., decarboxylation).

According to the above requirements it can be concluded that perfluorinated or the aromatic polymeric systems, such as polybenzimidazoles (PBI), poly(ether ether ketone)s, etc., functionalized with the corresponding acidic groups are featured with optimal properties for their exploration as polyelectrolyte membranes.

21.3 Fluorinated Proton-Conducting Membranes and FC on Their Basis

The first polymer electrolyte membrane for FC fitting all the considered conditions was developed in 1960s by Du Pont Co. on the basis of a copolymer comprising tetrafluoroethylene (TFE) and a fluorinated monomer containing sulfonyl-fluoride groups

$$\begin{array}{l} CF_2{=}CF \\ \quad | \\ \quad (OCF_2\text{-}CFY)_nOCF_2CFRSO_2Z \end{array}$$

where $Y = F$ or CF_3, $R = F$ or $C_mF_{2m+1}(m = 1 - 10), Z = F$, OH, OCH_3, NH_2, $n = 1 - 3$.

The developed functionalized monomer and copolymer were patented [6, 7] and the commercially production of the relating membrane was under the trade name Nafion was launched in 1972 [8].

The process developed by Du Pont Co. is based on the copolymerization of TFE with perfluoro(3,6-dioxo-4-methyloct-7-ene) is carried out in fluorocarbon solvents using radical initiators according to the following reaction

$$x\,CF_2{=}CF_2 + m\,\begin{array}{l} CF_2{=}CF \\ \quad | \\ \quad OCF_2\underset{\displaystyle CF_3}{\underset{|}{C}}FOCF_2CF_2SO_2F \end{array} \xrightarrow{R^\bullet} R[(CF_2CF_2)_x\begin{array}{l} CF_2CF]_nR \\ \quad\;\; | \\ \quad OCF_2\underset{\displaystyle CF_3}{\underset{|}{C}}FOCF_2CF_2SO_2F \end{array}$$

$x{=}16{-}12$; $n, m{=}1{-}2$

The studies on obtaining proton conducting membranes were also started by other leading companies suggesting and patenting different approaches to the synthesis of fluorinated functionalized comonomers and membrane copolymers on their basis, particularly, membrane copolymers under trade names Aciplex (Asahi Chemical Co.) [9], Flemion (Asahi Glass Technologies) [10] and Dowlex (Dow Chemical Co.) [11]. The structures and synthesis of the monomers useful for this application and copolymers obtained from these precursors are described in a recent review [12].

A new growth of research interest to proton conducting membranes for FC is observed in the early twenty-first century. Extended information on polymer electrolyte membranes is summarized in handbooks [13, 14]. Some new data insufficiently discussed in earlier publication, particularly relating to the thin structure, morphology and water retention properties of these membranes are presented in the reviews [3, 5, 14–22].

A significant complication in the development of commercial technologies for the production of Nafion and similar copolymers (Aciplex, Flemion etc.) is connected with a large difference between the copolymerization constants of TFE and monomers functionalized with sulfonyl-fluoride groups. These data unavailable for a long time were disclosed in a recent publication [23] relating to the copolymerization constants for TFE ($r_{TFE} = 9.0$) and perfluoro(3,6-dioxa-4-methyloct-7-ene)sulfonylfluoride available in Russia under the trade name FS-141 ($r_{FS\text{-}141} = 0.04$) as well as Alfey-Price parameters for FS-141 ($l = 2.23$, $Q = 0.019$). Thus, the difference in the monomer reactivities reached two orders of magnitude. For this reason the copolymerization is performed by loading the whole less active FS-141 to the reactor followed by feeding TFE and maintaining a constant pressure. Due to considerable growth of the polymerization system viscosity making impossible its stirring and obtaining the required copolymer composition the copolymerization is terminated at conversion degrees 20–25% followed by washing to remove the unreacted monomer. However, it is difficult to avoid losses of the expensive sulfonated monomer, that leads to the increase of the synthesized Nafion cost.

The resulting Nafion composition significantly affects its proton conducting and physico-mechanical properties.

In order to quantitatively assess the co-polymer composition in correlation with Nafion properties, the value 'equivalent weight' (EW) characterizing the molecular weight of a polymer chain fragment per sulfonic acid group is used. The experimental EW of a co-polymer lies in the interval between two limiting values. The upper one was associated with the percolation threshold, i.e., the minimum content of the ionogenic groups sufficient for ionic (proton) conductivity of a membrane as a result of formation of ionic clusters. The lower limit denotes the deterioration of physicomechanical properties of a polymer due to its substantial swelling and absorption of water, which can ultimately result in dissolution of Nafion in water. The data summarized in [3, 13, 14, 17–20] the best properties of Nafion membranes [24] are achieved at EW in the range 950–1,100 and according to [24] the optimal EW value for commercial Nafion membranes is about 1,100.

Recently it was shown [3, 22, 25] that the structure of the peroxide initiator the fragments of which are incorporated into the co-polymer as the terminal groups can substantially affect the thermal stability and durability of co-polymers and, correspondingly, the operational properties of the obtained PEM. It was experimentally confirmed that the use of perfluorinated peroxides as co-polymerization initiators allows the fabrication of membranes with enhanced thermal stability.

The copolymerization of TFE and perfluorosulfonyl monomers is usually carried out in fluorocarbon solvents (cooling solvents available in Russia under the trade name "Khladons") at 30–80°C depending on the type of the used initiator. The pressure of TFE in the system commonly maintained in the range of 0.3–0.8 MPa is not a critical parameter [6–12]. The applied pressure provides maintaining the TFE concentration in the liquid phase of the reaction system and consequently the monomer ratio necessary for the synthesis of the copolymer with the desired composition. The considered principles provide a background of the commercial processes for the production of perfuorinated membranes such as Nafion (Du Pont), Flemion (Asahi Glass Technology), Aciplex (Asahi Chemical), Dowlex (Dow Chemical). The application areas of these materials were discussed in the reviews [1, 3, 5, 8, 12–14, 16–20] The advantages of this class of membranes include their high chemical stability, mechanical strength and high proton conductivity in the temperature range of 40–90°C.

Another approach to obtaining copolymers of TFE with perfluorinated monomers containing sulfonyl fluoride groups is based on water-emulsion copolymerization patented in [26–30]. Most often, this process is performed using water-soluble initiators (e.g., persulfates or persulfate based redox initiating systems) and emulsion stabilizers such as perfluorinated emulsifiers, e.g., salts of perfluoropelargonic acid. The copolymers are obtained in the form of aqueous dispersions. The polymerization can be carried out up to ~50% conversion (and even higher, according to the patent [29]) of the sulfonic monomer. Despite the potential prospects of emulsion polymerization method, these patents are still not commercialized for the production of Nafion and similar polyelectrolytes.

To our point of view, a more promising approach can be based on recently patented [31, 32] method of TFE-FS-141 in an aqueous emulsion with radical initiators and a salt of a perfluorocarboxylic acid as the emulsion stabilizer providing a Nafion-like copolymer of the Nafion type with the optimal EW and satisfactory melt indices. Before the polymerization the system was emulsified in a rotor-stator homogenizer with the rotor speed of 8,000–12,000 rpm. The sulfonic monomer conversion degree reaches 70% without any change of EW values. This copolymerization technique is featured with increased safety due to the absence of fluorocarbon solvents and carrying out the process in aqueous media. Moreover, the increased conversion degree improves the economic performances of the process due to a more complete consumption of the expensive monomer FS-141.

The kinetic features and mechanism of aqueous emulsion TFE-FS-141 copolymerization are worthy of a detailed analysis. These features of the considered process are described in [33].

TFE-FS-141 copolymerization in water emulsion was performed in 0.26 and 1.5 L steel reactors maintaining a constant temperature 40°C, supplied TFE pressure automatically controlled on the level 0.5 MPa and FS-141:water volume ratio 1:12. The process was carried out using ammonium perfluorononaate (7.8 mmol/$L_{aqueous\ phase}$) as an emulsifier, potassium persulfate (PPS) – sodium metabisulfite red-ox system (4.6 mmol/L) as an initiator and phosphate buffer system for pH stabilization. Before copolymerization the reaction system was dispersed using a rotor-stator homogenizer at 1,000 rpm at ambient temperature within 15 min.

The preliminary dispersing provides aqueous emulsions of FS-141 with the particle size about 2–3 μm retaining stability within at least 1.5 h. An apparent increase of the droplet size observed using a Malvern Zetasizer Nano-ZS particle analyzer (affording particle size measurement in the range from 0.6 to 6 μm) is probably not caused by the flocculation since even in 1.5 h a simple shaking returns the emulsion into the initial dispersity condition (Fig. 21.1, plots 1–3). These data confirm a high importance of the preliminary emulsification in providing the emulsion copolymerization stability up to high FS-141 conversion degrees using a simple anchor stirrer in the reactor.

In order to characterize the changes in the reaction system during the polymerization the process was terminated at different time intervals, measured the latex particle dispersity using a Malvern Zetasizer Nano-ZS analyzer equipped with a He-Ne laser and determined the EW and MFI values for the obtained copolymer. The kinetic plots of TFE consumption during the copolymerization confirmed the process stability and reproducibility.

The effect of the monomer conversion degree upon EW (and consequently the resulting copolymer composition) is illustrated in Fig. 21.1. The obtained data show

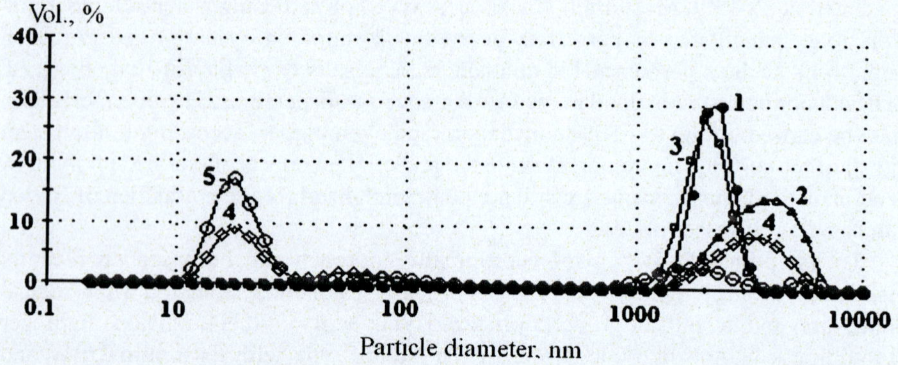

Fig. 21.1 Particle size distribution in FS-141 emulsion stabilized by ammonium perfluorononaate (7.8 mmol/L of aqueous phase) straight after preparation (*1*), in 45 min in a static mode (*2*), after the cell shaking in 90 min after the preparation (*3*) and in the polymerization system at FS-141 conversions 37% (*4*) and 52% (*5*)

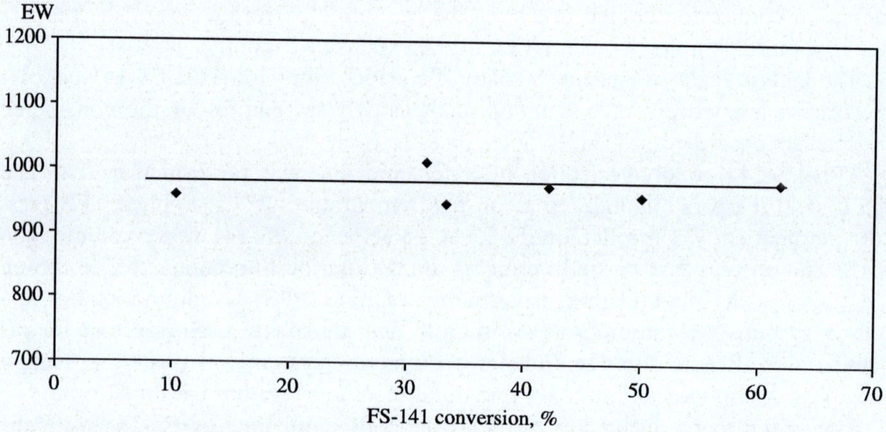

Fig. 21.2 Equivalent weight EW of the resulting copolymer as a function of FS-141 conversion in a 0.26 L reactor under identical process conditions

that both EW and copolymer composition remain almost constant up to high FS-141 conversions (60%) (Fig. 21.2).

The data in Fig. 21.1 show that copolymerization leads to the decrease in the volume part of initially formed monomer droplets and formation of polymer or polymer-monomer particles with the dispersity increased by about two orders. The part of the droplets with the highest dispersity (about several tens of nanometers) grows with the process time and conversion degree.

Although the observed significant dispersing of the polymerization system could be accounted for the micellar mechanism of the emulsion copolymerization, the characterization using Malvern analyzer did not revealed any micelles in the system that is probably determined by low ammonium perfluorononaate concentrations (close to the critical micelle formation concentration). Furthermore, a large difference in TFE and FS-141 reactivities should be also taken into consideration. Therefore, in order to obtain a copolymer with the required EW values the concentration ratio between the monomers should be shifted towards the excess of FS-141 that is hardly attainable in micellar systems. From the other hand, the implementation of the microemulsion copolymerization mechanism requires about one order higher dispersity for the initial emulsion system.

The above data allowed us to make the following conclusions regarding the mechanism and topochemical features of TFE-FS-141 copolymerization in an aqueous emulsion. The use of a water soluble initiator potassium persulfate provides the formation of primary radicals in the aqueous phase. Subsequently these radicals react with the dissolved TFE to transform into oligomeric radicals due to a higher water solubility and reactivity of TFE compared to FS-141. Highly reactive oligomeric radicals are immobilized onto dispersed FS-141 droplets containing the required amount of TFE and initiate the copolymerization yielding the copolymer with the required EW. The resulting copolymer is poorly soluble in FS-141 and leaves the monomer droplet forming a highly dispersed polymer particle. The process performance according to this mechanism depends on some critical parameters of the polymerization system, e.g., TFE pressure should not exceed certain threshold values to provide the required EW values of the resulting copolymers. Thus, TFE-FS-141 copolymerization is highly sensitive to the process conditions and requires their thorough selection and optimization.

The copolymers obtained by aqueous copolymerization technique were used for the preparation of membranes by liquid copolymer deposition from DMF solution. The electric (Volt-Amper) characteristics of the membranes were determined in an oxygen (air) – hydrogen fuel cell. The results shown in Fig. 21.3 suggest that the developed aqueous emulsion process affords membranes with the basic performances on the level of conventional membranes produced by Du Pont Co.

Studies of the morphological structure are important for perfecting the structure and operational characteristics of Nafion and analogous proton-conducting membranes. A historical overview of these studies is presented in [13]. A significant contribution to the concepts on the Nafion morphology was made by Gierke who was the first to put forward the cluster model that described the structure of this material [33]. Using information from modern methods of small-angle X-ray and neutron scattering, it was possible to analyze the structural anisotropy of the Nafion and determine the sizes and distribution of clusters, channels and hydrated sulfonic acid groups in these channels.

A "sandwich-like" structure element of the Nafion morphological organization is shown in Fig. 21.4 [34].

Presently it is generally accepted that the channels in Nafion are formed due to the phase separation between the originally hydrophobic polymer chain and the hydrophilic sulfonic acid groups bound to the chain. The proton transport proceeds through the water-containing channels; the molecules of water are either bound to the acidic groups or fill the channel space [14, 19, 20, 25, 34–36].

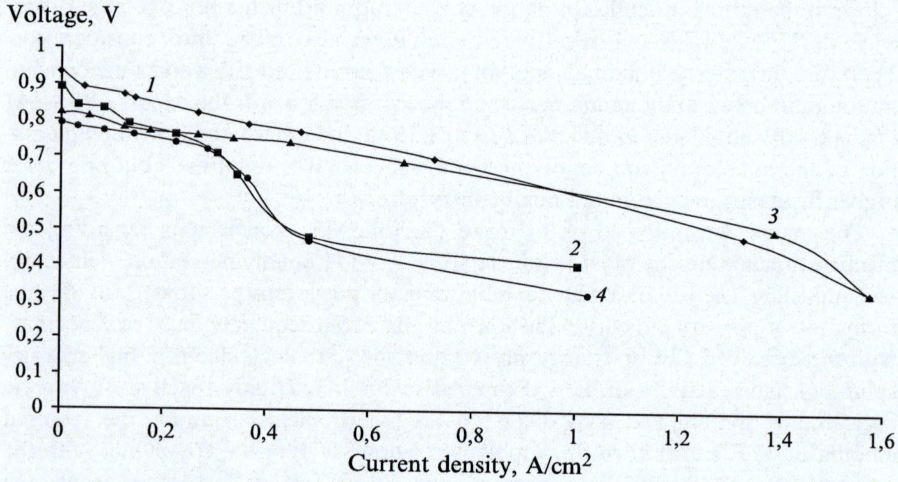

Fig. 21.3 Volt-Amper characteristic of the fuel cell involving the membrane obtained by aqueous emulsion copolymerization. Temperature 85°C, oxygen (air) and hydrogen collector Pt40/Cv15. The 50 μm thick membrane MF-4SK is obtained by deposition (pouring) of liquid copolymer. Ion exchange capacity 1.06 mmol/g. Performances using oxygen (*1*) and air (*2*); comparative data for 50 μm thick Nafion-212 membrane using oxygen (*3*) and air (*4*)

The level of proton conductivity depends on such factors as the density of distribution of acidic groups in the membrane, the structure and morphology of the membrane polymeric framework and the water content. The studies performed in order to provide a better understanding of the relationship between the structure and the proton conductivity of a membrane are presented in [37–49].

The microstructure of Nafion and analogous PEM was studied by small-angle X-ray and neutron scattering in a sufficiently wide range of polymer-solvent ratios [14, 19, 35]. It was noted [13] that for the nanophase distribution of PEM, the formation of interpenetrating domains is possible. In the same studies, it was found that the variations of transport characteristics of the Nafion channels can be controlled by changing the solvent during membrane washings or by affecting the orientation of the channels formed (for example, by the electric field).

The conductivity of Nafion membranes was studied by impedance spectroscopy as a function of temperature and sample pretreatment, in particular, hot-pressing [43]. Depending on the water content and the pore size, two modes of membrane conductivity were observed with a change in the activation energy in a temperature interval from –50°C to –15°C. Later it was shown that at higher temperatures the activation energy undergoes only slight changes remaining on the level about 130 meV at 25–75°C [44]. The effect of the pore size distribution on the membrane characteristics was also discussed [45].

A detailed characterization of the submicroheterogenic distribution of sulfonic acid groups and the interconnecting channels is also very important for the understanding of the proton conductivity mechanism in Nafion.

Recent data on the transport properties of channels in Nafion were reported by Roduner et al. [46] at the Europolymer Conference 2008. They studied the distribution

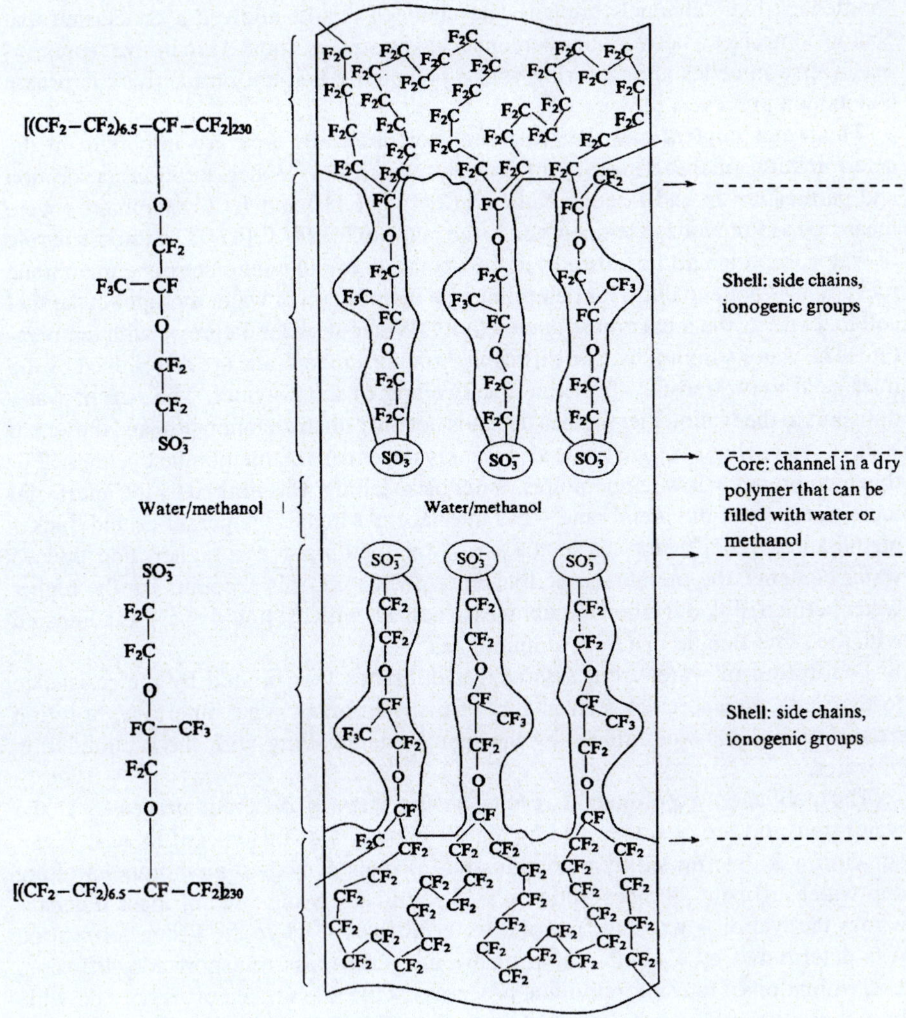

Fig. 21.4 "Sandwich-like" structural element of Nafion morphological organization [34]

of the proton conductivity across the membrane by electrochemical atomic force microscopy (EC-AFM) with nanosized probes switched to different channels in the membrane cross-section at a certain scanning rate. The obtained data were analyzed together with the results of determination of the Nafion microstructure by X-ray diffraction and small-angle neutron scattering. Measurements at different relative humidity have shown that the Nafion membrane surface was non-uniform and wide inactive ('dormant') surface areas exist between the active proton conducting channels. The profiles of active site distribution over the cross-section, their variations with time and the changes in the activity of a given channel with time (activation and decay) were studied. It was shown that at low humidity the proton conducting channels are predominantly formed by individual pores, while at high humidity associated "bottleneck-like" channels prevail. The obtained results allowed a conclusion that Nafion comprises a network of interconnected clusters of about 3 nm in size representing inverse micelles and featuring with a potentially non-uniform activity in proton conduction processes [46].

The water content and the water transfer processes are very important in the determination of the efficiency of Nafion membranes. Water sorption, desorption and permeation in and through Nafion 112, 115, 1,110 and 1,123 membranes were measured as functions of temperature in the range of 30–90°C [47]. The important role of water sorption and its transport in the operation of polymer electrolyte membrane fuel cell was pointed out. This determines the distribution of water throughout the fuel cell that affects the local proton conductivity. Water permeation grows with temperature. The water sorption by, and diffusion through, a membrane are determined by the interfacial mass transfer, diffusion and swelling of the polymer. The rate of water transport in the Nafion membranes depends not only on the temperature and the actual state of water (liquid or gas); it is also weakly dependent on membrane thickness. For thin membranes at low temperatures, water permeability was limited by the interfacial mass transport at the membrane – gas interface. At higher temperatures and thicker membranes, the diffusion resistance across the membrane was higher. For reduced water contents, the membrane permeability decreased; this accounts for the higher water permeability of Nafion membranes in contact with the liquid phase as compared with the same sample in a flow of humid gas.

Desorption of water from saturated membranes was limited by the resistance to the mass transport between phases at the membrane – gas interface. Sorption from a humid gas was limited by the membrane swelling with the accumulation of water.

The published experimental results on the rates of different processes at the water transport are characterized by big difference in diffusion coefficients. Water sorption was determined by a combination of rates of Nafion membrane swelling and water diffusion, the desorption was determined by the rates of mass transport across the Nafion – gas interface and water diffusion, while the water permeation was determined by a combination of the interface mass transport and diffusion. Determination of rates of individual processes at the mass transport is only possible by a thorough analysis of data on the permeability and sorption for membranes of different thicknesses and at different temperatures.

Thus, the proton conductivity of Nafion membranes and similar materials was directly related to the water content in ion channels of PEM [46] and accompanied by the water transport from the anode to the cathode. To compensate the water loss, water vapour was delivered with hydrogen. The water loss also determined the optimum temperature range of fuel-cell operation when common unmodified Nafion was used (up to 90°C). Limiting the working temperature interval revealed yet another drawback in the Nafion operation within a FC associated with poisoning of the platinum catalyst on the electrode with impurities contained in hydrogen. The trace amounts of CO (>10 ppm) in the hydrogen fuel are adsorbed on the catalyst surface thus reducing its efficiency. At temperatures above 130°C, the equilibrium of the CO and H_2 sorption on the platinum catalyst is shifted towards H2 for thermodynamic reasons, which prevents the catalyst poisoning. However, at higher temperatures (120–140°C), the proton conductivity of membranes decreases substantially due to Nafion dehydration.

Thus, notwithstanding several important advantages of proton-conducting membranes of the Nafion type (chemical resistance, mechanical stability, high proton conductivity), the search for optimal temperature modes of their operation with allowance of the retention of water content, the porosity and the structure of proton-conducting channels and the effect of impurities in fuel, requires modification of Nafion membrane structure and properties. The additional stimulus to further development of research in this direction were the high cost of this material and the demand for replacement of hydrogen by other kinds of fuel, particularly, alcohols.

21.4 Modification of Nafion Structure for the Optimization and Extension of Exploration Conditions

A common approach to improve the technical performances of the Nafion and similar membranes is based on their modification with inorganic components in order to enhance their water uptake ability and prevent their dehydration due to a high hydrophilicity and water retention ability of the modifying additives. This approach allows a substantial reduction of water loss at membrane exploration at elevated temperatures.

The modification of the Nafion structure and properties using inorganic additives such as SiO_2, heteropolyacids and their various combinations is described in numerous articles and reviews [4, 49–58]. Many of these publications report a possibility for efficient application of thus modified membranes at temperatures above 100°C (up to 140°C) and present possible mechanisms of the enhancing effect of these additives.

A particularly interesting approach is based on the incorporation of ultrafine SiO_2 surface-functionalized by covalent binding of organosilicon compounds bearing terminal SO_3H groups [59, 60] (e.g., by treatment of original silica nanoparticles with perfluoropropane-2,3-sultone). A membrane from Nafion filled with silica-SO_3H possessed enhanced proton conductivity dependent on the concentration of the added modifier (the maximum value was reached at the content of the modifier of 3%) and was equal to 0.12 S cm^{-1}. The modified membrane was also characterized by improved barrier properties.

It is also worthy to mention that the proton conductivity of Nafion can be improved by physical impact during the membrane casting, particularly, under the effect of an electric field, which favoured the alignment of ionic aggregations along the membrane thickness direction [61].

Another important direction represented by numerous studies relates to Nafion modification with polymer systems. Particularly, the enhancement of Nafion-like membrane physico-mechanical properties by styrene grafting is suggested using either radiation induced procedure [62] or performing the process in a supercritical CO_2 [63] followed by the grafted polystyrene sulfonation in a concentrated sulfuric acid. Another approach is based on Nafion modification with waterproof polymers (polyethylene, polyhexafluororopylene, copolymers of hexafluoropropylene with propylene or ethylene, etc.) [64]. This method is particularly suggested for FC involving methanol as a fuel. Other polymer systems useful for Nafion modification include polypyrrol [65, 66] and crosslinking additives [67]. The formation of two- and three-layered membranes with different EW values of separate modified Nafion layers are patented in [68]. Composite membranes based on Nafion in combination with some other polymers are described in [69–71].

The impressive progress in the field of the enhancement of Nafion and similar perfluorinated membranes in the recent decade is generally based on the following basic principles and achievements:

1. Characterization of various perfluorinated monomer systems for obtaining proton conducting membranes, including quantum-chemical studies, indicate that the compound discovered by Du Pont Co. in 1960s remains the optimal membrane material of this kind;
2. The optimal technological conditions for TFE copolymerization with sulfonyl fluoride monomers in freon solutions using perfluorinated peroxide initiators are defined;
3. New approaches to copolymerization in aqueous emulsion systems are suggested to provide the process safety and cost reduction;
4. The features of proton conducting membrane thin structure as well as the conditions of their effective functioning and prospects for increasing their proton conductivity and lifetime are revealed;
5. Approaches to structural modification of Nafion-like membranes with inorganic and polymeric compounds are suggested.

21.5 Prospects for the Application of Non-Fluorinated Polymer Systems as Proton Conducting Membranes for Fuel Cells

The review prepared by A. Glüsen and D. Stolten resents a comparative analysis of the development level and potential prospects for different polymer electrolyte membrane types including Nafion, sulfonated trifluoropolystyrene (BAM-3G), sulfonated poly(ether ether ketones) (SPEEK), polybenzimodazoles (PBI) and polyphosphazenes for different kinds of fuel cells, i.e. hydrogen-oxygen, direct methanol and high temperature FC.

For hydrogen-oxygen FC Nafion occupies the first place in both the R&D level and potential capacities. For direct methanol FC, the sulfonated polycondensation systems (SPEEK and polyphosphazenes) should be preferred, whereas for high-temperature systems, PBI is found to be the best option.

The results of experimental studies of various polycondensation polymeric membranes were considered in the review [18] with the comparison of their characteristics with those of Nafion. Generally, it was shown that in respect of proton conductivity, the majority of polycondensation polymers are inferior only to Nafion and it is only at high temperatures that the PBI-based membranes become superior.

In recent years, quite a number of reviews were published that touched upon the synthesis and properties of condensation polymeric systems as the proton-conducting materials. In a series of reviews by Rusanov et al. [16, 21, 72–76] approaches to the synthesis of many classes of condensation polymers [poly(benzimidazoles), poly (ether ether ketones), poly(ether sulfides), polyarylphosphazenes, polynaphthylimides] were described in detail as well as the polymer-analogous reactions of condensation systems. However, the analysis of the relationship between the proton conductivity and the chemical structure of polymeric systems under consideration and the prospects of optimizing their operational characteristics were given too briefly and insufficiently to draw any significant conclusions.

Another review [13] demonstrated the prospects for practical use and the properties of different sulfonated polyimides, sulfonated poly(benzimidazoles), sulfonated poly(aryl ether ketones), sulfonated polyphosphazenes and even sulfonated silicates. As a whole, it should be emphasized that condensation polymer membrane systems are characterized by high thermal stability (up to 500°C), tolerance with respect to admixtures of CO and CO_2 and low permeability towards different fuel types including methanol; they also surpass Nafion in mechanical strength. Thus, membranes based on the PBI-H_3PO_4 system remained stable at temperatures up to 500°C; their fairly high proton conductivity increases with an increase in temperature and phosphoric acid content. The conductivity of PBI containing 500 mol% of H_3PO_4 (i.e., 5 molecules of H_3PO_4 per PBI structural unit) at 190°C was $3.5 \cdot 10^{-2}$ S·cm^{-1}. In addition, the membrane permeability of these materials towards fuel components was one order of magnitude lower as compared with that of perfluorinated polymer electrolytes, and they surpassed Nafion-based membranes threefold in mechanical strength.

The first patents were granted to Pemeas Fuel Cell Technologies, which produced polybenzimidazole membranes for direct ethanol fuel cells. This company built a pilot plant for manufacturing membranes that was launched in 2004.122 A membrane developed by this company on the basis of poly(2,20-m-phenylene-5, 50-dibenzimidazole) under the name of Celtex-MEA was tested in a FC operating on a fuel system that contained 35% H_2, 64.8% N_2, 2,000 ppm CO and traces of CO_2 and H_2O. It exhibited efficiency in a temperature range of 120–200°C, tolerance to CO at concentrations above 50,000 ppm and had service life longer than 8,000 h. Yet another membrane version named Celtex-V was destined for direct methanol FC.

Recently the synthesis of a novel sulfonated polyimide membrane containing triazole groups was reported [77]. Its efficiency and high durability were demonstrated in a single fuel cell operating at 80°C for 5,000 h without any significant decrease in the ion-exchange capacity. However, the molecular weight of the membrane reduced by 10% consequently resulting in a decrease of its mechanical strength.

Nevertheless, no data on the commercialization of proton conducting membranes based on polycondensation polymer systems are available.

In order to reveal fundamental correlations between the structure, morphology and properties of the polymers useful as proton conducting membrane T.J. Peckham, J. Schmeisser, M. Rodgers and S. Holdcroft [78] performed a detailed analysis of proton conductivity as a function of acid groups content, effective proton mobility, type and rigidity of the polymer backbone and water content for the following five polymer systems: sulfonated poly(ether ketone) (SPEEK, **1**), poly (ethylenetetrafluoroethylene-graft-styrene sulfonic acid) (**2**), linear sulfonated polyimide (**3**), membrane based on sulfonated poly(a,b,b-trifluoro)styrene commercially produced by Ballard Power Systems under the trade name BAM (**4**) and Nafion (**5**) as the comparative material with the optimum combination of technical performances (reliability and service life), being also the best studied.

According to the chemical structure, the system **1** is an aromatic polymer, polyimide **3** can be assigned to polyaromatic rigid rod polymers and vinyl polymers **2** and **4** contain sulfonic acid groups in the lateral residues.

Since the trial-and-error method had not led to any substantial progress in the membrane development, the authors of the above study attempted to find a structure–property relationship for the considered polymer membrane systems and to propose on its basis the scientific principles of selecting PEM with optimal characteristics.

This analysis relied on the results of studying the formation of ionic channels for proton transport manifested by the phase separation of the hydrophobic polymer backbone from the hydrophilic part of the polymer with sulfonic acid groups. A In addition to the number of sulfonic acid groups, the features of their distribution thin the polymer matrix and the water content in the channels also significantly contribute into the membrane properties [79–81].

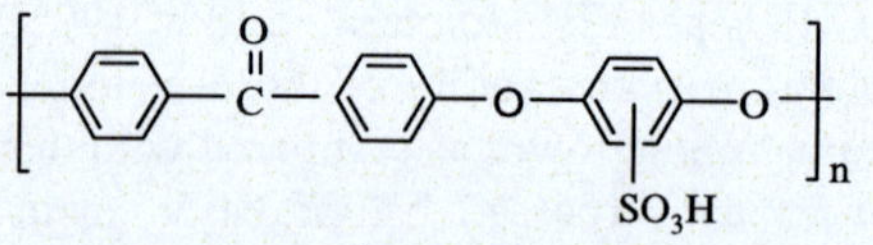

1

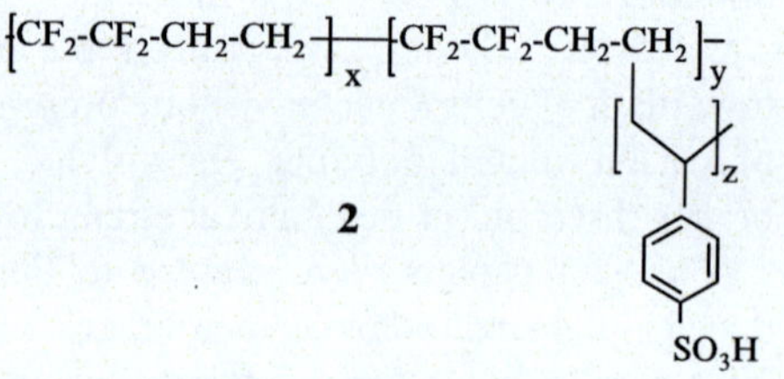

2

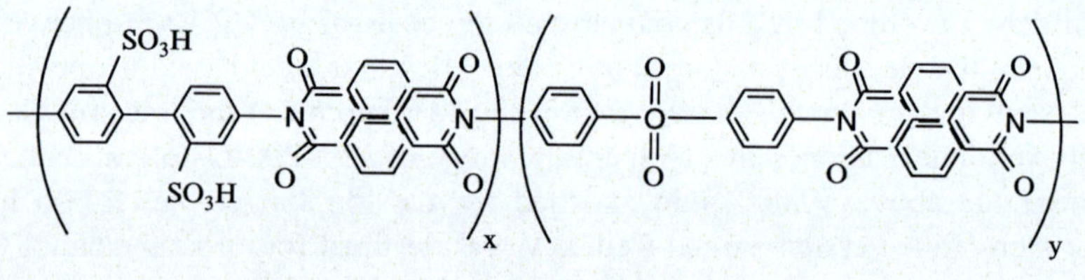

3

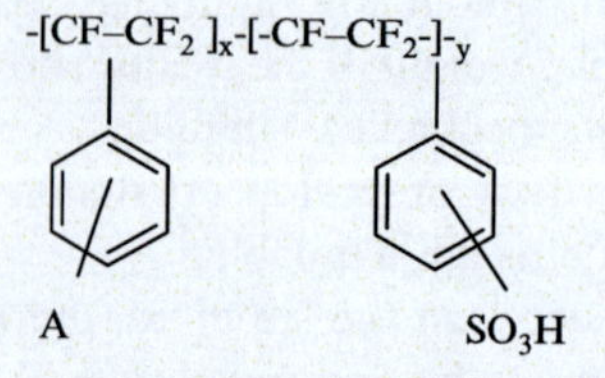

4

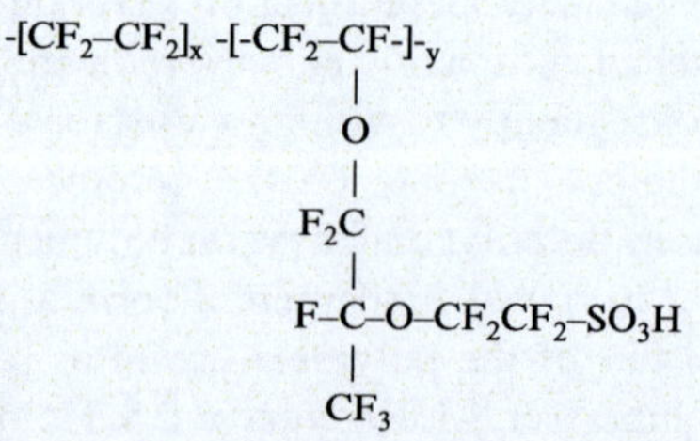

5

Let us consider several important results earlier reported in another study [19]. The comparison of the Nafion (**5**) and SPEEK (**1**) microstructures showed that in Nafion the water channels are wider and characterized by the higher connectivity. In the narrower SPEEK channels, protons have to be located closer to the SO_3 groups bound with the polymeric matrix and are subjected to stronger attractive forces enhanced by the basicity of polymer **1**. This explains the weaker mobility of protons in SPEEK as compared with Nafion.

The results of several published studies confirmed the aforementioned operation characteristics of these types of membranes. Thus the microphase separation was observed 130 in sulfonated polyimide membranes where the polyimide backbone bears sulfonic acid-terminated side alkoxy chains. The possibility of enhancing the proton conductivity of polyimides with sulfonic acid groups in the side chains was demonstrated [82]. A comparison of the microstructural characteristics of materials **2**, **4** and **5** has shown that whereas the latter (Nafion) showed clear indications of microphase separation in the form of ion aggregates, the majority of ion channels in polymer **2** were distributed uniformly, while the BAM membrane (4) demonstrated ion aggregation with uniform distribution and less developed localization of ionic domains [41, 43]. Generally, the system **2** is featured with a more pronounced microphase separation and higher conductivity compared with the system **4** [78].

The complex nature of the dependence of proton conductivity on the structural features of PEM, namely, on the domain structure, the sulfonic acid group concentration in the membrane, water content, proton effective mobility, was revealed [46, 78]. The relationship between these parameters turned out to be more complicated than that assumed in the earlier studies. The differences in the degree of conducting channels tortuosity for membranes with different distances between the neighboring SO_3H groups are shown in Fig. 21.5. The different sizes of channels and the presence of dead ends determine the deviations of the proton conductivity from its theoretical values.

A new direction in the development of PEMFC extensively reported in recent publications relates to hybrid polymer systems as a novel class of membrane materials. Since these systems are considered in detail in our recent review [3] we do not discuss them in the present paper.

Finally, it is necessary to consider the practical implementation of polymer membranes in FC useful for different purposes, e.g. in transport vehicles, submarines and automobile engines. The advances in these areas is determined by the progress in the studies on the enhancement of Pt catalysts efficiency, fuel cell structure and size as well as the improvement of hydrogen fuel storage systems and development of safe and low cost sources of pure hydrogen. These measures provided a significant cost decrease for the power generated by fuel cells.

These achievements resulted in the launch of commercial manufacture of 50–500 kW compact power sources involving on FC with Nafion based membranes. These units can be combined into batteries providing the power over 1,000 kW, e.g. for submarines [83]. According to the Project 212 a new 1,700 t volume submarine is equipped with FC batteries allowing autonomous power supply for about 2 weeks at the depth about 400 m. In addition to FC based

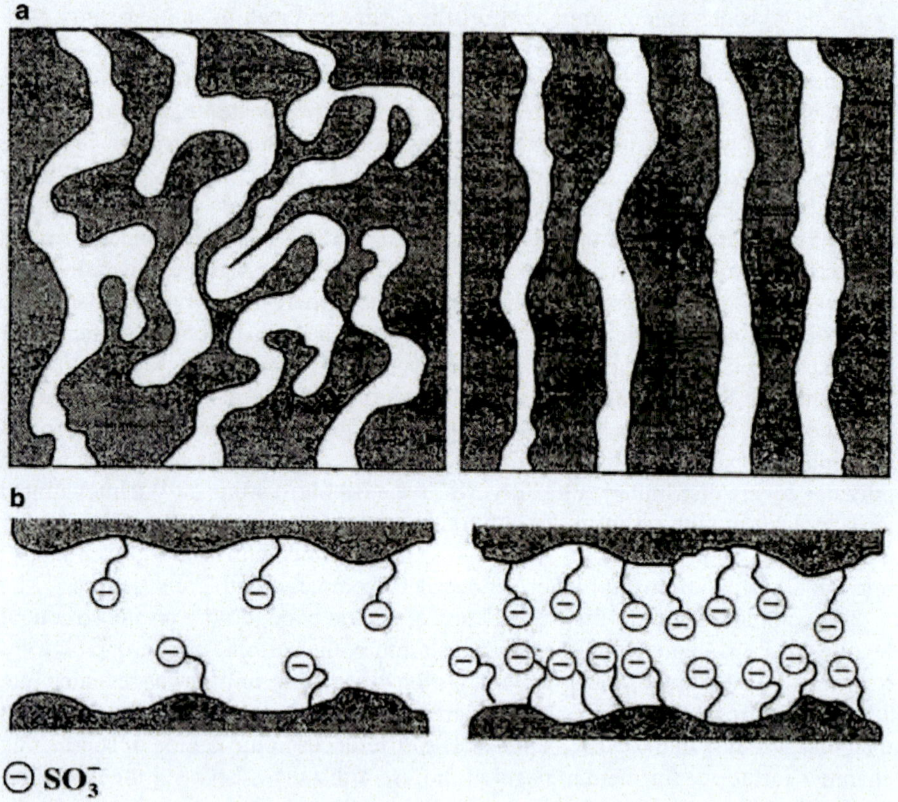

Fig. 21.5 Schematic presentation of aqueous domains in proton exchange membranes (*white*) with more (**a**) and less (**b**) tortuous channels. The spatial arrangement of acid groups within aqueous channels of the membranes with larger or smaller distances between the neighbouring sulfonic acid groups [78]

generator, the submarine is also equipped with a diesel engine and reserve accumulators. The fuel cells consume pure hydrogen generated in situ by controllable heating of metal hydrides. After exhaustion these hydrogen sources can be recovered by low temperature hydrogen absorption.

For the recent 2 years GM company performs field tests in the USA and Europe for 116 cars driven by crossover Equinox fuel cells using compressed hydrogen from 4.2 kg balloons. The application of this advanced compact FC instead of an internal combustion engine affords 322 km run per one balloon filling and the overall lifetime run increased from 80,000 to 130,000 km. Furthermore, the consumption of Pt catalyst in reduced from 80 to 30 g and expected to be further decreased to 10 g for the next generation of FC.

New hydrogen sources are also studied, particularly including a complex borane-nitrogen hydride (borazane) H_3NBH_3 developed by the Los-Alamos

National Laboratory and Alabama University [84]. This compound is solid at ambient conditions with the density 0.78 g/cm^3 and hydrogen content 20% wt. and can be recovered into the initial state after hydrogen yield [84].

In Denmark the company Amminex and Danmarks Tekniske Universitet developed another pure hydrogen source based on magnesium complexes with coordinated ammonium $Mg(NH_3)_6Cl_2$. This stable and safe solid substance yields about 9%wt. of hydrogen upon heating. The problem of exhaust purification from nitrogen oxides is completely resolved.

In this year a group of British companies presented in London a series of electrotaxies based on Fc and lithium battery engines. The applied 30 kW fuel cells provide a 402 km run by one hydrogen filling (3.7 kh H_2 compressed at 350 atm). The developed taxies are expected to be widely explored during the Olympic games in 2012.

21.6 Conclusions

The analysis of studies on the development of proton conducting polymer membranes made it possible to distinguish three periods of their evolution. The first period (in 1960s–1970s) was marked with bright achievements in the development of first fuel cells and fabrication of membranes of the Nafion type and FC thereof. After a of certain decay, a new rise in the activity of studies in the field of fuel cells and, correspondingly, in the development of polymeric membrane systems was observed in the 1990s.

Three groups of polymer systems suitable for the application in PEM for FC are clearly distinguished. The first group contained perfluorinated sulfonic acid containing polymers, i.e. Nafion, its analogs and different variations of these systems structurally modified according to various principles and approaches. Studies in this direction substantially advanced as regards their practical implementation, namely, at present, membranes of the mentioned type are already manufactured on the industrial scale and used in FC of different types.

The second group of polymer systems suitable for the manufacture of PEM comprised condensation polymers. This type of membranes are widely employed at present and considered to be very promising taking into account their thermal stability and the possibility of usage in a wide temperature range. Membranes based on polybenzimidazoles are produced on a pilot scale so that one can expect considerable extension of their application field in the nearest future.

The third group of polymer systems for membranes is represented by hybrid systems including hydrogels. The scientific basis substantiating the prospects of their application is only in the stage of development; however, the features of their structure, the accessibility and simplicity of their fabrication make expedient their further studies.

References

1. Press RJ, Santhanam KJ, Miri MJ, Bailey AV, Takacs GA (2009) Introduction to hydrogen technology. Wiley, Hoboken
2. European Commission. Eur 20719 EN – Hydrogen Energy and Fuel CellsÐA Vision of Our Future. Publications of European Committees, Luxemburg, 2003
3. Ivanchev SS, Myakin SV (2010) Polymer membranes for fuel cells: manufacture, structure, modification, properties. Russ Chem Rev 79:116
4. Zawodzinski TA, Derouin C, Radsinski S, Sherman RJ, Smith UT, Spenger TE, Gottesfeld S (1041) Water-uptake by and transport through Nafion(R) 117 membranes. J Electrochem Soc 1993:140
5. Bagotskii VS, Osetrova AM, Skundin AM (2003) Fuel cells: state-of-the-art and major scientific and engineering problems. Russ J Electrochem 39:919–934
6. Du Pont Co. (1962) Fluorocarbon sulfonyl fluorides. US Patent 3,041,317
7. Du Pont Co. (1966) Fluorocarbon vinyl ether polymers. US Patent 3,282,875
8. Yamabe M (1984) Advances in fluorine compounds technology, Moscow: Mir, 592 p. (trans. from Japanese into Russian)
9. Asahi Chemical Co. (1961) Fluorinated copolymers and cation exchange membrane and process for producing the same. GB Patent 2051831
10. Miyake N, Kita K, Honda M, Tanabe T (2008) R&D of Asahi Kasei aciplex membrane for PEM fuel cell applications. Electrochem Soc Trans 16:1219
11. Dow Chemical Co. (1990) Low equivalent weight sulfonic fluoropolymers. US Patent 4,940,525
12. Souzy R, Ameduri B (2005) Functional fluoropolymers for fuel cell membranes. Prog Polym Sci 30:644
13. Doyle M, Rajedran G (2003) Handbook of fuel cells fundamentals, technology and applications, vol 3, pt 3. Wiley, Chichester, p 351
14. Kreuer K-D (2003) Handbook of fuel cells: fundamentals, technology and applications, vol 3, part 3. Wiley, Chichester, p 420
15. Korovin VN (2004) Fuel cells. Alternat Energ Eko 10:8
16. Rusanov AL, Likhatchev DYu, Müllen K (2002) Proton-conducting electrolyte membranes based on aromatic condensation polymers. Russ Chem Rev 71:761
17. Glüsen A, Stolten D (2003) Membrane für polymerelektrolyt bronnstoffzellen. Chem Ing Tech 75:1591
18. Roziere J, Jones DJ (2003) Non-fluorinated polymer materials for proton exchange membrane fuel cells. Annu Rev Mater Res 33:503
19. Mauritz KA, Moore RB (2004) State of understanding of nafion. Chem Rev 104:4535
20. Kreuer K-D, Paddison SJ, Spohr E, Schuster M (2004) Transport in proton conductors for fuel-cell applications: simulations of elementary reactions and phenomenology. Chem Rev 104:4637
21. Rusanov AL, Likhatchev DYu, Kostoglodov PV, Müllen K, Klapper M (2005) Proton-exchanging electrolyte membranes based on aromatic condensation polymers. Adv Polym Sci 179:83
22. Shiraldi DA (2006) Perfluorinated polymer electrolyte membrane durability. Polym Rev 46:315
23. Odinokov AS, Bazanova OS, Sokolov LF, Barabanov VG, Timofeev SV (2009) Kinetics of copolymerization of tetrafluoroethylene with perfluoro(3, 6-dioxa-4-methyloct-7-ene) sulfonyl fluoride. Russ J Appl Chem+ 82:113
24. Aldrich Catalogue Handbook of Fine Chemicals, 1999-2000. Aldrich Chemical Co, Inc., Milwaukee, WI, 1999.
25. Ivanchev SS (2008) Fluorinated proton-conducting nafion-type membranes: the pars and the future. Russ J Appl Chem+ 81:529

26. Production of perfluorocarbon polymer having sulfonic acid type functional group. JP Pat. 62288614 (Asahi Glass Co., Ltd.) 1987.
27. Asahi Glass Co., Ltd. (1988) Production of perfluorocarbon polymer having sulfonic acid type functional group JP Patent 62288616
28. Asahi Glass Co., Ltd. (1988) Production of perfluorocarbon polymer having sulfonic acid type functional group. JP Patent 62288617, 62288617
29. Asahi Chemical Ind., Co. (1997) Perfluorocarbon copolymer containing functional groups and a method for producing it. US Patent 5,608,022
30. Du Pont Co. (2000) Free radical polymerization method for fluorinated copolymers. WO Patent 0052060
31. Boreskov Institite of Catalysis of the Siberian Branch of the Russian Acad. Sci. (2009) Method of obtaining perfluorinated copolymer containing functional groups. RU Patent 2348649C1
32. Boreskov Institite of Catalysis of the Siberian Branch of the Russian Acad. Sci. (2009) Method for producing perfluorinated functional groups containing copolymer. WO Patent 2009/082264 A1
33. Ivanchev SS (2010) Features of tetrafluoroethylene copolymerization with perfluoro (3,6-dioxa-4-methyloct-7-ene) sulfonylfluoride in aqueous emulsion system. Doklady Rus Acad Sci 2010 (in press)
34. Haubold H-G, Vad T, Jungbluth H, Hiller P (2001) Nano structure of NAFION: a SAXS study. Electrochim Acta 46:1559
35. Gabel G, Atkins P (2000) Structural evolution of water swollen perfluorosulfonated ionomers from dry membrane to solution. Polymer 41:5829
36. Rollet AL, Diat O, Gebel J (2002) A new insight into nafion structure. J Phys Chem B 106:3033
37. Slade S, Campbell SA, Ralph TR, Walsh FC (2002) Ionic conductivity of an extruded nafion 1100 EW series of membranes. J Electrochem Soc 149:A1556
38. Beatlie PD, Orfino FP, Basur VI, Zychowsky K, Ding J, Chuy C, Schmeisser J, Holdcroft S (2001) Conductivity and electrochemical ORR mass transport properties of solid polymer electrolytes containing poly(styrene sulfonic acid) graft chains. J Electrochem Chem 45:503
39. Ding J, Chuy C, Schmeisser J, Holdcroft S (2001) Solid polymer electrolytes based on ionic graft polymers: effect of graft chain length on nano-structured, ionic networks. Chem Mater 13:2231
40. Ding J, Chuy C, Holdcroft S (2002) Solid polymer electrolytes based on ionic graft polymers: effect of graft chain length on nanostructured ionic networks. Adv Funct Mater 12:389
41. Gao J, Lee D, Yang YS, Holdcroft S, Frisken BJ (2005) Self-assembly of surface-charged latex nanoparticles: a new route to the creation of continuous channels for ion conduction. Macromolecules 38:5854
42. Yang Y, Holdcroft S (2005) Synthetic strategies for controlling the morphology of proton conducting polymer membranes. Fuel Cells 5:171
43. Cappadonia M, Erning JW, Niaki SMS, Stimming U (1995) Conductance of Nafion 117 membranes as a function of temperature and water content. Solid State Ionics 77:65
44. Blumenthal G, Cappadonia M, Lehmann M (1996) Investigation of the proton transport in nafion® membranes as a function of direction, temperature and relative humidity. Ionics 2:102
45. Divisek J, Eikerling M, Mazin V, Schmitz H, Stimming U, Volfkovich YuM (1998) A study of capillary porous structure and sorption properties of nafion proton-exchange membranes swollen in water. J Electrochem Soc 145:2677
46. Roduner E, Hiesgen R, Alexandrova E (2008) Spatially resolved proton conductivity measurements and the structure of polymer electrolyte fuel cell membranes. In: Proceedings of europolymer conference 2008 (EUPOCn2008), Gargnano, Italy, 2008, p 17
47. Majsztrik PW, Satterfield MB, Bocarsly AB, Benzinger JB (2007) Water sorption, desorption and transport in nafion membranes. J Membr Sci 301:93

48. Chen H, Palmese GR, Elabd YA (2006) Membranes with oriented polyelectrolyte nanodomains. Chem Mater 18:4875
49. Antonucci PL, Arico AS, Creti P, Ramunni E, Antonucci V (1999) Investigation of a direct methanol fuel cell based on a composite Nafion®-silica electrolyte for high temperature operation. Solid State Ionics 125:431
50. Jung DH, Cho SY, Peck DH, Shin DR, Kim JS (2002) Performance evaluation of a Nafion/silicon oxide hybrid membrane for direct methanol fuel cell. J Power Sources 106:173
51. Adjemian VT, Lee SJ, Srinivasan S, Benzinger J, Bocarsly AB (2002) Silicon oxide nafion composite membranes for proton-exchange membrane fuel cell operation at 80–140°C. J Electrochem Sci 149:A256
52. Mioc U, Davidovic M, Tjapkin N, Colomban Ph, Novak A (1991) Equilibrium of the protonic species in hydrates of some heteropolyacids at elevated temperatures. Solid State Ionics 46:103
53. Tatsumisago M, Honjo H, Sakai Y, Minami T (1994) Proton-conducting silica-gel films doped with a variety of electrolytes. Solid State Ionics 74:105
54. Tazi B, Savadogo O (2000) Parameters of PEM fuel-cells based on new membranes fabricated from Nafion®, silicotungstic acid and thiophene. Electrochim Acta 45:4329
55. Nakajima H, Nomura S, Sugimoto T, Nishikawa S, Honma I (2002) High temperature proton conducting organic/inorganic nanohybrids for polymer electrolyte membrane. J Electrochem Soc 149:A953
56. Stati P, Arico AS, Baglio V, Lufrano F, Passalacqua E, Antonucci V (2001) Hybrid Nafion–silica membranes doped with heteropolyacids for application in direct methanol fuel cells. Solid State Ionics 145:101
57. Shao ZG, Joghea P, Hsing IM (2004) Preparation and characterization of hybrid Nafion–silica membrane doped with phosphotungstic acid for high temperature operation of proton exchange membrane fuel cells. J Membrane Sci 229:43
58. Arico AS, Baglio V, Di Blasi A, Creti P, Antonucci P, Antonucci V (2003) Influence of the acid–base characteristics of inorganic fillers on the high temperature performance of composite membranes in direct methanol fuel cells. Solid State Ionics 161:251
59. Lin YF, Yen CY, Ma CM, Liao SH, Lee CH, Hsiao YH, Lin HP (2007) High proton-conducting Nafion®/–SO_3H functionalized mesoporous silica composite membranes. J Power Sources 171:388
60. Su YH, Liu YL, Sun YM, Lai JY, Wang DM, Gao Y, Liu BL, Guiver MD (2007) Proton exchange membranes modified with sulfonated silica nanoparticles for direct methanol fuel cells. J Membr Sci 296:21
61. Lin HL, Yu TL, Han FH (2006) A method for improving ionic conductivity of Nafion membranes and its applications to PEMFC. J Polym Res 13:379
62. Nasef M, Hegazi ESA (2002) Preparation and applications of ion exchange membranes by radiation-induced hraft copolymerization of polar monomers onto non-polar films. Eur Polym J 38:87
63. Kim D, Sauk J, Byun J, Lee KS, Kim H (2007) Palladium composite membranes using supercritical CO_2 impregnation method for direct methanol fuel cells. Solid State Ionics 178:865
64. LaConti AB, Hamdan M, Kosek JA, Menezes T. Direct organic fuel cell proton exchange membrane and method of manufacturing the same. US Patent 116546A1
65. Lee HY, Kim JY, Park JH, Joe Y, Lee T (2004) Performance of polypyrrole-impregnated composite electrode for unitized regenerative fuel cell. J Power Sources 131:188
66. Smit MA, Ocampo A, Espinosa-Medina MA, Sabastian PJ (2003) A modified Nafion membrane with in situ polymerized polypyrrole for the direct methanol fuel cell. J Power Sources 124:59
67. Hoku Scientific, Inc. (2005) Composite electrolyte with crosslinking agents. US Patent 6,962,959
68. Dow Chemical Co. (1997) Composite fuel cell membranes. US Patent 5,654,109

69. Choi SW, Fu YZ, Ahn YR, Jo SM, Manthiram A (2008) Nafion-impregnated electrospun polyvinylidene fluoride composite membranes for direct methanol fuel cell. J Power Sources 180:167
70. Wu ZM, Sun GQ, Jin W, Wang Q, Hou H, Chan K-Y, Xin Q (2007) Use of in situ polymerized phenolformaldehydr resin to modife Nafion ® membrane for the direct methanol fuel cell. J Power Sources 167:309
71. Ainla A, Brandell D (2007) Nafion®–polybenzimidazole (PBI) composite membranes for DMFC applications. Solid State Ionics 178:581
72. Volkov VV, Mchedlishvili BV, Roldugin VI, Ivanchev SS, Yaroslavtsev AB (2008) Membranes and nanotechnologies. Nanotechnologies 3:656 (in Russian)
73. Rusanov AL, Solodova EA, Bulycheva EG, Abadie MJ, Voitekunas VYu (2007) Synthesis of polymers with protogenic groups by polymer-analogous transformations. Russ Chem Rev 76:1073
74. Dobrovol'skii YuA, Dzhannoli P, Lafitt B, Belomoina NM, Rusanov AL, Likhachev DYu (2007) Achievements in the field of proton- conductive portion electrolyte membranes. Russ J Electrochem+ 43:489
75. Rusanov AL, Kostoglodov PV, Abadie MJ, Voytekunas VYu, Likhatchev DYu (2008) Proton-conducting polymers and membranes carrying phosphonic acid groups. Adv Polym Sci 216:125
76. Rusanov AL, Bulycheva EG, Bugaenko MG, Voitekunas VYu, Abadie M (2009) Sulfonated polynanaphthylimides as proton-conducting membranes for fuel cells. Russ Chem Rev 78:53
77. Kabasawa A, Saito J, Yano H, Miyatake K, Uchida H, Watanabe M (1076) Durability of a novel sulfonated polyimide membrane in polymer electrolyte fuel cell operation. Electrochim Acta 2009:54
78. Peckham TJ, Schmeisser J, Rodgers M, Holdcroft S (2007) Main-chain, statistically sulfonated proton exchange membranes: the relationships of acid concentration and proton mobility to water content and their effect upon proton conductivity. J Mater Chem 17:3255
79. Shi ZQ, Holdcroft S (2005) Synthesis and proton conductivity of partially sulfonated poly ([vinylidene difluoride-co-hexafluoropropylene]-b-styrene) block copolymers. Macromolecules 38:4193
80. DeLuca NW, Elabd YA (2006) Nafion®/poly(vinyl alcohol) blends: effect of composition and annealing temperature on transport properties. J Membr Sci 282:217
81. Elabd YA, Napadensky E, Walker CW, Winey KI (2006) Transport properties of sulfonated poly(styrene-isobutylene-styrene) triblock copolymers at high ion-exchange capacities. Macromolecules 39:399
82. Yin Y, Fang JH, Watari T, Tanaka K, Kita H, Okamoto K (1062) Synthesis and properties of highly sulfonated proton conducting polyimides from bis(3-sulfopropoxy)benzidine diamines. J Mater Chem 2004:14
83. http://www/fuelcelltoday.com/FuelcellToday/Industry Directory/Industry Directory External/Industry Directory Display Company/04591 2234 00 html (2005)
84. Davis BL, Dixon DA, Garner EB, Gordon JC, Matus MH, Scott B, Stephens FH (2009) Efficient regeneration of partially spent ammonia borane fuel. Angew Chem Int Ed 48:6812

Chapter 22
Neutron Studies of Nanoscale Fullerenes and Fullerene Hydrides

V.A. Somenkov

Abstract Structure changes in amorphous phases of different modifications of carbon (diamond, fullerene) have been investigated by means of neutron and x-ray diffraction under irradiation and temperature. Polyamorphic transition (diamond-like – graphite-like phases) was found in irradiated diamonds by density change. It has been found that the amorphous fullerites subjected to high temperature (600–1,700°C) annealing undergo a polyamorphic transition from the molecular glass to the atomic glass, which is accompanied by the disappearance of fullerene halos at small scattering angles. We also studied of structure, sorption properties and thermal stability of intercalation compounds of crystalline and amorphous fullerenes C_{60} and C_{70} with saturated and unsaturated molecules. The analysis of structure by neutron scattering and precise thermogravimetry shows that sorption is determined by formation of solid solutions, superstructures or compounds on base of initial fullerenes.

Keywords Neutron diffraction · Amorphous carbon systems · Phase transition

22.1 Introduction

Phase transitions in disordered systems (amorphous solids, liquids, etc.), which sometimes are called polyamorphic transitions [1], occur at variations of thermodynamic parameters, such as pressure [2, 3] and temperature [4], and also under irradiation [5, 6]. Unlike the polymorphic transformations occurring in crystal systems in which the transitions are caused primarily by the change in the entropy, the polyamorphic transitions are associated first of all with the change in the internal energy (bonding type, coordination of nearest neighbors, etc.), particularly in the

V.A. Somenkov (✉)
RRC "Kurchatov Institute", Kurchatov sq. 1, 123182 Moscow, Russia
e-mail: somekov@usssph.kae.ru

S.Yu. Zaginaichenko et al. (eds.), *Carbon Nanomaterials in Clean Energy Hydrogen Systems - II*, NATO Science for Peace and Security Series C: Environmental Security 2,
DOI 10.1007/978-94-007-0899-0_22,

case of corresponding crystalline analogs. Crystals of simple materials undergo transitions from the molecular phase to the atomic phase, e.g., in iodine under high pressure [7] and in fullerenes at high temperatures and pressures [8–10]; in the latter case, the transition is accompanied by either polymerization or amorphization with the formation of nanodiamonds after rapid quenching [8]. In this work, we made an attempt to elucidate the possibility of such transitions occurring in amorphous systems by using carbon as an example and investigating the evolution of carbon systems under variations of temperature by diffraction methods.

22.2 Polyamorphic Transition in Irradiated Diamonds

A phenomenon of polyamorphism manifests itself in the existence of different structural forms of disordered state and can be expected for volume changes of different origin, particularly, in the irradiated crystals after its radiation amorphization. An increase in volume upon reactor irradiation is large and can reach several tens of percent for sufficiently high neutron fluences [11], which is equivalent to a "negative" pressure of several tens or even hundred of gigapascals.

In light of this circumstance, we use the neutron diffraction method in order to determine which structural changes occur in an irradiated diamond upon density change. As samples, we used natural-diamond powders with a mean particle size 14–20 μm to 0.5 mm that are irradiated in a beryllium block of an MR reactor, cooled by running water, up to a fluence of 1.51×10^{21} (175 days in a neutron flux of about 10^{14} $cm^{-2}s^{-1}$ with energies higher than 0.18 MeV). The irradiated powders turned out to be strongly inhomogeneous its densities being distributed in the range from 3.24 to 2.05 g/cm^3 (with an accuracy of no worse than 2%) [12]. The diffraction experiments were carried out on the DISK diffractometer [13] at the 4.5-MW IR-8 reactor. The wavelength of monochromatic neutrons was equal to 1.667 Å. Samples of various fractions with masses of 50–100 mg were investigated [14].

According to the results presented in Fig. 22.1, at density decrease, the diffraction lines of diamond broaden, the "tails" of diffraction lines overlap, and a "halo" corresponding to the formation of a fine-crystalline ("amorphous") material of the diamond-like type is formed. With a further decrease in density, the diffraction pattern exhibits a new halo, whose intensity increases gradually and whose position corresponds to the position of the first maximum on the defect pattern of irradiated graphite [15] or amorphous carbon (activated graphite) (Fig. 22.2a). The absence of small-angle scattering (for $q > 2 \times 10^{-2}$) from all samples indicates that they are homogeneous on a scale of 2–3 nm. The results can be treated as an evidence of a polyamorphic transition from diamond-like to graphite-like glass, which occurs when density decreases and which is likely associated with a decrease of the number of nearest neighbors in the first coordination sphere from four to three (in contrast to its increase at high pressures).

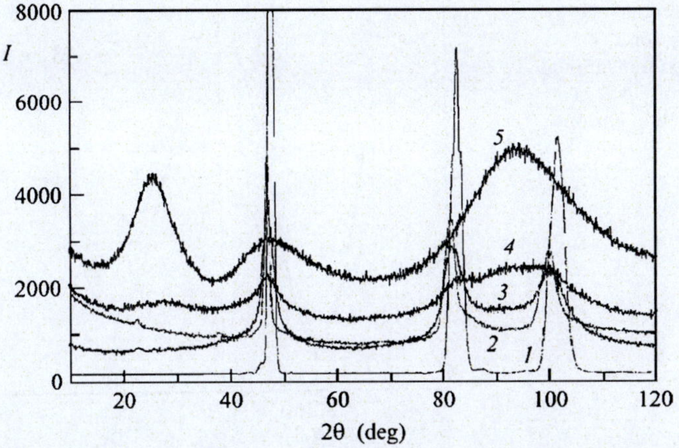

Fig. 22.1 Diffraction patterns showing the transition from a diamond-like glass to a graphite-like structure upon the change in density: (*1*) initial diamond, (*2*) 3.38, (*3*) 3.10, (*4*) 2.68, and (*5*) 2.11 g/cm^3

Fig. 22.2 (**a**) Diffraction patterns for low-density irradiated diamond (2.11 g/cm^3) and amorphous graphite and (**b**) resistivity vs. density

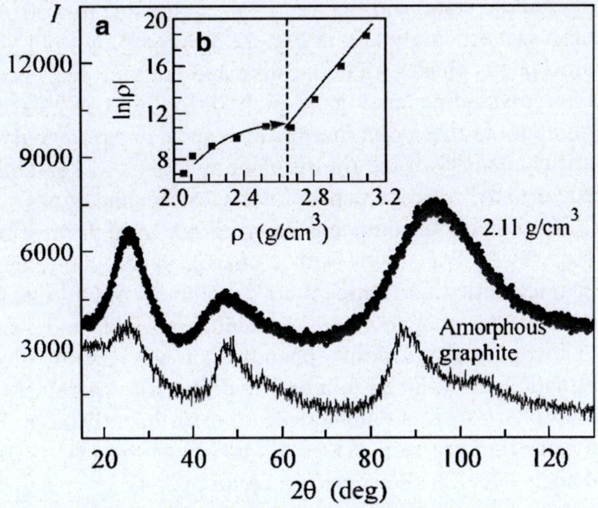

This transition is accompanied by a change in the resistivity of powders, which is measured by means of pressure contacts (Fig. 22.2b). The total change in resistivity is equal to six orders of magnitude in the density range under investigation and corresponds to a transition from the dielectric state to the metal one. The critical density, i.e., the density at which the transition occurs, is equal to $\rho \cong 2.7\text{–}2.9$ g/cm^3 according both to the diffraction measurements and to the electric-resistivity measurements.

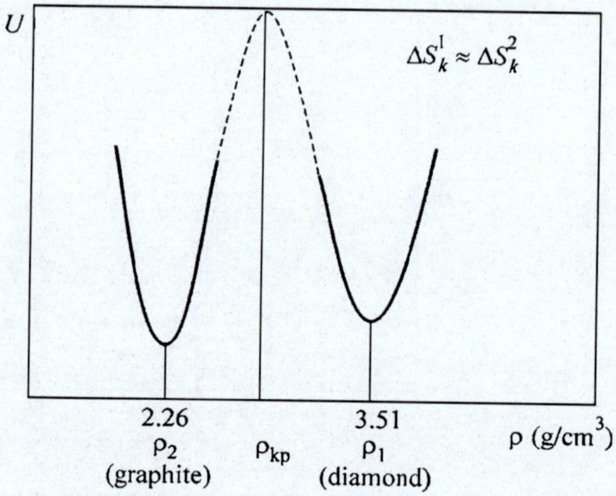

Fig. 22.3 Internal energy *U* vs. density for a polyamorphic transition at $\rho_{cr} \approx 2.7$ g/cm^3; ΔSk is the configuration entropy

Since the polyamorphic transition occurs between disordered phases at low temperatures, the role of entropy is not as large as in polymorphic transitions in a crystalline state, and the transition occurs due to a change in the internal energy, as is shown schematically in Fig. 22.3. According to Fig. 22.3, the transition between amorphous phases (diamond-like and graphite-like) is attributed to the existence of their crystalline analogs, which differ both in density and in the coordination number, so that polyamorphism is closely associated with polymorphism and the critical density of the transition corresponds to a saddle point and is approximately equal to the average density of crystalline analogues.

A similar phenomenon likely occurs in high pressure amorphous phases (SiO_2, H_2O, etc.). Transitions with a change in the coordination number (tetra–octa) are characteristic of a "lattice fluid" – solutions of hydrogen in metals [16] – where they occur upon change in temperature, pressure, and interstitial atom concentration [17, 18]. Finally, a similar phenomenon was recently observed upon the annealing of radiation-amorphized fullerene hydrides [6]. Therefore, it is not excluded that polyamorphic transitions are not rare upon sufficiently large change in density. In this case, the effect of pressure ($\Delta V < 0$) and irradiation ($\Delta V > 0$) makes it possible to change density over a wide range (to a factor of 2–4).

22.3 Polyamorphic Transition in Amorphous Fullerites C_{60}

The initial samples of 99.5%-purity C_{60} fullerenes fabricated by NeoTekProduct were prepared by high-temperature treatment of graphite followed by the isolation with organic solvents and a further chromatographic separation. The impurity composition of the samples was determined using spark mass spectroscopy at the Analytical-and-Certification Center of GIREDMET for the majority of the periodic table elements. It was established that main impurities are sulfur (0.09 wt%), aluminum (0.0006 wt%), and silicon (0.0008 wt%). The neutron diffraction patterns of the initial crystalline fullerenes correspond to the fcc lattice with the parameter $a = 14.16$ Å and provide good agreement with the experiment (the difference between the experimental and calculated intensities R_b did not exceed 4%).

Amorphous fullerites were prepared through mechanoactivation by grinding small weighed portions (~1.0–1.5 g) of the crystalline C_{60} fullerite in a Frisch-type mill at low grinding rates for long (to 120 h) times and fixing structural changes by neutron diffraction. Samples of amorphous fullerenes (C_{60} and C_{60}/C_{70} mixture) produced by application of mechanoactivation treatment in air and helium were received and their structural stability in relation to temperature and pressure influences were investigated. The results obtained shows that under mechanoactivation of fullerenes two processes occur – first, amorphization (at low milling velocities) with formation of a nanoscale fullerene-like amorphous phase and, second, graphitization (at high milling velocities).

Figure 22.4 shows that, as the grinding duration increases, the diffraction peaks, which are characteristic of the crystalline C_{60} fullerite, turn into wide fullerene halos at low mechanoactivation velocities, which are typical for amorphous (fine-grained) phases. In this case, the diffraction pattern ceases to change beginning from the grinding time of 40–60 h and the particle sizes estimated from the halo width are 2–4 nm. The samples thus prepared were annealed to high-temperature (up to 1,700°C) by step wise annealing in a vacuum furnace for 4 h in each cycle and, then, were cooled to room temperature.

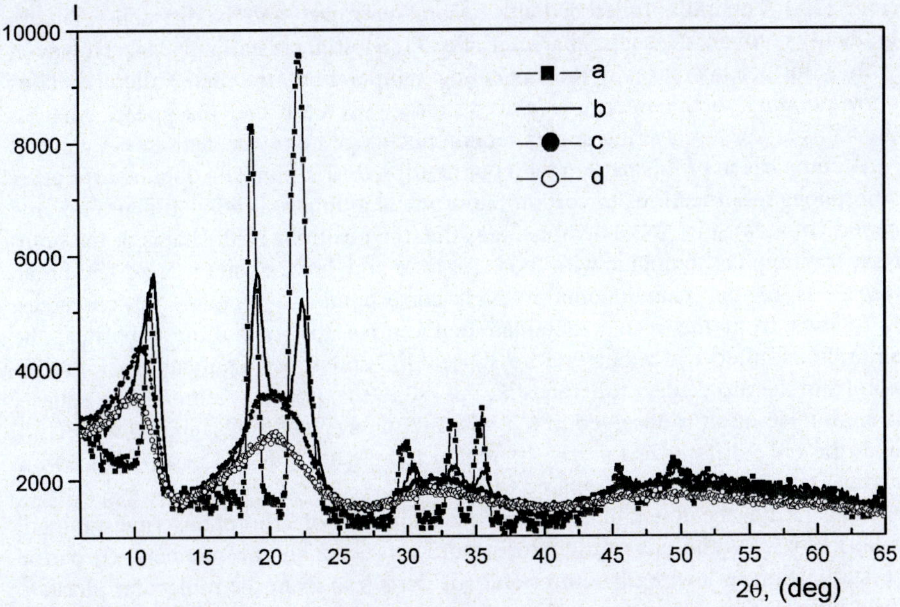

Fig. 22.4 Neutron diffraction patterns of the C_{60} fullerites for different times of grinding t = (*a*) 0, (*b*) 21, (*c*) 44, and (*d*) 58 h during the low-rate (<100 min^{-1}) mechanoactivation

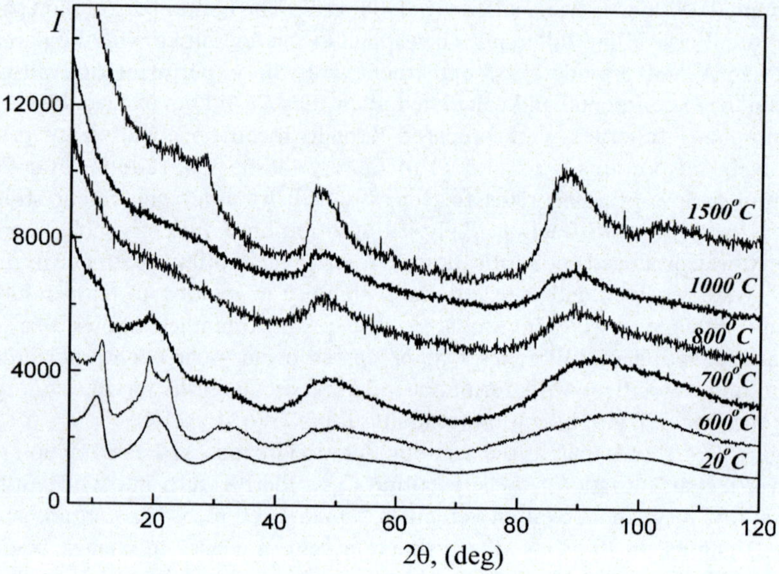

Fig. 22.5 Variations of the neutron diffraction patterns of the amorphous fullerites after annealing at temperatures $T = (1)$ 20, (2) 600, (3) 700, (4) 800, (5) 1,000, and (6) 1,500°C

As the annealing temperature increases, the diffraction pattern changes (Fig. 22.5). The initial fullerene halos become narrower, and the diffraction pattern gradually recovers the shape characteristic of the initial crystalline phase. However, with a further increase in the annealing temperature, the first fullerene halos corresponding to the intermolecular distances decrease and disappear, whereas the halos at large scattering angles remain unchanged or even increase.

A comparison of this pattern with our neutron diffraction data obtained for other amorphous modifications of carbon (amorphous graphite, irradiated diamond, and nanodiamonds) (Fig. 22.6) demonstrates that the positions of the halos of the high-temperature phase coincide with those for diamond, but they are more diffuse than the halos of the nanodiamond, which corresponds to smaller particle sizes. It follows from the results obtained that, unlike the crystalline fullerites, the amorphous fullerite at a temperature $T > 600$°C undergoes a polyamorphic transition from the molecular (fullerene-like) phase to the atomic (diamond-like) phase. A similar situation is observed in C_{60}–C_{70} mixtures (70 wt% C_{60} and 30 wt% C_{70}), with the only difference that the transition temperature in this case is somewhat higher (by 50–100°C) as compared to pure C_{60}.

Thus, it has been established that the evolution of amorphous (fine-grained) fullerites with variations in the temperature can occur in two manners: (i) partial crystallization at low temperatures and (ii) transition from the molecular phase to the atomic phase. This suggests that, when the particle sizes are small, the atoms at their boundaries can be bonded not only by weak (van der Waals) intermolecular bonds but also by strong (covalent) interatomic bonds, as is schematically shown in

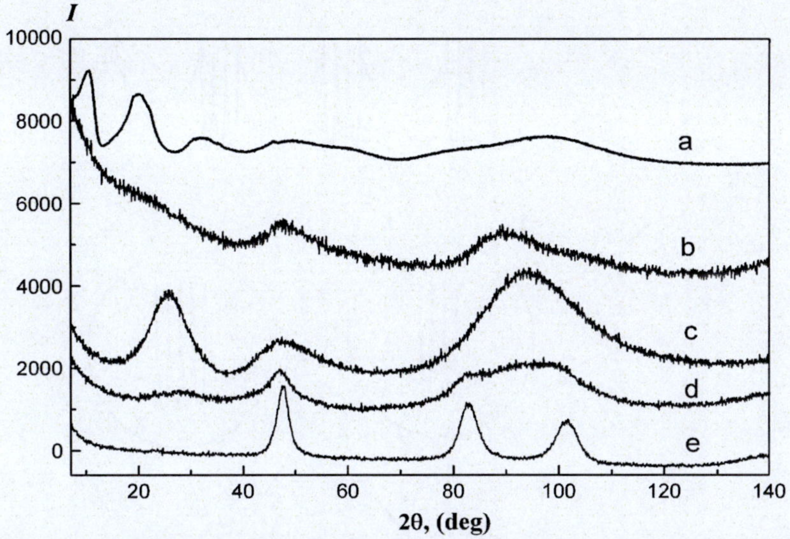

Fig. 22.6 Comparison of the neutron diffraction patterns of different amorphous carbon phases: (**a**) fullerene_like amorphous phase after mechanoactivation (molecular glass), (**b**) diamond_like amorphous phase after high-temperature annealing (atomic glass), (**c**) amorphous graphite, (**d**) irradiated amorphous diamond, and (**e**) nanodiamond

Fig. 22.7. Accordingly, it turned out that the sublimation temperature of the high-temperature amorphous phase is $T = 1{,}700°C$; i.e., it is almost twice as large as the temperature T_{evap} of the crystalline phase (870°C) and exceeds the melting temperatures T_m of many metals of practical interest (Fe, Ni, Al, Cu). The latter circumstance makes this phase promising for the use as additions in the preparation of nanomaterials. In order to elucidate the role played in this case by light atoms (O, N), which can enter into amorphous fullerites during a long-term mechanoactivation in air, we repeated the experiments in an inert atmosphere (He) and established that the same transition occurs during annealing. Thus, the particle size, rather than their composition, plays a decisive role in the transition.

22.4 Interaction of Crystline and Amorphous Fullerites with Hydrocarbon

Using neutron and X-ray diffraction the structure of the compound of fullerenes and hydrocarbons with hydrocarbon chloric derivatives was studied. The complexes production was performed by diffusion introduction of the molecules of methylene chloride (CH_2Cl_2), chloroform ($CHC1_3$), carbone tetrachloride (CCl_4), hexane (C_6H_{14}), heptane (C_7H_{16}) and octane (C_8H_{18}) into the space between fullerene molecules. It was detected creation of isomorphic compounds of the approximate composition C_{60}*2CCl_4, C_{60}*2$CHCl_3$, which have the simple

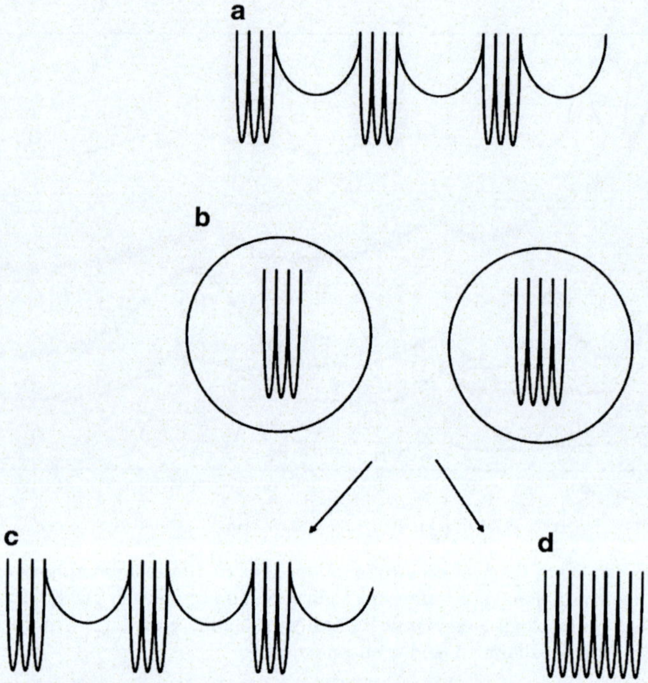

Fig. 22.7 Schematic diagram illustrating the phase transformations during temperature evolution of the amorphous fullerites: (**a**) potential relief in the fullerite crystals (molecular crystal), (**b**) nanoparticles in the amorphous fullerites (molecular glass), (**c**) recovery into the fine-grained state at low temperature (<600°C) annealing of the amorphous fullerenes (the onset of the crystallization of the molecular glass), and (**d**) polyamorphic phase transition during high-temperature (800–1,000°C) annealing of the amorphous fullerenes (transition from the molecular (fullerene-like) glass to the atomic (diamond-like) glass)

hexagonal crystal lattice with the lattice constants a = 10.12 Å, c = 10.64 Å for C_{60}*$2CCl_4$ and a = 10.01 Å, c = 9.94 Å for C_{60}*$2CHCl_3$. The ratio c/a increases at substitution of hydrogen by Cl.

Stability of the synthesized compounds was investigated and it was shown that the $C_{60}\cdot 2CCl_4$ samples lose CCl_4 near completely after holding during several weeks transforming into the nearly pure C_{60} with the slightly increased lattice parameter. In $C_{60}\cdot 2CHCl_3$ only some changes of diffraction reflex intensity are observed whereas $C_{60}\cdot 2CH_2Cl_2$ remains full stable even after holding during several months. All compounds lose hydrocarbon molecules at temperature increasing to 80°C. The obtained results show that the compound stability increases with hydrogen content increasing. Examining the neutron diffraction picture of compounds, which contain heptane (C_7H_{16}), octane (C_8H_{18}) and carbon tetrachloride (CCl_4), we can see additional (with respect to pure fullerene) peaks, which can be indexed using the same fcc lattice as pure fullerite has (Fig. 22.8). The analysis indicated that this picture is typical for superstructures with wave vector $\kappa = 2\pi b^*$ and with A_2X or A_4X_3 composition. Possible variants of these superstructures were analyzed within the framework of the concentration waves theory [19].

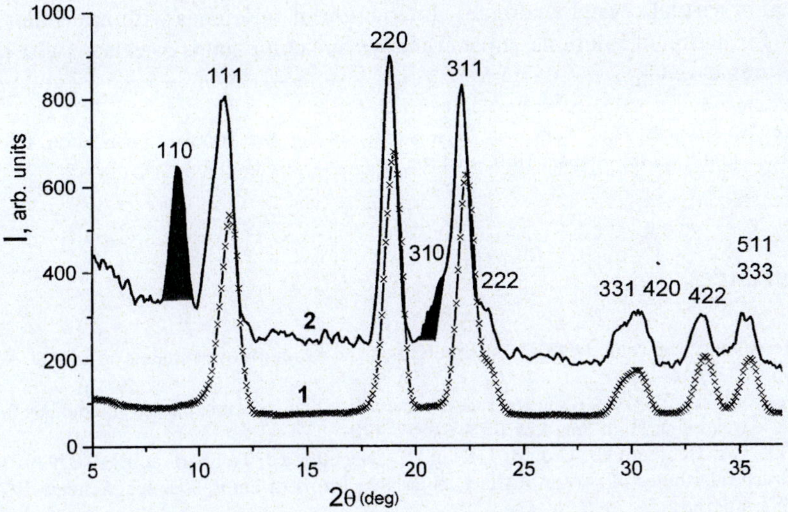

Fig. 22.8 Diffraction pattern of C_{60} with heptane: (*1*) – C_{60}, (*2*) – C_{60} + C_7H_{16}

Using the thermogravimetry method the dependence of desorption temperature on the composition of the compound being absorbed, including (CH_2Cl_2) – ($CHC1_3$) – (CCl_4) was received. Thermal stability of $C_{60} \cdot CH_{4-n}Cl_n$ compounds goes up together with chlorine atom quantity increase in the chlorine-replaced methane molecule. Extrapolating the received dependence we can expect the desorption temperature of the potential compound with CH_4 to be close to 50°C.

For the compound containing C_{60} and unsaturated halogen derivative – tetrachloroethylene C_2Cl_4 comparative analysis with in the framework of the concentration waves theory indicated that tetrachloroethylene compound represents another type of infusion superstructure as compared to the C_{60}· CCl_4. Additional superstructure peaks take place at the neutron diffraction picture. They correspond to the star of wave vector $\kappa = 2\pi b^*$ and to octahedral superstructure of the A_4X type, meanwhile CCl_4 compound corresponds to tetrahedral superstructure of the A_2X type. Precise derivative-graphical measurements confirm mole ratio fullerite/absorbate, which is 1/1 and it conforms to face-centered cubic lattice cavity occupation with C_{60}.

Mole ratio fullerene/absorbate, equal to 1/1, is characteristic for saturated molecules, while for unsaturated molecules it arrives at 1/2. This circumstance can be explained by specific interaction between π-electronic shells of unsaturated derivatives and the electronic structure of fullerenes and it can specify on possible using of fullerites for absorption and storage of unsaturated matters [20].

22.5 Conclusions

Polyamorphic phase transitions in the studied nanoscale carbon systems are associated with the appearance of phases, characteristic for high temperatures and high pressures, or with the absorption of gases. These phases correspond to the presence of real or virtual crystal analogues. Interaction of fullerenes with molecules leads to the formation of solutions, superstructures and compounds correlated with matrix structures C_{60}, C_{70}.

Acknowledgements This work has been supported by the Russian Foundation for Basic Research: grants 09-02-00464a, 10-02-00176a.

References

1. Benmore C, Siewenie J (2004) Polyamorphism and extreme environments on GLAD. Neutron New 15(3):16–18
2. Ponyatovsky EG (2000) Neutron scattering study of high-pressure phases obtained by thermobaric treatment. High Pres Res 17(3–6):157–170
3. Tulk CA, Benmore CJ, Urquidi J, Klug DD, Neuefeind J, Tomberli B, Egelstaff PA (2002) Structural studies of several distinct metastable forms of amorphous ice. Science 297:1320 (Washington)
4. Balagurov AM, Barkalov OI, Kolesnikov AI, Mironova GM, Ponyatovsky EG, Sinitsyn VV, Fedotov VK (1991) Neutron-diffraction study of phase transitions of high-pressure metastable ice VIII. JETP Lett 53(1):30
5. Agafonov SS, Glazkov VP, Nikolaenko VA, Somenkov VA (2005) Short-range order in irradiated diamonds. JETP Lett 81(3):122–124
6. Voronin VI, Glazkov VP, Goshchitsky BN, Somenkov VA, Teplykh AE (2002) Effect of reactor irradiation on the structure of fullerene hydrides. In: Abstracts of the XVII workshop on applications of neutron scattering in investigations of the Condensed State, Gatchina, Russia, 2002, p. 169 (in Russian)
7. Tonkov EYu (1988) High pressure phase transformations: a handbook, vol 2. Metallurgiya, Moscow, 358 p (Gordon & Breach, Philadelphia 1992) (in Russian)
8. Hirai H, Kondo K, Yoshizava N, Shiraishi M (1994) Amorphous diamond from C_{60} fullerene. Appl Phys Lett 64(14):1797–1799
9. Moseler M, Riedel H, Gumbsch P, String J, Mehlig B (2005) Understanding of the phase transformation from fullerite to amorphous carbon at the microscopic level. Phys Rev Lett 94:165503
10. Blank VD, Buga SG, Dubitsky GA, Serebryanaya NR, Popov MYu, Sundqvist B (1998) High-pressure polymerized phases of C_{60}. Carbon 36(4):319–343
11. Vance ER (1971) X-ray study of neutron irradiated diamonds. J Phys C Solid State Phys 4 (3):257–263
12. Kats MYa (1966) New methods in studies of minerals in gravitational field. Nauka, Moscow, p 162, in Russian
13. Glazkov VP, Naumov IV, Shilshtein SSh, Somenkov VA (1988) Superpositional many detector systems and neutron diffraction of microsamples. Nucl Instrum Method Phys Res A 264:367–374
14. Nikolaenko VA, Zabusov OO, Agafonov SS, Glazkov VP, Somenkov VA (2004) Neutron and X-ray investigations of phase transitions in diamond under irradiation. Extended abstracts, XVIII meeting on use of neutron scattering in studies of Condensed State, (in Russian) Zarechnyoe, Russia, p 93
15. Platonov PA, Chugunov OK, Alekseev SI, Vindryaevsky BA, Karpukhin VI, Tumanov YuP (1973) Study of radiation defects in irradiated pyrographite (in Russian). Preprint IAÉ-2266 – Institute of Atomic Energy, Moscow.
16. Somenkov VA, Shilstein SSh (1979) Phase transition of hydrogen in metals. Prog Mater Sci 24:267–268

17. Somenkov VA, Irodova AV, Shil'shtein SSh (1986) Phase transformations in hydrides of transition metals at high pressures (in Russian), Fizika Tverdogo Tela (Leningrad), 28:3200 (Phys Solid State, 28:1804)
18. Fukai Yu (1995) Formation of superabundant vacancies in metal hydrides at high temperatures. J Alloy Compd 231(1–2):35–40
19. Agafonov SS, Glazkov VP, Somenkov VA, Filippov AA (2005) Interaction of fullerenes and nanotubes with hydrocarbons and their derivatives. In: Abstracts book 7th Biennial International Workshop "Fullerenes and atomic clusters", St-Petersburg, Russia, June 27–July 1 2005, p 144
20. Agafonov SS, Somenkov VA, Filippov AA (2009) Structure of intercalates C_{60} and C_{70}. In: Abstracts of 9th Biennial International Workshop "Fullerenes and Atomic Clusrers" (IWFAC 2009), St-Peterburg, Russia, 6–10 July 2009 p 108

Chapter 23
Electronic Structures of Fullerene C_{60} Derivative: DFT Modeling

D.V. Lopatin

Abstract In present paper we report results of computer simulation of electronic structure of the molecular complexes – fullerene C_{60} and donors: TMPDA, Bz_4BTPE, LMG and LCV. It is shown, that each donor molecules forms several types shortened van-der-Waals contacts with adjacent C_{60} spheres. All complexes have energy spectrum typical for semiconductors. The molecular orbital spectrum indicates that molecular complexes have more fine electronic structure than the isolated fullerenes C_{60} and the donor molecules. Displacement and increase in number of lines in the spectrum of molecular complexes in comparison with the spectrum of individual C_{60} and the donor molecules take place.

Keywords Computer simulation · Electronic property and structure · Fullerene · Complex · Van-der-Waals contact · Energy spectrum

23.1 Introduction

In recent years fullerene-based materials have brought up steadily increasing attention both in academic and industrial researches. It is of great interest to use fullerene and fullerite not only by itself but also as modified derivatives, that enhance the properties of fullerenes and extend the application possibility. Recent developments suggest that fullerene-based materials could be successfully used in a wide range of areas such as IT devices, solar cells, clinical diagnostics, pharmaceuticals, environmental and energy industries. It seems interesting to study the electron-optical properties of photoactive fullerene C_{60}-based molecular complexes characterized by a high quantum efficiency of charge transfer and long-lived charge separation [1, 2]. In present paper we report results of computer simulation of electronic structure of the molecular complexes fullerene C_{60} and

D.V. Lopatin (✉)
Tambov State University, 392000 Tambov, Russia
e-mail: lopatin@tsu.tmb.ru

S.Yu. Zaginaichenko et al. (eds.), *Carbon Nanomaterials in Clean Energy Hydrogen Systems - II*, NATO Science for Peace and Security Series C: Environmental Security 2,
DOI 10.1007/978-94-007-0899-0_23,

organic donors: TMPDA (N,N,N′,N′-tetramethyl-p-phenylenediamine), Bz4BTPE (tetrabenzo(1,2-bis[4 H-thiopyran-4-ylidene]ethene)), LCV (Leuco Crystal Violet), LMG (Leucomalachite Green).

23.2 Methods

Initial data preparation and intermediate transformations were carried out using original software. Optimization of geometry of complexes was determined in several ways. For initial calculations we used a method of molecular mechanics Universal Force Field (UFF) [3]. UFF method is realized in software packages Gaussian 03 [4] or ArgusLab [5]. It allows one to refine configuration quickly and, thus, to reduce total calculation time. The other approaches are semi-empirical method PM3 (Parametric Method 3) [6] with various parameterization in three quantum-chemical packages PC-GAMESS [7] or Gaussian 03. Density functional theory (DFT) calculations were performed using the Gaussian 03. Electronic structure were calculated B3LYP (hybrid functional – Becke's three-parameter formulation [8] and gradient corrected correlation functional – Lee, Yang and Parr [9]) with the 6-311++g(2d, 2f) basis set. Tight SCF convergence criteria (10^{-6} to 10^{-8} a.u.) were used for all calculations.

23.3 Results

Modeling results for isolated molecules correspond to the X-ray diffraction data for fullerene C_{60} based molecular complexes with volumetric substitutes indicating that donor molecules are bent in such compounds [10]. The bend of donor molecule skeleton provides its more dense conformity to the spherical C_{60} molecule that increases the number of energetically favorable van-der-Waals contacts. Each donor molecules forms several types shortened van-der-Waals contacts with adjacent C_{60} spheres (Table 23.1). These distances are corresponding to the X-ray diffraction analysis of real crystals [11, 12] (see references in Table 23.1). It is worth mentioning that the data on structure LCV·C_{60} are received for the first time and absent in Cambridge Crystallographic Data Centre [13].

Computer modeling of C_{60}, Bz4BTPE·C_{60}, LMG·C_{60}, LCV·C_{60} and TMPDA·C_{60} molecules electronic structure based on non-empirical DFT (B3LYP/6-311++g(2d, 2f)) calculations allowed to define molecular orbitals energy E and destiny of states (DOS) spectrum. Figure 23.1 shows calculated spectrum for complex near Fermi level. According to Fig. 23.1 all complexes have energy spectrum typical for semiconductors with band-gap $\Delta E = 1.265$–1.731 eV. For the donor isolated molecules $\Delta E = 2.9$–4.7 eV that considerably exceeds (on 1.2–3.2 eV) the same parameter for the molecular complex. The HOMO-LUMO gap for investigated complexes is less than for fullerene C_{60} molecule. Data of researches of optical properties real crystals [11, 12, 14, 15] confirms results of modeling (Table 23.2).

Table 23.1 Structural data

	Van-der-Waals contacts in molecule		Van-der-Waals contacts in 3D cluster	
	Type	Range, nm	Type	Range, nm
TMPDA · C_{60}	C(TMPDA) · C (C_{60})	0.2962–0.3112; 0.301 Ref. [11]	–	
	C(CH_3) · C(C_{60})	0.2849		
	H(CH_3) · C(C_{60})	0.2635;0.263 Ref. [11]		
	N(TMPDA) · C (C_{60})	0.3185		
Bz_4BTPE · C_{60}	C(C_9SH_4) · C(C_{60})	0.3130–0.3594;	0.340–0.353 Ref. [12]	–
	H(C_6H_4) · C(C_{60})	0.2809–0.3102		
	S(C_9SH_4) · C(C_{60})	0.3577 0.357 Ref. [12]		
LMG · C_{60}	C(LMG) · C(C_{60})	0.3196–0.3410 0.322–0.341 Ref. [11]	C(LMG) · C(C_{60})	0.3208–0.3410; 0.322–0.341 Ref. [11]
	H(CH_3) · C(C_{60})	0.2664	H(CH_3) · C(C_{60})	0.2686
	N(LMG) · C(C_{60})	0.2996; 0.299–0.320 Ref. [11]	N(LMG) · C(C_{60})	0.2984; 0.2986; 0.299–0.320 Ref. [11]
	–	–	C(C_{60}) · C(C_{60})	0.2842–0.3493; 0.300–0.338 Ref. [11]
LCV · C_{60}	C(LCV) · C(C_{60})	0.3244–0.3395	C(LCV) · C(C_{60})	0.3216–0.3413
	C(CH_3) · C(C_{60})	0.2832	C(CH_3) · C(C_{60})	0.2803–0.2896
	H(CH_3) · C(C_{60})	0.2623	H(CH_3) · C(C_{60})	0.2662
	N(LCV) · C(C_{60})	0.3060	N(LCV) · C(C_{60})	0.2863–0.3055
	–	–	C(C_{60})·C(C_{60})	0.2838–0.3498

Comparison of C_{60}, TMPDA·C_{60}, LMG·C_{60}, LCV·C_{60} and Bz4BTPE·C_{60} energy spectrum in UV–vis range indicates that molecular complexes have more fine electronic structure than the isolated fullerenes C_{60} and the donor molecules (see Fig. 23.2). Displacement and increase in number of lines in the spectrum of molecular complexes in comparison with the spectrum of individual C_{60} and the donor molecules take place.

Main causes of distinctions in spectrum of complexes and the isolated molecules connected with significant increase in number of optical transitions and their shift in the molecular complex forming process. There are two reasons of such changes. The first, distortion of C_{60} skeleton geometry in complexes leads to the symmetry reduction that effects on selection rules and the intermolecular excitation energy (shift and splitting of valence and unoccupied levels being degenerated for C_{60} with I_h-symmetry).

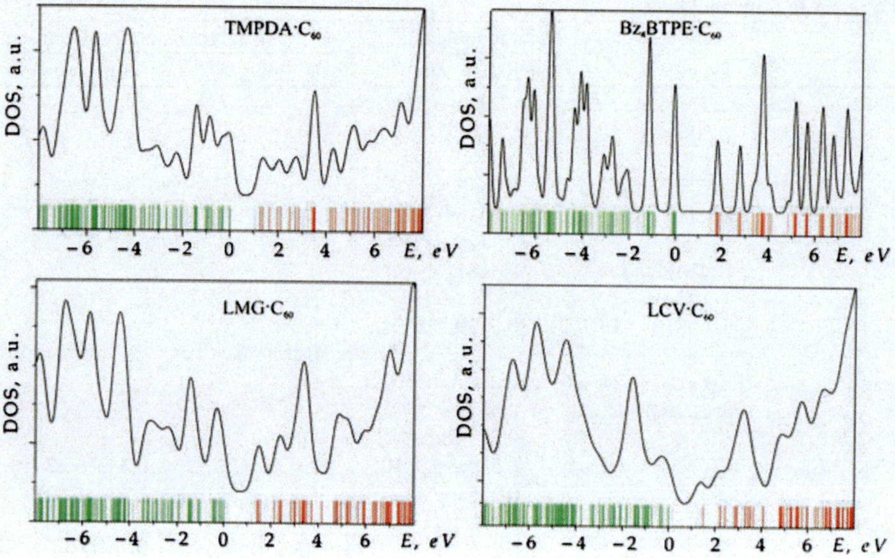

Fig. 23.1 DOS spectrum of molecular complexes TMPDA·C_{60}, Bz4BTPE·C_{60}, LMG·C_{60} and LCV·C_{60}

Table 23.2 Energy of optical transitions

Energy, eV												Ref. (method)
Bz_4BTPE · C_{60}												
1.55–1.77	1.96			2.16					3.72		4.81	[12] absorption
				2.29								[12] photoconductivity
	2.00			2.21		2.63			3.57			
1.73	1.85					2.75			3.74		4.81	DFT/B3LYP
						2.73			3.75		4.74	
TMPDA · C_{60}												
1.26		2.1						2.7				[11] absorption
		2.29						2.72			2.96	[14] photoconductivity
1.26		2.10			2.50			2.74			2.83	DFT/B3LYP
											2.90	
LCV · C_{60}												
		1.7			2.1			2.7			3.5	[11] absorption
1.49		1.71										[15] photoconductivity
1.49		1.67			2.23			2.84			3.45	DFT/B3LYP
		1.75										
LMG · C_{60}												
1.48			2.38				3.36			4.85		DFT/B3LYP

The second, in the complexes molecular orbitals are formed due to the π-electrons of phenyl fragments of Bz4BTPE, LMG, LCV donors (or benzene ring of TMPDA) in contrast to individual molecules. Electrons of phenyl rings carbon atoms participate in valence orbitals forming process due to rings considerable

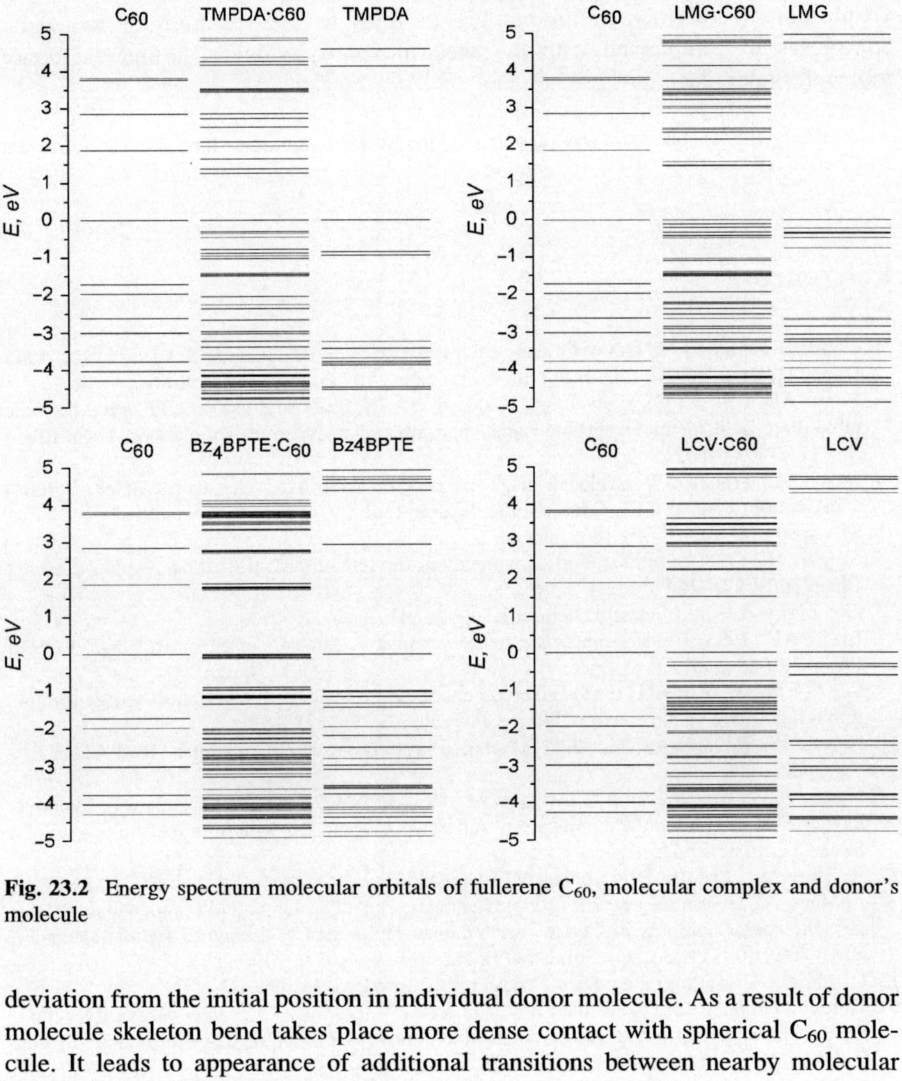

Fig. 23.2 Energy spectrum molecular orbitals of fullerene C_{60}, molecular complex and donor's molecule

deviation from the initial position in individual donor molecule. As a result of donor molecule skeleton bend takes place more dense contact with spherical C_{60} molecule. It leads to appearance of additional transitions between nearby molecular orbitals of C_{60} and donor molecules.

23.4 Conclusions

Electronic structure of complexes was investigated by B3LYP/6-311++g(2d, 2f) method. The optimized structures are shown bend of donor molecule skeleton provides its more dense conformity to the spherical C_{60} molecule that increases the number of energetically favorable van-der-Waals contacts. All complexes have energy spectrum typical for semiconductors with band-gap 1.265–1.731 eV. Energy spectrum of such type molecular complex was shown fine electronic structure.

Displacement and increase in number of lines in the spectrum of molecular complexes in comparison with the spectrum of individual C_{60} and the donor molecules take place.

Acknowledgements The work was supported by the Russian foundation for basic research (grant 10-02-00763a).

References

1. Hoppe H, Sariciftci NS (2004) Organic solar cells: an overview. J Mater Res 19(7):1924–1945
2. Hoppe H, Sariciftci NS (2008) Polymer solar cells. Adv Polym Sci 214:1–86
3. Rappe AK, Casewit CJ, Colwell KS, Goddard WA III, Skiff WM (1992) UFF, a full periodic table force field for molecular mechanics and molecular dynamics simulations. J Am Chem Soc 114:10024–10035
4. Frisch MJ, Trucks GW, Schlegel HB, Scuseria GE, Robb MA, Cheeseman JR et al (2004) Gaussian 03 revision E.01. Gaussian Inc, Wallingford
5. Thompson M. http://www.arguslab.com
6. Stewart JJP (1989) Optimization of parameters for semiempirical methods. Appl J Comput Chem 10(2):221–264
7. Granovsky AA. http://classic.chem.msu.su/gran/gamess/index.html
8. Becke AD (1993) Density-functional thermochemistry. The role of exact exchange. J Chem Phys 98:5648–5652
9. Lee C, Yang W, Parr RG (1988) Development of the Colle-Salvetti correlation-energy formula into a functional of the electron density. Phys Rev B 37(2):785–789
10. Neretin IS, Slovokhotov YL (2004) Chemical crystallography of fullerenes. Russ Chem Rev 73(5):455–486
11. Konarev DV, Kovalevsky AY, Litvinov AL, Drichko NV, Tarasov BP, Coppens P, Lyubovskaya RN (2002) Molecular complexes of fullerenes C_{60} and C_{70} with saturated amines. J Solid State Chem 162(2):474–485
12. Konarev DV, Lopatin DV, Rodaev VV, Umrikhin AV, Khasanov SS, Saito G, Nakasuji K, Litvinov AL, Lyubovskaya RN (2002) Synthesis, crystal structute and photoconductivity of new molecular complex of C60 with tetrabenzo (1, 2-bis[4 H-thiopyran-4-ylidene]ethane): Bz4BTPEC60. J Phys Chem Solids 66(5):711–715
13. Cambridge Crystallographic Data Centre. http://www.ccdc.cam.ac.uk.
14. Golovin YuI, Lopatin DV, Rodaev VV, Konarev DV, Litvinov AL, Lyubovskaya RN (2007) On the photoconductivity of layered molecular complex of fullerene C_{60} with saturated amine TMPDA. Phys Status Solidi R 1(2):56–58
15. Golovin YuI, Lopatin DV, Rodaev VV, Konarev DV, Litvinov AL, Lyubovskaya RN (2006) Photoconductivity of crystalline molecular complex of fullerene C_{60} with amine LCV. Phys Status Solidi B 243(11):78–80

Chapter 24
New Substances: Red Carbon Suboxide, Red N-doped Fullerene $(C_{50}N_{10})O_3H_{10}$ and Red Carbon

A. Kharlamov, G. Kharlamova, O. Khyzhun, and N. Kirillova

Abstract New process of thermal transformation of organic molecules is elaborated. First fullerene C_{60} and at its hydrides are synthesized at polycondensation of molecules of benzene and at temperatures excluding sublimation of carbon. New closed heteroatomic molecule $C_{50}N_{10}O_4H_{10}$ is formed during polycondensation of pyridine molecules. Red molecular crystals of this substance are easy dissolved in alcohol and water, but badly in benzene and toloene. The atoms of carbon in a molecule $C_{50}N_{10}O_4H_{10}$ are found in two nonequivalent states: (C_2)–C–N and (C_2)–C–C. Atom of nitrogen in this molecule is found in the supercoordinated state: four bonds with atoms of carbon in pyridine net and one bond is realized with surface atoms of oxygen (3 bonds N–O–H) and hydrogen (7 bonds N–H). First carbon suboxide $(C_{>5,4}O)$ and "red carbon" are obtained.

Keywords Closed heteroatomic molecule $C_{50}N_{10}O_4H_{10}$ · Red carbon · Carbon suboxide · Supercoordinated nitrogen · Polycondensation reaction

24.1 Introduction

Today after general enthusiasm to unique closed molecules (fullerenes and single-walled nanotubes) as well as to 0D (nanoparticles and nanophases) and 1D (nanostructures) objects new brand of scientific pilgrimage has appeared. Graphene as 2D object and its oxide attract the increasing attention. Again as before many researchers and especially "chemical modelers" (who simulate virtual or already really existing objects) try by any way to touch this brand: "for the first time to repeat" 100 times already repeated synthesis of substance or "for the first

A. Kharlamov (✉), G. Kharlamova, O. Khyzhun, and N. Kirillova
Frantsevich Institute for Problems of Materials Science of NAS,
3 Krzhyzhanovsky Str., 03142 Kiev, Ukraine
and
Kiev National Taras Shevchenko University, 64 Volodimirska Str, 03001 Kiev, Ukraine
e-mail: akharlamov@ukr.net

S.Yu. Zaginaichenko et al. (eds.), *Carbon Nanomaterials in Clean Energy Hydrogen Systems - II*, NATO Science for Peace and Security Series C: Environmental Security 2,
DOI 10.1007/978-94-007-0899-0_24,

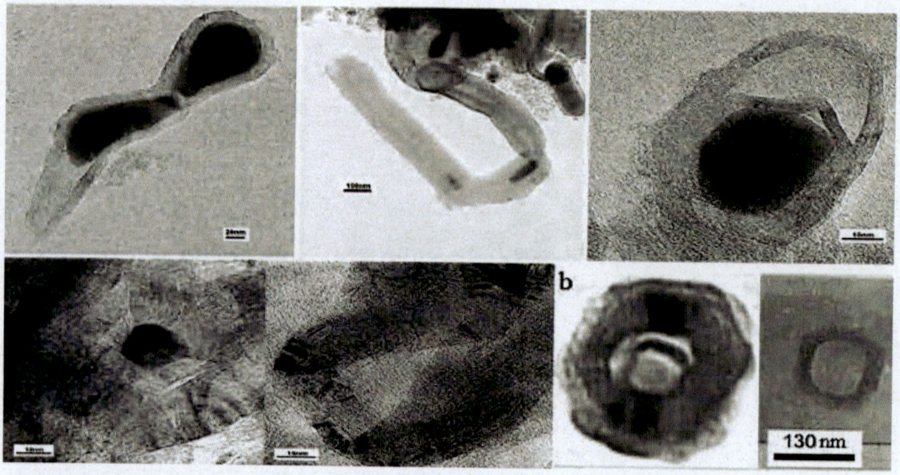

Fig. 24.1 Unusual nanostructures of carbon [1–4]

time to predict" already really existing structure. We participated in similar studies also [1–4]. By means of electronic microscopy we tried to find in products of pyrolysis this unique solid-state molecule (single-walled nanotube) or even nanostructure with original morphology. As it is possible to see (Fig. 24.1), structures, really, original were found out but for study of their properties the reproduced results are required.

Despite of such concentrated approach to study of new and unique nanoobjects the question concerning mechanisms of formation of molecules of fullerenes and single-walled carbon nanotubes and also anisotropic nanostructures of many other substances remains till now open. (Very difficult to imagine that nanostructures presented on Fig. 24.1 grow according to generally accepted VLS mechanism [5–7] on metal nanoparticle). Understanding of the reasons of so contrast distinction in electrophysical properties between two graphene isomers [8] (armchair and the zigzag) cylindrical solid-state molecule of carbon is absent also. However the knowledge and understanding of the mechanism is a way to creation of new routes of reactions and, consequently, synthesis of new substances, which alongside with graphene can be also attractive to a wide spectrum of the experts.

Before we offered the alternative mechanism to usual (VLS to the mechanism), according to which the stage of decomposition of molecules of reagent up to carbon on particle of the catalyst is not stipulated as intermediate. On the contrary, nanoparticles of metal as the centres of nucleation of reaction only activate the initial molecules capable to the further polycondensation with formation of closed single-walled graphene nets and multi-walled structures. Due to huge exoeffect at formation of a number of bonds C–C particles of metal melt and some particles are located inside a solid-state molecule or nanostructure. This reaction is one of many

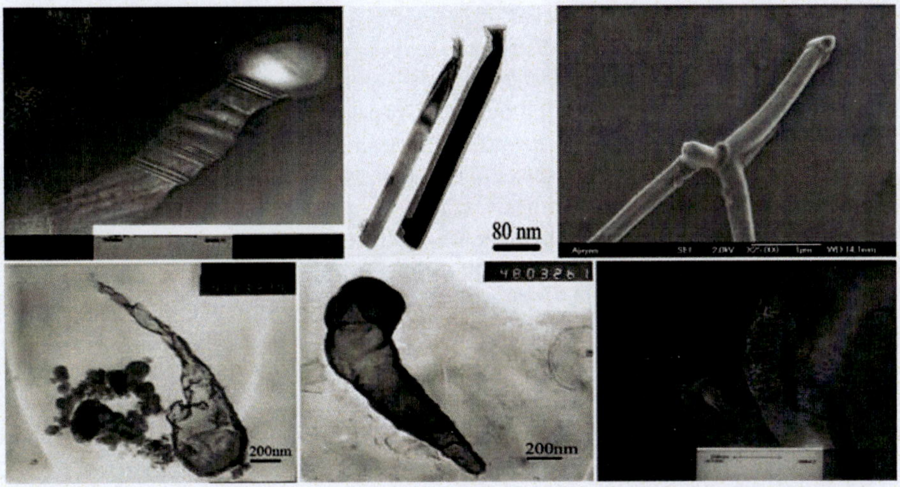

Fig. 24.2 Unusual nanostructures of silicon carbide, silicon and boron nitride (*bottom*) [9–13]

reactions developed by us as a method of low-temperature (1200°C) exothermal (unisothermal) nanothynthesis. Exothermal nanothynthesis is reactions of self-accelerated growth of anosotropic nanostructures from atoms (molecules) of evaporating initial substances. The reaction arises on nanocenter and then considerably is accelerated due to raising (for the account of exoeffect of reaction) velocity of sublimation of initial powdery reagents. By means of this method for the first time such important (in the practical relation) very hard substances as boron and silicon carbides, nitride and suboxide boron as well as boride and boroncarbide phases aluminium were synthesized as nanostructures and nanosized particles (Fig. 24.2).

The method of low-temperature exothermal synthesis is very simple and less energy-intensive in comparison to analogue. As initial reagents (the same to a usual industrial method) are used powdery simple substances (boron, silicon, carbon and aluminium). However opposite to a method of analogue the mixture of powders is used in free (not pressed) kind and temperature of synthesis much lower. Powdery product is formed (instead of strongly sintered as in a method of analogue) and consists from anisotropic nanostructures of various morphology, length and diameter. At the fixed direction of heat removal it is possible to create oriented growth of more homogeneous nanostructures. Essentially increased (in comparison with equilibrium pressure) "reactionary" sublimation of initial reagents (silicon, boron and carbon) is carried out locally, only in microvicinities of much more high-temperature (in comparison with fixed temperature in reactor) nanocenter of growth nanostructure. The mainly directed flow of atoms on nanocenter of growth occurs at the expense of their fast annihilation at formation of a product. Nanocenter melting and "reactionary" sublimation of initial reagents in its vicinities occur owing to exoeffect of reaction. However, sintering of a product (as in a method of analogue) does not occur since heat removal in the not pressed initial mixture is optimum: superfluous exoeffect is easily distributed and is extinguished in interpartial space.

It was accepted earlier to consider that the interaction between solid reagents is solid-phase: the reaction is carried out on diffusive mechanism, where the stage of diffusion of reagents through a layer of a formed sintered product is limiting. We believe that the interaction between solid reagents is always made according to the mechanism of exothermal synthesis which at uncontrollable heat removal (in particular at use of the pressed reagents mixture) passes in a mode of thermal explosion (so-called self-spreading high-temperature synthesis).

Consequently, if polycondensation mechanism at exothermal nanosynthesis of graphene nets, really, is realized, fullerenes (and not just nanotubes and onions) can be obtained from molecules of benzene at its thermal transformation. Today it is accepted to consider that fullerenes are formed only from carbon atoms (or liquid linear and ring clusters) generated at superhigh temperatures (>4.000°C) of evaporation of graphite.

In this work we present perfectly new substances and describe experimental results to confirm the reality of polycondensation mechanism of formation of carbon structures.

24.2 Experimental Results and Discussion

24.2.1 Investigation by Mass-Spectrometric Analysis of Products of Benzene Policondensation

The research is carried out with use as initial reagents of benzene (C_6H_6) as pyridine (C_5H_5N). The molecule of pyridine is aromatic heteroanalogue of benzene. Because of greater basicity and greater mobility of atoms of hydrogen (especially in orto and para places) pyridine can be more active reagent in reaction of polycondensation. The synthesis from molecules of pyridine the closed molecule, fullerene-like nitrogen – carbon containing will be bright confirmation of realization of polycondensation mechanism. However, main objective of research is to create a new type of reactions in order to obtain new substances.

The experiences on thermal transformation of benzene and pyridine are carried out in original reactionary conditions which are distinct from typical process of pyrolysis. As products, as a rule, new substances are formed or substances which are not characteristic for usual process of pyrolysis. So, it is established that in mass – spectra both positive and negative ions of toluene extract of a product of polycondensation (Fig. 24.3) there are most intensive peaks of ions from m/z 720, 696, 672. Just these three ions are characteristic for high-temperature (received at evaporation of carbon material) fullerene C_{60} and clusters C_{56} and C_{58} as products of disintegration (as it is accepted to consider) at its ionization. At mass – spectra of negative ions in addition there are also peaks of ions from m/z 648 and 624, which were

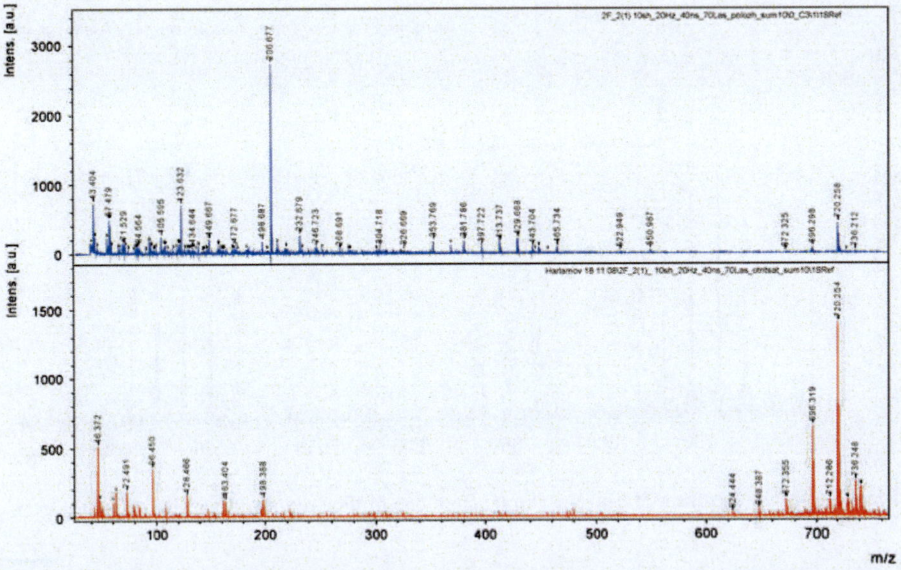

Fig. 24.3 Mass-spectrum of toluene extract from the product of thermal transformation of benzene

not found out at all earlier. Consequently, these ions (or clusters C_{52} and C_{54}) are characteristic for low-temperature fullerene. (It is possible, that the ions as satellites of fullerene C_{60} correspond not to clusters of its disintegration but to individual fullerenes C_{56} and C_{58}, C_{52} and C_{54}). At mass – spectra there are also peaks of ions, which can correspond to hydrides of fullerenes of composition $C_{60}H_8$, $C_{60}H_{16}$ and $C_{60}H_{20}$. All revealed ions with the greater intensity are shown in mass – spectra of negative ions.

Is remarkable that at mass – spectra of benzene extracts of some products there are peaks not only ions, characteristic for low-temperature fullerene. So three groups of lines (Fig. 24.4) are found out which can correspond to hydrides of fullerenes C_{60} and C_{70}:

$$C_{60}H_8,\ C_{60}H_{16} \text{ и } C_{60}H_{20}$$
$$C_{70}H_7,\ C_{70}H_{21},\ C_{70}H_{28}$$

In a spectrum there are also peaks of ions with value m/z, which can correspond to clusters:

$$C_{82}H_{17},\ C_{82}H_{31},\ C_{82}H_{45},\ C_{82}H_{59}.$$

The periodicity in change of values m/z in next clusters composes 14 units (or group CH_2, or atom of nitrogen).

Thus, fullerenes C_{60} and C_{70} and also their hydrides are formed in reactionary conditions excluding sublimation of carbon. The growth of the closed polymeric molecules is realized with participation of benzene molecules.

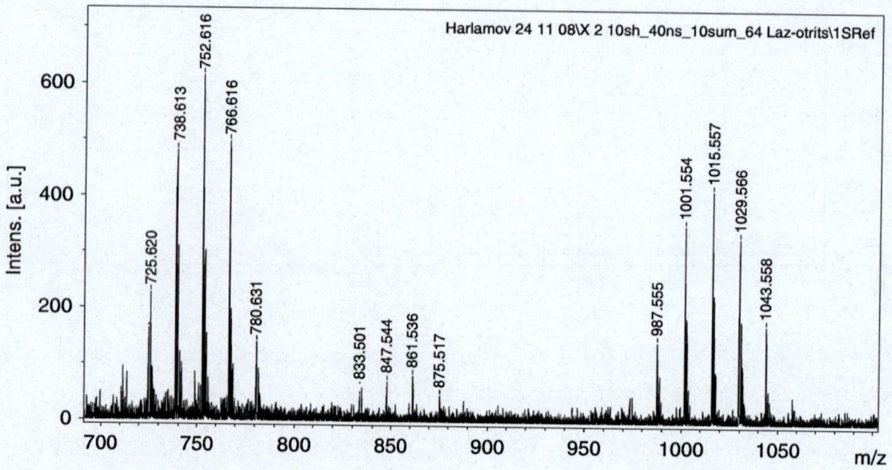

Fig. 24.4 Mass-spectrum of benzene extract from one of the product of benzene transformation

24.2.2 Synthesis of Closed Fullerene-like C, N–Containing Molecule during Pyridine Policondensation

Pyridine in a regime of primary polycondensation is transformed to one of products which is extracted by means of water and ethanol. This substance versus fullerene C_{60} in the unpolar solvent (for example, benzene) is dissolved only slightly. Its concentrated the alcohol solution has brightly red colour. This substance from alcohol (or water) solution can be precipitated as bright of red crystals (Fig. 24.5) or nanodisperse dark-red powder with a various degree of crystallinity (Fig. 24.6).

According to the chemical analysis this substance consists of nitrogen (17.1%), oxygen (5.5%), hydrogen (1.2%) and carbon (72.2% mass.) and, hence, its formula composition is $C_{6,01}N_{1,22}O_{0,344}H_{1,2}$. Study of this substance by means of XPS has shown (Fig. 24.7) that it contains carbon, oxygen and nitrogen. Binding energy of N1s and O1s is equal 399.6 and 531.7 eV accordingly. However XPS – spectrum C1s has two maxima: 285.0 and 286.6 eV that can testify to presence in the given substance of two kinds of atoms of carbon.

Alcohol solution of $C_{6,01}N_{1,22}O_{0,344}H_{1,2}$ was investigated by means of chromatographic and mass – spectrometric analyses. In mass – spectra there are four lines with mass numbers: 802.6; 413.0; 364.8 and 339.9. Most intensive is the first peak and, hence, molecular formula of $C_{6,01}N_{1,22}O_{0,344}H_{1,2}$ is $C_{50}N_{10}O_3H_{10}$ with molecular mass 798. (Difference between 802.6 and 798 there is, probably, because of heavy isotopes of some elements).

Raman spectrum of $C_{50}N_{10}O_3H_{10}$ is rather complex and contains nine lines (Fig. 24.8). The most intensive lines (1.375 and 1.500 cm^{-1}) are characteristic for atoms of carbon in sp^3 – and sp^2 – states accordingly. The line of atom of carbon

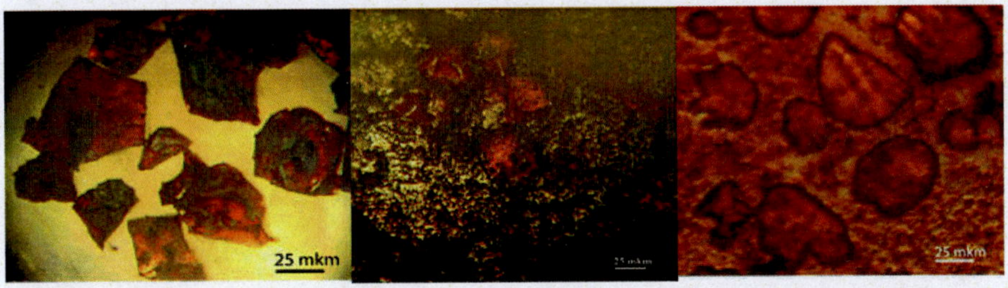

Fig. 24.5 Images of optical microscopy in transmitted light of crystals of new substance

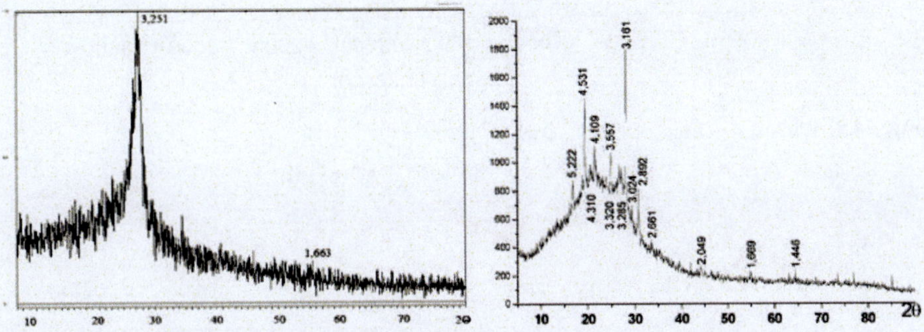

Fig. 24.6 X-ray patterns of new substance of different degree of crystallinity

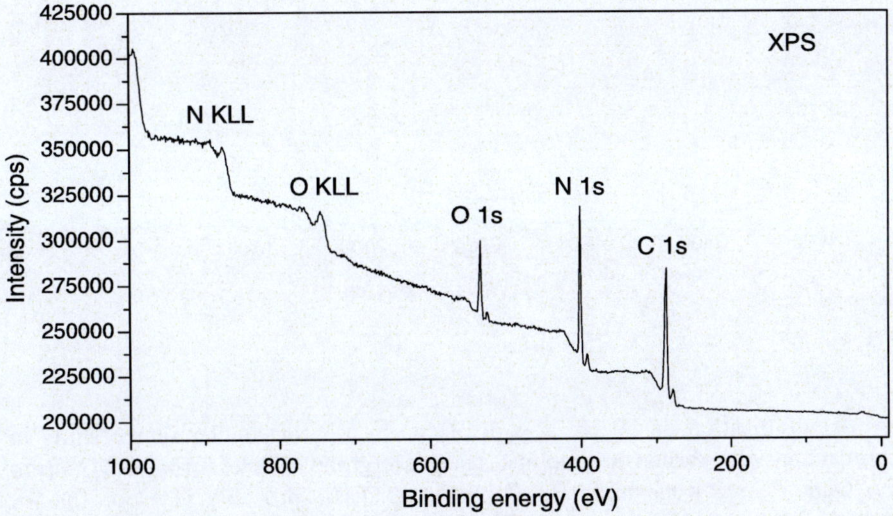

Fig. 24.7 XPS-spectrum of new substance, $C_{6,01}N_{1,22}O_{0,344}H_{1,2}$

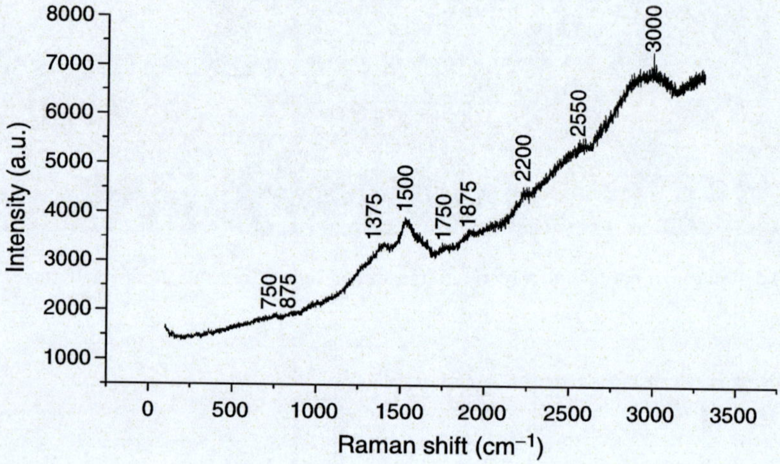

Fig. 24.8 Raman spectrum of $C_{50}N_{10}O_4H_{10}$

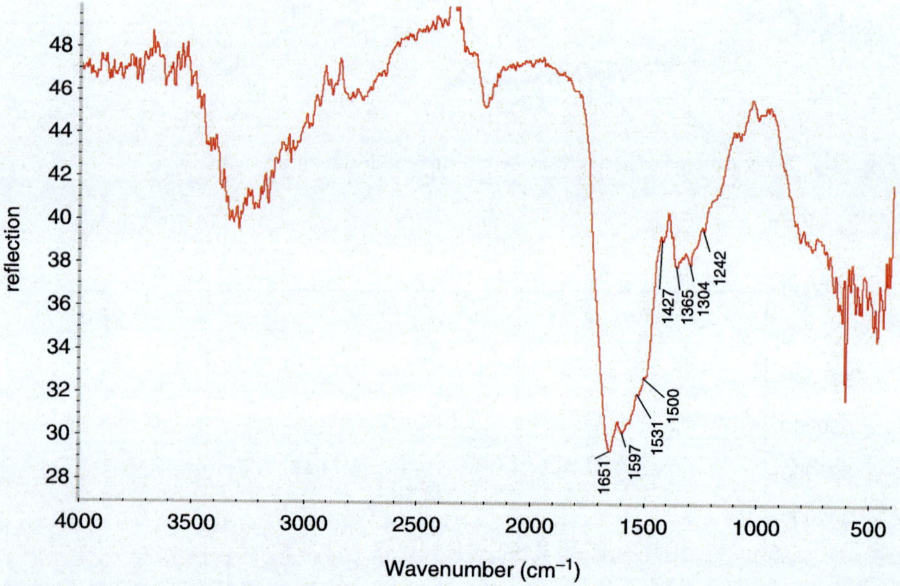

Fig. 24.9 IR – spectrum of $C_{50}N_{10}O_4H_{10}$

sp^2 – hybridization on 90 cm^{-1} is displaced in area of smaller raman shifts in comparison with carbon in graphite. The basic group of lines (from 1.242 up to 1.670 cm^{-1}) in IR – spectrum (Fig. 24.9) $C_{50}N_{10}O_3H_{10}$, probably, is responsible for skeletal fluctuations from bonds C–C which, usually, are characteristic for aromatic hydrocarbons.

On the basis of the represented results we believe that the molecule $C_{50}N_{10}O_3H_{10}$ has a closed fullerene-like structure with a construction element of the molecule of pyridine. The polycondensation of pyridine molecules (as well as of benzene molecules at formation of fullerene) results in formation of pyridine (or N-graphene) net.

Because of high exoeffect of formation of great number of the new connected bonds C–C in pyridine structure both atom of nitrogen, and atom of carbon can be in extremely exited (down to formation of a carbon radical) states. The atom of nitrogen with the nearest radical carbon forms fourth covalent bond N–C, using one electron of not divided pair of electrons. Residual electron of atom of nitrogen promotes to it the transfer in the supercoordinated state at the expense of formation of bonds N–O–H or N–H on a surface of pyridine net.

24.2.3 Shortly about a Suboxide of Carbon and "Red Carbon"

Here we also represent the first results concerning synthesis of such new substances as red suboxide of carbon ($C_{>5,4}O$) and "red carbon" (both substances of red colour (Fig. 24.10). (Now it is known 4 gaseous oxide of carbon (CO, CO_2, C_3O_2 and $C_{12}O_9$) and solid oxide of graphite ($C_{<2}O$)).

Carbon suboxide is formed at catalytic oxidation of graphite by oxygen. It should be remind that oxide of graphite is formed at treatment of graphite by such strong oxidizers as a nitric and sulfuric acid as well as perchlorate or permanganate of potassium. Suboxide of carbon has variable composition with the contents of oxygen from 13 up to 20% mass (oxide of graphite contains not less than 40% mass of oxygen). At long exposure in water suboxide of carbon loses crystalinity (Fig. 24.11). According to XPS – spectra (Fig. 24.12) suboxide of carbon alongside with carbon and oxygen contains also iron, which easily leaves at processing a sample by a hydrochloric acid.

In alcohol suboxide of carbon is decomposed with formation as one of products amorphous (Fig. 24.11) substance (red colour), which in accord XPS – spectrum (Fig. 24.12) contains only carbon. However in IR – spectrum of red carbon (as against IR – spectrum initial suboxide of carbon) there are lines (2.956,

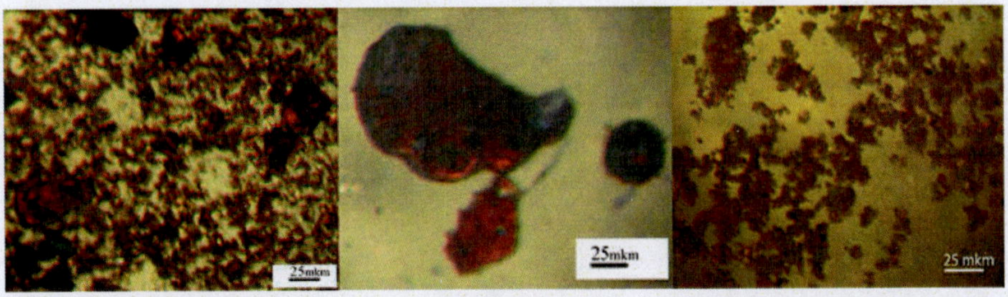

Fig. 24.10 Images of optical microscopy in transparent light of carbon suboxide and "red carbon"

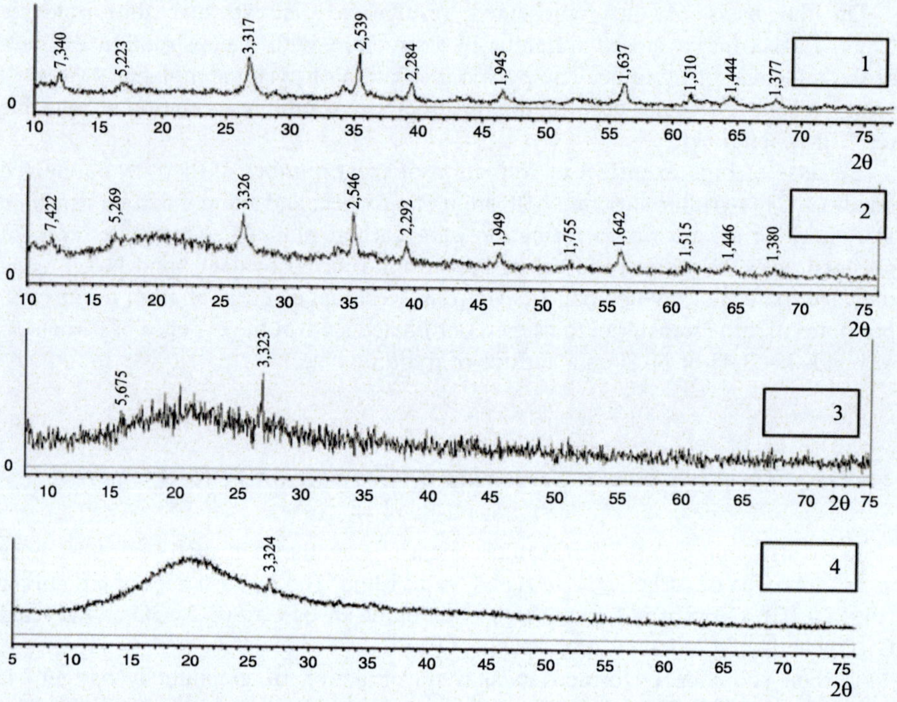

Fig. 24.11 X-ray patterns of $C_{>5,4}O$ initial (1), after water treatment (2,2) and "red carbon" (4)

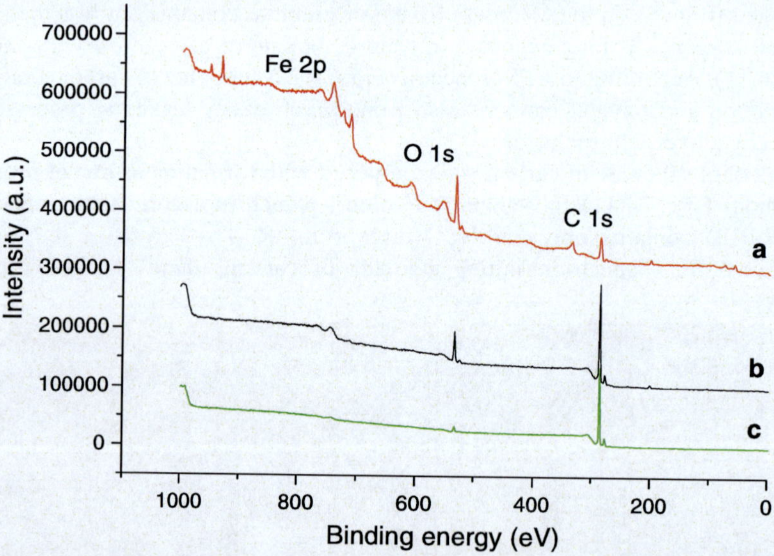

Fig. 24.12 XPS -spectra of carbon suboxide initial (**a**), after hydrochlonic acid (**b**) and "red carbon" (**c**)

2.919 and 2848 cm^{-1}), which usually are characteristic for valent C–H fluctuations. These lines are absent also in IR-spectrum of a sample of carbon suboxide after its treatment in water. Is remarkable, that groups of lines from 1.260 up to 1720 cm^{-1}, which are characteristic for skeletal fluctuations of bonds in aromatic hydrocarbons, in IR – spectra and carbon suboxide and red carbon are very similar.

24.3 Conclusions

Thus, first is established, that fullerene C_{60} is formed in conditions excluding of sublimation of carbon. The formation C_{60} is carried out in reactionary conditions favorable for polycondensation (but not for destruction) of molecules of benzene. Partially hydrogenated fullerenes C_{60} ($C_{60}H_8$, $C_{60}H_{16}$, $C_{60}H_{20}$) and C_{70} ($C_{70}H_7$, $C_{70}H_{21}$, $C_{70}H_{28}$) are also products of reaction of dehydrogenated polymerization of benzene.

For the first time is established, that at polycondensation of pyridine as aromatic analogue of benzene the closed fullerene-like molecule $C_{50}N_{10}O_4H_{10}$ is formed. New substance as red molecular crystals is dissolved in alcohol and water. The atom carbon in a molecule $C_{50}N_{10}O_4H_{10}$ is in two different states: (C_2)–C–C and (C_2)–C–N. It is possible, that 10 atoms of nitrogen in heteroatomic molecule are in the supercoordinated state: four bonds with atoms of carbon in pyridine net and one bond are realized with superficial atoms of oxygen (3 bonds N–O–H) and atoms of hydrogen (7 bonds N–H). All three substances have been synthesized in several gram quantities.

References

1. Kharlamov AI, Loythenko SV, Kirillova NV, Kaverina SV, Fomenko VV (2004) Toroidal nanostructures of carbon. Single-walled 4 –, 5 – and 6 hedrons and nanorings. Rep Acad Sci Ukraine 1:95–100, Russian
2. Kharlamov AI, Kirillova NV, Ushkalov LN (2006) Simultaneous growth of spheroidal and tubular carbon structures during the pyrolysis of benzene. Theor Exp Chem 42(2):90–95
3. Kharlamov AI, Ushkalov LN, Kirillova NV, Fomenko VV, Gubareny NI (2006) Synthesis of onion nanostructures of carbon at pyrolysis of aromatic hydrocarbons. Rep Acad Sci Ukraine 3:97–103, Russian
4. Kharlamov AI, Kirillova NV (2009) Fullerenes and fullerenes hydrides as products of transformation (polycondensation) of aromatic hydrocarbons. Rep Acad Sci Ukraine 5:112–120, Russian
5. Baker RTK, Barber MA, Harris PS, Feates FS, Waite RJ (1972) Nucleation and growth of carbon deposits from the nickel catalyzed decomposition of acetylene. J Catal 26:51–62
6. Lobo LS, Trimm DL, Figueiredo JL (1973) Kinetics and mechanisms of carbon formation from hydrocarbons on metals In: Hightower JW (ed) Proceedings of 5th international congress on catalysis, North Holland, Amsterdam, 2:1125–1135
7. Tibbetts GG (2000) Nucleation and growth of carbon filaments and vapor-grown carbon fibers In: Biro LP, Bernardo CA, Tibbetts GG and Lambin Ph (ed) Carbon filaments and nanotubes: common origins, differing applications? Series E: Applied Sciences 372:63–73

8. Kharlamov AI, Kirillova NV, Skripnichenko AV, Gubareni NI, Fomenko VV (2010) Nanochemical peculiarities of nanostructures, nanophases and nanoparticles. Rep Acad Sci Ukraine 4:100–108, Russian
9. Kharlamov AI, Kirillova NV, Karachevtseva LA, Kharlamova AA (2003) Low-temperature reactions between vaporizing silicon and carbon. Theor Exp Chem 39(6):374–379
10. Kholmanov I, Kharlamov AI, Milani P et al (2002) A simple method for the synthesis of silicon carbide nanorods. J Nanosci Nanotechnol 2(5):453–456
11. Kharlamov AI, Kirillova NV (2002) Gas-phase reactions of formation of silicon carbide nanofilaments from silicon and carbon powders. Theor Exp Chem 38(1):59–63
12. Kharlamov AI, Kirillova NV, Loytchenko SV et al (2002) Synthesis of elongated nanostructures of silicon carbide from powdery silicon and carbon. Rep Acad Sci Ukraine 10:98–105
13. Kharlamov AI, Kirillova NV, Karachevtseva LA, Fomenko VV, Bondarenko ME (2006) Vapor-gaseous process of low-thermal (1200°C) transformation of polycrystalline silicon to its highly orient anisotropic particles. Rep Acad Sci Ukraine 12:48–55

Chapter 25
Source of Ultraviolet Radiation with Field-Emission Cathode Made of Nanostructured Carbon Material

I.V. Ehmenina, E.P. Sheshin, and N.N. Chadaev

Abstract This article considers possibility of creation of basically new sources of ultraviolet radiation based on field-emission under action of electrons emitted from nano-structured cathode.

Keywords Spectral characteristic · Field-emission · Nanostructured cathode · Carbon fibre · Ultraviolet radiation intensity

25.1 Introduction

Nowadays there is a lot of extremely important for the laser technique, medicine, ecology and petrochemistry photochemical technologies and schemes for which inexpensive, effective and compact sources of ultra-violet radiation are necessary.

But unfortunately, widely known now ultraviolet sources have a number of essential lacks, such as: bulkiness of a design, the small area of a radiating surface, low efficiency, high cost, and use of ecologically harmful substance – mercury. Therefore working out of new methods of receiving ultraviolet on the basis of last achievements in the field of optoelectronics, for the purpose of creation of a light source possessing high light efficiency, the great durability, and also being as much as possible ecological both in manufacture, operation and recycling, is necessary.

In the given work the technique of reception of the ultraviolet, based on the phenomenon of field-emission has been developed.

This technique consists in use of ultra-violet phosphor. The field-emission light source is a vacuum lamp with an electronic gun and the screen covered with ultraviolet phosphor. Electrons are accelerated by anode voltage, under the influence of this high energy electrons phosphor shine [1, 2] (Fig. 25.1).

I.V. Ehmenina (✉), E.P. Sheshin, and N.N. Chadaev
The Moscow Institute of Physics and Technology (The State University),
Institute lane, 9 Dolgoprudnyj, 141700 Moscow, Russia
e-mail: ehmenina@mail.ru

S.Yu. Zaginaichenko et al. (eds.), *Carbon Nanomaterials in Clean Energy Hydrogen Systems - II*, NATO Science for Peace and Security Series C: Environmental Security 2,
DOI 10.1007/978-94-007-0899-0_25,

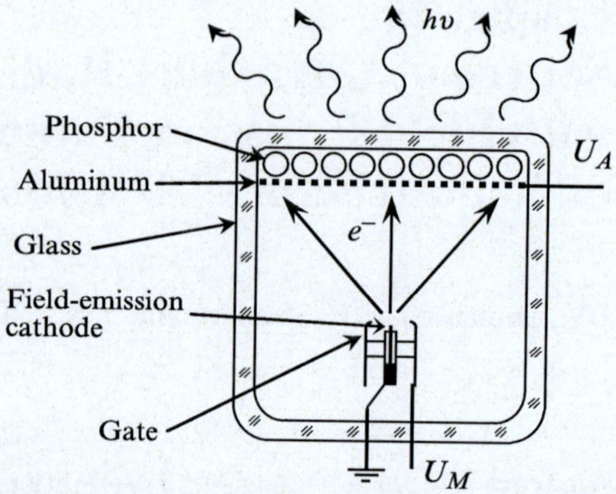

Fig. 25.1 Work circuit diagram of field-emission light source

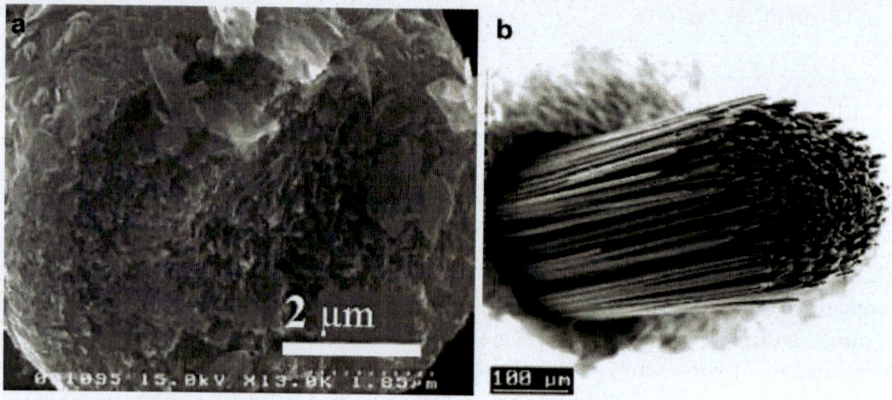

Fig. 25.2 Face surface of a carbon fibre (**a**) and the field-emission cathode from a bunch of fibres (**b**)

In our laboratory we use a bunch of carbon tubes [3, 4] (Fig. 25.2) as field-emission cathode.

Excitation of phosphor by electronic bunch is difficult process in which we can distinguish a number of consecutive stages:

1. Penetration of electrons in a crystal lattice of phosphor and formation in it the cascade secondary electrons as a result of unelastic collisions. The part of them is lost as a result of secondary emission.
2. Excitation of the luminescence centers by electrons.
3. Release of the absorbed energy in the form of radiative (luminescence) or radiationless transitions (energy loss for heating phosphor). The ratio between probabilities of these transitions characterizes efficiency of phosphor if consider losses of primary and secondary electrons as a result of reemission.

Doubtless advantages of an field-emission source of UV are ecological compatibility, a wide range of working temperatures, high stability to mechanical vibrations and voltage fluctuations in a network, low inertance (time of "electric" inclusion of the cathode does not exceed 10^{-8} s) and high durability. Also it is necessary to notice that thanks to application of the field-emission cathode the source has no heated parts.

25.2 Results and Discussion

For definition of characteristics of the future field-emission source of ultra-violet radiation, researches of spectra ultra-violet phosphor have been done (Fig. 25.3).

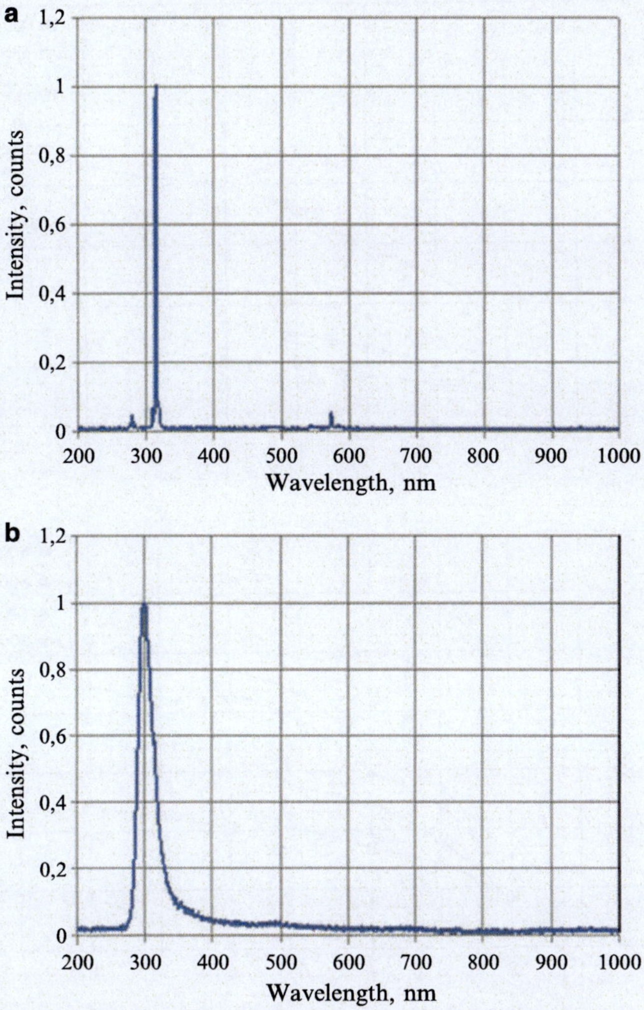

Fig. 25.3 Spectral distribution of radiation of some ultra-violet phosphors with different chemical compounds

Spectral characteristics of phosphors with various chemical compounds and the sizes of grains were studied, dependences of intensity of radiation on the applied voltage and power have been revealed (Fig. 25.4).

For measurement of phosphor's parameters they have been put on an anode glass plate. The anode was put in diode construction with the field-emission cathode from a bunch of carbon fibers. Phosphors were tested at anode voltage 2 ÷ 10 kV and an anode current 40 ÷ 450 μA.

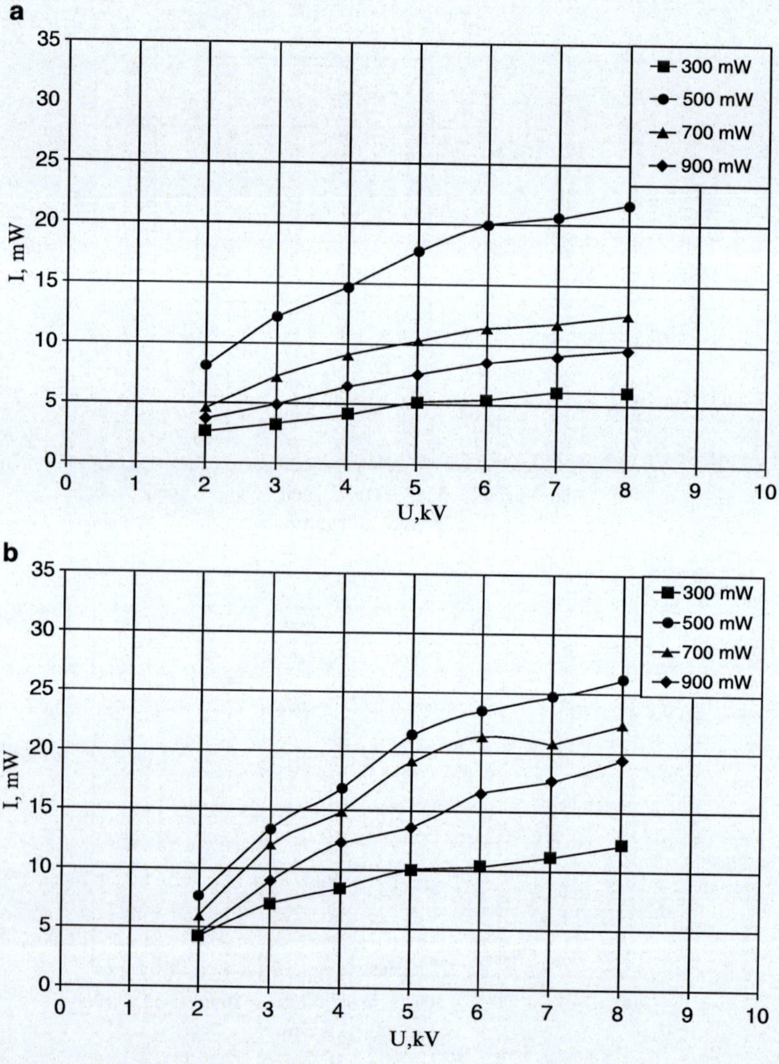

Fig. 25.4 Dependence of intensity of UV radiation of phosphor with the size of grains 10 nm (**a**) and 5 nm (**b**) on anode voltage at different applied powers

Apparently from Fig. 25.4 for an ultra-violet phosphor with larger size of grains the sharp increase in intensity of an ultra-violet luminescence is observed at increase in a current of an electronic bunch (the curve corresponding to the applied power 900 mW is located much above than previous ones) whereas for the same phosphor with smaller size of grain similar jump of intensity is not observed (all curves are equally removed from each other).

25.3 Conclusions

Researches have shown that spectral characteristics of investigated phosphors have emissive peak on wavelengths of 295–310 nm that corresponds to a near ultra-violet range, and also that phosphors with smaller size of grains not only are better put on the anode and have more uniform luminescence, but also possess the stablest characteristics at various applied voltage and power.

References

1. Sheshin EP, Suvorov AL, Bobkov AF, Dolin DE (1994) Light source on the basis Of multitip field emission cathode from carbon materials. In: Abstract of 7th international vacuum microelectronics conference, Grenoble, France, 1994, pp 423–426
2. Leshukov M, Chadaev N, Sheshin E et al (2002) Light sources on the basis of field emission with cathodes from carbon fibres. Proceedings of conference "Carbon: fundamental problems of a science, materials and technology", Moscow (in Russian)
3. Sheshin EP (2001) Structure of surfaces and field-emissive properties of carbon materials, MIPT, p 287 (in Russian)
4. Sheshin EP (1999) Field emission of carbon fibers. Ultramicroscopy 79:101–108

Chapter 26
Hydrogen Desorption Temperature and Its Storage in Cylindrical and Plane Graphene-Based Carbon Nanostructures. A Comparative Analysis with Nanocrystallites-Based Carbon Nanostructures

A.P. Soldatov and O.P. Parenago

Abstract The new technique of synthesis of new carbon nanostructures is proposed: the oriented carbon nanotubes with graphene walls (OCNTG). To synthesize these structures the consistent covering of surface of pores of ultrafiltration inorganic membranes with monolayers of graphenes was held. These monolayers were formed during pyrolysis of the definite quantity of methane. This resulted in formation of OCNTG inside the nano-size pores ($D_{av.} = 50$ and 90 nm). Formation of the monolayer of graphenes was identified with the help of X-ray photoelectron spectroscopy (XPS). The depth of covering with monolayer was controlled with scanning electron microscope (SEM). Investigation of regularities of adsorption, storage and desorption of hydrogen in OCNTG and plane graphene-based carbon nanostuctures (PGNS) were carry out. It was shown that quantity of the adsorbed hydrogen reached ~14.0% from mass of OCNTG and ~4.2% from mass of PGNS. Adsorption of hydrogen in OCNTG and PGNS was identified for the first time using thermogravimetric analysis (TGA) coupled with mass-spectrometric analysis, and it was found that its desorption at atmospheric pressure occurs at temperature of ~175°C. Investigation of adsorptivity to hydrogen of three carbon structures was held which are: nanosized pyrocarbon crystallites (NPC), theirs superposition (an NPC superposition is *n* sequentially formed pyrocarbon nanocrystallites each consisting of *m* monolayers) and OCNTG, which showed that this property is characteristic only for the latter. The new effect of hydrogen variation of performance (HVP) was found; this effect consists in that fact that hydrogen adsorbed in OCNTG influences in transport properties of membranes decreasing their performance on liquids in 4–26 times which fact is the indirect confirmation of its high activity which rides probably on dissociative mechanism of hydrogen adsorption.

A.P. Soldatov (✉) and O.P. Parenago
Topchiev Institute of Petrochemical Synthesis RAS,
Leninsky av. 29, 119991 Moscow, Russian Federation
e-mail: Soldatov@ips.ac.ru

S.Yu. Zaginaichenko et al. (eds.), *Carbon Nanomaterials in Clean Energy Hydrogen Systems - II*, NATO Science for Peace and Security Series C: Environmental Security 2,
DOI 10.1007/978-94-007-0899-0_26, © Springer Science+Business Media B.V. 2011

Keywords Graphene · Carbon nanotube · Monolayer · Pyrocarbon · Pyrolysis · Hydrogen adsorption · Storage and desorption · Dissociative mechanism

26.1 Introduction

Now, much attention is paid to the synthesis and investigations of materials composed of nanosized particles. Among these works, studies of carbon nanostructures (nanocrystallites, nanotubes, nanofibers, graphenes, etc.) [1–3], which have extraordinary adsorption and electron emission properties [4–6], hold a special place.

The main factors which determine these properties of nanomaterials are multiplicity of interfaces, existence of dislocations, vacancies and so forth [7, 8] which are determined by particular qualities of their structure and construction. So it is interesting to synthesize and investigate of hybrid carbon nanostructures which might perform new promising properties. Of great interest are the synthesis and study of new hybrid carbon nanostructures, namely, oriented graphene-based carbon nanotubes (OGCNTs). This work continues the studies on modification of porous structures of inorganic membranes with nanosized pyrocarbon crystallites (NPCs) [9–15]. These structures are called hybrid because they are carbon nanotubes with walls formed from graphene sheets.

Synthesis of OGCNTs was performed by successive deposition of one or several carbon monolayers composed of graphene sheets to the pore surface of oxide membranes (Al_2O_3, SiO_2, TiO_2, MgO, etc.) by means of topochemical dehydrogenation of methane or its homologues. One of the unique and practically important properties of carbon nanotubes is their ability to adsorb and store hydrogen [16, 17]. In this context, it is of interest to elucidate whether OGCNTs deposited on the pore surface are capable of adsorbing hydrogen.

In this paper, we report the results of a study on the synthesis procedure and identification of single- and multiwall carbon nanotubes in membrane pores, as well as their hydrogen-adsorption properties and influence of hydrogen on permeability of membranes. At the same time perform a comparative analysis of the surface adsorption properties of various carbon structures (NPCs, their superposition, and OCNTGs).

26.2 The Depth of Covering Surface of Pores by Carbon Nanostructures

Ultrafiltration composite membranes "TRUMEN" (TiO_2 + MgO and TiO_2 + Cr_2O_3 on porous steel) were used in this work for synthesis of OCNTG; the depth of the selective layer was ~20 μm. Investigation of their porous structure using dynamic desorption porometry (DDP) [18] showed that $D_{average} = 90$ and

50 nm correspondingly, porosity was ~25–35%, the surface of pores was ~1.7–2.2 m^2. PGNS were synthesized on Al_2O_3 grade SCS-9 with surface area ~800 m^2/g.

Forming of OCNTG and PGNS was performed at 800°C on the plant with quartz reactor ($D_{inner} = 54$, L = 990 mm) and electric oven using supply line methane with CH_4 percentage of ~99% as a pyrolized reagent. For programming of temperature on different stages and for assignment of heating and cooling rates the microprocessor controller of temperature with 12 regulation zones was used.

The process thermocouple which was located inside the oven was calibrated by means of control thermocouple which was placed in the reaction area with the help of the devoted reactor tip. Its regulus was fixed at the distance $\leq$1.0 mm from the surface of membrane. Membrane was settled in the fixative frame [14] and was placed in the process zone of the oven. The set of calibration experiments which were performed in the conditions of the real process (process vacuum and delivery of methane) made it possible to find the exact correlation between process and control thermocouples.

Formation of OCNTG was performed by means of step be step plating surface of pores with monolayers of graphenes. Experiments were performed according to the scheme: heating up to 800°C → isothermal soaking (plating with monolayer) → cooling down to 500°C or to the room temperature → heating up to 800°C etc.

SEM investigation of chips of membranes was performed with the field emissive raster electron microscope JSM-6700F with add-on for energy-dispersive spectrometry (EDS) JED-2300F (JEOL, Japan). The apparatus was equipped with so called "cold" cathode and new "semi-in-lens" objective lenses with electron filtration. Analysis was performed with accelerating voltage of 15 kV, emission current of 10 mkA, primary electron bunch current of 5 nA and process distance of 8 mm.

Add-on JED-2300F was used for qualitative and quantitative analysis using EDS technique. In the investigation quantitative analysis with reference spectra of elements was used (function "differential filter + least-squares method + ZAF method").

It is evident that the formation of OGCNTs in the membrane pores should be carried out under Knudsen diffusion conditions (deposition over the whole pore depth). The λ value [19] estimated at an operating pressure of 4.9 kPa showed that the ratio $\lambda/D = 94$ and 52 for the membranes are used.

The SEM images of chip the initial membrane and after deposition of NPCs are presented in Fig. 26.1. In Fig. 26.1a, the SEM image of the cleaved surface of the pristine membrane clearly shows the selective layer with a fine-grained structure and the substrate composed of bulky. After deposition of NPCs, dark carbon areas appear in the SEM image of the cleaved surface (Fig. 26.1b), and the deposition depth corresponds to the thickness of the selective layer. Indeed, as is easily seen, the bottom border of carbon fragments coincides with the border between the selective layer and the substrate. It can also be seen that the membrane has a discrete structure in this area, and that dark NPC areas alternate with lighter bands of the selective oxide layer. This is due to the fact that the line of fracture of the membrane after cooling in liquid nitrogen passes not only through pores, but also through pore walls, whose cleaved surface has structure of the pristine membrane.

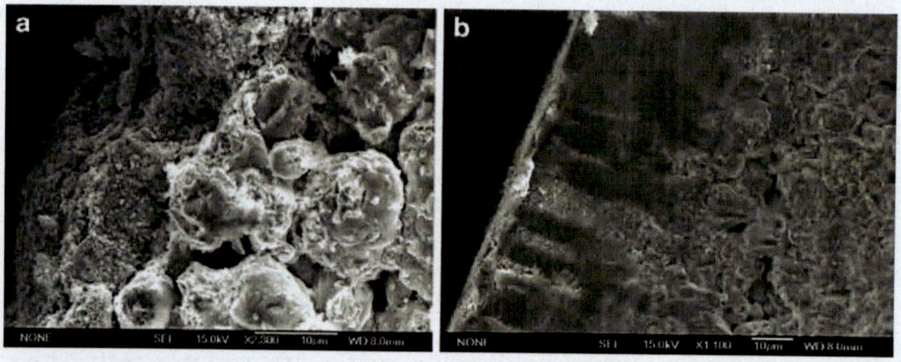

Fig. 26.1 Photographs of cleavage surfaces of TRUMEM ultrafiltration membranes: (**a**) initial and (**b**) with a PNC layer deposited at $p = 4.9$ kPa

It is evident that plating with NPC at 4.9 kPa takes place for the whole depth of pores. Hence we can form OCNTG in the pores of membranes if we can work through the technique of plating with monolayers of NPC and identification of the latter which were precipitated on the surface of pores.

26.3 The Formation of a Graphene Monolayer

Previously [10], we showed that NPCs deposited on the membrane have the perfect crystal structure of graphite. They have a hexagonal lattice with following parameters: $d_{002} = 3.368$ A, $L_c = 40.0$ nm, and $L_a = 80.0$ nm. Therefore, layer-by-layer deposition of pyrocarbon monolayers on the pore surface results in the formation of single- and multiwall carbon nanotubes oriented normally to the membrane surface, their walls being composed of graphene fragments (OGCNTs). The pyrocarbon mass required to form a monolayer can be determined by the equation given in [10]. For the membranes used, this mass is 0.0026 g.

The formation of a graphene monolayer on the pore surface of membranes was identified by XPS. Investigation was performed on the spectrometer PHI 5500 ESCA (Perkin Elmer) using Mg Kα radiation (hν = 1253.6 eV) with 300 W/14 kV power. Pressure of the residual gases in the measuring camera was 8–9 $\times$ 10^{-10} Tore. Fine spectrums were got with analyzer transmission energy of 11.75 eV and frequency of data acquisition of 0.1 eV/step. Data processing was performed with approximation with non-linear least-squares method using Gauss-Laurence function. Correction of charging effects was performed by means of calibrating of binding energies scale relative to carbon $C1_s - 285.0$ eV.

To change the depth of analysis in the surface zone spectrums were got at different angles of the sample to the axis of analyzer [20]. For purification of surface the ion gun was used in the blow off mode with energy of ions 150 eV, raster 2 $\times$ 2 mm^2, partial pressure for argon 12 mPa.

The results of these measurements are shown in Fig. 26.2. We compared the spectra of the pristine membrane (sample 2), the membrane heated at the temperature of pyrocarbon deposition under vacuum (sample 3), and the membrane with a deposited graphene monolayer (0.0028 g, sample 1) (Fig. 26.2). The spectra were analyzed taking into account positive charging of the sample surface due to electron emission. For this, the binding energies were determined from the peak positions shifted by the value ΔE, which was found from the peak position of carbon adsorbed on the pristine membrane (sample 2), because the state of carbon in others samples after heat treatment (sample 3) or after grapheme monolayer deposition (sample 1) is unknown.

In the spectrum of sample 2, the shift of the adsorbed carbon peak from the peak maximum of oxygen in TiO_2 (529.75 eV for O1s) was determined (Fig. 26.2a), and this value was used to determine ΔE for samples 1 and 3.

An analysis of the spectra showed that the titanium peaks were identical in the spectra of all samples (Fig. 26.2b) and corresponded to titanium dioxide.

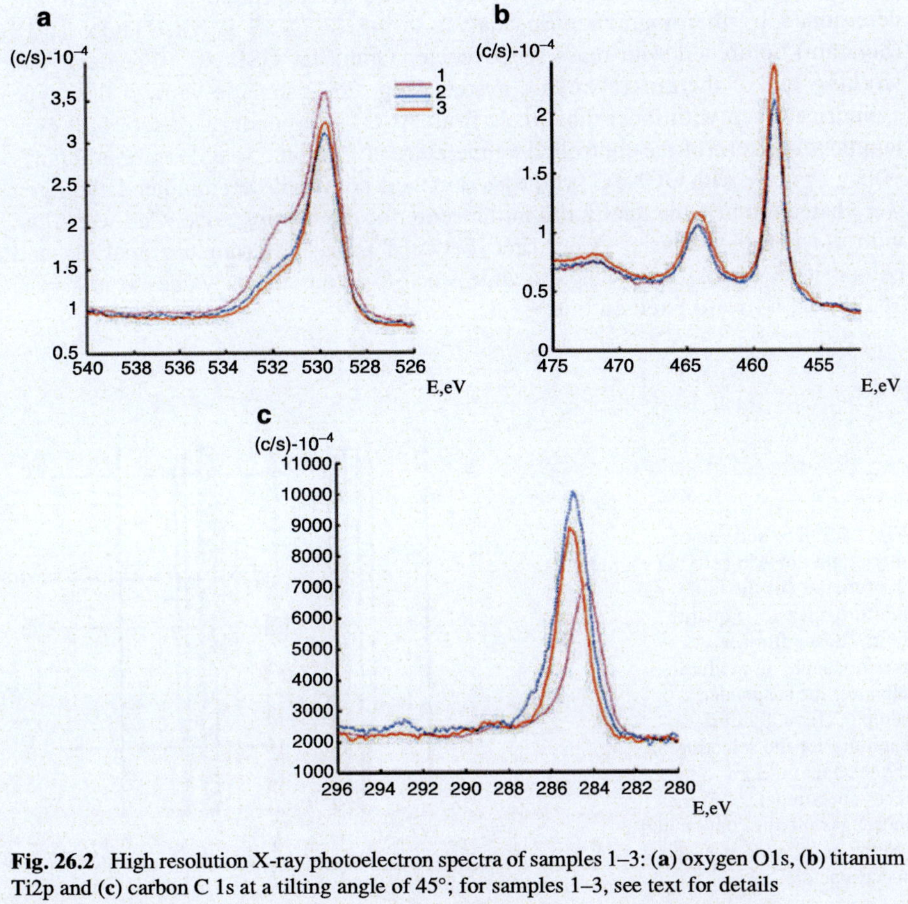

Fig. 26.2 High resolution X-ray photoelectron spectra of samples 1–3: (**a**) oxygen O1s, (**b**) titanium Ti2p and (**c**) carbon C 1s at a tilting angle of 45°; for samples 1–3, see text for details

The carbon peak positions were identical only for samples 2 and 3 and corresponded to adsorbed carbon, whereas sample 1 exhibited a carbon peak that was shifted to lower energies (284.3 eV) (Fig. 26.2c). This is indicative of the presence of a film of pure carbon on the surface of sample 1.

Thus, the procedure developed yields new carbon nanostructures, namely OCNTGs, in pores of inorganic membranes. The formation of OCNTGs was proved by SEM and XPS.

26.4 Adsorption and Storage of H_2 in OCNTG and PGNS. Temperature Desorption of H_2

Quantity of hydrogen which was adsorbed in OCNTG and PGNS was determined using gravimetric method to hundredth parts of mg. The initial pressure of hydrogen at saturation was 10–13 MPa. Temperature of desorption of hydrogen was determined by thermogravimetric analysis using SETSYS EVOLUTION 16/18 (Setaram) combined with mass spectrometer OmniStar GSD 301 (Pfeiffer). The working sell of thermogravimetric system (Fig. 26.3) include vertical flow-type cylindrical oven with inner tube made from Al_2O_3 with inner diameter of 20 mm; length of the zone of the controlled temperature is ~30 mm. Samples of membranes ~0.5 g by mass with OCNTG synthesized in their pores with accumulated hydrogen were hanged up on the quartz rod in the center of the heating zone. Cell was filled with He before analysis. Delay time between mass spectrometric analysis and composition of gas in the reaction zone was ~8 s. Ion current values at $m/e = 2$, 18, 28, 32, 44 were fixed on line.

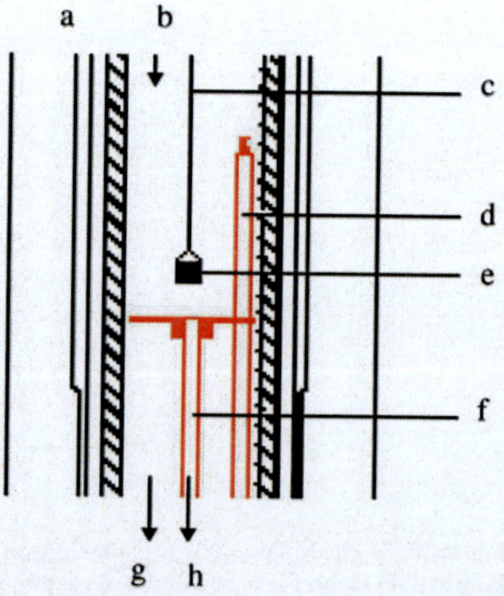

Fig. 26.3 The scheme of a thermogravimetric cell: (**a**) the furnace, (**b**) the initial gas mixture inlet, (**c**) a quartz fibre, (**d**) the furnace thermocouple in an alumina tube, (**e**) the suspended sample, (**f**) an alumina capillary for the selecting of the reactant mixture to the mass-spectrometer, (**g**) the initial gas mixture outlet, and (**h**) the gas flow to the mass-spectrometer

Table 26.1 Adsorption of H_2 in OCNTG (the samples 1–10) and PGNS (the samples 11–12) – P = 11.0–12.5 MPa

No[a]	m_c[b], mg	m_{H2}[b], mg	A[c], %	A_{25}[c], %	A_{60}[c], %	$A_{уд}$, mg/mono-layer
1	2.4	0.28	11.7	11.2		0.257
2	1.4	0.07	5.0	5.0	5.0	0.077
3	5.3	0.18	3.4	3.4		0.075
4	18.6	0.94	5.4	5.4	5.1	0.104
5	19.6	0.45	2.3	2.3	2.3	0.056
6	2.0	0.28	14.0	2.9	2.0	0.237
7	2.2	0.26	11.8	11.0	10.5	0.107
8	2.4	0.13	5.4	5.4		0.119
9	6.2	0.12	1.9	1.9		0.043
10	2.0	0.15	7.5	7.4		0.143
11	2.4	0.10	4.2			
12	3.6	0.08	3.4			

[a] To prepare the samples 1–5 the initial membranes were used with D_{pores} = 50 nm, for samples 6–10 – 90 nm; for samples 11–12 were used Al_2O_3 grade SCS-9

[b] m_C и m_{H2} – mass of carbon placed on the surface of pores and mass of adsorbed hydrogen correspondingly

[c] A, A_{25} and A_{60} – quantity of adsorbed hydrogen (% of mass of carbon) directly after saturation, after 25 days and after 60 days correspondingly

Table 26.1 shows the results for adsorption and storage of hydrogen in OCNTG (samples 1–10) and PGNS (samples 11–12), which were received via gravimetric method. In the last column the specific adsorption of hydrogen to one-wall tube is shown. Samples 1 and 6 are characterized with maximum quantity of this value but duration of storage of hydrogen in first is very low: already after 25 days its quantity reduces from 14.0% to 2.9%. Maximal absolute quantity of hydrogen which was 0.94 mg was adsorbed by nine-wall OCNTG (sample 4) which is equal to ~10.5 ml (at normal conditions) or 4.84 10^{-4} mol.

From data in the table it is also follow that specific adsorption of hydrogen for sample No. 4 with nine-layer OCNTG and for sample No. 7 with monolayer tubes is virtually the same. Apparently for these samples hydrogen is adsorbed by each carbon nanotube in approximately the same quantity. Therefore hydrogen molecules diffuse through monolayer OCNTG and interlaminar spaces and are adsorbed on the surface of tubes during accumulation of hydrogen. Quantitative distribution of hydrogen on the layers is probably sufficiently uniform although some dispersion exists.

Let us mark out sample No. 6 which was found to have the large capacity to hydrogen which fact is connected probably to its porous structure. Increased adsorption of hydrogen in the common carbon nanotubes occurs in that places where there are some structure defects which were formed during synthesis [21]. In our case OCNTG which are formed in membrane pores exactly repeat its geometry. Of course that the higher factor of pore tortuosity (τ), the more complex is the configuration of the nanotube formed in the pores, and the more it has points and knots with defective or strained structure which are places with heightened adsorptive potential. Probably the porous structure of sample No. 6 is characterized with higher quantity of τ. The indirect confirmation of this fact is sufficiently fast and significant (~80% from the adsorbed quantity) desorption of hydrogen comparing with other samples already at 20–22°C.

It is necessary to study the regularities of hydrogen desorption from OCNTG and PGNS depending on temperature. Thermogravimetric and mass-spectrometric analysis were used for this purpose. Figure 26.4a and 26.4c shows the results received during investigation of the sample of membrane with $D_{pores} = 50$ nm and sample of Al_2O_3 on whose porous surface OCNTG and PGNS were synthesized and 7.3% (to mass of OCNTG) and 4.2% (to mass of PGNS) of hydrogen were accumulated in them. One can see that change of ion current (2) for $m/e = 2$ (hydrogen) conforms to 170–180°C – temperature which can be correlated with process of desorption of H_2 form OCNTG and PGNS. This process goes with decrease of mass of sample (1) which probably occurs as a result of desorption of hydrogen and adsorbed water.

Let us note the existence of the second peak of ion current at $m/e = 2$ which appears at 500–550°C and accompanied with significant decrease of mass of the sample (Fig. 26.4a). Its appearance is a result of interaction of water steam that is always found in the measuring cell (air condensate, water from the samples etc.) with OCNTG carbon according to the reaction: $C + 2H_2O = 2H_2 + CO_2$ which starts at ~550°C [22]. Confirmation of this are the results of the investigation of the sample with OCNTG formed in its pores without adsorption of hydrogen (Fig. 26.4b). One can see that peak of ion current at $m/e = 2$ at 170–180°C is absent. One more confirmation of that fact that the noted reaction takes place in the measuring sell is change of the mass quantity of water (Fig. 26.5, curve 2). One can see that up to ~250°C this value is almost constant but at higher temperatures it decreases because of desorption. Probably at temperatures >500°C together with desorption of water interaction of carbon with water steam takes place.

Very important for our investigation is the question about mechanism and energy of hydrogen adsorption in OCNTG; the detailed investigation of this problem will be presented in the next publications. However we'd like to note that temperature of H_2 desorption (170–180°C) shows that energy of its binding with OCNTG is significantly higher that energy of hydrogen bond. It is known [23] that energy of hydrogen bond for water molecules which form 4-molecular associates at 0°C it rather high and is equal to 25 kJ/mol. However at boiling point and at atmospheric pressure water contains lesser than 1% of dimers [24] i.e. hydrogen bonds are destroyed almost entirely. At the same time in OCNTG hydrogen is stored up to significantly higher temperature – 170–180°C. In the case of dissociative adsorption of H_2 in OCNTG the forming bonds are weaker than typical chemical C-H bonds; in the case of nondissociative adsorption the interaction under review is much more stronger than physical sorption (Van der Waals forces, energy of desorption is of the same order as enthalpy of evaporation of liquid hydrogen, i.e. ~0.9kJ/mol [25, 26]). In the case of dissociative adsorption mechanism of hydrogen spillover [27] is possible while in the second case structurally-adsorptive mechanism is possible.

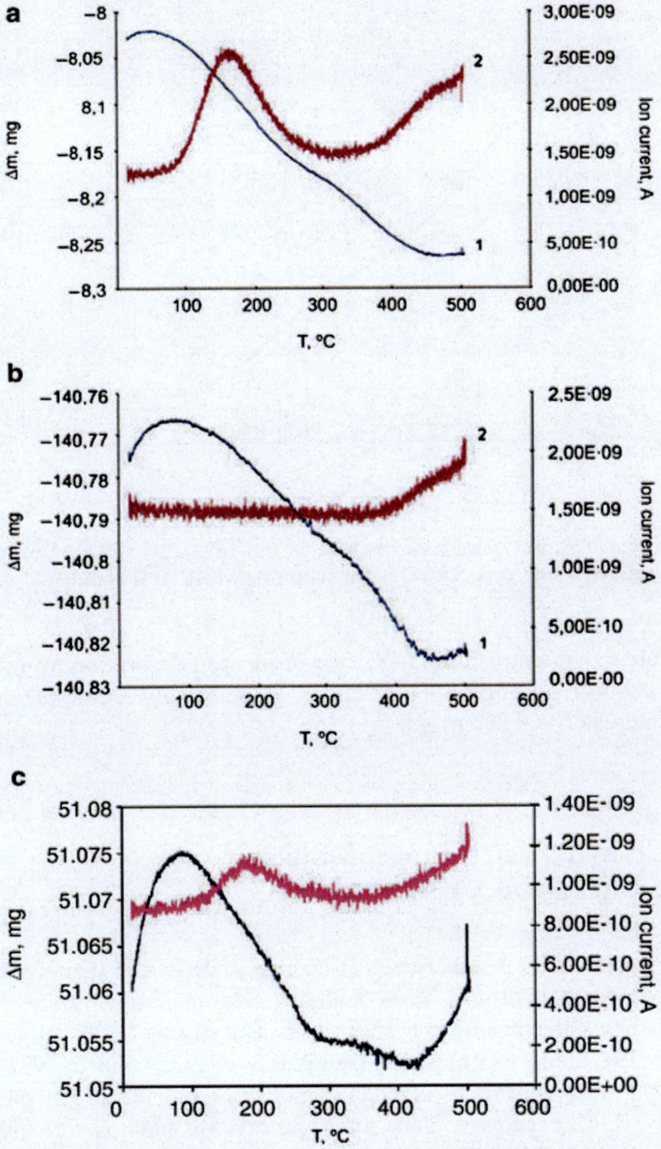

Fig. 26.4 Change of mass (*1*) and ion current (*2*) at $m/e = 2$ for membrane sample in whose pores OCNTG is synthesized where H_2 is adsorbed (**a**) and for sample where there is no adsorbed H_2 (**b**) and for sample of Al_2O_3 (grade SCS-9) on which surface PGNS is synthesized where H_2 is adsorbed (**c**)

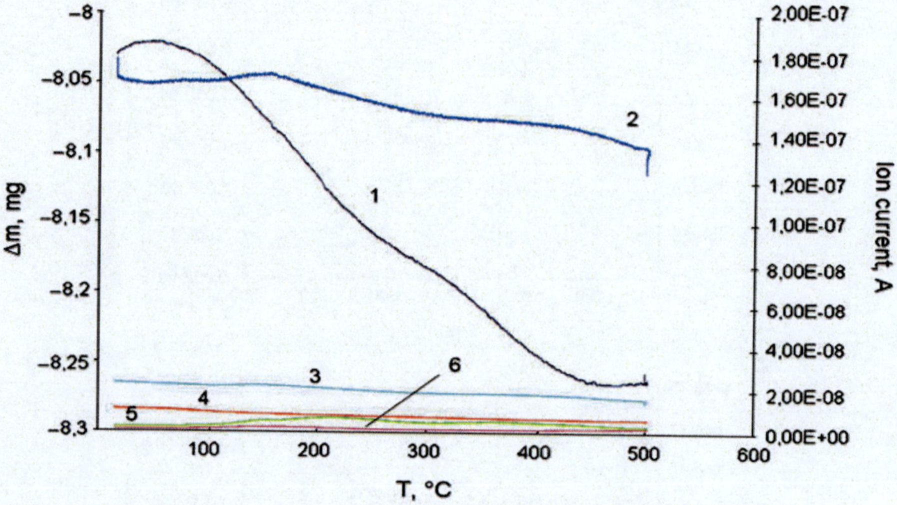

Fig. 26.5 Change of mass (*1*) and ion current (*2*) at m/e = 18 (*2*), 28 (*3*), 32 (*4*), 44 (*5*) an 2 (*6*) for membrane sample in whose pores OCNTG is synthesized where H_2 is adsorbed

Let us notice that temperature interval of hydrogen desorption from OCNTG is fully corresponds to hydrogenation reactions which usually occur at comparatively low temperatures (20–200°C) [28].

26.5 NCP and Theirs Superposition. An Uncoupled Layer Effect

It was necessary to find whether such structures as NCP and theirs superposition were able to adsorb hydrogen while studying hydrogen adsorption in OCNTG. Table 26.2 shows the results for adsorption and storage of hydrogen in NCP (samples 15–19), theirs superposition (samples 3–6, 9–11) and OCNTG (samples 1–2, 7–8, 12–14) which were received via gravimetric method and by thermogravimetric analysis. One can see that ability to adsorb and to store hydrogen is characteristic only to OCNTG.

Table 26.3 presents selected data on the superposition of NPCs and its influence on the electrophysical properties of the surface and membrane permeability. It gives the number of superpositions *n* on the porous surface of membranes, between which, the sample was cooled until methane dehydrogenation stopped, as mentioned above; the number of monolayers *m* in each superposition calculated by the procedure of [9] and the superposition time. An analysis of the results given in Table 26.3 revealed a new effect regarding the even or odd total number of monolayers in the NPC superposition. Let us trace these changes for samples 1–5.

Table 26.2 Adsorption of H_2 in NCP (samples 15–19), in NCP superposition (samples 3–6, 9–11) and in OCNTG (samples 1–2, 7–8, 12–14) – P = 11.0–12.5 MPa

#[a]	m_c[b], mg	m_{H2}[b], mg	A[c], %	A_{25}[c], %	A_{60}[c], %	$A_{spec.}$, mg/mono-layer
1	2.0	0.28	14.0	2.9	2.0	0.237
2	2.2	0.26	11.8	11.0	10.5	0.107
3	7.9	–	–			
4	9.8	–	–			
5	12.7	–	–			
6	10.9	–	–			
7	2.4	0.13	5.4	5.4		0.119
8	6.2	0.12	1.9	1.9		0.043
9	6.8	–	–			
10	7.0	–	–			
11	8.6	–	–			
12	1.4	0.07	5.0			
13	18.6	0.94	5.4	5.4	5.1	0.104
14	19.6	0.45	2.3	2.3	2.3	0.056
15	18.6	–	–			
16	7.9	–	–			
17	9.8	–	–			
18	12.7	–	–			
19	10.9	–	–			

[a]For preparation of samples 1–8, 16–17 the initial membranes were chosen of D_{pores} = 90 nm, for samples 9–15, 18–19 – 50 nm

[b]m_C и m_{H2} – mass of carbon placed on the surface of pores and mass of adsorbed hydrogen correspondingly

[c]A, A_{25} and A_{60} – amount of adsorbed hydrogen (% of mass of carbon) directly after saturation, after 25 days, and after 60 days correspondingly

About 1.5 monolayers were deposited on sample 1 during 3 min. In the second superposition, another 3.5 monolayers were added (sample 2); i.e., two superpositions resulted in 5 monolayers. After that, the deposition rate decreased drastically, and after the next 32 min, only ~0.2 monolayers were deposited (sample 3).

Earlier [11], we studied in detail the kinetics of thetopochemical dehydrogenation of methane on membranes with a selective layer with a similar chemical composition. The kinetic evaluation of the results of the superposition revealed that for samples 1 and 2, they were fully consistent with the data of [11], but the rate constant of samples 3–5 was more than one order of magnitude lower. In other words, the NPC superposition occurs in agreement with the kinetics of methane decomposition if $m_1 + \ldots + m_n$ is even, but is drastically decelerated if it is odd.

Thus, the kinetics of methane dehydrogenation in continuous deposition of NPCs is not valid for superposition. Indeed, after the first two superpositions, sample 7 has a total of 4 layers, and further superposition follows the reaction kinetics (2.2 and 2.4 monolayers, Table 26.3).

The same tendency is observed for membranes with D = 90 nm, whose selective layer consists of TiO_2 + MgO. Thus ~3 monolayers were deposited in the first

Table 26.3 Superposition of NPCs on Trumem membranes with $D = 50$ nm (samples 1–7) and 90 nm (8–11). Changes in the electrophysical properties of the pore surface (ξ, σ) and permeability (Q, ml/(s cm^2 Torr)), $T_{dep} = 800°C$, $p_{dep} = 4.7–4.9$ kPa

Sample	$-\xi \times 10^3$, V	$-\sigma \times 10^4$, C/cm^2	Q × 10^5, (ethanol)	Q × 10^5, (dodekan)	Q × 10^5, (H_2O)	n, un.	m, un.	τ, min
I_{50}	20.0	47.5	2.9	2.9	4.4	–	–	–
1	3.2	5.6	5.5	3.7	4.4	1	1.4	3
2	3.5	5.1	3.2	8.0	–	2	1.4 + 3.4	3 + 32
3	4.0	6.3	2.5	2.2	–	3	1.4 + 3.4 + 0.21	3 + 32 + 32
4	4.2	6.0	2.5	2.2	–	3	1.4 + 3.4 + 0.14	3 + 32 + 32
5	3.5	6.1	–	–	–	3	1.2 + 1.5 + 0.3	3 + 32 + 32
6	3.0	6.5	–	–	–	4	1.2 + 1.5 + 0.3 + 2.3	3 + 32 + 32 + 32
7	2.8	6.4	3.8	6.4	6.3	2	1.5 + 2.5 + 2.2 + 2.4	3 + 32 + 32 + 32
I_{90}	27.4	6.5	9.2	6.5	14.2	–	–	–
8	2.5	5.4	6.6	10.1	–	4	1.6 + 3.5 + 1.2 + 1.8	3 + 32 + 32 + 32
9	3.3	5.6	11.5	15.8	–	1	2.8	35
10	3.0	5.1	11.0	15.5	–	2	2.8 + 1.2	35 + 35
11	2.7	4.9	11.1	19.7	–	3	2.8 + 1.2 + 2.8	35 + 35 + 35

I_{50} and I_{90} are the starting membrane samples with $D_{pore} = 50$ and 90 nm, respectively; n is the number of superpositions in the NPCs formed; m is the number of carbon monolayers in each superposition; and τ is the deposition time of each superposition

superposition (sample 9), ~1 monolayer in the second (sample 10), and again ~3 monolayers in the third (sample 11). The amount of deposited carbon changes in a similar way depending on the total number of previously deposited monolayers in all superpositions, decreasing if the number is even (the rate constant decreases by more than one order of magnitude) and further deposition follows the reaction kinetics if it is odd.

For samples 6 and 8 with four superpositions, an odd number of monolayers (approximately three for 6 and five for 8) were deposited after two superpositions. After the first superposition, 1.2 monolayers were deposited on 6 and 1.6 on 8; i.e., the deposition on 6 is closer to the monolayer (odd) type and hence only 1.5 monolayers were deposited on its surface after the second superposition. For 8, the deposition is closer to the two_layer (even) type and 3.5 monolayers were deposited (Table 26.3). After two superpositions, both samples had an odd number of monolayers, namely, ~3 (exactly 2.7) for 6 and ~5 (exactly 5.1) for 8. That is why the third superposition was small for both samples: 0.3 monolayers for 6 and 1.2 for 8.

These results strongly suggest that there is an uncoupled layer effect in NPC superposition. Evidently, superposition of an even number of monolayers occurs with their coupling accompanied by a certain stabilizing redistribution of electron density, formation of a three-dimensional structure, etc. Therefore the properties of a superposition with an even number of monolayers are close to those of NPCs. When forming a superposition with an odd number of monolayers, we obtain a structure with a two-dimensional upper layer, whose interaction with lower-lying coupled layers is rather limited. In this layer, considerably greater mobility of the electron cloud is possible along with other effects inherent in two-dimensional structures. They seem to be the reason for the NPC superposition effect which we discovered for the first time.

An indirect support for the suggested interpretation may be the electron microscopy data for samples with an NPC superposition and nanosized crystallites. The latter process is understood to be the conventional continuous deposition of pyrocarbon nanostructures. As is known [29], in high-resolution electron microscopy for dielectric samples with a surface charge (starting membrane), the secondary electron emission coefficient decreases and the surface starts to be changed. This charging leads to image distortion and diffusion.

Figure 26.6 shows the photographs of the starting membrane (top) and a membrane with deposited pyrocarbon nanocrystallites (bottom) (at magnification $\times 10^5$). It can be seen that the image of the starting membrane became very unclear and diffuse. When pyrocarbon nanocrystallites were deposited, the surface became conducting, and its charge decreased by approximately one order of magnitude (Table 26.3); this is why the image remained clear and contrastive (Fig. 26.6). Figure 26.7 presents the photograph of sample 2, for which m is odd, after two superpositions. A threefold superposition does not form on it (samples 3 and 4, Table 26.3). On the photograph at $\times 10^5$ magnification, it can be seen that the image of this sample starts to lose contrast and becomes slightly diffuse. Evidently, the properties of its surface differ significantly from those of NPCs and the differences

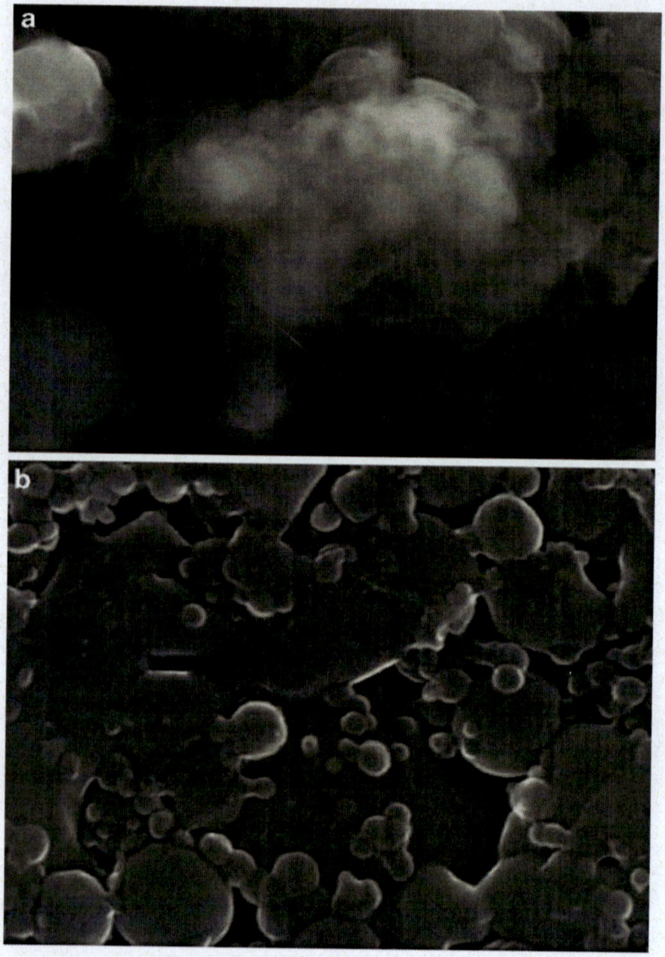

Fig. 26.6 Photographs of the surface of (**a**) the starting membrane and (**b**) the membrane with deposited pyrocarbon crystallites (magnification $\times 10^5$)

are due to the electron density redistribution, surface charge effects, etc., which should be evaluated by quantum-mechanics methods.

The changes in the electrophysical characteristics of membranes such as the pore surface and permeability for different liquids (Table 26.3) give certain information, but do not reflect the mechanism of the processes quite adequately. The surface density of the charge and the ξ potential decreased substantially (by approximately one order of magnitude) when the pore surface was covered even with one NPC monolayer for membranes with a selective layer of TiO_2 + Cr_2O_3 or titanium oxide with magnesium oxide (Table 26.3). In addition, the permeability of the nonpolar (dodecane) fluid increased considerably for samples 2 and 11 with an uncoupled upper layer (Table 26.3).

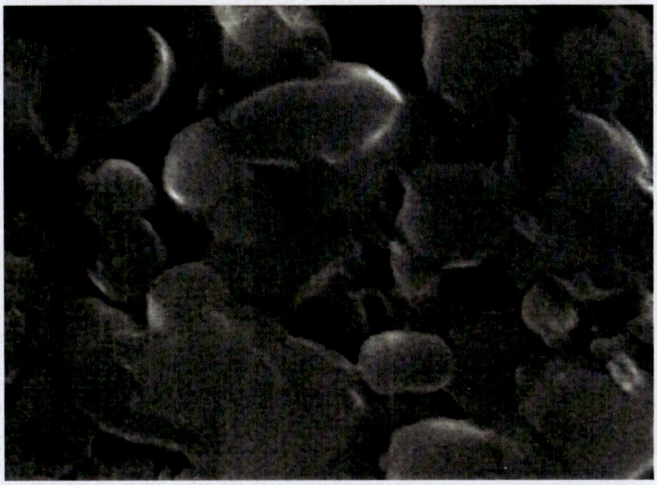

Fig. 26.7 Photograph of the surface of a membrane with a twofold superposition and odd *m* (magnification $\times 10^5$)

26.6 Influence of Adsorbed H_2 on Trans-membrane Transport

Very important question is the problem of reactivity of H_2 which is accumulated and stored in OCNTG. Investigation of this parameter is the next stage of the study, and results will be soon received. However the indirect confirmation of high activity of adsorbed in OCNTG hydrogen is the effect of hydrogen variation of performance (HVP) of membranes with OCNTG synthesized and hydrogen accumulated in their pores.

Table 26.4 and Fig. 26.8 show the results of the investigation which demonstrate that performance of membranes with D = 50 nm and 90 nm increases after formation of OCNTG (without H_2) in their pores (Table 26.4, samples 1 and 2). Performance for all fluids decreases in 4–26 times after adsorption 0.7–5.1% of hydrogen.

In the pores of sample 6 two-layer OCNTG were formed, and on their surface there was ~5.1% H_2 taking into consideration the abovementioned uniform distribution of H_2 in the tubes.

Dependence of permeability on quantity of adsorbed hydrogen shown in Fig. 26.8 has minimum corresponding to 3.6% of H_2. There are two possible reasons of this fact: either at this quantity of adsorbed hydrogen HVP effect is maximal or, that is most likely, in two-layer OCNTG there exchange stabilization interactions between H_2 molecules from different tubes, and influence of hydrogen on permeability decreases.

The significant fact is that performance of membranes with adsorbed H_2 with D_{pores} = 90 nm turns to be similar and even sometimes almost in an order less than

Table 26.4 Effect of hydrogen adsorption in OCNTG on the electrophysical characteristics of the pore surface and efficiency of membranes

Sample	$-\zeta \times 10^3$, V	$-\sigma \times 10^4$, C/m^2	A, %	$Q \times 10^5$, ml/s·sm^2· Torr (H_2O) (H_2O)	$Q \times 10^5$ (ethanol)	$Q \times 10^5$ (dodecane)
I_{50}	20.0	47.5	–	4.4	2.9	2.9
1	2.6	6.1	–	7.9	6.8	5.3
I_{90}	27.4	66.5	–	14.2	9.2	6.5
2	2.2	5.1	–	18.5	13.6	9.2
3	8.3	17.2	0.7	2.2	2.4	2.8
4	8.1	18.9	1.9	1.7	2.1	2.3
5	7.0	16.1	3.6	0.59	0.76	0.43
6	7.3	17.0	5.1	1.1	2.0	2.1

A is the amount of adsorbed hydrogen (% of the mass of deposited carbon) during the efficiency measurement; (1, 2) the samples with monolayer OCNTGs but without adsorbed H_2; (3–5) the samples with H_2 adsorbed in the single-walled OCNTGs; (6) the sample with H_2 adsorbed in the two-walled OCNTG

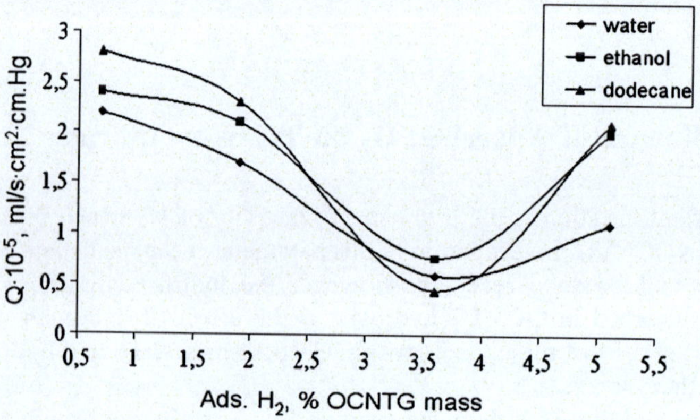

Fig. 26.8 Change of performance for membranes with $D_{pores} = 90$ nm versus quantity of hydrogen adsorbed in OCNTG which are formed on the surface of their pores

those of membranes with $D_{pores} = 50$ nm which don't contain accumulated hydrogen (Table 26.4).

Very interesting is also influence of H_2 adsorption on electro-physical properties of porous surface. The electric surface properties of membranes were investigated using method of flow potential with the help of chlorine-silver (Ag-AgCl) electrodes [30]. In present investigation the potential drop ΔE was studied as the function of pressure ΔP on membrane during flowing of 0.01 M solution of KCl through membrane. The flow potential ($\Delta E/\Delta P$) was used to determine the quantity of ξ-potential using Smoluhovski equation; value of the latter was used in Gui equation for determination of charge density on porous surface. The sign of the surface charge was determined using slope ratio of $\Delta E/\Delta P$.

One can see from Table 26.4 that ξ-potential and charge of the surface decrease ~ in an order after OCNTG formation comparing with the initial membrane. However after saturation of OCNTG with hydrogen both ξ-potential and charge of the surface increase in ~3–4 times. It confirms indirectly that hydrogen in two-dimensional carbon structure is rather active and takes part in exchange or coordination interactions which are probably the result of dissociative adsorption mechanism. Also important that HVP effect is revealed both at flowing though the membrane of liquids with significantly high dipole moment of molecules which is equal for water and ethanol $\mu = 6.1$ and 5.7 Coulomb m and with zero dipole moment as well – dodecane. Let us notice that data for sample 3 (Table 26.4) are the mean values for tree parallel measures.

26.7 Conclusions

Thus, new carbon nanostructures were synthesized, namely, oriented single- and multiwall carbon nanotubes with walls formed of graphenes and plane graphene-based carbon nanostuctures. They turned out to be able to adsorb and store molecular hydrogen and their adsorption capacity is in line or larger than similar value of traditional carbon nanotubes. This ability of OGCNTs can be of interest for solving the problem of hydrogen storage for fuel cells. These structures are also good candidates for another end-use application: membrane nanoreactors, in which one of the reagents (hydrogen) will be stored in OGCNTs and will desorb as required to enter into the reaction. At the same time, we can to create autonomous and compact energy elements using PGNS.

Results on adsorption and storage of H_2 in OCNTG formed in pores let one to consider membrane as the integrity of ~25 10^{10} of nano-size reactors in which ~4.84 10^{-4} mol of hydrogen is accumulated. It is shown for the first time that desorption of hydrogen from OCNTG occurs at 170–180°C which is typical for many hydrogenation reactions.

To summarize, we have studied for the first time the effect of the interfaces of nanosize substances on their properties using pyrocarbon structures as an example. Indeed, on passing from NPCs to their superposition, we obtain a structure with an *n*-fold increase in the interfaces and the uncoupled layer effect is observed. These structures, however, do not possess the hydrogen adsorption ability. Subsequent *n*-fold increase in the interfaces leads to two-dimensional OCNTGs, which are capable of accumulating and preserving hydrogen. Introducing the S_{surf}/V criterion that characterizes the surface to volume ratio of a nanostructure, we can see that it increases and tends to infinity for two-dimensional OCNTGs. Its numerical value can probably serve as a nanodimensionality factor, showing that the object acquired new specific properties.

Effect of hydrogen variation of performance (HVP) for membranes with OCNTG synthesized in pores with hydrogen accumulated in that OCNTG which was found for the first time can be considered as the indirect confirmation of high activity of H_2 adsorbed in two-dimensional structure. The registered increase of surface charge and of ξ-potential is connected probably with dissociative mechanism of hydrogen adsorption (mechanism of hydrogen spillover).

This work was supported in part by the Russian Foundation for Basic Research (RFBR), Grant No. 10-03-00659-a.

References

1. Sun X, Li R, Stansfield B, Dodelet J-P et al (2007) Controlled synthesis of pointed carbon nanotubes. Carbon 45(4):732–737
2. Harris PJ (2007) Solid state growth mechanisms for carbon nanotubes. Carbon 45(2):229–239
3. Gao L, Zhou X, Ding Y (2007) Effective thermal and electrical conductivity of carbon nanotube composites. Chem Phys Lett 434(4–6):297–300
4. Zuttel A, Sudan P, Mauron P et al (2004) Model for the hydrogen adsorption on carbon nanostrutures. Appl Phys A 78(7):941–946
5. Hagen A, Moos G, Talalaev V, Hertel T (2004) Electronic structure and dynamics of optically excited single-wall carbon nanotubes. Appl Phys A 78:1137–1145
6. Zhang M, Atkinson KR, Baughman RH (2004) Multifunctional carbon nanotube yarns by downsizing an ancient technology. Science 306:1358–1361
7. Wang WH, Hong TH, Kuo CT (2007) Super growth of vertically aligned SWCNTs using self-assembled nanoparticles from $CoCrPtO_x$ ultra-thin film. Carbon 45(1):97–102
8. Bachtold A, Hadley P, Nakanishi T et al (2003) Logic circuits based on carbon nanotubes. Physica E 16(1):42–46
9. Soldatov AP, Rodionova IA, Shkolnikov EI et al (2004) Pyrocarbon modification of composite inorganic membranes. Russ J Phys Chem A 78(9):1659–1664
10. Soldatov AP, Rodionova IA, Parenago OP (2006) Influence of pyrocarbon modification on physico-chemical characteristics of surface of pores and on transport properties of inorganic membranes. Russ J Phys Chem A 80(3):500–506
11. Soldatov AP, Berezkin VV, Gontar IV et al (2008) Knudsen diffusion procedure of methane in the pores of inorganic membranes: kinetics and depth of plating with pyrocarbon, its influence on transport properties. Russ J Phys Chem A 82(6):1124–1130
12. Soldatov AP, Syrtsova DA, Parenago OP (2008) The depth of nanocrystallites of pyrocarbon deposition in pores of ultrafiltration membranes and its influence on their efficiency. Russ J Phys Chem A 82(11):1903–1907
13. Soldatov AP, Parenago OP (2008) Carbon nanotubes from graphenes in the pores of inorganic membranes. In: Baranowsky B, Zaginaichenko SY, Schur DV, Skorokhod VV, Veziroglu A (eds) Carbon nanomaterials in clean energy hydrogen systems, NATO-OTAN. Springer, The Netherlands, pp 225–232
14. Soldatov AP, Shkolnikov EI, Rogailin MI et al. (2002) Technique of modification of porous structure of inorganic membranes with pyrocarbon. Russian Federation Patent 2179064 (in Russian)
15. Soldatov AP, Parenago OP (2007) Carbon nanotubes from graphenes in the pores of inorganic membranes, Book of Abstracts of X ICHMS'2007, Sudak-Crimea-UKRAINE, 22–28 Sept 2007, pp 758–759
16. Ning GQ, Wei F, Luo GH et al (2004) Hydrogen storage in multi-wall carbon nanotubes using samples up to 85 g. Appl Phys A 78(7):955–959
17. Shiraishi M, Takenobu T, Kataura H et al (2004) Hydrogen adsorption and desorption in carbon nanotube systems and its mechanisms. Appl Phys A 78(7):947–953
18. Shkolnikov EI, Elkina IB, Volkov VV (1999) Method of analysis of porous structure. Russian Federation Patent 2141642 (in Russian)

19. Mulder M (1999) Introduction in membrane technology. Mir, Moscow, pp 212–266 (in Russian)
20. Moulder JF, Stickle WF, Sobol PE (1992) In: Chastain J (ed) Handbook of x-ray photoelectron spectroscopy. Physical Electronics, Eden Prairie, 439 p
21. Cheng YM, Yang QY, Liu C (2001) Hydrogen storage in carbon nanotubes. Carbon 39(10):1447–1454
22. Slavinski MP (1952) Physico-chemical properties of elements. Metallyrgizdat, Moscow, 276 p (in Russian)
23. Goronovski IT, Nazarenko YP, Nekryach EF (1974) Short chemical hand-book. Naukova Dumka, Kiev, p 766 (in Russian)
24. Sinyukov VV (1976) Structure of monoatomic liquids, water and water solutions of electrolytes. Science, Moscow, 238 p (in Russian)
25. Zuttel Z, Eklund PC (2002) Hydrogen storage in carbon nanostructures. Int J Hydrogen Energy 27(2):203–212
26. Murata R, Yang RT (2002) Adsorption mechanism of supercritical hydrogen in internal, and interstitial nanospaces of single-wall carbon nanohorn assembly. J Phys Chem B 106:11131–11138
27. Lueking A (2004) Carbon-metal composites: hydrogen spillover to enhance hydrogen storage. Transactions of the VIII World Renewable Energy Congress, Denver, p 112
28. Geits BK, Ketzir J, Shuit G (165) Chemistry of catalytic processes. Science, Moscow, p 165
29. Schimmel G (1972) Technique of electron microscopy. Nauka, Moscow, p 187 (in Russian)
30. Berezkin VV, Volkov VI, Kiseleva OA (2003) Charge of pores of tracking membranes from polyethylene terephtalate. Colloid J 65:129–132

Chapter 27
Temperature Ferroelastic Phase Transition in Hydroxyapatite. Hydroxyl Solubility, Configuration Heat-Capacity, Hysteresis Effect, Elasticity Modulus

Z.A. Matysina, S.Yu. Zaginaichenko, D.V. Schur, and N.A. Shvachko

Abstract The statistical theory of ferroelastic-paraelastic phase transition has been elaborated in this paper. The equation of thermodynamic equilibrium state has been examined and the temperature of transition between phases has been estimated. The calculation of temperature dependence of hydroxyl solubility in crystal has been carried out. The manifestation hysteresis effect has been considered. The temperature dependence of elastic compliance and elasticity modulus has been evaluated and the verity of the rule of "negative two" and Curie-Weiss law has been found. The dependence configuration heat capacity on temperature has been calculated, its extremality has been ascertained and the abrupt change of heat capacity has been defined in the point of phase transition. The established regularities are consistent qualitatively with literature experimental data.

Keywords Molecular-kinetic theory · Hydroxyapatite structure · Ferroelastic phase transition · Solubility of hydroxyl · Hysteresis effect · Elastic compliance · Heat capacity

27.1 Introduction

The investigation of apatites and apatite-like materials attracted attention of scientists for a long time. Their researches began as early as the 30th years [1–10], the structure, properties of these materials, temperature transformations, character of such transformations in apatites were studied and the possibility of practical use of

---

Z.A. Matysina
Dnepropetrovsk National University, 72 Gagarin str, Dnepropetrovsk 49000, Ukraine

S.Yu. Zaginaichenko (✉), D.V. Schur, and N.A. Shvachko
Institute for Problems of Materials Science of NAS of Ukraine,
Department # 67, 3 Krzhyzhanovsky str, Kiev 03142, Ukraine
e-mail: shurzag@materials.kiev.ua

S.Yu. Zaginaichenko et al. (eds.), *Carbon Nanomaterials in Clean Energy Hydrogen Systems - II*, NATO Science for Peace and Security Series C: Environmental Security 2,
DOI 10.1007/978-94-007-0899-0_27,

such materials was discussed by various investigators. The apatites aroused considerable interest especially in the last decade [11–82].

Apatites and apatite-like materials have occupied the prominent place in many branches of science and production due to their unique properties. The range of possibilities of their practical application [14, 15, 22–24, 32, 33, 42, 43, 49–53, 58, 59, 65, 66, 76–82] has been extended more and more in acousto-electronics, acoustooptics and other technical branches, in stomatology, orthopaedy. Apatites have been used for the creation of miniature devices on visualization of acoustic fields, for the transformation of electric signals into audible signals (and conversely), for the creation of biocomposites as ecological sorbents with a view to bury the radioactive wastes, for the realization of highly sensitive methods of control of environment quality, for the use in medicine as the dentures and bone materials that reach the level of natural biocompatibility. The apatites are perspective crystals for the production of oxide materials with proton conduction in consequence of the easy introduction of hydrogen atoms into their structural free sites.

Therefore, there is a need to perform an in-depth, comprehensive investigation of their atomic and electronic structure, the physical and chemical properties, phase transformations and other processes going on in them with the change of external terms, to study the effect of impurities and different atoms intersubstitution on the properties and processes in apatite materials.

At present time more than 100 apatites and apatite-like materials have already been investigated and described [26, 34, 63].The apatites form the group of minerals with the general formula [63]

$$\mathrm{Me_{10}(XO_4)_6Y_2 \equiv Me_5(XO_4)_3Y},$$

where

$$\mathrm{Me} = \mathrm{Ca^{2+}, Cd^{2+}, K^{+}, Na^{+}, Ba^{+}, Mg^{2+}, Mn^{2+}, Ni^{2+}, Pb^{2+}, Sr^{2+}, Th^{3+}}\ \text{and others,} \tag{27.1}$$

$$\mathrm{XO_4 = PO_4^{3-}, CO_4^{2-}, CoO_4^{2-}, CrO_4^{3-}, Mn_4^{2-}, SO_4^{2-}, SiO_4^{3-}, VO_4^{3-}}\ \text{and others,}$$

$$\mathrm{Y = F^{-}, Cl^{-}, OH^{-}, O^{2-}, CO_3^{2-}}\ \text{and others.}$$

The apatites are easily susceptible to the processes of hydrotation and dehydrotation. In addition, for apatites a deficit on the Me component is often characteristic, for example, on a calcium [20, 38, 40, 41, 47, 48, 74]. The hydrogen atoms as a result of introduction into the crystal can substitute both the hydroxyl groups OH and vacant sites of calcium atoms. In this case at the deficiency of calcium the formula of hydroxyapatite becomes $Ca_{5-x}H_x(PO_4)_3OH$. The presence of outside electric stress assists movement of hydrogen atoms on the structural channels of crystal along the principal screw or along a triad axis, creating proton conductivity [83–85].

The hydroxyapatite $Ca_5(PO_4)_3OH$ is under investigation in the present paper and we examine the revealed in it ferroelastic phase transition of paraelastic-ferroelastic type [9, 10] at the temperature of 484.5 K [8, 22, 45, 54, 61, 83, 84, 86, 87] manifested in the ordered displacement of hydroxyl groups OH, we calculate the solubility of hydroxyl OH that escape from the crystal lattice at the temperature of 1,073 K with no change of structure in accordance with experimental data (since the melting point is high and equal to 1,614–1,622 K [83, 84, 88]) and elucidate also the possibility of hysteresis effect manifestation that can shows itself in crystals at the first-kind phase transition under the directed outside mechanical stress, evaluate the elastic compliance and modulus of elasticity, estimate the configuration heat capacity, its temperature dependence, special feature of this dependence near-by the temperature of phase transition.

Experimental investigation of heat capacity of hydroxyapatite made possible to discover the peak on its temperature dependence in the vicinity of ferroelastic-paraelastic phase transition at the $T_o = 484.5$ K (Fig. 27.1). The average heat capacity of such crystals was measured in the temperature range from 298 K to1298 K in the papers [56, 57] and the corresponding magnitudes of $C_p = 530 J \cdot mol^{-1} \cdot K^{-1}$ and $C_p = 635 J \cdot mol^{-1} \cdot K^{-1}$ were obtained that are in agreement with heat capacity values of Fig. 27.1, but such peak was not found in these works.

Study of curves of thermographic (TG) and differential thermal analysis (DTA) [5, 48, 74, 83] makes possible to reveal also both the temperature of phase transition and temperature of hydroxyl loss. The DTA curves of Fig. 27.2 for the aqueous solution of hydroxyapatite [74] illustrate the exothermal peaks at the temperatures area near 757°C of hydroxyl OH loss, the endothermal peak at the structural ferroelastic phase transition in the point of 210°C and also the exothermal peaks corresponding to the decrease of concentration of water mass in the temperature

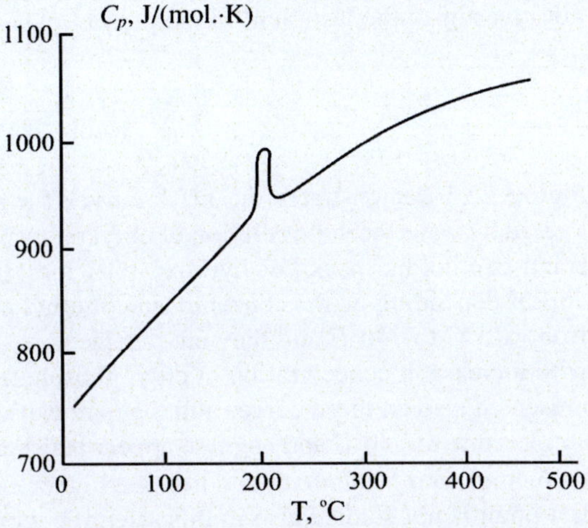

Fig. 27.1 Experimental plot of temperature dependence of hydroxyapatite heat capacity [89]. The peak is observed at the ferroelastic-paraelastic structural phase transition at $T_o \approx 210°C$

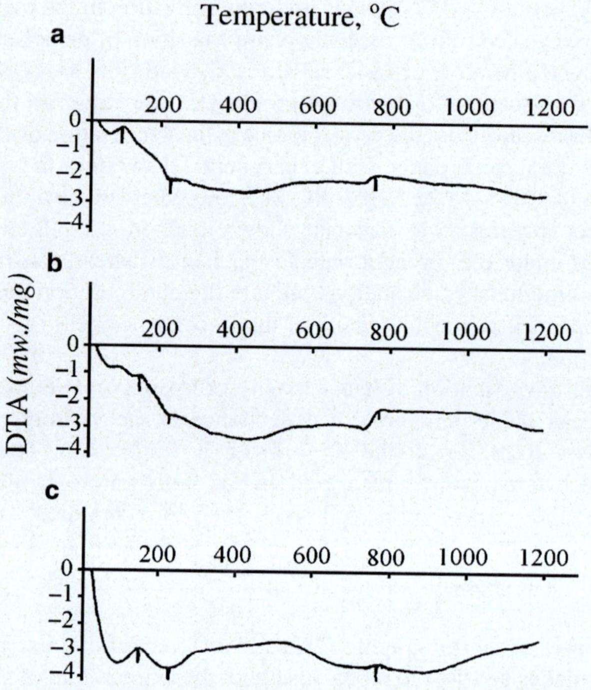

Fig. 27.2 DTA curves of water solution of hydroxyapatite [74] without calcium deficit (with Ca/P molar ratio equal to 1.67 (**a**)) and with calcium shortfall (with Ca/P ratios 1.6 (**b**) and 1.5 (**c**)) illustrated the exothermic peaks with decreasing water concentration at the points T = 98°C, 154°C, 155°C, the exothermic peaks of hydroxyl OH loss over the temperature region T = 757°C and also the small endothermic peaks at the structural ferroelastic-paraelastic phase transition over the temperature area T ≈ 210°C

range from 98°C to 155°C depending on the degree of calcium shortfall. We does not investigate the last point in the present paper.

$$C_p, \frac{\mathrm{J}}{(\mathrm{mol.} \cdot \mathrm{K})}$$

Figure 27.3 demonstrates the DTA curves for specimens of hydroxyapatite [90] treated by water-acetone solution of ethyl phosphate of different concentration. The small exothermal peaks of hydroxyl OH loss are seen on these curves and they appear depending on the concentration of ethyl phosphate in the temperature range from 615°C to 740°C and they are displaced in the area of more low temperatures with increase in concentration of ethyl phosphate. The deep endothermal peaks are presented also on these curves and they are caused by the loss of water mass at the temperature of 350°C and begin to appear at the concentration of ethyl phosphate as $0.46 \cdot \mathrm{mol} \cdot \mathrm{dm}^{-3}$ (curve d) and become higher as this concentration is greater. The last point is not studied also in this scientific work.

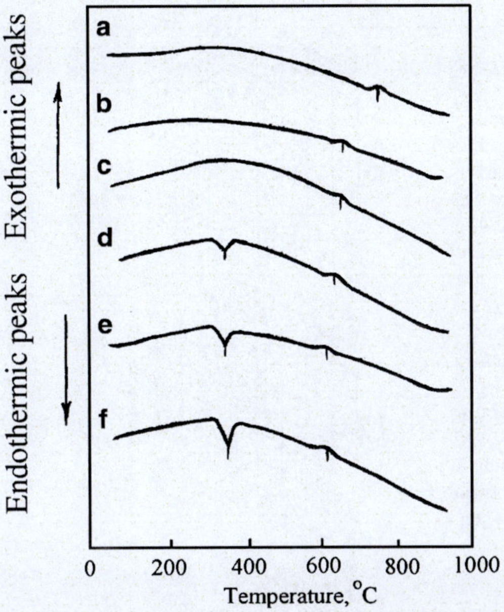

Fig. 27.3 DTA curves of calcium hydroxyapatites specimens [90] treated with water-acetone solutions of ethyl phosphates (EP) at different concentrations of the last as 0; 0.09; 0.18; 0.46; 0.68; 0.91 mol·dm^{-3} (curves a, b, c, d, e, f, respectively). The curves exhibit endothermic DTA peaks of water mass losses in the temperatures range T = 615–740°C in dependence on EP concentration

To solve the above-listed problems, in the present investigation calculations of Helmholtz free energy and Gibbs thermodynamic potential for hydroxyapatite $Ca_5(PO_4)_3OH$ are performed on the basis of the molecular-kinetic concept, their dependence on temperature, order parameter and energetic constants of interatomic and intermolecular interactions in crystal is found. The method of average energies is used ignoring the correlation in substitution of lattice positions in crystal by atoms and molecular groups and taking into account only interactions between the nearest structural blocks and also considering the square-law dependence of ordering energy on order parameter and hydroxyl concentration.

27.2 Structure of Hydroxyapatite

The structure of hydroxyapatite $Ca_5(PO_4)_3OH$ is shown in Fig. 27.4 in projection to a planar plane (0001) perpendicular to the principal screw axis z for the values z Z = 0, c/4, c/2, 3c/4 [6, 16, 34, 80]. The apatites of the same type have the same structure at the replacement of the hydroxyl OH^- by fluorine F^- and chlorine Cl^- [17, 35, 83, 84].

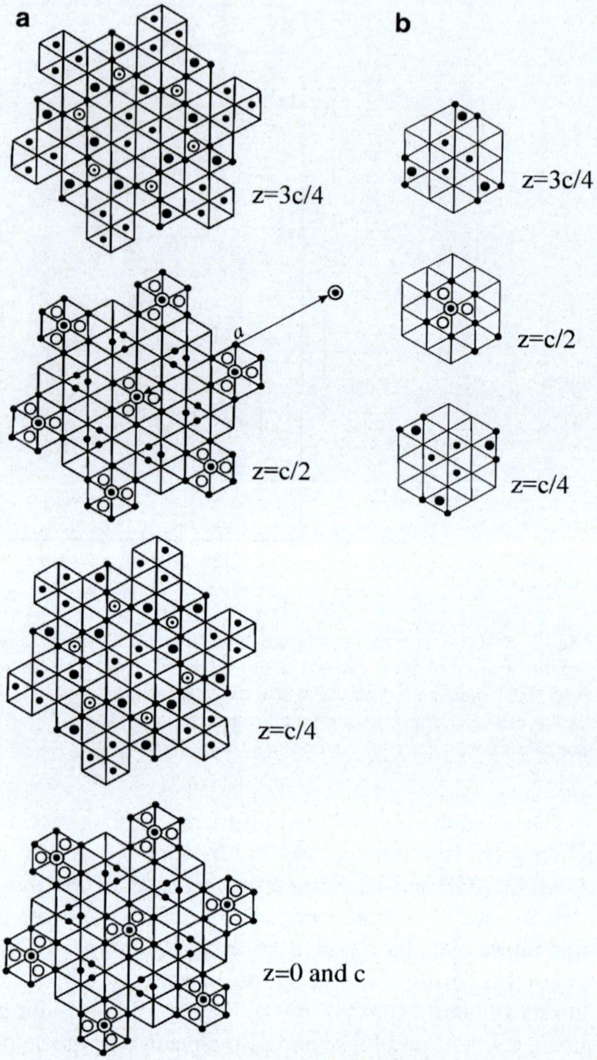

Fig. 27.4 Structure of hydroxyapatite $Ca_5(PO_4)_3OH$, *a*, c – crystal lattice constants. (**a**) projection of crystal lattice on the planar planes (0001) perpendicular to the principal screw axis z for z = 0, c/4, c/2, 3c/4, c; (**b**) hydroxyl OH (z = c/2) and its neighbouring atoms of calcium Ca (z = c/2), oxygen O (z = c/4, 3c/4) and phosphorus-oxygen PO_4 tetrahedrons (c/4, 3c/4) entered into the interacting hydroxyl complexes; ○⊙ – atoms of calcium Ca′, Ca″, ● – atoms of phosphorus P, •◉ – atoms of oxygen O and hydroxyl OH.

The parameters of a crystal lattice, presented in different references, are not too differ. The values of lattice constants *a* and *c* for hydroxyapatite $Ca_5(PO_4)_3OH$ are given in Table 27.1 from different papers.

All apatites are distinguished for their ability to the isomerous replacements both in the anionic and in the cationic sublattices [62], as noted in [1]. As this takes place, even slight changes of concentration and kind of dopants can essentially change the physical and chemical properties of material with retention of crystal structure [26, 34, 91]. The realization of solid solutions of apatites with the different nature of

Table 27.1 Parameters of a crystal lattice of hydroxyapatite in Å

a	9.432		9.4176		9.3642		9.4238	
c	6.881	[8, 14, 72, 83, 84]	6.8814	[91]	6.8811	[31]	6.8854	[45]
a	9.418		9.4214		9.4302		9.422	
c	6.884	[62]	6.8814	[62]	6.8911	[63]	6.884	[74]

components Me, X, Y and also presence of vacancies in cationic and anionic sublattices are possible [27, 83, 84, 92].

We consider monocrystals, although the polycrystals realization is possible, crystallites of which depending on hydroxil concentration are microcrystallites in synthetic material and with decreasing OH concentration up to 20% they are reformed in nanocrystallites in biological apatite [58]. Notice that hydroxyapatite of calcium is an inorganic component of bone fabric of animals and man [87], i.e. is the important biomaterial.

The phosphorus-oxygen tetrahedrons PO_4, hydroxyls OH and calcium-oxygen groups $Ca''O_9$ are the characteristic structural units for calcium hydroxyapatite. The hydroxyls OH are arranged in the centre of hexagonal prisms along the principal screw axis at c/2 intervals. They are somewhat displaced (~0.35 Å) upward and downward from the planar planes corresponding to z = 0 and c/2 [35, 83, 84]; in doing so their surrounding by oxygen and calcium atoms for z = c/2 is rotated by 60° as compared to the surrounding for z = 0.

The calcium atoms in crystal occupy two crystallographic different positions [43, 44, 83, 93], these atoms are named as Ca′ and Ca″. Each hydroxyl OH is surrounded by three atoms of calcium Ca′ in planes with z = 0, c/2 and *c*. Atoms Ca″ take up triple axes in planes with z = c/4 and 3c/4. They are in the centre of trigonal prisms and are surrounded by nine atoms of oxygen forming the group $Ca''O_9$, oxygen atoms in which are located on the middle of all edges of prisms. The trigonal prisms are connected with each other by the bases so that the lower base of one prism is the upper base of the next prism and oxygen atoms in these bases are common for both prisms. The oxygen atoms of $Ca''O_9$ group in a planar planes with z = c/4 and 3c/4, surrounding the Ca″ atoms, enter into the composition of the nearest three phosphorus-oxygen tetrahedrons PO_4. In addition, the oxygen atoms, surrounding the hydrohyls OH, are a part of these tetrahedrons and these atoms are located in the planar planes with z = 0, c/2 and c. Belonging of oxygen atoms to structural units as OH, PO_4, $Ca''O_9$, their common character provide enough stable structure of apatite.

The high structural stability of hexagonal apatites is exhibited also by virtue of the high mobility of ions that provides their self-ordering [62].

The ordered state of crystal is realized at the rather low temperatures (below 480 K) and it manifests itself in alternation of displacements of hydroxyl OH, upper and lower, in the planar planes with z = 0, c/2, c about a vertical axis z. With increase in temperature the displacements of hydroxyls OH in each channel can be both upper and lower and at the temperature of 484.5 K the order in alternation of their displacements disappears, the low-temperature monoclinic phase transforms into high-temperature hexagonal phase [54]. The realization of binary mixture of these phases is possible in the neighbourhood of the phase transition temperature [45]. The high-temperature treatment of hydroxyapatite leads to its decay, the β phase of three calcium-phosphate (β-TCP) is formed [40, 41, 94, 95].

27.3 The Order Parameters. Energetic Parameters

The ordered state of crystal is set at the expense of interaction between hydroxyls OH that tend to be displaced in some channel along the axis z and in another channel opposite to the axis z.

The number of positions of hydroxyl OH is symbolized by N. The totality of these positions is given the name of hydroxyls lattice. We divide this lattice into two sublattices: in one sublattice the displacement of all hydroxyls is upper in fully-ordered state, in another sublattice, in contrast, this displacement is lower along the axis z and these displacements are collinear. The hydroxyl orientations, both upwards and downwards, with reference to the crystal lattice are equal and so the numbers of sites in sublattices are the same.

In the partially ordered state in each sublattice the displacements can be both upper and lower, but in first sublattice they are predominantly the upper, in the second – mainly the lower.

The probabilities of these displacements are determined the following formulae

$$P_U^{(1)} = N_U^{(1)}/N_1,\ P_U^{(2)} = N_U^{(2)}/N_2,\ P_L^{(1)} = N_L^{(1)}/N_1,\ P_L^{(2)} = N_L^{(2)}/N_2, \qquad (27.2)$$

where N_1, N_2 are the numbers of sites of the first and the second sublattices,

$$N = N_1+N_2, \quad N_1=N_2 = N/2. \qquad (27.3)$$

$N_U^{(1)}$, $N_U^{(2)}$, $N_L^{(1)}$, $N_L^{(2)}$ are the numbers of hydroxyls in sublattices with the upper and lower displacements, respectively.

In this case the probabilities (27.2) satisfy the relations as follows:

$$P_U^{(1)} + P_L^{(1)} = 1, \quad P_U^{(2)} + P_L^{(2)} = 1. \qquad (27.4)$$

We introduce in examination the order parameters in sublattices as follows:

$$\zeta_1 = P_U^{(1)} - P_L^{(1)}, \quad \zeta_2 = P_U^{(2)} - P_L^{(2)}. \qquad (27.5)$$

At the full disorder in each sublattice the number of the upper displacements is equal to the number of the lower displacements, i.e. $P_U^{(1)} = P_L^{(1)}$ and $P_U^{(2)} = P_L^{(2)}$, therefore $\zeta_1 = \zeta_2 = 0$. In the case of full order in sublattices $P_U^{(1)} = 1$, $P_L^{(1)} = 0$, $P_U^{(2)} = 0$, $P_L^{(2)} = 1$, in this case $\zeta_1 = 1$, $\zeta_2 = -1$, i.e. the order parameters can vary within the limits

$$-1 \leq \zeta_1 \leq 1, \quad 1 \geq \zeta_2 \geq -1. \tag{27.6}$$

The change of sign of order parameter on opposite mean that the roles of sublattices are interchanged.

The probabilities (27.2) taking into account the relations (27.4) and (27.5) are expressed in terms of parameters as follows:

$$P_U^{(1)} = \frac{1+\zeta_1}{2}, \quad P_L^{(1)} = \frac{1-\zeta_1}{2}, \quad P_U^{(2)} = \frac{1+\zeta_2}{2}, \quad P_L^{(2)} = \frac{1-\zeta_2}{2}. \tag{27.7}$$

The interaction between hydroxyls is in reality realized between hydroxyl complexes, the composition of each is made up from: hydroxyl OH, three atoms of calcium Ca′ at distance r_o from the first, six atoms of oxygen (in three above and below hydroxyl) at distance r' and six phosphorus-oxygen tetrahedrons PO_4 at distance r''.

The distances r_o, r', r'' are represented by lattice parameters *a*, *c* by the following formulae

$$r_o = a\Big/8\sqrt{3}, \quad r' = \sqrt{c^2 + \frac{a^2}{12}}\Big/4, \quad r'' = \sqrt{c^2 + \frac{7a^2}{12}}\Big/4. \tag{27.8}$$

Each hydroxyl complex has six nearest complexes in one planar plane at the distance r and two complexes above and below at the distance r*. The distances r and r* are equal to

$$r = a\sqrt{13}/8, \quad r^* = c/2. \tag{27.9}$$

For lattice parameters presented in many papers, for example in [8] (see Table 27.1), the distances r_o r', r'', r and r* are equal to

$$r_o = 0{,}681\,\text{Å}, \; r' = 1{,}850\,\text{Å}, \; r'' = 2{,}491\,\text{Å}, \; r = 4{,}251\,\text{Å}, \; r^* = 3{,}4405\,\text{Å}. \tag{27.10}$$

On the condition of equality of sites in sublattices of different type these hydroxyl complexes each in site of definite type have at the distance r four sites of another type and two sites of same type, but at the distance r* – two sites of identical to that type.

27.4 Calculation of Numbers of Interacting Hydroxyl Pairs

We take into account in calculations only interaction between the nearest hydroxyl groups.

The hydroxyl complexes with upper and lower displacements of hydroxyl are denoted by K_U and K_L. In this case the pair interaction K_UK_U and K_LK_L is realized at a distances of r and r* correspondingly, but K_U K_L interaction – at slightly greater or smaller distances as $r \pm \Delta, r^* \pm \Delta'$.

We introduce the following symbols for energies of these interactions:

$$V(r), V(r+\Delta), V(r-\Delta), V(r^*), V(r^*+\Delta'), V(r^*-\Delta'). \tag{27.11}$$

The energy of K_U, K_L complexes consists of energies of hydroxyl OH interaction with the nearest three calcium atoms Ca′, six oxygen atoms O and six phosphorus-oxygen tetrahedrons PO_4 and these energies are designated as $\upsilon_c(r_0)$, $\upsilon_O(r')$, $\upsilon_p(r'')$.

Using the taken symbols, we get the formula for the energy of hydroxyl complexes

$$V(r) = 3\upsilon_c(r_o) + 6\upsilon_o(r') + 6\upsilon_p(r''). \tag{27.12}$$

It is evident that

$$|V(r-\Delta)|>|V(r)|>|V(r+\Delta)|, \cdots |V(r*-\Delta')|>|V(r*)|>|V(r*+\Delta')|. \tag{27.13}$$

Also we introduce the following symbols for the numbers of complex pairs K_UK_U, K_LK_L, K_UK_L at different distances:

$$\begin{aligned} N(r) &= N_{UU}(r) + N_{LL}(r), \\ N(r+\Delta) &= N(r-\Delta) = N_{UL}(r \pm \Delta), \\ N(r^*) &= N_{UU}(r^*) + N_{LL}(r^*), \\ N(r^*+\Delta') &= N(r^*-\Delta') = N_{UL}(r^* \pm \Delta'). \end{aligned} \tag{27.14}$$

The calculation of these numbers respectively by the a priori probabilities (27.2) gives the following formulae:

$$\begin{aligned} N(r) &= N\left[2\left(P_U^{(1)}P_U^{(2)} + P_L^{(1)}P_L^{(2)}\right) + \frac{1}{2}\left(P_U^{(1)^2} + P_U^{(2)^2} + P_L^{(1)^2} + P_L^{(2)^2}\right)\right], \\ N(r\pm\Delta) &= \frac{1}{2}N\left[2\left(P_U^{(1)}P_L^{(2)} + P_U^{(2)}P_L^{(1)}\right) + P_U^{(1)}P_L^{(1)} + P_U^{(2)}P_L^{(2)}\right], \\ N(r^*) &= \frac{1}{2}N\left(P_U^{(1)^2} + P_U^{(2)^2} + P_L^{(1)^2} + P_L^{(2)^2}\right), \\ N(r^*\pm\Delta') &= \frac{1}{2}N\left(P_U^{(1)}P_L^{(1)} + P_U^{(2)}P_L^{(2)}\right) \end{aligned} \tag{27.15}$$

and in view of formulae (27.7) – by the order parameters ζ_1, ζ_2 in sublattices:

$$\begin{aligned} N(r) &= N\left(\frac{3}{2}+\zeta_1\zeta_2+\frac{\zeta_1^2+\zeta_2^2}{4}\right), \\ N(r\pm\Delta) &= \frac{1}{2}N\left(\frac{3}{2}-\zeta_1\zeta_2-\frac{\zeta_1^2+\zeta_2^2}{4}\right), \\ N(r^*) &= \frac{1}{2}N\left(1+\frac{\zeta_1^2+\zeta_2^2}{2}\right), \\ N(r^*\pm\Delta') &= \frac{1}{2}N\left(1-\frac{\zeta_1^2+\zeta_2^2}{2}\right). \end{aligned} \tag{27.16}$$

These formulae will be taken into consideration in the computation of free energy.

27.5 Free Energy

The free energy of the crystal is calculated by the known formula

$$F = U - kT\ln W, \tag{27.17}$$

where U is the configuration internal energy, W is the thermodynamic probability, T is absolute temperature, k is Boltzmanns constant.

The internal energy, determined by the sum of interaction energies of all neatest hydroxyl complexes, will be equal

$$\begin{aligned} U = &- N(r)V(r) - N(r+\Delta)V(r+\Delta) - N(r-\Delta)V(r-\Delta) \\ &- N(r^*)V(r^*) - N(r^*+\Delta')V(r^*+\Delta') - N(r^*-\Delta')V(r^*-\Delta'), \end{aligned} \tag{27.18}$$

where energies are taken with the opposite sign.

From the formulae (27.16) for the pairs numbers it follows that the configuration energy will consist of three components: constant value independent of order parameter, component proportional to $\zeta_1\zeta_2$ and a third component with square dependence on order parameters. So the configuration internal energy can be written as follows

$$U = -N\left[V + V'\,\zeta_1\,\zeta_2 + V''(\,\zeta_1^2 + \zeta_2^2)\right], \tag{27.19}$$

where V, V′, V″ energies are defined by the sum of interaction energies (27.11).

Thermodynamic probability is determined by the number of energetically distinguishable states of crystal, i.e. by the number of energetically distinctive distribution of hydroxyl complexes over their positions and calculated by the combinatorial rule by formula

$$W = \frac{N_1!}{N_U^{(1)}!N_L^{(1)}} \times \frac{N_2!}{N_U^{(2)}!N_L^{(2)}}. \qquad (27.20)$$

The natural logarithm of thermodynamic probability in view of Stirling formula $\ln X! = X(\ln X - 1)$ for large X numbers and formulae (27.2) can be written as

$$\ln W = -\frac{1}{2}N\left[\left(P_U^{(1)}\ln P_U^{(1)} + P_L^{(1)}\ln P_L^{(1)} + P_U^{(2)}\ln P_U^{(2)} + P_L^{(2)}\ln P_L^{(2)}\right)\right]. \qquad (27.21)$$

Substituting the a priori probabilities (27.7), we find the logarithm of thermodynamic probability in terms of order parameter

$$\ln W = -\frac{1}{4}N\left[(1+\zeta_1)\ln\frac{1+\zeta_1}{2}+(1-\zeta_1)\ln\frac{1-\zeta_1}{2}+(1+\zeta_2)\ln\frac{1+\zeta_2}{2}+(1-\zeta_2)\ln\frac{1-\zeta_2}{2}\right]. \qquad (27.22)$$

Taking into account expressions (27.19) and (27.22) and in consideration for one lattice site the free energy (27.17) can be written as

$$f=\frac{F}{N}=-V-V'\zeta_1\zeta_2-V''\left(\zeta_1^2+\zeta_2^2\right)$$
$$+\frac{1}{4}kT\left[(1+\zeta_1)\ln\frac{1+\zeta_1}{2}+(1-\zeta_1)\ln\frac{1-\zeta_1}{2}+(1+\zeta_2)\ln\frac{1+\zeta_2}{2}+(1-\zeta_2)\ln\frac{1-\zeta_2}{2}\right]. \qquad (27.23)$$

The derived expression (27.23) for free energy of hydroxyapatite shows its dependence on temperature T, order parameters ζ_1, ζ_2 and the energetic constants V, V', V''.

27.6 Equations of Thermodynamic Equilibrium. Order-Disorder Transition Temperature

At the equilibrium state the free energy is minimal and in this case the conditions for minimum of free energy are written as follows:

$$\frac{\partial f}{\partial \zeta_1} = 0, \quad \frac{\partial f}{\partial \zeta_2} = 0. \qquad (27.24)$$

After substitution of free energy f into (27.24) we find the relations for order parameters

$$kT\ \ln\frac{1+\zeta_1}{1-\zeta_1} = 4(V'\ \zeta_2 - V''\ \zeta_1),$$
$$kT\ \ln\frac{1+\zeta_2}{1-\zeta_2} = 4(V'\ \zeta_1 - V''\ \zeta_2). \quad (27.25)$$

It is easy to verify that with ζ_1 in place of $\pm\ \zeta_2$ Eqs. 27.25 transforms one into another. Therefore, we can restrict our consideration to the one of equations and shall believe that $\zeta = \zeta_1 = -\zeta_2$ (or $\zeta = \zeta_1 = \zeta_2$). This gives the relation

$$kT\ \ln\frac{1+\zeta}{1-\zeta} = 2\ \omega\ \zeta, \quad (27.26)$$

where

$$\omega = \begin{cases} 2(2V'' - V') & \text{at}\ \ \zeta_1 = -\ \zeta_2, \\ 2(2V'' + V') & \text{at}\ \ \zeta_1 = \zeta_2 \end{cases} \quad (27.27)$$

is the energy of hydroxyls ordering in crystal.

Considering $\zeta \to 0$ in Eq. 27.26 after expansion of logarithm as a series in small ζ, we find the temperature of order-disorder transition in the form

$$kT_o = \omega_o, \quad (27.28)$$

where ω_o is the constant component of energy ω. The ordering temperature depends on the distance between structural units and consequently on order parameter, in most cases with appearance of order the parameters of crystal lattice decrease moderately. In the simplest case the dependence $\omega = \omega(\zeta)$ is square

$$\omega = \omega_o + \alpha\zeta^2. \quad (27.29)$$

On this basis, the free energy for one site of lattice at $\zeta = \zeta_1 = -\zeta_2$ can be written as follows:

$$f = -V - (\omega_o + \alpha\,\zeta^2)\,\zeta^2 + \frac{1}{2}kT\left[(1+\zeta)\ln\frac{1+\zeta}{2} + (1-\zeta)\ln\frac{1-\zeta}{2}\right]. \quad (27.30)$$

In the condition for minimum of free energy $\partial f/\partial\,\zeta = 0$ we find the equation of thermodynamic equilibrium of system

$$kT\ \ln\frac{1+\zeta}{1-\zeta} = 2(\omega_o + 2\alpha\zeta^2)\zeta. \quad (27.31)$$

In the neighbourhood of phase transition temperature, when $\zeta \to 0$ and $\ln\frac{1+\zeta}{1-\zeta} \approx 2\,\zeta$, we get the approximate formula for the temperature dependence of order parameter

$$\zeta^2 = \left(\frac{T}{T_o} - 1\right) \Big/ 2\frac{\alpha}{\omega_o}. \qquad (27.32)$$

As is evident from this relation, the ζ^2 dependence on temperature is nonlinear. In the majority of cases for different types of phase transitions of the second kind the formula (27.32) is in agreement with experimental data and theoretical calculation results of another authors [96–98].

27.7 The Hydroxyl Solubility

The presence of vacancies on the sites of hydroxyl group OH and also the high mobility of the last will provide that hydroxyls escape the crystal with increase in temperature. Experimentally, the full loss of hydroxyl from the crystal manifests itself at the temperature of 1,073 K and it is accompanied by the appearance of small peak on the DTA curves (Figs. 27.2 and 27.3).

We shall calculate the solubility of hydroxyl OH in the hydroxyapatite crystal under examination the area of high temperatures when the order in alternation of hydroxyl displacements is nonexistent ($\zeta = 0$) [99].

From the N sites of hydroxyls the part is occupied by group OH, another part is vacant, i.e.

$$N = N_{OH} + N_V,\ c = N_{OH}/N,\ c_V = N_V/N,\ c + c_V = 1, \qquad (27.33)$$

where N_{OH}, N_V are the number of hydroxyls and vacant sites and c and c_V are their concentrations correspondingly.

The numbers of hydroxyl pairs at the distances r, rr $\pm\,\Delta, r^*, r^* \pm \Delta'$ are equal to

$$N(r) = 3Nc^2,\ N(r \pm \Delta) = \frac{3}{2}Nc^2,\ N(r^*) = Nc^2,\ N(r^* \pm \Delta') = \frac{1}{2}Nc^2. \qquad (27.34)$$

From these formulae it is apparent that the configuration internal energy is proportional c^2 and having regard to variation of energetic parameter with the square dependence of order parameter on concentration we find the configuration energy in the following form

$$U = -NVc^2 = -N\left(V_o + \alpha c^2\right)c^2. \qquad (27.35)$$

The thermodynamic probability is calculated by the formula

$$W = \frac{N!}{N_{OH}!N_V!} \quad (27.36)$$

and it gives the natural logarithm of W in the form

$$\ln W = -N[c\ln c + (1-c)\ln(1-c)]. \quad (27.37)$$

Substituting values of U (27.35) and ln W (27.37) in Eq. 27.17 and adding the component with hydroxyl activity, we find free energy as follows:

$$f = \frac{F}{N} = -V_o\left(1+\frac{\alpha}{V_o}c^2\right)c^2 + kT[c\ln c - (1-c)\ln(1-c)] - kTc\ln\lambda, \quad (27.38)$$

where λ is the activity of hydroxyl OH.

The solubility is determined by equilibrium concentration *c* from the condition of minimum of free energy ($\partial f/\partial c = 0$) and this gives the relation

$$kT\ln\frac{c}{\lambda(1-c)} = 2\left(V_o + 2\alpha c^2\right)c, \quad (27.39)$$

that can be rewritten in the form

$$c = \left[1+\frac{1}{\lambda}\exp\frac{-2(V_o+2\alpha c^2)c}{kT}\right]^{-1}. \quad (27.40)$$

For small concentration *c* the formula (27.39) is simplified to

$$kT = 2\left(V_o + 2\alpha c^2\right)c\Big/\ln\frac{c}{\lambda}. \quad (27.41)$$

Figure 27.5 gives the plot of temperature dependence of hydroxyl OH solubility, constructed by formula (27.39) for energetic parameters V = 3.33 eV, $\alpha = -2.05$ eV, $\lambda = 0.05$ (they are estimated from experimental fact that over the

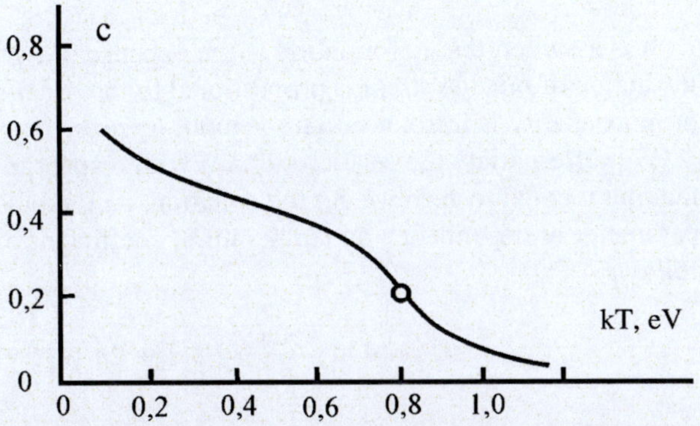

Fig. 27.5 The calculation plots for solubility of hydroxyl OH in hydroxyapatite as a function of temperature. The *circle* on the curve corresponds to the extreme point of dependence dc/dT

temperature region of 1,073 K hydroxyl escape the crystal (c → 0)). It is seen from Fig. 27.5 that with increased temperature the hydroxyls concentration decreases sharply at first and further more smoothly and at temperature of kT > 1 eV the curve become asymptotic to the *c* zero value.

A peculiarity of c(T) dependence maniifests itself in the fact that in the temperature area, when kT ≈ 0.5 eV, the derivative dc/dT is found to be extremal (as marked off by circle on the curve in Fig. 27.5). This occurrence points to the disturbance of uniform increase of temperature of specimen at the DTA research and small-size peak makes its appearance on the experimental DTA curves (Figs. 27.2 and 27.3). The peak is presented in Fig. 27.3 in the temperature region of 0.5–0.7 eV in dependence on concentration of ethyl phosphate in water-acetone solution, by which hydroxyl specimens are treated.

27.8 Hysteresis Effect

The hysteresis effect can present in condensates with first-kind phase transitions, if outside directed mechanical stress is acted on specimen.

We shall investigate the Gibbs thermodynamic potential to reveal the possibility of manifestation of hysteresis effect. The studies of hysteresis effect for ferro of H4 and $G5_1$ structures were conducted by us in the papers [100–103].

Thermodynamic potential for one hydroxyl at $\zeta = \zeta_1 = -\zeta_2$, taking into account the square dependence of ordering energy on the order parameter under the action of the outside oriented mechanical stress σ collinear to the displacements of hydroxyl groups OH and considering the proportionality of groups OH kT, eV shift to the order parameter, is determined by the following equation (the dimensionless units, i.e. in the ratio to ω_o)

$$\Psi = -\upsilon - (1 + \delta\zeta^2)\zeta^2 + \frac{1}{2}\theta\left[(1+\zeta)\ln\frac{1+\zeta}{2} + (1-\zeta)\ln\frac{1-\zeta}{2}\right] - \sigma\zeta, \quad (27.42)$$

where the following designations are used

$$\upsilon = V/\omega_o, \ \delta = \alpha/\omega_o, \ \theta = T/T_0. \quad (27.43)$$

It is assumed that deformation at the expense of hydroxyls displacements under the action of outside stress is proportional to the order parameter, the coefficient of proportionality is introduced into symbol of stress σ.

We differentiate the relation (27.42) with respect to the order parameter and by making it equal to zero we get the equation, defined the equilibrium value of order parameter in dependence on temperature, coefficient δ and the outside stress σ, as follows

$$\theta \ln\frac{1+\zeta}{1-\zeta} = 2(1 + 2\delta\,\zeta^2)\zeta + \sigma. \quad (27.44)$$

Figures 27.6 and 27.7 illustrate the plots of temperature dependence of order parameter, constructed by formula (27.44) for the cases $\alpha \neq 0$, $\sigma = 0$ (Fig. 27.6) and $\alpha = 0$, $\sigma \neq 0$ (Fig. 27.7).

It is evident from Fig. 27.6 that at $\alpha = 0$ and $\alpha < 0$ with a rise in temperature the order parameter decreases progressively, reaches zero at $T = T_L = T_0$, i.e. in this case we have second-kind phase transition. At $\alpha \neq 0$ this reduction is more appreciable. If $\alpha > 0$, the dependence $\zeta = \zeta(T)$ can be extreme. The order

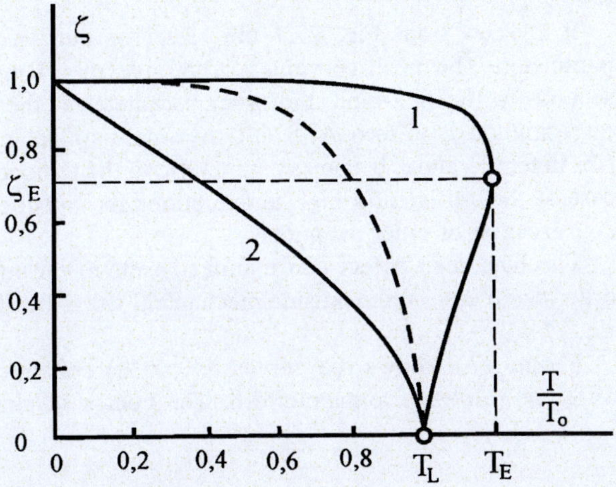

Fig. 27.6 Theoretical plots of temperature dependence of order parameter constructed by formula (27.44) for $\alpha = 0$ (the dashed line) and $\alpha = \pm\, 0.5$ (the solid lines 1 and 2, respectively) in the absence of outside oriented mechanical stress. The extreme point is marked off by circle on the curve 1

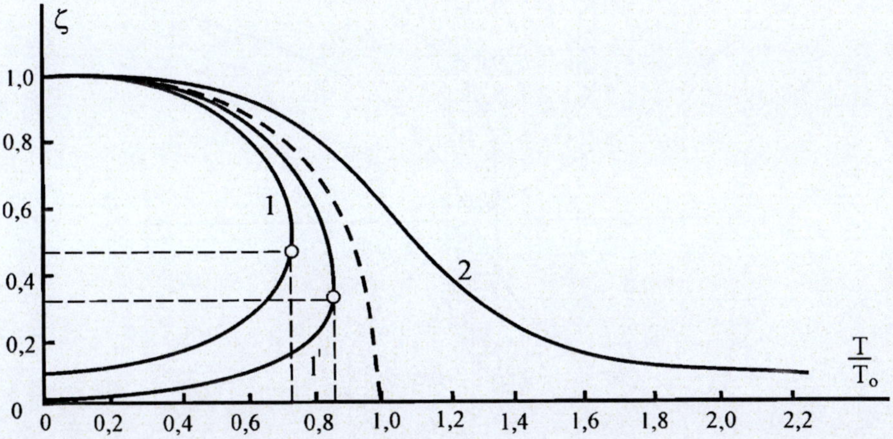

Fig. 27.7 Theoretical plots of temperature dependence of order parameter constructed by formula (27.44) and determined the effect of external oriented mechanical stress on the magnitude of order parameter at $\alpha = 0$ and$\sigma = 0$ (the dashed curve) and also for $\alpha = 0$ and $\sigma = \pm\, 0.3$ (curves 1 and 2, respectively). The curve $1'$ is constructed for $\alpha = 0$ and $\sigma = 0.1$. The extreme points are marked off by circles on the curves 1 and $1'$

parameter in this situation falls moderately at first to the temperature T_0 and then reduces abruptly to zero. In this case

$$T_L < T_0 < T_E, \tag{27.45}$$

where T_L is the temperature of loss of stable state, T_0 is the temperature of phase transition, T_E is the temperature corresponding to the extreme point. We have the first-kind phase transition. Consequently, the energetic parameter δ can change the kind of phase transition. The ordering temperature T_0 is determined by condition $\psi(\zeta) = \psi(0)$.

It is seen from Fig. 27.7 the effect of outside mechanical stress on order parameter. The positive values of σ give the change of phase transition from second- to the first-kind, the order parameter at the temperature below T_E goes discontinuously to zero. At $\sigma < 0$ the order parameter reduces progressively with a rise in temperature, becomes asymptotic to the temperature axis at $T \to \infty$. In this case at the sufficiently high temperatures the outside stress is favourable for the conservation of order parameter.

The hysteresis effect can manifest itself in the nonmonotonic dependence of order parameter on the outside mechanical stress σ. This dependence is defined by Eq. 27.44.

Figure 27.8 shows the curves $\zeta = \zeta\ (\sigma)$ constructed by formula (27.44) for different values of temperature θ. The course of change of ζ with increased or

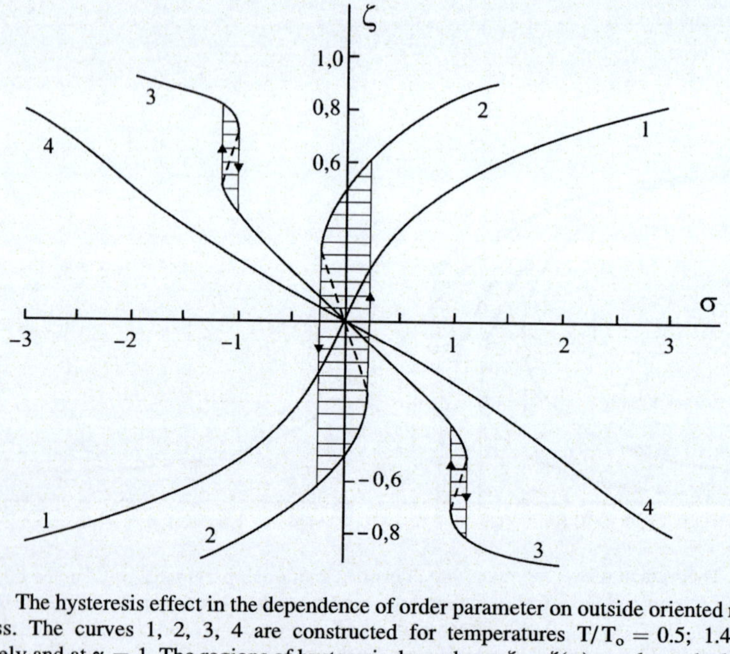

Fig. 27.8 The hysteresis effect in the dependence of order parameter on outside oriented mechanical stress. The curves 1, 2, 3, 4 are constructed for temperatures $T/T_o = 0.5$; 1.4; 2; 2.6 respectively and at $\alpha = 1$. The regions of hysteresis dependence $\zeta = \zeta(\sigma)$ are shown by hatching. The *dotted portions* of the curves correspond to the unstable state

decreased stress σ is marked off by an arrows. The sharp decrease or increase of order parameter present in these portions of curves. The manifestation both of one and two hysteresis loops is possible, as is seen from this figure. In these cases the order-disorder phase transition will be the transition of first kind.

27.9 Elastic Compliance. Modulus of Longitudinal Elasticity

The elastic compliance is characterized by the change in the order parameter ζ corresponding to the unit change of stress σ and is determined by derivative $S = \partial\zeta/\partial\sigma$ and modulus of longitudinal elasticity is the reciprocal value $E = \partial\sigma/\partial\zeta$ defined the variations in stresses under change of order parameter.

In the neighbourhood of phase transition temperature, when ζ is small and $\ln\frac{1+\zeta}{1-\zeta} \approx 2\,\zeta$, we shall rewrite the formula (27.44) at follows

$$E = 2\left(\frac{T - T_0}{T_0} - 2\delta\zeta^2\right)\zeta. \tag{27.46}$$

Equation 27.46 in view of (27.32) allow us to get the formula for elasticity modulus

$$E = 2\left(\frac{T - T_0}{T_0} - 6\delta\zeta^2\right) = \begin{cases} 2x \text{ at } \zeta = 0, \\ -4x \text{ at } \zeta^2 = \frac{T - T_0}{T_0}/2\delta, \end{cases}$$

where

$$x = \frac{T - T_0}{T_0}. \tag{27.47}$$

From (27.47) it follows that in this case the rule "negative two" as

$$E_{\zeta\neq 0}/E_{\zeta=0} = -2 \tag{27.48}$$

and the Curie-Weiss law (proportionality of elasticity modulus to temperature) are valid.

Differentiating (27.44) with respect to ζ we can get more exact formula for elasticity modulus

$$E = 2\left(\theta\frac{1}{1-\zeta^2} - 6\delta\zeta^2 - 1\right). \tag{27.49}$$

If the value ζ^2 (for small ζ) is ignored in denominator of this formula as compared to one, we obtain the result of formula (27.47).

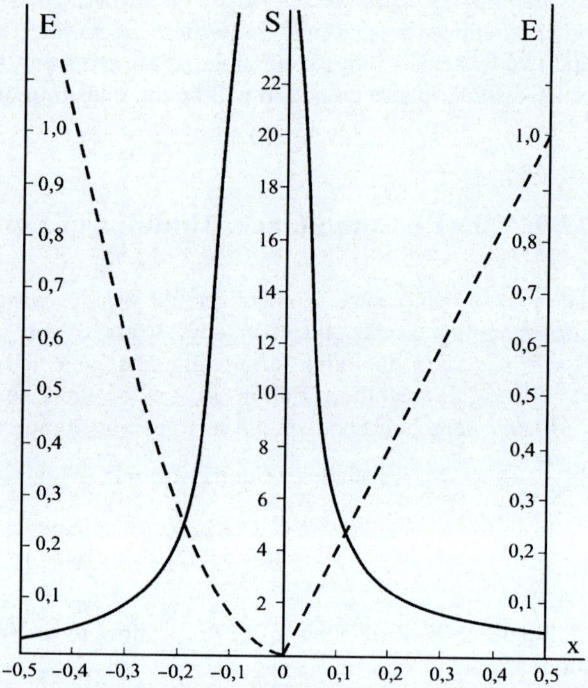

Fig. 27.9 Theoretical plots of temperature dependence of elastic compliance S (*the full curves*) and elasticity modulus E (*the dotted curves*). The curves for $x > 0$ ($T > T_o$) correspond to the disordered state ($\zeta = 0$) and forx < 0 ($T < T_o$) – to the ordered state

Substituting ζ^2 from (27.32) in Eq. 27.49, we derive the following expression for the elasticity modulus at any values of ζ

$$E = \begin{cases} 2x & \text{at } x > 0\ (\zeta = 0), \\ -4x\left[1 + \dfrac{1+x}{2(x - 2\,\delta)}\right] & \text{at } x < 0\ (\zeta \neq 0). \end{cases} \tag{27.50}$$

The plots of temperature dependence of elastic compliance S and elasticity modulus E, constructed using formula (27.50) for $\delta = 0.2$, are shown in Fig. 27.9. The curves for $x > 0$ ($T > T_0$) correspond to the disordered state, for $x < 0$ ($T < T_0$) – to the ordered state. The dependence E(x) in the ordered state for small x differ from linear, but from $x = -0.3$ and on this dependence approximates to the linear. For great x (in absolute value) the dependence S(x) is so that the curve approaches to zero, but with $\pm\, x \rightarrow 0$ it tends to infinity.

27.10 Configuration Heat Capacity

The heat capacity is determined from the expression for internal configuration energy

$$u = -\,\upsilon - \left(1 + \delta\zeta^2\right)\zeta^2 \tag{27.51}$$

according to the following formula

$$C = \frac{du}{d\theta} = -\left(1 + 2\delta\zeta^2\right)\frac{d\zeta^2}{d\theta}, \tag{27.52}$$

for which the derivative d $\zeta^2/d\,\theta$ is estimated from the equation of thermodynamic equilibrium (27.44). We find this derivative in the form

$$\frac{d\zeta^2}{d\theta} = -3\zeta^2\left(1 + 2\delta\zeta^2\right)\Big/\left(\frac{1}{1-\zeta^2} - \frac{1 + 6\delta\zeta^2}{\theta}\right). \tag{27.53}$$

Substituting (27.53) in Eq. 27.52, we find the heat capacity in dependence on order parameter and temperature as follows

$$C = 3\zeta^2\left(1 + 2\delta\zeta^2\right)^2\Big/\left(\frac{1}{1-\zeta^2} - \frac{1 + 6\delta\zeta^2}{\theta}\right). \tag{27.54}$$

Taking into account θ from the equilibrium equation (27.44) at $\sigma = 0$

$$\theta = 2\zeta\left(1 + 2\delta\zeta^2\right)\Big/\ln\frac{1+\zeta}{1-\zeta}, \tag{27.55}$$

we can construct the plots defined the dependences of heat capacity on order parameter $C = C\,(\zeta)$ and on temperature $C = C\,(\theta)$.

Figure 27.10 illustrates these plots constructed by formulae (27.54), (27.55) for different values of δ. It can be seen from Fig. 27.10 that positive values of δ lead to a rise of heat capacity and as much, as δ is greater.

It turns out that the dependence $C = C(\theta)$ is extremal and this dependence reaches an extremum in the neighbourhood of ordering temperature. The relative temperature θ_m corresponding to a maximum of heat capacity is approximately equal to

$$\theta_m = 1 - \frac{1}{3}\,\zeta^2, \tag{27.56}$$

i.e. for small value ζ the temperature θ_m approaches the temperature θ_0 but is less than one, as shown in Fig. 27.10b,c.

The maximum value of heat capacity is equal to

$$C_m = \begin{cases} 4.29 \text{ at } \delta = 0, \\ 5.19 \text{ at } \delta = 0{,}1, \\ 6.04 \text{ at } \delta = 0{,}15. \end{cases} \tag{27.57}$$

We find the jump of heat capacity at $\theta_0 = 1$ as follows

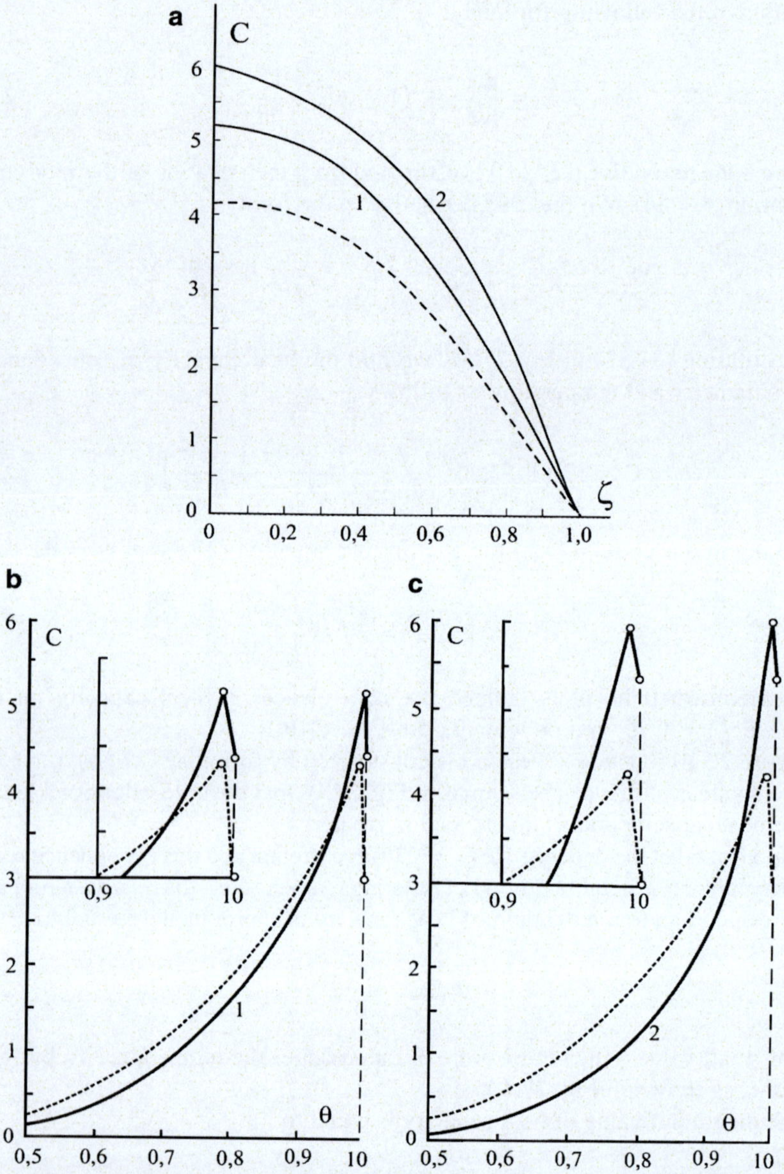

Fig. 27.10 The calculation plots for the dependence of configuration heat capacity of hydroxyapatite on order pameter (**a**) and on relative temperature (**b**) and (**c**) constructed by formula (27.54) for the values $\alpha_o = 0$ (*the dotted curves*) and $\alpha_o = 0.1$; 0.15 (curves 1 and 2, respectively). The extreme points and points of discontinuous change of heat capacity at $\zeta \to 0$ are marked off by *circles*

$$\Delta C = C_{T_o-0} - C_{T_o+0} = \begin{cases} 3 & \text{at } \delta = 0, \\ 4{,}29 & \text{at } \delta = 0{,}1, \\ 5{,}45 & \text{at } \delta = 0{,}15. \end{cases} \quad (27.58)$$

As is evident from the foregoing, the value δ tend to increase both the maximum heat capacity C_m and the heat capacity jump ΔC.

27.11 Conclusions

The elaborated statistical theory for hydroxyapatite makes it possible for us to provide an explanation and substantiate its properties: the hydroxyl solubility in the crystal, the hysteresis effect, the compliance, the heat capacity. These properties have been derived by the calculation and investigation of thermodynamic functions: Helmholtz free energy and Gibbs thermodynamic potential.

The equation of thermodynamically equilibrium state of the system has been developed and we have determined the ordering temperature and evaluated the temperature dependence of order parameter by this equation. For ζ^2 near the ordering temperature T_o the last-mentioned is linear and agreeing with experimental data and with the result of another theories for the second-kind phase transitions [96]. The calculated ordering temperature T_o is defined by the interaction energies of hydroxyl complexes. The existence of ferroelastic-paraelastic phase transition has been illustrated by DTA curves at the temperature of $T_o = 483$ K in Fig. 27.2.

The carried out calculations of solubility of hydroxyl groups OH in apatite have shown the monotonic dependence of hydroxyl concentration that decreases with increased temperature. These dependence has a special feature: the presence of inflection point at the temperature of $kT \approx 0.8$ eV. This circumstance is responsible for the extremality of temperature dependence of derivative of solubility with respect to temperature and thus it disturbs the uniform increase of specimens temperature on their heating that brings into existence the peak at the DTA curves (Figs. 27.2 and 27.3). Experimentally such peak appeared over the temperature area as 0.5–0.7 eV.

The investigation of thermodynamic potential, including the component on crystal substantiation deformation at the expense of mechanical stress, allows, firstly, the evaluation of influence of mechanical stress on order parameter and, secondly, the substantiation of possibility of hysteresis effect manifestation that is pointed out as experimental factor in many literature sources [104]. At this takes place, the manifestation both of one and two hysteresis loops is possible for different values of temperatures.

The calculations of elastic compliance and modulus of longitudinal elasticity from equation of thermodynamic equilibrium have shown that at small values of order parameter the rule of "negative two" and Curie-Weiss law are performed. Over the wide range of temperatures the dependence of elasticity modulus on temperature has been found to the non-linear for the ordered state.

We have fulfilled the calculation of configuration heat capacity and its temperature dependence that, firstly, is found to be extremal and, secondly, permits the estimation of heat capacity jump in the point of order-disorder phase transition. Both of these facts manifest themselves experimentally, as shown in Fig. 27.1.

The following plots have been constructed in the paper: the dependence of hydroxyl OH solubility on temperature (Fig. 27.5), the temperature dependence of order parameter for different values of thermodynamic parameters α and σ (Figs. 27.6 and 27.7), the order parameter as a function of mechanical stress (Fig. 27.8), the temperature dependences of compliance and elasticity modulus (Fig. 27.9), the heat capacity depending on order parameter and temperature (Fig. 27.10).

The calculation results have been compared with experimental data for hydroxyapatite and this has shown their qualitative agreement.

References

1. Mehmel M (1984) Uber die struktur des apatits. Zeit Krist 75:323–331
2. Náray-Szabó S (1930) The structure of apatite. Zeit Krist 75:387–398
3. Beevers CA, Mc. Intyre DB (1946) The atomic structure of fluor-apatite and its relation to that of tooth and bone material. Mineralog Mag 27:254–257
4. Ormont BF (1950) Structures of inorganic substances. Gostekhizdat, Moscow, 968 p (in Russian)
5. Silverman SR, Fuyat RK, Weiser JD (1952) Quantitative determination of calcite associated with carbonate-bearing apatite. Am Mineralogist 37:211–222
6. Posner AS, Perloff A, Diorio AF (1958) Refinement of the hydroxyapatite structure. Acta Cryst 11:308–309
7. Hayashi S (1960) Analysis of the crystal structure of fluorapatite by nuclear magnetic resonance. J Chem Soc Japan 81:540–542
8. Kay MJ, Young RA, Posner AS (1964) Crystal structure of hydroxyapatite. Nature 204 (12):1050–1052
9. Aizu K (1969) Possible species of "ferroelastic" crystals and of simultaneously ferroelectric and ferroelastic crystals. J Phys Soc Japan 27(2):387–396
10. Aizu K (1970) Possible species of ferromagnetic, ferroelectric and ferroelastic crystals. Phys Rev B2(3):754–772
11. Morgan H, Wilson RM, Elliott JC, Dowker SEP, Anderson P (2000) Preparation and characterization of monoclinic hydroxyapatite and its precipitated carbonate apatite intermediate. Biomaterials 21:617–627
12. Rakovan JF, Hughes JM (2000) Strontium in the apatite structure: strontian fluoride and belovite - (Ce). Can Miner 38:839–845
13. Sampath Kumar TS, Manjubala I, Gunasekaran J (2000) Synthesis of carbonated calcium phosphate ceramics using microwave irradiation. Biomaterials 21:1623–1629
14. Mathew M, Takagi S (2001) Structures of biological minerals in dental research. J Res Natl Inst Stan technol 106(6):1035–1044
15. Torrent-Burgues J, Rodriguez-Clemente R (2001) Hydroxyapatite precipitation in a semi-batch process. Cryst Res Technol 36:1075–1082

16. Gomez-Morales I, Torrent-Burgues J, Boix T, Fraile B, Rodriguez-Clemente R (2001) Precipitation of stoichiometric hydroxyapatite by a continuous method. Cryst Res Technol 36(1):15–26
17. Leroy N, Bres E (2001) Structure and substitutions in fluorapatite. Eur Cell Mater 2:36–48
18. Koutsopoulos S (2001) Kinetic study on the crystal growth of hydroxyapatite. Langmuir 17:8092–8097
19. Hadrich A, Lautie A, Mhiri T (2001) Vibrational study and fluorescence bands in the FT-Raman spectra of $Ca_{10-x}Pb_x(PO_4)_6(OH)_2$ compound. Spectrochim Acta A 57:1673–1681
20. Raynaud S, Champion E, Bernache-Assollant D, Thomas P (2002) Calcium phosphate apatites with variable Ca/P atomic ratio I Synthesis, characterization and thermal stability of powders. Biomaterials 23:1065–1072
21. Koutsopoulos S (2002) Synthesis and characterization of hydroxyapatite crystals: a review study on the analytical methods. J Biomed Mater Res 62:600–612
22. Shpak AP, Karbovskii VL, Trachevskii VV (2002) Apatites. Akademperiodika, Kiev, 414 p (in Russian)
23. Elliott JC (2002) Calcium Phosphate Biominerals. In: Kohn MJ, Rakovan J, Hughes JM (eds) Phosphates: geochemical, geobiological and materials importance, vol 48, Reviews in mineralogy and geochemistry. Mineralogical Society of America, Washington, DC, pp 427–454
24. Chaumont J, Soulet S, Krupa JC, Carpena J (2002) Competition between disorder creation and annealing in fluorapatite nuclear waste forms. J Nucl Mater 301:122–128
25. Gross KA, Berndt CC (2002) Biomedical Application of Apatites. In: Kohn MJ, Rakovan J, Hughes JM (eds) Phosphates: geochemical, geobiological and materials importance, vol 48, Reviews in mineralogy and geochemistry. Mineralogical Society of America, Washington, DC, pp 631–672
26. Pan Y, Fleet ME (2002) Compositions of the Apatite-Group Minerals: Substituting Mechanisms and Controlling Factors. In: Kohn MJ, Rakovan J, Hughes JM (eds) Phosphates: geochemical, geobiological and materials importance, vol 48, Reviews in mineralogy and geochemistry. Mineralogical Society of America, Washington, DC, pp 13–50
27. Hughes JM, Rakovan J (2002) The Crystal Structure of Apatite, $Ca_5(PO_4)_3$(F, OH, Cl). In: Kohn MJ, Rakovan J, Hughes JM (eds) Phosphates: geochemical, geobiological and materials importance, vol 48, Reviews in mineralogy and geochemistry. Mineralogical Society of America, Washington, DC, pp 1–12
28. Murugan R, Sampath Kumar TS, Panduranga Rao K (2002) Fluorinated bovine hydroxyapatite: preparation and characterization. Mater Lett 57:429–433
29. John M, Rakovan H, Rakovan J (2002) The crystal structure of apatite $Ca_5(PO_4)_3$(F, OH, Cl). Rev Mineralog Geochem 48(1):1–12
30. Shpak AP, Karbovsky VL, Vakhney AG, Kluyenko LP (2002) Spectral and quantum-mechanical study of electronic structure of ultradispersed calcium hydroxyapatite. Chem Phys Technol Surface 7–8:57
31. Calderin L, Stott MJ (2003) Electronic and crystallographic structure of apatites. Phys Rev B.67:134106
32. Shpak AP, Karbovskii VL, Tracheskii VV, Didenko RV, Kurgan NA (2003) Diagnostics of apatitelike structures based on alkali-earth metals. Metallophys Mod Technol 25 (10):1279–1301
33. Pinchuk ND, Ivanchenko LA (2003) Technological processes of production of calcium phosphate biomaterials. Poroshk Metall 7–8:36–52 (in Russian)
34. White TJ, Zhili D (2003) Structural derivation and crystal chemistry of apatites. Acta Crystallogr B59:1–16
35. Calderin L, Stott MJ (2003) Electronic and crystallographic structure of apatites. Phys Rev B B67(34106):1–7
36. Hidouri M, Bouzouita K, Kooli F, Khattech I (2003) Thermal behaviour of magnesium-containing fluorapatite. Mater Chem Phys 80:496–505

37. Monteiro MM, Rocha NCC, Rossi AM, Soares GA (2003) Dissolution properties of calcium phosphate granules with different compositions in simulated body fluid. J Biomed Mater Res A65:299–305
38. Mavropoulos E, Rossi AM, da Rocha NCC, Soares GA, Moreira JC, Moure GT (2003) Dissolution of calcium-deficient hydroxyapatite synthesized at different conditions. Mater Charact 50:203–207
39. Kim SR, Lee JH, Kim YT, Riu DH, Jung SJ, Lee YJ, Chung SC, Kim YH (2003) Synthesis of Si, Mg substituted hydroxyapatites and their sintering behaviors. Biomaterials 24:1389–1398
40. Pena J, Vallet Regi M (2003) Hydroxyapatite, tricalcium phosphate and biphasic material prepared by a liquid mix technique. J Eur Ceram Soc 23:1687–1696
41. Kasten P, Luginbuhl R, Van Griensven M, Barkhausen T, Krettek C, Bohner M, Bosch U (2003) Comparison of human bone marrow stromal cells seeded on calcium-deficient hydroxyapatite, β-tricalcium phosphate and demineralized bone matrix. Biomaterials 24:2593–2604
42. Shi J (2004) From dental enamel tu synthetic hydroxyapatite-bazed biomaterials. Ph. D. thesis, University of Hamburg
43. Tadic D, Epple MA (2004) Thorough physicochemical characterisation of 14 calcium phosphate-based bone substitution materials in comparison to natural bone. Biomaterials 25:987–994
44. Pasteris JD, Wopenka B, Freeman JJ et al (2004) Lack of OH in nanocrystalline apatite as a function of degree of atomic order: implication for bone and biomaterials. Biomaterials 25:229–238
45. Markovic M, Fowler BO, Tung MS (2004) Preparation and comprehensive characterization of a calcium hydroxyapatite. Reference material. J Res Natl Inst Stand Technol 109(6):553–568
46. Webster TJ, Massa-Schlueter EA, Smith JL, Slamovich EB (2004) Osteoblast response to hydroxyapatite doped with divalent and trivalent cations. Biomaterials 25:2111–2121
47. Siddharthan A, Seshadri SK (2004) Sampath Kumar T.S. Microwave accelerated synthesis of nanosized calcium deficient hydroxyapatite. J Mater Sci Mater Med 15:1279–1284
48. Wang T, Dorner-Reisel A, Müller E (2004) Thermogravimetric and thermokinetic investigation of the dehydroxylation of a hydroxyapatite powder. J Eur Ceram Soc 24:693–698
49. Karbovskii VL (2005) Electronic and atomic structure of apatites and apatitelike structures. Ph.D. thesis, Kiev. Instiutute of Metal Physics of NAS of Ukraine, 313 p (in Russian)
50. Brik AB, Shpak AP, Karbovskii VL et al (2005) ESR of nanoscale particles in biological and synthetic carbonate containing apatite. Mineral J 27(1):5–26 (in Russian)
51. Wopenka B, Pasteris JD (2005) A mineralogical perspective on the apatite in bone. Mater Sci Eng C25:131–143
52. Danilchenko SN, Kulik AN, Bugay OM et al (2005) Determination of content and localization of magnesium in bioapatite of bone. J Appl Spectr 72(6):821–826 (in Russian)
53. Miro S, Grebille D, Chateigner D, Pelloquin D, Stroquert JP, Grob JJ, Costantini JM, Studer F (2005) X-ray diffraction study of damage inducted by swift heavy ion irradiation in fluorapatite. Nucl Instrum Methods Phys Res B227:306–318
54. Hochrein O, Kniep R, Zahn D (2005) Atomistic simulation study of the order/disorder (monoclinic to hexagonal) phase transition of hydroxyapatite. Chem Mater 17(8):1978–1981
55. White T, Ferraris C, Kim J, Madhavi S (2005) Apatite – an adaptive framework structure. Rev Miner Geochem 57(1):307–401
56. Cruz FJAL, Lopes JNC, Calado JCG, Piedade MEM (2005) A molecular dynamics of the thermodynamic properties of calcium apatites. 1. Hexagonal phases. J Phys Chem B 109 (51):24473–24479
57. Cruz FJAL, Lopes JNC, Calado JCG (2006) Molecular dynamics of the thermodynamic Properties of Calcium Apatites. 2. Monoclinic Phases. J Phys Chem 110(9):4387–4392
58. Leventouri Th (2006) Synthetic and biological hydroxyapatites: crystal structure questions. Biomaterials 27:3339–3342

59. Danilchenko SN, Kulik AN, Pavlenko PA, Kalinichenko TG et al (2006) Thermally activated diffusion of magnesium from bioapatite crystals. J Appl Spectr 73(3):437–443
60. Glimcher MJ (2006) Bone: Nature of the calcium phosphate crystals and cellular, structural, and physical chemical mechanisms in their formation. Rev Mineral Geochem Mineralog S Am 64:223–282
61. Shin WJ, Wang JW, Wang MC, Hon MH (2006) A study on the phase transformation of the nanosized hydroxyapatite synthesized by hydrolysis using in situ high temperature x-ray diffraction. Mater Sci Eng C 26:1434–1438
62. Danilchenko SN (2007) Structure and properties of calcium apatites from viewpoint of biomineralogy and biomaterials science. Visnyk SimGU Ser Phys Math Mech 2:33–59 (in Russian)
63. Dorozhkin SV (2007) Calcium orthophosphates. J Mater Sci 42:1061–1095
64. Suvorova EI, Petrenko PP, Buffat PA (2007) Scanning and transmission electron microscopy for evaluation of order/disorder in bone structure. Scanning 29:162–170
65. Dorozhkin SVA (2007) hierarchical structure for apatite crystals. J Mater Sci Mater Med 18:363–366
66. Brik AB, Danilchenko SN, Radchuk VV et al (2007) Thermoactivated changes of properties of biogenic and synthetic carbonatecontaining apatites by data of X-ray diffraction and ESR. Miner J 29(2):32–47 (in Russian)
67. Li ZY, Lam WM, Yang C, Xu B, Ni GX, Abbah SA, Cheung KMC, Luk KDK, Lu WW (2007) Chemical composition, crystal size and lattice structural changes after incorporation of strontium into biomimetic apatite. Biomaterials 28:1452–1460
68. Landi E, Tampieri A, Celotti G, Sprio S, Sandri M, Logroscino G (2007) Sr-substituted hydroxyapatites for osteoporotic bone replacement. Acta Biomater 3:961–969
69. Dagang G, Kewei X, Yong H (2008) The influence of Sr doses on the *in vitro* biocompatibility and *in vivo* degradability of single-phase Sr-incorporated HAP cement. J Biomed Mater Res A 86:947–958
70. Siva Rama Krishna D, Siddharthan A, Seshadri SK, Sampath Kumar TS (2007) A novel route for synthesis of nanocrystalline hydroxyapatite from eggshell waste. J Mater Sci Mater Med 18:1735–1743
71. Lee SJ, Yoon YS, Lee MH, Oh NS (2007) Nanosized hydroxyapatite powder synthesized from eggshell and phosphoric acid. J Nanosci Nanotechnol 7:4061–4064
72. Zhang S (2008) CO_2 Laser induced structural changes of dental enamel. Dissertation on Doctor Degree of Natural Sciences, University of Hamburg, 115 p
73. Chinthaka Silva GW, Ma L, Hemmers O, Lindle D (2008) Micro-structural characterization of precipitation-synthesized fluorapatite nano-material by transmission electron microscopy using different sample preparation techniques. Micron 39:269–274
74. Siddharthan A, Sampath Kumar TS, Seshadri SK (2009) Synthesis and characterization of nanocrystalline apatites from eggshells at different Ca/P rations. Biomed Mater 4:1–9
75. Ma G, Liu XY (2009) Hydroxyapatite: hexagonal or monoclinic? Cryst Growth Des 9 (7):2991–2994
76. Magdich LN, Molchanov VYa (1978) Acoustooptical devices and their applications. Sov. Radio, Moscow, 306 p (in Russian)
77. Yu Gulyaev V (1980) Acoustoelectronic apparatus for communication systems and information handling. In: Kotelnikov VA (ed) Problems of modern radio engineering and electronics. Nauka, Moscow, pp 297–319 (in Russian)
78. Balakshii VI, Parygin VN, Chirkov LE (1985) Physical principles of acousto-optics. Radio i Svyaz, Moscow, 323 p (in Russian)
79. Suchanek W, Yoshimura M (1998) Processing and properties of hydroxyapatite-based biomaterials for use as hard tissue replacement implants. J Mater Res 13(1):94–117
80. Song QM, Wang CY, Wen SL (1998) Electron structure of a grain boundary and a crystal of calcium phosphate bioceramic (hydroxyapatite). Philos Mag A Phys Condens Matter Struct Defect Mech Prop 77(5):1309–1321

81. Kanazawa T (ed) (1989) Inorganic phosphate materials. Elsevier Science, Amsterdam
82. Zyman Z, Ivanov I, Glushko V et al (1999) Preparation and properties of inhomogeneous hydroxyapatite ceramics. J Biomed Mater Res 46:135–140
83. Dudnik EF, Duda VM, Dyachenko L (1998) Crystal chemistry features of hyfrogen introduction in oxide crystals with apatite-like structure. Visnyk DGU Fizika 1(3):141–149 (in Russian)
84. Dudnik EF, Duda VM (1999) Protonic conductivity and ferroelasticity in materials with apatite-like structure. Ferroelectrics 233(1–2):121–127
85. Royce BS (1973) The defect structure and ionic transport properties of calcium apatite. J Phys 34:11–12, C9-327-332
86. Orlovskii VP, Zakharov NA, Ivanov AA (1996) Structural transition and dielectric characteristics of high-purity hydroxyapatite. Inorg Mater 32(6):654–656
87. Orlovskii VP, Zakharov NA, Ivanov AA (1996) Structural transition and dielectric characteristics of high-purity hydroxyapatite of calcium. Inorg Materials 32(6):736–739 (in Russian)
88. Azimov SY, Ismatov AA, Fedorov NF (1990) Apatites and their rare-earth analogues. Fan, Tashkent,150 p (in Russian)
89. Sharpataya GA, Fedoseev AD, Bogacheva VV, Orlovskii VP (1995) The heat capacity of $Ca_{10}(PO_4)_6(OH)_2$ from 300 to 1070 K. Russ J Inorg Chem 40(4):590–593
90. Ishikawa T, Tanaka H, Yasukawa A, Kandori K (1995) Modification of calcium hydroxyapatite using ethyl phosphates. J Mater Chem 5(11):1963–1967
91. Elliott JC (1994) Structure and chemistry of the apatites and other calcium orthophosphates. Studies in inorganic chemistry 18. Elsevier, Amsterdam, 389 p
92. Tsuda H, Arends J (1994) Orientational micro-Raman spectroscopy on hydroxyapatite single crystals and human enamel crystallites. J Dent Res 73(11):1703–1710
93. Wen SL (1989) Human enamel structure studied by high resolution electron microscopy. Electron Microsc Rev 2(1):1–16
94. Mortier A, Lemaitre J, Rouxhet PG (1989) Temperature-programmed characterization of synthetic calcium-deficient phosphate apatites. Thermochim Acta 143:265–282
95. Deaza PN, Guitian F, Santos C (1997) Vibrational properties of calcium phosphate compounds. 2. Comparison between hydroxyapatite and β-tricalcium phosphate. Chem Mater 9:916–922
96. Smirnov AA (1966) Molecular-kinetic theory of metals. Nauka, Moscow, 488 p (in Russian)
97. Matysina ZA (1978) Molecular-kinetic theory of ordered solid solutions. DGU, Dnepropetrovsk, p 119 (in Russian)
98. Matysina ZA, Zaginaichenko SYu, Schur DV (2005) Orders of different type in crystals and phase transformations in carbon materials. Nauka i obrazovanie, Dnepropetrovsk, 524 (in Russian)
99. Matysina ZA, Zaginaichenko SYu, Schur DV (2006) Solubility of impuritites in metals, alloys, intermetallic compounds, fullerites. Nauka i obrazovanie, Dnepropetrovsk, 514 p (in Russian)
100. Matysina ZA, Chumak VA (2000) Statistic theory of deformational ordering in crystals of sheelite structure. Dep. N 1766-B00. (M. VINITI. 2000. 24 p.). Izv Vuzov Fizika 9:112 (in Russian)
101. Matysina ZA, Chumak VA (2001) Deformation hystresis and elastic compliance of crystals with structure H4 near Curie point Ukr. Fiz Zhurn 46(9):957–959 (in Russian)
102. Matysina ZA, Zaginaichenko SYu, Eremenko AM (2002) Phase transformations in elastic crystals of the structure G_{51}. Proc. of Int. conf. "Science for Materials in the Frontier of Centuries: Advantages and Challenges". Kiev. IPM of NASU, November 4-8, 2002, P. 208–209
103. Matysina ZA, Eremneko AM (2005) Investigation of phase transformation in ferroelastics of structure G_{51}. Deformation hysteresis, configuration heat capacity. Dep.N 100-B2004. (M.: VINITI. 23 p.). Izv Vuzov Fizika 1:94 (in Russian)
104. Parker SP (Editor in Chief) (1983) McGraw-Hill encyclopedia of physics. Mc-Graw-Hill Book Company, New-York

Chapter 28
The Theory of Phase Transformations and Heat Capacity in Crystals of Fluorofullerenes

S.Yu. Zaginaichenko, Z.A. Matysina, and D.V. Schur

Abstract The theory of phase transition of order–disorder type on the molecular-kinetic grounds has been developed for the mixture of fluorofullerenes $C_{60}F_{48}$, $C_{60}F_{36}$: transition from the ordered body-centered tetragonal (bct) structure to the disordered face-centered cubic (fcc) structure. The free energies of these phases have been found, their dependence on temperature, composition of material, the degree of ordering and energetic constants has been determined. The temperature of transition between phases has been calculated. The constitution diagram has been constructed and it defines the temperature and concentration areas of bct, fcc phases formation and also the region of realization of both bct and fcc phases. The configuration heat capacity of bct phase and its temperature dependence $C_p(T)$ has been defined. The peak-shaped increase of heat capacity in the neighbourhood of the temperature of phase transition has been estimated and this is in agreement with experimental data.

Keywords Fluorofullerene · Phase transition · Constitution diagram · Heat capacity · Temperature · Order parameter

Nomenclature

F_i [J]	Free energy of phases
E_i [J]	Configuration internal energy
υ_{ij} [J]	Interaction energy of fluorofullerenes
k [J/K]	Boltzmann constant
c_i [–]	Concentration of phases

S.Yu. Zaginaichenko (✉), Z.A. Matysina, and D.V. Schur
Institute for Problems of Materials Science of NAS of Ukraine,
Krzhyzhanovsky str. 3, Kiev 03142, Ukraine
e-mail: shurzag@ipms.kiev.ua

S.Yu. Zaginaichenko et al. (eds.), *Carbon Nanomaterials in Clean Energy Hydrogen Systems - II*, NATO Science for Peace and Security Series C: Environmental Security 2,
DOI 10.1007/978-94-007-0899-0_28,

C_p [J/K·mol]	Heat capacity
T [K]	Temperature
T_o [K]	Temperature of phase transition
η [–]	Order parameter in distribution of fluorofullerenes
a [μm]	Lattice parameter
N [–]	Number of lattice sites
X_i [–]	Composition of mixture
Φ_i [–]	Flurofullerenes of two types (i = 1, 2)

28.1 Introduction

The intensive research studies of fullerenes and features of their formation of last years led to the findings of thousands new compounds possessing the unique physicochemical properties. The fluorofullerenes are of great interest for scientific groups. Among the inorganic derivatives of fullerenes they have the high thermodynamic stability in comparison, for example, with chlorides, bromides, oxides and manifest much wide variability. The attachment of fluorine atoms to C_{60} molecules affect much their physical and chemical properties. This brings up the prospects of production of new materials with unusual properties and the awakening of interest to the fluoride derivatives of fullerenes.

The solid-phase fluorination of C_{60} crystals by the molecular flow of fluorine F_2 brings primarily into existence the molecules $C_{60}F_{36}$ which possess enhanced stability [1–7]. The atomic bond C-F in them is sufficiently strong. The next fluorine atoms have the less bonding force with molecules C_{60}. Depending on temperature, pressure and duration of interaction of fluorine molecules flow with fullerite the addition of 12 supplementary fluorine atoms gives the formation of fluorofullerene $C_{60}F_{48}$ [3, 7–10]. The last 12 fluorine atoms migrate readily over the carbon frame, redistribute in condensate and even break away [11–13], resulting in rearrangement of structure of fluorofullerene molecules, formation of structures of different type, production of distinct their isomers and consequently change of physical and chemical properties. The exothermicity of fluorination reaction and temperature increase provides the intensity of regrouping of fluorine atoms.

The extensive experimental and theoretical investigation of physicochemical and thermodynamic properties of fluorofullerenes begins and in the most of research papers just more stable fluorofullerenes $\Phi_1 = C_{60}F_{48}$, $\Phi_2 = C_{60}F_{36}$ are being examined. The laboratories perform research on their crystalline and electronic structure, degree of stability, mass composition, the availability of isomers of different level symmetry, phase transitions in condensate with the external pressure and temperature variations, carry out the estimation of phase transition temperature, measure the lattice constants, determine their bulk modulus, conduct the plotting

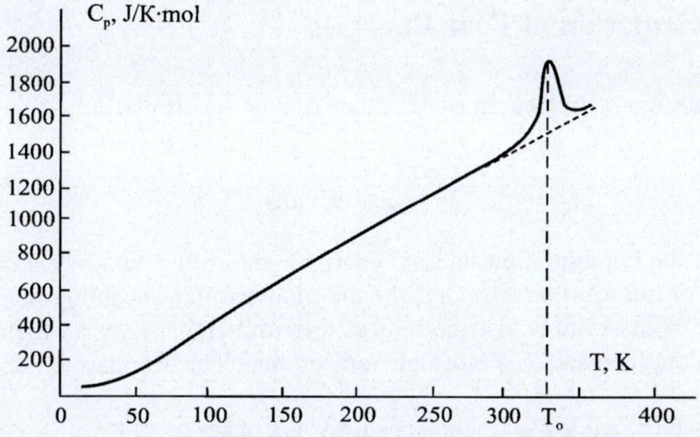

Fig. 28.1 Experimental plot of temperature dependence of heat capacity of $C_{60}H_{48}$ fluorofullerenes [22, 30, 39]. To is the temperature of structural phase transition bct → fcc. The *dotted line* shows the dependence $C_o(T)$ in the absence of phase transition

of constitution diagram with indication of temperature and pressure regions of realization of fullerite, fullerene polymers, fluorofullerenes and amorphous phase, evaluate and calculate the heat of fluorofullerenes formation, the enthalpy of their forming, thermal coefficient of expansion, heat capacity [12, 14–39].

Experimental investigation of heat capacity C_p of fluorofullerene Φ_1 establish the anomaly in its temperature dependence [22, 30, 39], there is the peak in the curve $C_p(T)$ in the temperature region $T_0 \sim 330$ K (Fig. 28.1). The further determination of crystalline and molecular structure of condensate demonstrates that at this temperature the phase transition of order–disorder type takes place. The fluorofullerene structure is changed from the ordered body-centered tetragonal (bct) to the disordered face-centered cubic (fcc) [21, 23]. The anomalous change of lattice parameter [23] is observed at this temperature, it seems reasonable to say that it is caused by realization of phase transition.

Theoretical investigation of fluorofullerenes, the development of statistical theory of transition between phases at the temperature T_0, the construction of constitution diagram of the system being studied, the elucidation of temperature dependence of fluorofullerite heat capacity, the explanation and justification of appearance of possible anomaly in this dependence at the temperature T_0 are of our interest.

To solve this problem we calculate the free energy of condensate. In theoretical calculations the mixture of fluorofullerenes Φ_1, Φ_2 is considered and according to experimental data this system has ordered bct crystalline lattice at temperatures below T0 and disordered fcc lattice at temperatures higher than T_0. The method of average energies [40], the approximation of account of interaction energies of nearest fluorofullerene pairs [40, 41] and the model of symmetric spherically rigid spheres are used in calculations.

28.2 Calculation of Free Energies

The computation of free energies of bct and fcc phases is performed by the known formula

$$F_i = E_i - kT\ln G_i, \tag{28.1}$$

where E_i is the configuration internal energy equal to the sum of energies of pair interaction of fluorofullerenes, G_i is the thermodynamic probability determined by the amount of discernible distributions of fluorofullerenes over all their positions, k is Boltzmanns constant, T is absolute temperature. The bct phase is named as first, the fcc phase – as second.

At first we calculate the free energy F_1 of bct phase.

The configuration energy is equal to

$$E_1 = -N_{11}\upsilon_{11} - N_{22}\upsilon_{22} - N_{12}\,\upsilon_{12}, \tag{28.2}$$

where N_{ij}, υ_{ij} are the numbers of pairs and energies of interaction with opposite sign of fluorofullerenes $\Phi_1 = C_{60}F_{48}$, $\Phi_2 = C_{60}F_{36}$ (i, j = 1, 2).

The distance between the nearest fluorofullerenes, at which energy υ_{ij} is determined for bct structure, is equal to

$$r_1 = \sqrt{2a_1^2 + a_2^2/2}, \tag{28.3}$$

where α_1, α_2 are parameters of bct lattice.

The numbers of pairs of the nearest fluorofullerenes $\Phi_1\Phi_1$, $\Phi_2\Phi_2$, $\Phi_1\Phi_2$ are defined by expressions

$$\begin{aligned} N_{11} &= \frac{1}{2}zNP_1^{(1)}P_1^{(2)}, \quad N_{22} = \frac{1}{2}zNP_2^{(1)}P_2^{(2)}, \\ N_{12} &= \frac{1}{2}zN\left(P_1^{(1)}P_2^{(2)} + P_1^{(2)}P_2^{(1)}\right), \end{aligned} \tag{28.4}$$

where z = 8 is the coordination number, N is the number of all sites (fluorofullerenes), $P_i^{(\alpha)}$ are the a priori probabilities of substitution of bct lattice sites of $\alpha = 1$, 2 type with fluorofullerenes of $i = \Phi_i$ $(i = 1, 2)$ kind. For the studied structure they are equal to

$$P_1^{(1)} = c_1 + \frac{1}{2}\eta, \quad P_1^{(2)} = c_1 - \frac{1}{2}\eta, \quad P_2^{(1)} = c_2 - \frac{1}{2}\eta, \quad P_2^{(2)} = c_2 + \frac{1}{2}\eta, \tag{28.5}$$

η is the order parameter in fluorofullerenes distribution over the sites of crystal bct lattice

$$\eta = 2\left(P_1^{(1)} - c_1\right). \tag{28.6}$$

In view of relations (28.5) we rewrite the numbers N_{ij} as

$$N_{11} = \frac{1}{2}zN\left(c_1^2 - \frac{1}{4}\eta^2\right), \quad N_{22} = \frac{1}{2}zN\left(c_2^2 - \frac{1}{4}\eta^2\right),$$
$$N_{12} = zN\left(c_1c_2 + \frac{1}{4}\eta^2\right). \tag{28.7}$$

Then the configuration energy E_1 is determined by formula

$$E_1 = E_{O1} - \frac{1}{8}zN\,\omega_1\eta^2, \tag{28.8}$$

where augend for E_{O1} energy

$$E_{O1} = -\frac{1}{2}zN(c_1^2\upsilon_{11} + c_2^2\upsilon_{22} + 2c_1c_2\upsilon_{12})$$
$$= -\frac{1}{2}zN(c_1\upsilon_{11} + c_2\upsilon_{22} + c_1c_2\,\omega_1) \tag{28.9}$$

is independent of energy order and

$$\omega_1 = 2\upsilon_{12} - \upsilon_{11} - \upsilon_{22} \tag{28.10}$$

is the ordering energy of first phase.

The thermodynamic probability G_1 is calculated according to the rules of combinatorics

$$G_1 = \frac{N_1!}{N_1{}^{(1)}!N_2{}^{(1)}!}\cdot\frac{N_2!}{N_1{}^{(2)}!N_2^{(2)}!}, \tag{28.11}$$

where N_1, N_2 are the numbers of lattice sites of $\alpha = 1, 2$ type $N_i^{(\alpha)}$and are the numbers of fluorofullerenes of i kind on the sites of α type

$$N_i^{(\alpha)} = N_iP_i^{(\alpha)},\ i = 1, 2;\ \alpha = 1, 2. \tag{28.12}$$

Taking into consideration formulae (28.5) and using Stirling formula $\ln X! = X(\ln X-1)$ for large X numbers, we find the thermodynamic probability

$$lnG_1 = -\frac{1}{2}N\left[\left(c_1+\frac{1}{2}\,\eta\right)ln\left(c_1+\frac{1}{2}\,\eta\right)+\left(c_1-\frac{1}{2}\,\eta\right)ln\left(c_1-\frac{1}{2}\,\eta\right)+\right.$$
$$\left.+\left(c_2-\frac{1}{2}\,\eta\right)ln\left(c_2-\frac{1}{2}\,\eta\right)+\left(c_2+\frac{1}{2}\,\eta\right)ln\left(c_2+\frac{1}{2}\,\eta\right)\right]. \tag{28.13}$$

Substituting configuration energy E_1 (28.8) and thermodynamic probability G_1 (28.13) in equation (28.1), we define the free energy of bct phase calculated for one site of crystal lattice as follows

$$f_1 = F_1/N = e_1 - \omega_1\eta^2 + \frac{1}{2}kT\Delta_1, \tag{28.14}$$

where the following designations are used

$$\Delta_1 = \left(c_1+\frac{1}{2}\,\eta\right)ln\left(c_1+\frac{1}{2}\,\eta\right)+\left(c_1-\frac{1}{2}\,\eta\right)ln\left(c_1-\frac{1}{2}\,\eta\right)+$$
$$+\left(c_2-\frac{1}{2}\,\eta\right)ln\left(c_2-\frac{1}{2}\,\eta\right)+\left(c_2+\frac{1}{2}\,\eta\right)ln\left(c_2+\frac{1}{2}\,\eta\right), \tag{28.15}$$

$$e_1 = E_{O1}/N = -4(c_1\upsilon_{11} + c_2\upsilon_{22} + c_1c_2\,\omega_1). \tag{28.16}$$

The formulae (28.14)–(28.16) determine the dependence of free energy of bct phase on temperature, order parameter, phase composition (concentrations c_1, c_2) and energetic parametres υ_{11}, υ_{22}, ω_1.

The calculation of free energy F_2 of the second fcc phase is similar to the foregoing

$$f_2 = F_2/N = e_2 + kT\Delta_2, \tag{28.17}$$

where

$$\Delta_2 = c_1 lnc_1 + c_2 lnc_2, \tag{28.18}$$

$$e_2 = E_2/N = -6\left(c_1^2\upsilon'_{11} + c_2^2\upsilon'_{22} + 2c_1c_2\upsilon'_{12}\right), \tag{28.19}$$

and $z_2 = 12$ is taken into account for fcc phase. The υ'_{ij}energies of interaction of the nearest fluorofullerenes in fcc phase are determined at the distances

$$r_2 = a\big/\sqrt{2}, \tag{28.20}$$

a is the parameter of fcc lattice.

According to formulae (28.17)–(28.19), the free energy of second fcc phase depends on temperature, concentrations c_1, c_2 and energetic parametres υ'_{11}, υ'_{22}, υ'_{12}.

28.3 Temperature of Phase Transition. Equations of Thermodynamic Equilibrium. Constitution Diagram

The phase transition bct → fcc occurs at the temperature $T = T_o$ when free energies of both phases are equal

$$f_1 = f_2. \tag{28.21}$$

We equate the two expressions (28.14), (28.17) and find the temperature T_o of phase transition

$$kT_o = \left(e_1 - e_2 - \omega_1\eta^2\right) \Big/ \left(\frac{1}{2}\Delta_1 - \Delta_2\right). \tag{28.22}$$

The equilibrium value of order parameter η in (28.22) is estimated from the condition of thermodynamic equilibrium, i.e. by equation

$$\partial f_1 / \partial \eta = 0 \tag{28.23}$$

and this gives the relation

$$kT\,ln\frac{\left(c_1 + \frac{1}{2}\eta\right)\left(c_2 + \frac{1}{2}\eta\right)}{\left(c_1 - \frac{1}{2}\eta\right)\left(c_2 - \frac{1}{2}\eta\right)} = 8\,\omega_1\eta. \tag{28.24}$$

Considering $\eta \to 0$ in (28.24) and straightforward rearranging gives the temperature of ordering of the first bct phase

$$kT_C = 8c_1c_2\,\omega_1. \tag{28.25}$$

The equilibrium value of order parameter depends on temperature and concentrations c_1, c_2. The maximum order ($\eta = 1$) is realized for the infinitely low temperatures and concentrations corresponding to the stoichiometric composition $c_1 = c_2 = 0.5$.

For bct phase of stoichiometric composition the equation (28.24) in terms of (28.25) takes the form

$$ln\frac{1+\eta}{1-\eta} = 2\frac{T_c}{T}\eta. \tag{28.26}$$

For the numerical assessment of temperatures of phase transitions T_o of bct → fcc type and T_C of order–disorder type the evaluation of energetic parametres e_1, e_2, ω_1 should be made. These parametres are estimated approximately using the experimental data for the temperature T_o of phase transition bct → fcc ($T_o \approx 330$ K or $kT_o = 0.028$ eV). This estimation shows that

$$e_1 = -0.01 \text{ eV}, \quad e_2 = -0.005 \text{ eV}, \quad \omega_1 = 0.01 \text{ eV}. \tag{28.27}$$

As an example we evaluate this temperature T_o for the maximum value of order parameter defined by equalities

$$\eta_m = \begin{cases} 2c_1 & \text{at} \quad c_1 \leqslant 0.5, \\ 2c_2 & \text{at} \quad c_1 \leqslant 0.5. \end{cases} \tag{28.28}$$

It is evident that $\eta_m = 1$ for the stoichiometric composition of the system. In this case $kT_C = 0.02$ eV, i.e. the temperatures T_o and T_C are close to each other.

The value Δ_1 (28.15) for order parameter η_m (28.28) takes the following form

$$\Delta_1 = \begin{cases} 2c_1 ln2c_1 + (c_2 - c_1)ln(c_2 - c_1) & \text{at} \quad c_1 \leqslant 0.5 \\ 2c_2 ln2c_2 + (c_1 - c_2)ln(c_1 - c_2) & \text{at} \quad c_1 \geqslant 0.5. \end{cases} \tag{28.29}$$

Furthermore, the calculation of free energies f_1, f_2 is carried out by formulae (28.14), (28.17) using the energetic parametres (28.27), magnitude η_m (28.28), values of Δ_1 (28.29), Δ_2 (28.18).

Figure 28.2 gives the plots of concentration dependences of free energies constructed for different temperatures. These graphical representations of free energies make it possible to estimate the temperature T_o of phase transition bct → fcc and to construct the constitution diagram of the system under study. The concentration and temperature regions of realization of one- and twophase condensates are determined in the method of common tangents to the curves $f_1(c_1), f_2(c_1)$.

The constitution diagram of the system being studied is presented in Fig. 28.3. It is constructed by the use of method of common tangents to the curves $f_1(c_1)$, $f_2(c_1)$ and by the intersection points of these curves. It is seen from this figure that at low temperatures the ordered bct phase is bound to be realized from the mixture of fluorofullerenes Φ_1, Φ_2 in accordance with experimental data. The phase transition

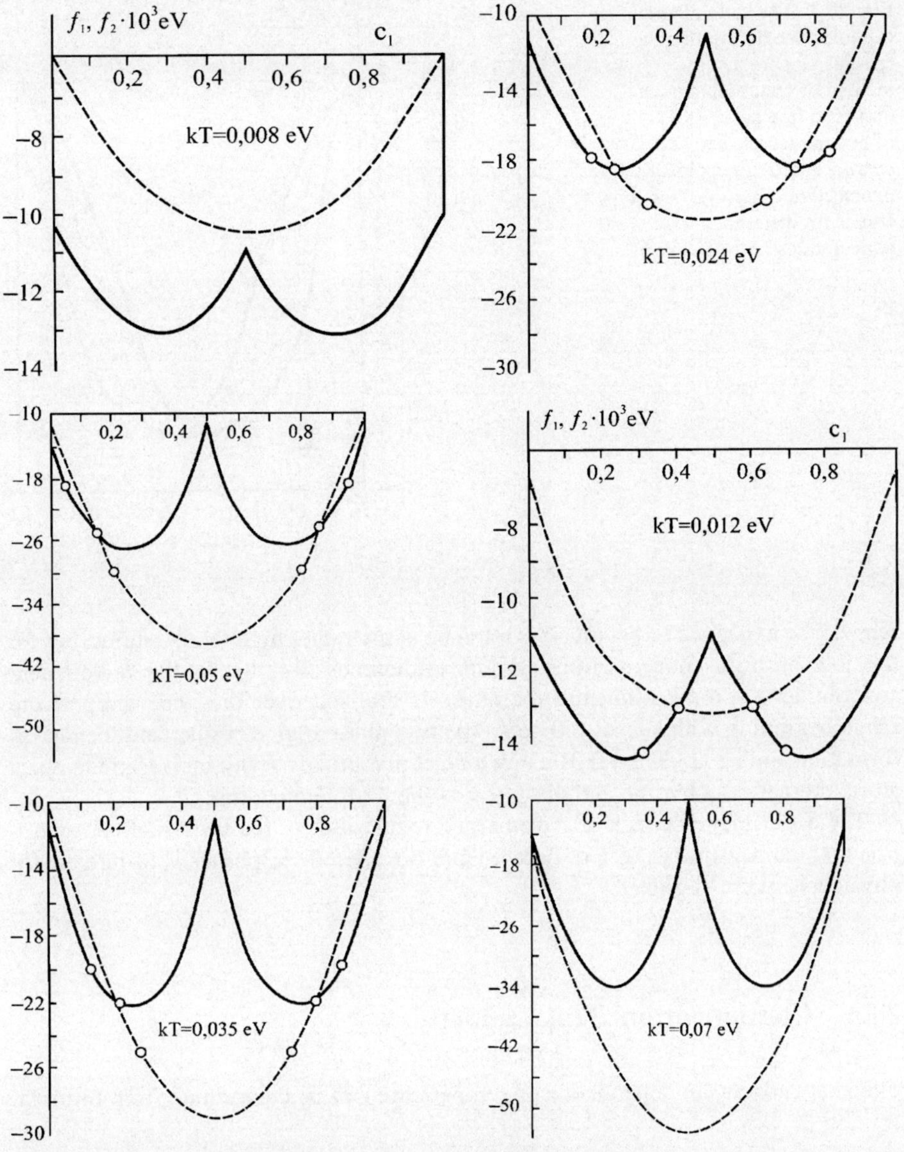

Fig. 28.2 The design plots of concentration dependence of free energies of ordered bct phase (*unbroked curves*) and disordered fcc phase (*dotted curves*) of flurofullerenes constructed for different temperatures. The intersection points of $f_1(c_1), f_2(c_1)$ functions and points of common tangent to them are marked off by the *open circles* on the curves

into disordered state of fluorofullerenes Φ_1, Φ_2 mixture with fcc lattice takes place with increase in temperature that also corresponds to experimental data. As the temperature increases, the concentration range of fcc phase formation broaden and at the sufficiently high temperatures the fcc phase is realized over all concentration

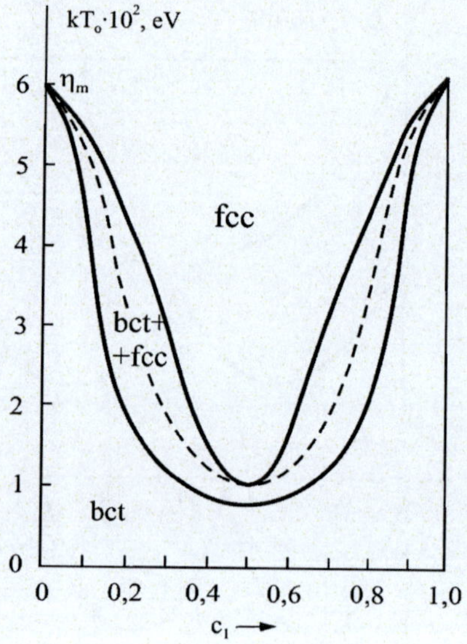

Fig. 28.3 The state diagram of molecular fluorofullerene crystals constructed by method of common tangent curves to free energies f_1, f_2 of bct and fcc phases (*full curves*) and by the points of intersection of $f_1(c_1), f_2(c_2)$ curves for different temperatures (*dotted curve*)

range. The existence of bct phase is retained at the rather high temperatures, but for the low or high concentrations of fluorofullerenes Φ_1, Φ_2. In the region near stoichiometric composition the fcc phase is realized over the wide temperature range beginning with $kT_o > 0.01$ eV. The two-phase regions of bct and fcc phases formation appear at $kT_o \approx 0.01$ eV, which at first broaden with increase in temperature, thereafter converge and disappear at the high temperatures.

The experimental check of the emerged regularities of realization of concentration and temperature ranges of ordered and disordered fcc phases is of interest for physicochemical engineers.

28.4 Configuration Heat Capacity

The heat capacity in dependence on temperature can be determined from formula

$$C_V = \frac{\partial E_1}{\partial T} = -2N\,\omega_1 \eta \frac{\partial \eta}{\partial T} \tag{28.30}$$

in which the order parameter η and its temperature derivative $\partial\eta / \partial T$ should be determined from the equation of thermodynamic equilibrium (28.24). The configuration heat capacity of disordered fcc phase is equal to zero, because the E_2 energy is temperature independent.

The calculation of configuration heat capacity with account of formula (28.25) gives the result

$$\frac{C_V}{kN} = ln^2 \frac{(c_1 + \frac{1}{2}\eta)(c_2 + \frac{1}{2}\eta)}{(c_1 - \frac{1}{2}\eta)(c_2 - \frac{1}{2}\eta)} \Bigg/ 4\left[\frac{c_1}{c_1^2 - \frac{1}{4}\eta^2} + \frac{c_2}{c_2^2 - \frac{1}{4}\eta^2} - \frac{1}{\eta} ln \frac{(c_1 + \frac{1}{2}\eta)(c_2 + \frac{1}{2}\eta)}{(c_1 - \frac{1}{2}\eta)(c_2 - \frac{1}{2}\eta)}\right]. \tag{28.31}$$

This formula defines the dependence of configuration heat capacity of bct phase on order parameter $C_V = C_V(\eta)$ and on temperature $C_V = C_V\,(T)$ taking into consideration the relations (28.24), (28.25).

The step of heat capacity in the point of phase transition bct → fcc is equal to

$$\Delta C = (C_V)_{T_C - 0} - (C_V)_{T_C + 0} = (C_V)_{T_C - 0}, \tag{28.32}$$

because $(C_V)_{T_C + 0} = 0$ (see Fig. 28.4a).

In consequence of the performed calculations the expression for step of heat capacity takes the form

$$\Delta \frac{C}{kN} = \frac{3}{2} \cdot \frac{c_1 c_2}{c_1^3 + c_2^3}. \tag{28.33}$$

This formula gives $\Delta \frac{C}{kN}$ values for different concentrations c_1.

$$\Delta \frac{C}{kN} = \begin{cases} 1.5 \text{ at } c_1 = 0.5, \\ 1.286 \text{ at } c_1 = 0.4, \\ 0.851 \text{ at } c_1 = 0.3, \\ 0.462 \text{ at } c_1 = 0.2, \\ 0.185 \text{ at } c_1 = 0.1, \end{cases} \tag{28.34}$$

i.e. the heat capacity step decreases with deviation of composition of bct phase from stoichiometry.

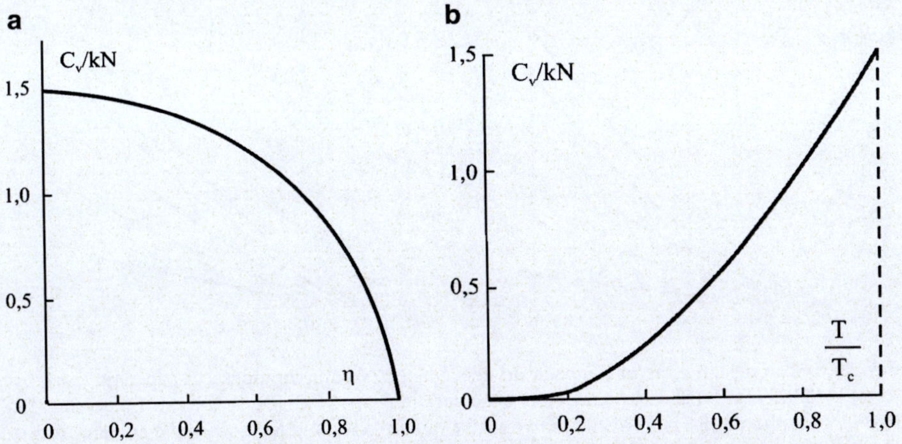

Fig. 28.4 The plots of configuration heat capacity of bct phase of stoichiometric composition as a function of order parameter (**a**) and temperature (**b**)

In the case of bct phase of stoichiometric composition the formula (28.31) is simplified

$$\frac{C_V}{kN} = \eta(1 - \eta^2) ln^2 \frac{1+\eta}{1-\eta} \Big/ 2\left[2\eta - (1 - \eta^2) ln \frac{1+\eta}{1-\eta}\right]. \tag{28.35}$$

In Fig. 28.4 the plots of configuration heat capacity dependence on order parameter (a) and on temperature (b) are shown for bct phase of stoichiometric composition. From these curves one can see that heat capacity falls with a decrease in order parameter and increases with a rise in temperature.

The heat capacity for the mixture of two phases (two-phase region at the constitution diagram) is determined by formula

$$\frac{C_V}{kN} = X_1 \left(\frac{C_V}{kN}\right)_{bct} + X_2 \left(\frac{C_V}{kN}\right)_{fcc} = X_1 \left(\frac{C_V}{kN}\right)_{bct}, \tag{28.36}$$

because the configuration heat capacity of fcc phase is equal to zero. In this formula the values X_1, X_2 define the composition of bct and fcc phases mixture, in this case

$$X_1 + X_2 = 1, \ldots 0 \leqslant X_1, X_2 \leqslant 1. \tag{28.37}$$

The temperature dependences of configuration heat capacity for different values of concentration c1 (a) and of composition X_1 of the first bct phase (b) are given in Fig. 28.5. The character of dependences $C_V(T_1)$ is identical for both cases (a) and (b). The heat capacity and its step at $T = T_C$ show a decrease with decreasing value c_1 or X_1. The theory gives results presented in Fig. 28.5 in qualitative agreement with experimental data (Fig. 28.1) for the heat capacity dependence on temperature for the ordered bct phase. The decrease of heat capacity C(T) in a gradual manner at

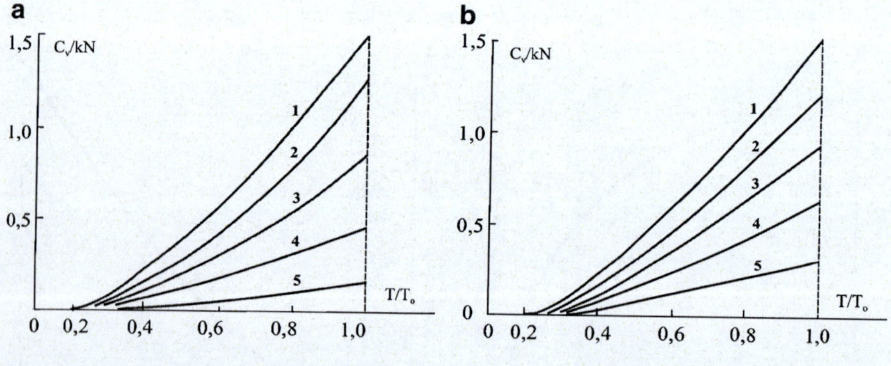

Fig. 28.5 The design plots of temperature dependence of configuration heat capacity: (**a**) – by formula (28.31) for fct phase and different concentrations $c_1 = 0.5; 0.4; 0.3; 0.2; 0.1$ (curves 1–5); (**b**) – by formulae (28.31), (28.36) for two-phase mixture (bct and fcc), when the content of first phase is $X_1 = 1; 0.8; 0.6; 0.4; 0.2$ (curves 1–5)

$T > T_C$ on the experimental plot can be caused by the presence of short-range ordering, which is not taken into account in our calculations.

28.5 Conclusions

The developed statistical theory makes possible to justify and to provide an explanation for the phase transition of order–disorder type in the mixture of fluorofullerenes $C_{60}F_{48}$, $C_{60}F_{36}$ which is revealed experimentally at the temperature of 330 K when transformation of ordered bct phase into disordered fcc phase is realized.

The calculation of free energies of both phases has been carried out, the evaluation of phase transition temperature in dependence on composition of phases and order parameter has been performed.

The construction of constitution diagram has been made for determination of temperature and concentration regions of constituent phases in the formation of pure bct and fcc phases as well as their mixture. It follows from this diagram that the ordered bct phase has been realized at low temperatures, the temperature rise has stimulated the formation of disordered fcc phase. Both theoretical conclusions are in agreement with experimental data.

The calculation of configuration heat capacity of bct phase has shown its peak-shaped increase in the neighbourhood of phase transition temperature that also corresponds to experimental observations. The evaluation of heat capacity step in the point of phase transition has been carried out and the value of this step decreases with deviation of fluorofullerenes mixture composition from stoichiometry.

References

1. Selig H, Lifshitz C, Peres T et al (1991) Fluorinated fullerenes. J Am Chem Soc 113:5475–5476
2. Gakh AA, Tuinmann AA, Adcock JL et al (1994) Selective synthesis and structure determination of $C_{60}F_{48}$. J Am Chem Soc 116:819–820
3. Clare BW, Kepert DL (1999) The structures of $C_{60}F_{36}$ and new possible structures for $C_{60}H_{36}$. J Mol Struct Theochem 466(1–3):177–186
4. Meier MS (2000) In: Kadish KM, Ruoff RS (eds) Fullerenes: chemistry, physics and technology. Wiley-Interscience, New York, p 129
5. Bagryantsev VF, Zapolskii AS, Boltalina OV et al (2000) Reaction of fullerenes with molecular fluorine. J neorganicheskoy khimii 45(7):1121–1127
6. Avent AG, Taylor R (2002) Fluorine takes a hike: remarkable room-temperature rearrangement of the C1 isomer of C60F36 into the C3 isomer via a 1, 3-fluorine shift. Chem Commun 22:2726–2727
7. Sidorov LN, Yurovskaya MA, Borschevskii AY et al (2005) Fullerenes. Ekzamen, Moscow, p. 688 (in Russian)

8. Boltalina OV, Sidorov LN, Bagryantsev VF et al (1996) Formation of $C_{60}F_{48}$ and fluorides of higher fullerenes. J Am Chem Soc 2:2275–2278
9. Privalov VI, Boltalina OV, Galeva NA, Taylor R (1998) Structure of crystalline $C_{60}F_{48}$ by NMR data with rotation under angle, Doklady RAN. Seriya khimicheskaya 360(4):499–502 (in Russian)
10. Troyanov SI, Troshin PA, Boltalina OV et al (2001) Two isomers of $C_{60}F_{48}$: an indente fullerene. Angew Chem Int Ed 40(12):2285–2287
11. Papina TS, Kolesov VP, Lukyanova VA et al (1999) The standard molar enthalpy of formation of fluorofullerene $C_{60}F_{48}$. J Chem Thermodyn 31(10):1321–1328
12. Papina TS, Kolesov VP, Lukyanova VA et al (2000) Enthalpy of formation and C-F bond enthalpy of fluorofullerene $C_{60}F_{36}$. J Phys Chem B 104(23):5403–5405
13. Gakh AA, Tuinmann AA (2001) 'Fluorine dance' on the fullerene surface. Tetrahedron Lett 42 (41):7137–7139
14. Tuinman A, Gakh A, Adcock J, Compton R (1993) Hyperfluorination of buckminsterfullerene. J Am Chem Soc 115:5885–5886
15. Kniaz K, Fischer JE, Selig H et al (1993) Fluorinated fullerenes: synthesis, structure and properties. J Am Chem Soc 115(4):6060–6064
16. Fowler PW, Sandall JPB, Taylor R (1997) Structural parallels in hydrogenated and fluorinated [60] - and [70] – fullerenes. J Am Chem Soc 2:419–423
17. Boltalina OV, Galeva NA, Markov VYu et al (1997) A mass spectrometric study of $C_{60}F_{48}$. Mendeleev Commun 5:169–212
18. Clare BW, Kepert DL (1997) An analysis of the 94 possible isomers of $C_{60}F_{48}$ containing a three-fold axis. J Mol Struct Theochem 389(1–2):97–103
19. Mitsumoto R, Araki T, Ito E et al (1998) Electronic structures and chemical bonding of fluorinated fullerenes studied by NEXAFS, UPS, and Vacuum-UV adsorption spectroscopies. J Phys Chem A 102(3):552–560
20. Taylor R (1998) Progress in fullerenes fluorination. Russ Chem Bull 47(5):823–832
21. Kawasaki S, Aketa T, Touhara H et al (1999) Crystal structures of the fluorinated fullerenes $C_{60}F_{36}$ and $C_{60}F_{48}$. J Phys Chem B 103(8):1223–1225
22. Druzhinina AI, Galeva NA, Varushchenko RM et al (1999) The low temperature heat capacities of fluorofullerenes. J Chem Thermodyn 31(11):1469–1482
23. Kawasaki S, Okino F, Touhara H (2000) Crystal structures and phase transformations of the fluorinated fullerenes. Mol Cryst Liq Cryst 340:629–633
24. Boltalina OV, Galeva NA (2000) Direct fluorination of fullerenes. Russ Chem Rev 69 (7):661–674
25. Gakh AA, Tuinman AA (2001) The structure of $C_{60}F_{36}$. Tetrahedron Lett 42(41):7133–7135
26. Slanina Z, Uhlik F, Boltalina OV, Kolesov VP (2002) Isomeric $C_{60}F_{36}$ (g) species: computed structures and heats of formation. Phys Solid State 44(3):511–512
27. Kawasaki S, Yao A, Okino F et al (2002) High pressure phases of fullerenes, hydrofullerenes and fluorofullerenes. Mol Cryst Liq Cryst 386(1):109–114
28. Yao A, Matsuoka Yu, Komiyama S et al (2002) Structural properties of fluorinated fullerenes at high pressures and high temperatures. Solid State Sci 4(11–12):1443–1447
29. Avent AG, Clare BW, Hitchcock PB et al (2002) $C_{60}F_{36}$: there is a third isomer and it has C_1 symmetry. Chem Commun 20:2370–2371
30. Druzhinina AI, Varuschenko RM, Boltalina OV, Sidorov LN (2002) Selective synthesis of C_{60} F_{48} Proceedings of All-Russian conference on chemical thermodynamics, Saint-Petersburg, Paper I-P 28 (in Russian)
31. Gakh AA, Romanovich AYu, Bax A (2003) Thermodynamic rearrangement synthesis and NMR structures of C_1, C_3 and T isomers of $C_{60}F_{36}$. J Am Chem Soc 125(26):7902–7906
32. Rau JV, Cesaro SN, Boltalina OV et al (2004) Raman and infrared spectroscopic study of $C_{60}F_{18}$, $C_{60}F_{36}$ and $C_{60}F_{48}$. Vibrational Spectrosc 34(1):137–147

33. Boltalina OV, Strauss SH (2004) Fluorofullerenes. In: Schwarz JA, Contescu CI, Putyera K (eds) Dekker encyclopedia of nanoscience and nanotechnology, vol 2. Dekker, New York, pp 1175–1190
34. Papoular RJ, Allouchi H, Dzyabchenko AV et al (2006) High-resolution x-ray powder diffraction structure determination of $C_{60}F_{48}$. Fullerenes Nanotubes Carbon Nanostruct 14 (2–3):279–285
35. Popov A, Senyavin V, Boltalina OV et al (2006) Infrared, Raman and DFT spectroscopic studies of $C_{60}F_{36}$ and $C_{60}F_{48}$. J Phys Chem A 110:8645–8652
36. Bulusheva LG, Okotrub AV, Shnitov VV et al (2009) Electronic structure of $C_{60}F_{36}$ studied by quantum-chemical modeling of experimental photoemission and x-ray absorption spectra. J Chem Phys 130:014704
37. Sheka EF (2009) Step-wise computational synthesis of fullerene C60 derivatives. 1. fluorinated Fullerenes $C_{60}F_{2k}$ In: Lecture notes in computer science, Sloot P.M.A et al (ed.), Berlin: Springer, Part 1, pp 1–32
38. Mikoushkin VM, Shnitov VV, Bryzgalov VV et al (2009) Core electron level structure in $C_{60}F_{18}$ and $C_{60}F_{36}$ fluorinated fullerenes. Tech Phys Lett 35(3):256–259
39. Boltalina OV, Sidorov LV Buckminsterfullerene, higher fullerenes and their endo and fluorinated derivatives. Russ Chem Rev (accepted for publication)
40. Matysina ZA, Zaginaichenko SY, Schur DV (2005) Orders of various type in crystals and phase transitions in carbon materials. Nauka i obrazovanie, Dnepropetrovsk, 524 p (in Russian)
41. Schur DV, Matysina ZA, Zaginaichenko SY (2007) Carbon nanomaterials and phase transformations in them. Nauka i obrazovanie, Dnepropetrovsk, 680 p (in Russian)

Chapter 29
Heat Stability of Me-C Nanocomposites

E.I. Golovko, Al.D. Zolotarenko, D.V. Schur, S.Yu. Zaginaichenko, A.P. Pomytkin, E.P. Rudakova, O.V. Milto, and Z.A. Matysina

Abstract The results of investigation of thermal resistance on air of Me-C nanocomposites containing Fe, Ni and Fe-Ni mixtures have been given in the present paper.

Thermal analysis of samples produced by arc-discharge method due to the evaporation of pure graphite and its mechanical mixtures with ferromagnetics (Fe, Ni, Fe-Ni) have been performed using the derivatograph Q-1500D in conditions of dynamic heating from room temperature to 1,000°C.

The results of mass change (TG), the rate of mass change (DTG), heat changes (enthalpy change, DTA) for synthesis products during the heating process have been obtained. The results of thermal investigation of products produced by joint evaporation of graphite and ferromagnets, the X-ray phase analysis for graphite samples with ferromagnets before and after oxidation and the temperature dependence of TG, DTG, DTA curves for samples produced by evaporation of pure graphite and mechanical mixtures (C + Fe; C + Ni; C + Fe + Ni) have been presented.

On the basis of the performance of experiments it is necessary to notice that smooth mass decreasing of soot in the low temperature range occurs due to amorphous carbon burning. Under heating till 600°C fullerene-liked nanostructures oxidation occurs, and during the following temperature growth the oxidation of graphitized mass proceeds and possible multi-wall carbon nanotubes begins.

The investigations have shown that produced Me-C nanocomposites are oxidized in higher temperature range than the pure soot. Increasing of upper interval of temperature oxidation may be explained by ferromagnetics (catalysts of nanotubes

E.I. Golovko, Al.D. Zolotarenko, D.V. Schur, S.Yu. Zaginaichenko (✉),
A.P. Pomytkin, E.P. Rudakova, and O.V. Milto
Institute for Problems of Materials Science of NAS of Ukraine,
Krzhyzhanovsky str. 3, Kiev 03142, Ukraine
e-mail: shurzag@materials.kiev.ua

Z.A. Matysina
Dnepropetrovsk National University, 72 Gagarin str, Dnepropetrovsk 49000, Ukraine

S.Yu. Zaginaichenko et al. (eds.), *Carbon Nanomaterials in Clean Energy Hydrogen Systems - II*, NATO Science for Peace and Security Series C: Environmental Security 2,
DOI 10.1007/978-94-007-0899-0_29,

growth) and multywall nanotubes oxidation. Decreasing of the temperature of oxidation beginning can be explained by formation of amorphous soot-liked phase in the samples.

Keywords Fullerite · Carbon nanotube · Arc evaporation · Oxidation · Ferromagnet · Temperature

29.1 Introduction

At present time metal carbon nanocomposites have attracted considerable attention due to their unique physicochemical properties. Particular emphasis has been placed on Me-C nanocomposites based on metals of Fe group. The presentation of research results into thermal resistance on air of Me-C nanocomposites containing Fe, Ni and Fe-Ni mixtures has been performed in the present paper.

In the course of tests the method of arc discharge in gaseous phase (ADG) has been used for production of Me-C composites based on iron, nickel and their mechanical mixtures.

This paper describes also the processes occurring in the space between the electric arc and the reactor wall (near electrode area) and on the reactor wall itself.

29.2 Investigation Methods and Equipment

The product was synthesized using arc plasma -chemical installation with vertical location of the reactor which had mobile cathode. The mechanical mixture of graphite and metals powders was added into the anode along its axis. Graphite of MPG-7 type was used as material of evaporated cores.

Me-C nanocomposite material was taken out of soot by magnetic separation of parietal soot suspension in hydrocarbons.

As noted before, the charged carbon particles are kept by electromagnetic field of the interelectrode space and practically cannot leave for the near electrode area. Therefore, on consideration of peculiarities of the processes occurring in the gaseous phase and on the reactor walls, their insignificant contribution can be neglected, but account must be taken of the neutral particles.

Relation between the number of neutral and charged particles that are generated by graphite evaporation will depend on the conditions of the performance of the technological process. On the basis of the theoretical analysis and the study of the composition and morphology of the products formed on the reactor walls, the process of nanostructures formation in the gaseous phase and on the reactor walls by arc evaporation of graphite can be represented as the schematic model shown in Fig. 29.1 [1].

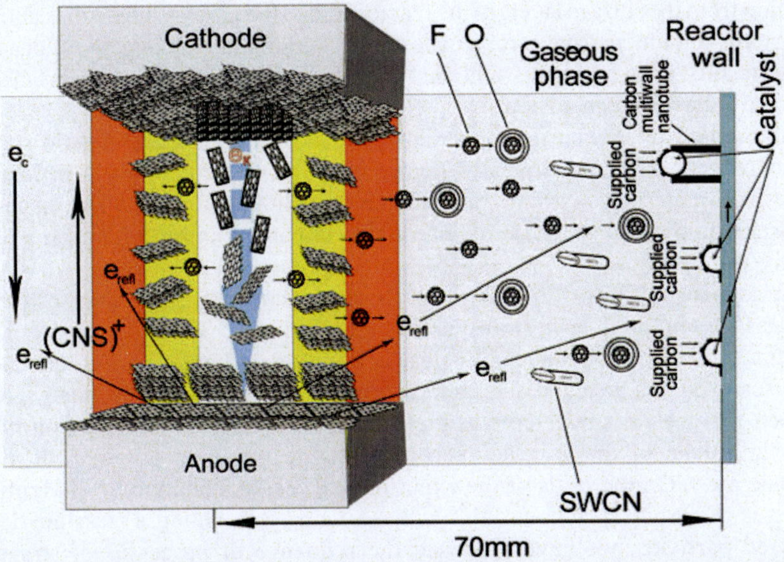

Fig. 29.1 The schematic diagram of carbon nanostructures formation during arc synthesis: F are fullerenes, O are onions, SWCN are single-wall carbon nanotubes; e_c is electrons flow of a power up to 8 kW (U = 25–30 V, I = 300 A) that moves from cathode to anode; e_{refl} are the reflected electrons; $(CNS)^+$ are carbon nanostructures with positive charge; CNT are carbon nanotubes

Under the present experimental conditions, in the arc evaporation of graphite the portion of carbon vapor, which consists of neutral particles of plasma stream, moves under the action of both gradients (ΔT and ΔP) from the center of the arc column to the periphery along the arc radius. This carbon vapor escapes the arc area (interelectrode space) at a velocity of more than 20–25 m/s., reaches the reactor wall within 0.003 s. and cools down to the room temperature. For this period, a number of processes occur; their duration varies starting from fractions of a nanosecond. The formed products concentrate on the reactor walls and are called "wall soot".

As particles move away from the axis of the electric arc column, their temperature and concentration decrease, their geometrical dimensions increase, the diffusion rate is lowered, so that the number of collisions with other molecules per time unit decreases. Different compounds and structures can be formed under change of energy and the number of carbon reagents collisions in the gaseous phase. The forming fullerenes can transform into onions and into nanotubes and another structures in the presence of a catalyst.

The extraction studies have shown that metals can catalyze the process of fullerene molecules destruction and graphite-like structures formation on their basis.

The production of third bodies is quite possible in the gaseous phase due to the fact that the wall temperature is 550–600°C. The electromagnetic radiation and electron beams generated by electric arc of a power up to 8 kW. favour the metal atoms transition from the surface layers of the reactor wall to the gaseous phase thus allowing these atoms to influence the carbon nanostructure formation in the gaseous phase.

The results of experiments and analysis of the obtained data are sufficient to allow the following conclusions.

The method of carbon nanomaterials synthesis defines the energetic state of starting reagents and, consequently, the peculiarities of formation and structure of synthesized materials. Carbon nanostructures, including fullerene-like materials, are formed in the vapour phase by interaction of carbon atoms with each other or with the third body.

The sequence of transformations which carbon undergoes in the nanostructure synthesis by any method is shown schematically in [2]. So, to synthesize a new structure, reagents are produced by destruction of a carbon or carbon-containing precursor. As this takes place, interaction of carbon atoms or their groups between themselves under certain thermodynamic conditions brings into existence the nuclei of the specified carbon structure (carbine, graphite, diamond, fullerene etc.). On the basis of the experimental results obtained by studying the processes of carbon nanomaterials synthesis using S.P. Gubin's classification of dispersed particles, we have proposed the sequence of processes of structures formation occurring in carbon materials when the system goes from the separate atoms through the clusters to the nanoparticles and further to the massive samples [3].

Duration of carbon nanomaterials synthesis defines their amount or the change in their geometrical dimension, but not the physical and chemical nature determined by the process thermodynamics at the nucleation stage. Interaction at the atomic level (nucleation) proceeds relatively fast, within fractions of a nanosecond. To obtain the product of a specified dispersion, i.e. the material consisting of the particles of certain geometrical dimensions and structure and having certain properties, the duration of interaction needs to be controlled at each level of structure formation.

A nucleus can be constructed from the chains of different lengths and degrees of branching, cycles and polyhedrons. Its frame can be skeleton and the combination of the above-listed structural elements. Increasing number of atoms in the cluster frame (nuclearity) results in the increasing number of methods for their connection. When nuclearity is higher than 20, according to the thermodynamics and geometry the spherical spatial structure is the most favourable; this is observed in the case of fullerene clusters of carbon.

Investigations of micro- and nanostructures of produced composites have been carried out using transmission electron microscope (TEM). The spectrum analysis of extracts of obtained product has been performed by the method of UV-VIS spectroscopy by spectrophotometer SF-2000.

Thermal analysis of samples produced by the arc-discharge method through the evaporation of pure graphite and its mechanical mixtures with ferromagnets (Fe, Ni, Fe-Ni) has been carried out with the derivatograph Q-1500D in conditions of dynamic heating in air from the room temperature to 1,000°C.

29.3 Results and Discussion

TEM-investigations have shown that Me-C composites consist substantially of metallic particles covered by carbon (Fig. 29.2, a – Ni; b – Fe). As is obvious from figure, metallic particles of 1–30 nm in diameter are completely enclosed in multi-layer carbon capsules. These metallic particles have quasi-spherical morphology.

The X-ray investigations have shown that nickel nanoparticles covered by carbon shells are predominantly the monocrystalline metal (of FCC lattice). Iron particles form carbide Fe_3C.

It should be pointed out that by the action of external magnetic field on the suspension of produced Me-C composites their particles move along strength lines

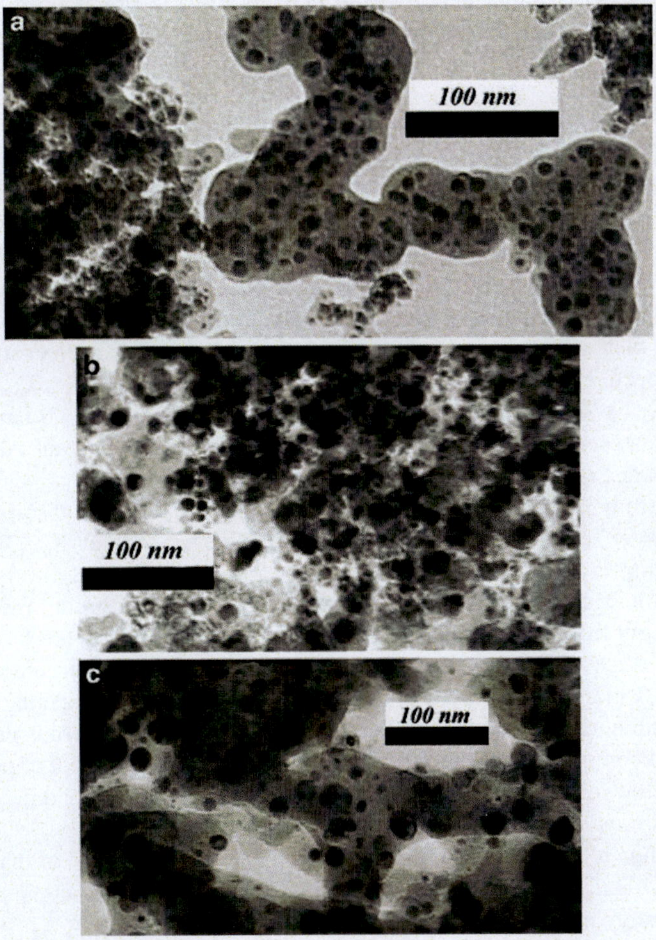

Fig. 29.2 Metallic particles of Ni (**a**), Fe (**b**) and Fe + Ni (**c**) covered by carbon film

of the field. The process of separation of these composite particles is based on this property. The fraction of the very low-mobility contains the composite with encapsulated particles of 5–10 nm in diameter.

TEM-investigations have been carried out before and after extraction of a component from the synthesis products. Pictures of B.2(X) samples of fullerene-containing soot have engaged one's attention (Fig. 29.3). The soot particles, encapsulated as thin web of fibrous crystals of 5 nm in diameter, have been clearly seen in Fig. 29.3a after soot washing in ethanol. After the same sample washing in toluene during 5 min, the following toluene outlet, and at last filling the system by ethanol, TEM picture has changed slightly (Fig. 29.3b). The most of the smallest crystals have disappeared, although some of them have stayed. In their instead more large crystals have appeared of length $l = 20–150$ nm and of diameter d = 5–20 nm. These appeared fullerite crystals are the products of process of fullerene molecules salting-out by alcohol from toluene solution residue. A well washed insoluble graphite-like nanoparticles remain after the prolonged extraction (Fig. 29.3c).

The same pictures have been observed for Ni-C (B.7), Fe-C (B.5), (Fe + Ni)-C (B.9) samples of composites.

The extracts of B.2(X), B.5(Fe), B.7(Ni) and B.9(Fe + Ni) samples, which contain presumably mixture of fullerenes C_{60} and C_{70}, have been analyzed using electron-absorption UV-VIS–spectroscopy according to the method developed at our department of the Institute for Problems of Materials Science of NASU.

The kinds of the resulting spectrums are given in Figs. 29.4 and 29.5.

Analysis of the spectrums Sp.1 and Sp.2 have shown that B.2(X) and B.5(Fe) extracts respectively have good detected adsorption bands (AB) with $\lambda_{max} = 287{,}9$ nm and $\lambda_{max} = 407{,}2$ nm corresponding to C_{60} and $\lambda_{max} = 336{,}0$ nm and $\lambda_{max} = 472$ nm corresponding to C_{70} respectively. This clearly demonstrates that both of solutions consist of C_{60} (84%) and C_{70} (16%) mixture.

The spectrum Sp.3 of B.7(Ni) probe solution has good detected absorption bands with $\lambda_{max} = 287{,}9$ nm and $\lambda_{max} = 336$ nm, however, AB at $\lambda_{max} = 407$ nm (corresponding to C_{60}) and $\lambda_{max} = 472$ nm (corresponding to C_{70}) are absent in the spectrum. The spectrum Sp.4 determined for probe solution B.9(Fe + Ni) differs strongly from the previous ones. Instead of two AB there are brightly split AB at $\lambda_{max} = 287{,}9$ nm and $\lambda_{max.} = 291{,}9$ nm and three slightly detected AB at $\lambda_{max} = 335{,}8$ nm, $\lambda_{max} = 353{,}2$ nm, $\lambda_{max} = 376{,}1$ nm. These results show that soluble component of synthesis products of composite nanomaterials depends on the chemical composition of precursor. The third soluble component of products of arc evaporation for mechanical mixtures of graphite with various components can be both endo- and exocomponents of fullerenes.

The findings of an investigation into mass change (TG), the rate of mass change (DTG), heat changes (enthalpy change, DTA) for synthesis products in the course of heating are given in Fig. 29.6 and Table 29.1.

The results of X-ray phase analysis for the samples before and after oxidation, which have been received using the diffractometer DRON-3 M, are represented in the Table 29.2.

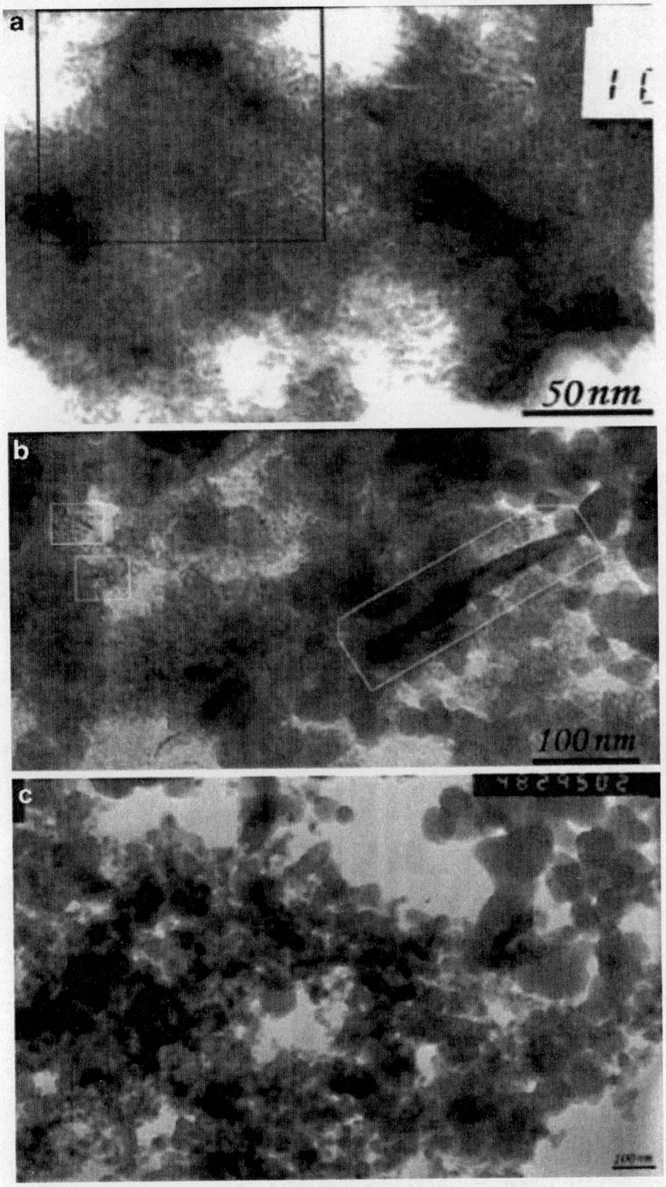

Fig. 29.3 Fullerene-containing soot for the sample B.2(X)

The investigation which have been carried out show that oxidation of graphitized parietal soot occurs in the temperature range 329–778°C. The broad asymmetric peak with $T_{max} = 684°C$ corresponds to this process on the DTG curve. The DTG curve nature in the temperature range ~500–665°C points to the formation of a

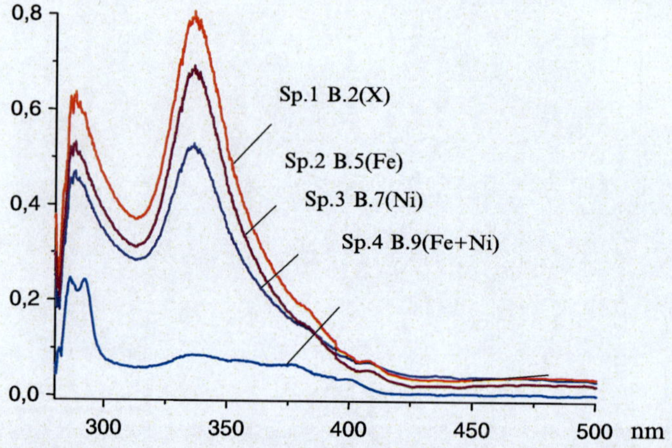

Fig. 29.4 The spectrums of absorption UV-VIS spectroscopy for B.2(X), B.5(Fe), B.7(Ni), B.9 (Fe + Ni) extracts

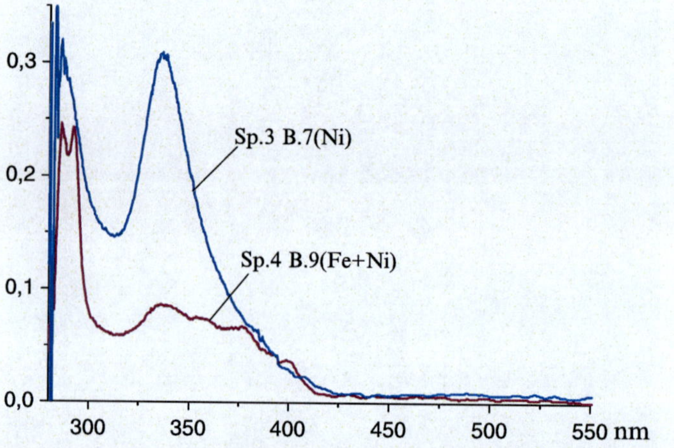

Fig. 29.5 The spectrums of absorption UV-VIS spectroscopy for B.7(Ni) and B.9(Fe + Ni) extracts

number of nanostructures with various thermal resistance. The moderate broad exothermic peak is noticed in this temperature range on the DTA curve. It overlaps with more large exothermic peak with the maximum intensity at 709°C (Fig. 29.6a; Tables 29.1 and 29.2).

It has been known from literature data that amorphous carbon burns up in the temperature range 300–500°C. Fullerenes inflame at 550°C. The oxidation temperatures of one-wall and multi-wall carbon nanotubes equal 650 and 750°C respectively. The graphitized particles interaction with air oxygen occurs at ~800°C [4, 5].

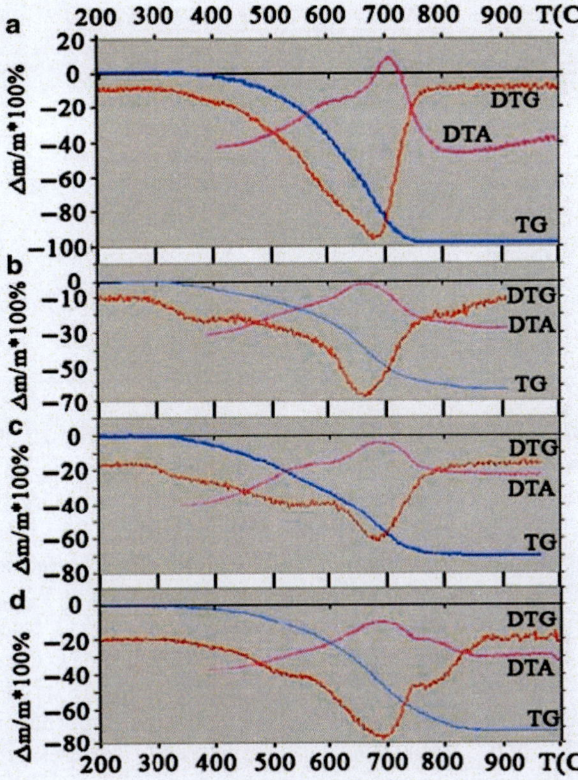

Fig. 29.6 The temperature dependence of TG, DTG, DTA curves for samples produced by evaporation of pure graphite (**a**) and mechanical mixtures C + Fe (**b**); C + Ni (**c**); C + Fe + Ni (**d**)

Table 29.1 The results of thermal investigation of products produced by joint evaporation of graphite and ferromagnets

No	Material	Temperature range of oxidation, °C	Mass Decrease %	DTG $T_{1shoulder}$ (°C)	DTG $T_{2shoulder}$ (°C)	DTG T_{1max} (°C)	DTG T_{2max} (°C)	DTA
1.	$C_{(soot)}$	329–778	98			684		709
2.	C + Fe	209–886	61	386		668		668
3.	C + Ni	209–849	70.4	346	530	683		692
4.	C + Fe + Ni	206–894	72.3		505	690.5	776	697

On the basis of obtained data it should be mentioned that the smooth decreasing of soot mass in the low temperature range occurs due to amorphous carbon burning. The oxidation of fullerene-liked nanostructures, on heating till 600°C, occurs and as the temperature increases later on the oxidation of graphitized mass and possible of multy-wall carbon nanotubes begins.

The products containing ferromagnetics begin to oxidize at lower temperatures (206–209°C) and this process takes place in broader temperature range.

Table 29.2 The results of X-ray phase analysis for graphite samples with ferromagnets before and after oxidation

No	Material	Phase composition before oxidation	Phase composition after oxidation
1.	$C_{(soot)}$	Rombohedral distortion of graphite	–
2.	C + Fe	Soot (amorphous phase, haloon $2\theta \approx 12$–$22°$), graphite traces, (line of intermediate intensity, d = 3,36A°), α-Fe (VCC) and γ-Fe FCC- lattices	Main phase – Fe_2O_3, amorphous phase, (haloon $2\theta \approx 14$–$24°$)
3.	C + Ni	Soot Ni(C)- solid solution(interplanar spacing are increased), traces of NiC.	Main phase – NiO, amorphous phase, (halo on $2\theta \approx 16$–$25°$)
4.	C + Fe + Ni	$Fe_{0.64}$ $Ni_{0.36}$ (cubic crystal system), soot, traces of graphite with rombohedral distortion	Main phase – $Ni_{1.43}$ $Fe_{1.7}O_4$(FCC), NiO, amorphous phase, (haloon $2\theta \approx 12$–$24°$)

The decrease of the product masses constitutes 61–72.3% (Fig. 29.6b–d; Table 29.1, no 2–4). Under oxidation of the product containing Fe the shoulder is observed on the DTG curve at temperature about 386°C and large asymmetric peak is seen at $T_{1max} = 668°C$ (Fig. 29.6b; Table 29.1, no 2). The DTG curve of oxidation for the product with Ni in parallel with the low temperature shoulder at $T_{1shoulder} = 346°C$ (Table 29.1, no 3) contains the second shoulder at $T_{2shoulder} = 530°C$. Temperature of the large asymmetric peak equals $T_{1max} = 683°C$ as in the case of oxidation of the product with Fe (Fig. 29.6c; Table 29.1, no 3). The shoulder presence at 505°C only (Table 29.1, no 4) the displacement of the large asymmetric peak to higher temperature range (690°C), the appearance of one more little peak at 776°C (Fig. 29.6d; Table 29.1, no 4) are special features of high temperature behavior for the product containing Fe-Ni.

Under oxidation of the samples with ferromagnets the occurrence of shoulders on the DTG curves at 346 and 386°C is evidently associated with the burning of amorphous carbon. More intricate character of TG, DTG, DTA curves of these samples oxidation (in comparison with pure soot) at higher temperatures is associated with formation of solid products during heating due to the metal oxidation at temperatures above 300°C. According to the X-ray phase analysis oxides Fe_2O_3, NiO, complex oxide $Ni_{1,43}Fe_{1,7}O_4$, amorphous halo at $2\Theta \approx 12$–24^0 (Table 29.2) are found in the residue of the samples that have been oxidized.

The appearance of wide peak ($T_{max} = 668$–$690°C$) (Fig. 29.6b–d; Table 29.1, no 2–4) on the DTG curve as in the case of soot oxidation points to the availability of carbon nanotubes in products with ferromagnets.

29.4 Conclusions

The conventional schematic model of carbon nanostructures formation at the arc evaporation of graphite in gaseous (neutral) phase has been suggested. The basis for this model is the behaviour of particles in electromagnetic field at the extremely high gradients of pressure and temperature on an arc radius.

One of the most important conclusions following from the experimental observations can be considered the fact that all micro- and macrovolumes of carbon nanomaterials are formed at the stage of nucleation, i.e. a nanostructural product consists of nuclei of different structures.

The Me-C nanocomposites (where Me = Fe, Ni, Fe + Ni) have been produced and they have structures like the nucleus-cover.

It has been noticed the response of nanocomposites to the magnetic field action.

The initial fullerite crystals, which are contained in soot before extraction process, have been shown.

The self-descriptiveness and the high of the speed performance method developed by these authors of the qualitative and quantitative analysis of the soluble nanomaterials have been noted.

The carried out investigations have shown that produced Me-C nanocomposites are oxidized in a wider temperature range than the pure soot. Increasing of upper interval of temperature oxidation can be explained by oxidation of ferromagnets (catalysts of nanotubes growth) and multiwall nanotubes. Decreasing of the temperature of oxidation beginning can be attributed to the formation of amorphous soot-liked phase in the samples being studied.

References

1. Zolotarenko AD (2009) The peculiarities of synthesis of carbon nanostructures and their hydrogen capacity. Ph.D. Dissertation for degree in chemical sciences, Kiev
2. Schur DV, Zaginaichenko SY, Skorokhod VV (2005) On the mechanism of carbon nanostructures formation. In: Extended abstracts of IX international conference on hydrogen materials science and chemistry of carbon nanomaterials (ICHMS'2005), Sevastopol, pp 534–537
3. Schur DV, Zaginaichenko SYu, Lysenko EA, Golovchenko TN (2007) The forming peculiarities of C_{60} molecule. Carbon nanomaterials in clean energy hydrogen systems, NATO Science Series, Netherlands: Springer pp 53–66
4. Golovko EI, Bogolepov VA, Zaginaichenko SYu, Lysenko EA, Schur DV et al (2005) The use of thermogravimetric analysis for certification of nanostructural materials. Nanosyst Nanomater Nanotechnol 3(3):633–643 (in Russian)
5. Eletskii AV (2002) Carbon nanotubes and their emissivity. Phys Usp 172(4):401–439 (in Russian)

Chapter 30
$[C_{76}]$ – Fullerenes: Enumeration of Isomer Substitutions in Terms of Apical, Edge and Face Differentiation

V.M. Smolyakov, D.V. Sokolov, D.Yu. Nilov, V.V. Grebeshkov, and D.M. Fedin

Abstract The discovery of fullerenes in quasicrystals having a symmetry of icosahedral point groups stimulated the interest to the usage of symmetry point groups having axes of the 5th order incompatible with the symmetry of infinite crystal lattice. Of interest are the crystals obtained from the $[C_{60}]$–$[C_{100}]$ molecules. These molecules may form crystals (fullerides) together with other chemical elements and radicals. The structures having carbon layers forming lengthy hollow cylinders (carbon nanotubes ending with half-spherical cap) are distorted halves of fullerene molecule. Some nanotubes have open ends. Nanotubes combine the properties of a molecule with one-dimensional solid body. They are characterized by a large variety of properties perspective for applications. A method is proposed of the property prediction for the series of isomer substitutions of basic structure making use of the polygonal number splitting of the Pascal triangle (Smolyakov et al., Abstract of the XVIII Mendeleev conference on general applied chemistry, vol 2, Moscow, 23–28 Sept 2007, p 524; Nilov et al., Fullerenes C_{60} – C_{80}: enumeration of exohedral substitution isomers and methods of thermodynamical properties estimation, In: Abstracts of VIII international conference solid state chemistry and contemporary micro- and nanotechnologies, Kislovodsk, 14–19 Sept 2008, pp 366–368 (in Russian); Smolyakov, About construction of additive schemes calculation of alkanes properties in the third approximation, Calculation methods in physical chemistry, KSU, Kalinin, pp 39–68, 1988; Smolyakov et al. C_{60} and C_{78} fullerenes: identification of exohedral substitution isomers. In: Abstracts of IX national conference on crystal growth, Moscow, 16–20 Oct 2000, pp 650 (in Russian); Yu et al., *Vestnik Tver State Univ Ser Chem* 2(30):87–94, 2007 (in Russian); Smolyakov et al. Rare Mater Technol 28:626–636, 2009). With the aid of the analysis of molecule structural elements On the basis of the analysis of fullerene molecule structural elements a formula is obtained for the calculation

---

V.M. Smolyakov (✉), D.V. Sokolov, D.Yu. Nilov, V.V. Grebeshkov, and D.M. Fedin
Department of Physical Chemistry, Tver State University, Russia,
Sadovy pereulok, 35, Tver 170002, Russia
e-mail: smolyakov@inbox.ru

S.Yu. Zaginaichenko et al. (eds.), *Carbon Nanomaterials in Clean Energy Hydrogen Systems - II*, NATO Science for Peace and Security Series C: Environmental Security 2,
DOI 10.1007/978-94-007-0899-0_30, © Springer Science+Business Media B.V. 2011

of fullerene thermodynamical properties. Calculations are made for C^{o}_{p298K}, ΔH_{subl}, of gaseous clusters $[C_{60}]$–$[C_{100}]$, not studied experimentally.

Keywords Fullerenes · Enumeration of isomers · Symmetry group · Mathematical models

30.1 Introduction

The chiral and achiral isomers substitution $[C_{76}]$–fullerene (Fig. 30.1) is given with the $\boldsymbol{T}_d$. Symmetry formulas Z and generating functions for the number of isomers on the basis of Poya theorem; isomers distribution over the $\rho(m)$ families and depending on the number $m = 76$ of the substitution sites are presented.

30.2 Substitution of the $[C_{76}]$-Fullerene Molecule Over Vertices

The symmetry group operations $\boldsymbol{T}_d$. (E, $4C_3$, $4C_3^2$, $3C_2^1$, $3S_4^1$, $3S_4^3$, $6\sigma_d$). The cyclic indices corresponding to the symmetry operations are: $E \Rightarrow f_1^{76}$, $4C_3^1 \Rightarrow 4f_1^1 f_3^{25}$, $4C_3^2 \Rightarrow 4f_1^1 f_3^{25}$, $3C_2^1 \Rightarrow 4f_2^{38}$, $3S_4^1 \Rightarrow 3f_4^{19}$, $3S_4^3 \Rightarrow 3f_4^{19}$, $6\sigma_d \Rightarrow 6f_1^6 f_2^{35}$. Exchanging of possible substitution sites, recorded in cycles of symbols of the form $f_l^{\alpha}, f_m^{\beta}$, where α, β – number of cycles, and l, m – their length. The sum of all $f_l^{\alpha}, f_m^{\beta}$ gives the cyclic indice of the group to calculate the isomers substitution properties without regard to the chirality [1].

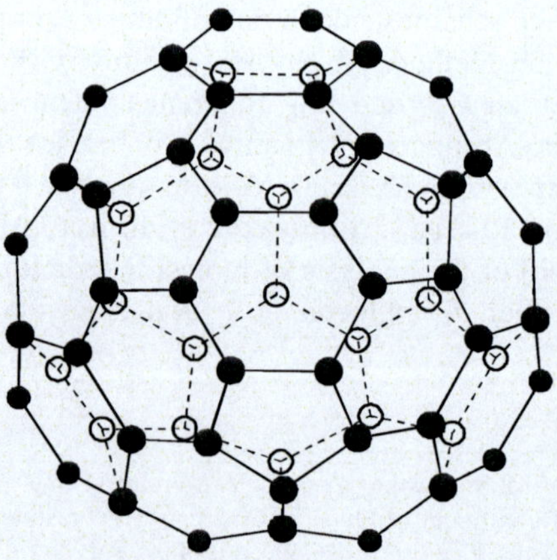

Fig. 30.1 Structure of molecule $[C_{76}]$– fullerene

Table 30.1 Isomer numbers of some X-substituted [C_{76}]-fullerenes over vertices calculated by formulas (30.2)–(30.4)

Family	Isomer number		
	Accounting for enantiomers	Chiral pairs	Chiral
$h^{76}h0$	1	0	1
$h^{75}h^{1}$	7	2	3
$h^{74}x^{2}$	247	111	25
$h^{73}x^{3}$	5,875	2,880	115
$h^{72}x^{4}$	107,107	53,265	577
$h^{71}x^{5}$	1,539,570	768,716	2,138
$h^{70}x^{6}$	18,220,554	9,106,278	7,998
$h^{69}x^{7}$	182,182,650	91,078,480	25,690
$h^{68}x^{8}$	1,571,342,085	785,631,126	79,833
$h^{67}x^{9}$	11,872,224,525	5,936,000,105	224,315
$h^{66}x^{10}$	79,544,021,063	39,771,708,331	604,401
$h^{65}x^{11}$	477,263,364,265	238,630,923,567	1,517,131
$h^{64}x^{12}$	2,585,177,255,039	1,292,586,806,131	3,642,777
$h^{63}x^{13}$	12,727,023,055,500	6,363,507,391,310	8,272,880
$h^{62}x^{14}$	57,271,606,866,864	28,635,794,434,992	17,996,880
$h^{61}x^{15}$	236,722,628,710,860	118,361,295,665,902	37,379,056
$h^{60}x^{16}$	902,505,034,086,408	451,252,479,773,322	74,539,764
$h^{59}x^{17}$	3,185,311,841,735,700	1,592,655,849,506,780	142,722,140
...	...	...	...
$h^{39}x^{37}$	559,657,232,158,858,000,000	279,828,616,043,845,000,000	71,168,166,300
$h^{38}x^{38}$	574,385,054,066,608,000,000	287,192,526,997,243,000,000	72,123,443,700

$$Z_2^V(T_d) = 1/24(f_1^{76} + 8f_1^1 f_3^{25} + 3f_2^{38} + 6f_4^{19} + 6f_1^6 f_2^{35}). \qquad (30.1)$$

Making use of the substitutions $f_l^\alpha = (h^l + x^l + y^l + \ldots)^\alpha$ on the basis of Poya theorem for the formula (30.1) obtains function $h^\nu + Ah^{\nu-1} + Bh^{\nu-2}x^2 + \ldots$, having coefficients (1, A, B, ...) equal to the number of isomers substitution of the type $h^\alpha x^\beta y^\gamma$... (see Tables 30.1–30.6). It is found that [C_{76}]-fullerene substitutes decompose into $\rho(m)$ = **3,689** families. For the rotation group ***T*** (operation σ_d, и S_n are excluded) the cyclic index, corresponding to the total number of chiral and achiral isomers of [C_{76}]-fullerene, can be written as

$$Z_1^V(T) = 1/12(f_1^{76} + 8f_1^1 f_3^{25} + 3f_2^{38}). \qquad (30.2)$$

$$Z_{Achir}^V(T_d) = 1/12(6f_4^{19} + 6f_1^6 f_2^{35}). \qquad (30.3)$$

$$Z_{Chiral\ pairs}^V(T_d) = 1/24(f_1^{76} + 8f_1^1 f_3^{25} + 3f_2^{38} - 6f_4^{19} - 6f_1^6 f_2^{35}). \qquad (30.4)$$

Table 30.2 Isomer numbers of some XYZ-substituted [C_{76}]-fullerenes over vertices calculated by formulas (30.2)–(30.4)

Family	Isomer number Accounting for enantiomers	Chiral pairs	Achiral
$h^{73}xyz$	35,150	17,545	60
$h^{72}x^2yz$	1,282,975	641,180	615
$h^{71}x^3yz$	30,791,400	15,394,620	2,160
$h^{71}x^2y^2z$	46,187,100	23,090,670	5,760
$h^{70}x^4yz$	546,547,350	273,267,630	12,090
$h^{70}x^3y^2z$	1,093,094,700	546,536,310	22,080
$h^{70}x^2y^2z^2$	1,639,654,704	819,801,762	51,180
$h^{69}x^5yz$	7,651,662,900	3,825,812,550	37,800
$h^{69}x^4y^2z$	19,129,157,250	9,564,520,560	116,130
$h^{69}x^3y^3z$	25,505,543,400	12,752,733,900	75,600
$h^{69}x^3y^2z^2$	38,258,314,500	19,129,054,070	206,360
$h^{68}x^6yz$	87,994,123,350	43,996,985,550	152,250
$h^{68}x^5y^2z$	263,982,370,050	131,990,992,560	384,930
$h^{68}x^4y^3z$	439,970,616,750	219,985,097,850	421,050
$h^{68}x^4y^2z^2$	659,956,146,570	329,977,549,230	1,048,110
$h^{68}x^3y^3z^2$	879,941,233,500	439,970,231,750	770,000
$h^{67}x^7yz$	854,800,055,400	427,399,813,500	428,400
$h^{67}x^6y^2z$	2,991,800,193,900	1,495,899,354,390	1,485,120
$h^{67}x^5y^3z$	5,983,600,387,800	2,991,799,551,300	1,285,200
...	...	...	...
$h^{64}x^6y^4z^2$	35,830,547,113,605,000	17,915,273,429,587,300	254,430,400
$h^{64}x^6y^3z^3$	47,774,062,763,027,700	23,887,031,297,345,200	168,337,304
...	...	...	...
$h^{20}x^{19}y^{19}z^{18}$	$6.81694 \cdot 10^{41}$	$3.40847 \cdot 10^{41}$	0
$h^{19}x^{19}y^{19}z^{19}$	$7.17572 \cdot 10^{41}$	$3.58786 \cdot 10^{41}$	0

Table 30.3 Isomer numbers of some X-substituted [C_{76}]-fullerenes over edges calculated by formulas (30.5) and (30.6)

Family	Isomer number Accounting for enantiomers	Chiral pairs	Achiral
h^{114}	1	0	1
$h^{113}x$	10	3	4
$h^{112}x^2$	551	255	41
$h^{111}x^3$	20,092	9,926	240
$h^{110}x^4$	556,472	277,496	1,480
$h^{109}x^5$	12,234,376	6,113,676	7,024
$h^{108}x^6$	222,251,626	111,109,369	32,888
$h^{107}x^7$	3,428,918,856	1,714,392,540	133,776
$h^{106}x^8$	45,861,703,074	22,930,589,768	523,538
$h^{105}x^9$	540,147,973,500	270,073,053,738	1,866,024
$h^{104}x^{10}$	5,671,552,781,202	2,835,773,203,696	6,373,810
$h^{103}x^{11}$	53,621,945,580,840	26,810,962,620,780	20,339,280
$h^{102}x^{12}$	460,255,025,630,106	230,127,481,698,305	62,233,496
$h^{101}x^{13}$	3,611,231,684,226,600	1,805,615,751,876,350	180,473,900
$h^{100}x^{14}$	26,052,457,099,470,000	13,026,228,298,220,900	503,028,200
...	...	...	...
$h^{58}x^{56}$	$1.2683 \cdot 10^{32}$	$6.34148 \cdot 10^{32}$	594,475,150,812,905,000
$h^{57}x^{57}$	$1.29055 \cdot 10^{32}$	$6.45273 \cdot 10^{32}$	810,647,932,926,689,000

Table 30.4 Isomer numbers of some X-substituted [C_{76}]-fullerenes over pentagonal faces calculated by formulas (30.7) and 30.8

Family	Isomer number: Accounting for enantiomers	Chiral pairs	Achiral
h^{12}	1	0	1
$h^{11}x$	1	0	1
$h^{10}x^2$	7	2	3
h^9x^3	21	8	5
h^8x^4	45	18	9
h^7x^5	66	28	10
h^6x^6	86	38	10

Table 30.5 Isomer numbers of some X-substituted [C_{76}]-fullerenes over haxagonal faces calculated by formulas (30.9) and (30.10)

Family	Isomer number: Accounting for enantiomers	Chiral pairs	Achiral
h^{28}	1	0	1
$h^{27}x$	3	0	3
$h^{26}x^2$	35	11	13
$h^{25}x^3$	279	118	43
$h^{24}x^4$	1,735	807	121
$h^{23}x^5$	8,190	3,956	278
$h^{22}x^6$	31,510	15,466	578
$h^{21}x^7$	98,694	48,808	1,078
$h^{20}x^8$	259,259	128,714	1,831
$h^{19}x^9$	575,631	286,413	2,805
$h^{18}x^{10}$	1,094,149	545,089	3,971
$h^{17}x^{11}$	1,789,515	892,167	5,181
$h^{16}x^{12}$	2,535,981	1,264,855	6,271
$h^{15}x^{13}$	3,120,264	1,556,634	6,996
$h^{14}x^{14}$	3,343,908	1,668,324	7,260

Table 30.6 Distribution of X-substituted [C_{76}]-fullerenes [1] over type for substitutions over pentagonal and hexagonal faces taking into account (30.7)–(30.10)

Substitution type	Isomer number for substitution: Over pentagonal faces: Chiral[a]	Over pentagonal faces: Achiral[a]	Over hexagonal faces: Chiral[a]	Over hexagonal faces: Achiral[a]
Mono-	0	1	0	3
Di-	4	3	22	13
Three-	16	5	236	43
Tetra-	36	9	1,614	121
Penta-	56	10	7,912	278
Hexa-	76	10	30,932	578

[a] Chiral and achiral isomer numbers are related by expressions: $Z^F_{Chir}(T_d) = 2\{Z^F_1(T) - Z^F_2(T_d)\}$ and $Z^F_{Achir}(T_d) = Z^F_1(T) - Z^F_{Chir}(T_d)$

30.3 Substitution of the [C_{76}]-Fullerene Molecule Over Edges

Cyclic indices for the calculation of [C_{76}]-fullerene isomers substitution over edges (with and without accounting for enantiomery) were found to be

$$Z_2^E(T_d) = 1/24(f_1^{114} + 8f_3^{38} + 3f_1^2 f_2^{56} + 6f_2^1 f_4^{28} + 6f_1^8 f_2^{53}), \tag{30.5}$$

$$Z_1^E(T) = 1/12(f_1^{114} + 8f_3^{38} + 3f_1^2 f_2^{56}). \tag{30.6}$$

30.4 Substitution of the [C_{76}]-Fullerene Molecule Over *Pentagonal* Faces

Substitutions over *pentagonal* faces correspond to cyclic indices

$$Z_2^F(T_d) = 1/24(f_1^{12} + 8f_3^4 + 3f_2^6 + 6f_4^3 + 6f_1^2 f_2^5), \tag{30.7}$$

$$Z_1^F(T) = 1/12(f_1^{12} + 8f_3^4 + 3f_2^6). \tag{30.8}$$

30.5 Substitution of the [C_{76}]-Fullerene Molecule Over *Hexagonal* Faces

Substitutions over *hexagonal* faces correspond to indices

$$Z_2^F(T_d) = 1/24(f_1^{28} + 8f_1^1 f_3^9 + 3f_2^{14} + 6f_4^7 + 6f_1^6 f_2^{11}), \tag{30.9}$$

$$Z_1^F(T) = 1/12(f_1^{28} + 8f_1^1 f_3^9 + 3f_2^{14}). \tag{30.10}$$

30.6 Numerical Calculations of ΔH_{subl}, C^o_{p298} by the Additive Scheme

Numerical calculations of ΔH_{subl}, C^o_{p298} for gaseous [C_{60}]–[C_{100}]- fullerenes are performed [2] by the additive scheme (Table 30.7)

$$P_{fullerene} = n_C x_0 + n_{C_6} x_2 + n_{edge} x_3 + n_F x_4, \tag{30.11}$$

Table 30.7 Thermodynamical properties[a] of carbon clusters C_{60}–C_{100} calculated by (30.11)

Fullerene	$\Delta H_{субл}$ kJ/mole	C^o_{p298} J/mole·K	Fullerene	$\Delta H_{субл}$ kJ/mole	C^o_{p298} J/mole·K
C_1	716.67	20.838	C_{80}	216.9	671.7
C_2	830.9	43.208	C_{82}	220.22	689.02
C_{60}	183.7	498.5	C_{84}	223.54	706.34
C_{62}	187.02	515.82	C_{86}	226.86	723.66
C_{64}	190.34	533.14	C_{88}	230.18	740.98
C_{66}	193.66	550.46	C_{90}	233.5	758.3
C_{68}	196.98	567.78	C_{92}	236.82	775.62
C_{70}	200.3	585.1	C_{94}	240.14	792.94
C_{72}	203.62	602.42	C_{96}	243.46	810.26
C_{74}	206.94	619.74	C_{98}	246.78	827.58
C_{76}	210.26	637.06	C_{100}	250.1	844.9
C_{78}	213.58	654.38			

[a] Experimental values for the calculation of fullerene properties by (30.11) are taken from [3]

where n_c and n_{edge} is the number of carbon atoms and C—C bonds in the cluster; n_{C_6} is the number of six-membered cycles; n_F is the total number of five- and six-membered cycles, and x_0, x_2, x_3, x_4 are the scheme parameters, obtained by rms of experimental [3] P values.

30.7 Conclusions

Point groups are applicable for the description of the position symmetry, i.e. the environment symmetry of one or another crystal space point, in particular, the symmetry of the atom (or molecule) position in the crystal structure. Since crystal faces are nodal grids the crystal lattice symmetry manifests itself in the crystal habitus (external shape). Inasmuch as crystal edges are nodal rows, the Z edge substitution is used. Unfortunately point groups having different geometrical meaning are isomorphous (e.g., $\boldsymbol{D}_{3h}$, $\boldsymbol{D}_{3d}$, $\boldsymbol{D}_6$). Conversely, parallelohedron structure is not uniquely described by the symmetry formula Z. The systematization of isomers of edge and face fullerene [C_{60}]–[C_{100}] substitutes was performed by us with the object of deriving other still unknown polyhedrons from these fullerenes. The presented combinatorial notions may be useful for the construction of mathematical models on the basis of splitting Pascal triangle polygonal numbers [2, 4–8] for the estimation of physico-chemical properties of fullerene and fulleride derivatives.

References

1. Smolyakov V, Sokolov D, Nilov D, Grebeshkov V, Fedin D (2009) Fullerenes C60–C80: enumeration of exohedral substitution isomers and methods of thermodynamical properties estimation. Rare Mater Technol 28:626–636

2. Nilov DY, Sokolov DV, Fedin DM, Grebeshkov VV, Smolyakov VM (2008) Fullerenes C_{60}–C_{80}: enumeration of exohedral substitution isomers and methods of thermodynamical properties estimation. In: Abstracts of VIII international conference solid state chemistry and contemporary mcro- and nanotechnologies, Kislovodsk, 14–19 Sept 2008, pp 366–368 (in Russian)
3. Kolesov VP, Pimenova SM, Pavlovich VK et al (1996) Enthalpies of combustion and formation of fullerene C_{60}. J Chem Thermodyn 28:1121–1125
4. Smolyakov VM, Sokolov DV, Nilov DY (2007) Fullerenes, carboranes, adamantane, arenas: the generation of isomers and evaluation of their properties on the graph model on the basis of Pascal's triangle. Abstracts of the XVIII Mendeleev conference on general and applied chemistry, vol 2, Moscow, 23–28 Sept 2007, p 524
5. Smolyakov VM (1988) About construction of additive schemes calculation of alkanes properties in the third approximation, Calculation methods in physical chemistry. KSU, Kalinin, pp 39–68
6. Smolyakov VM, Sokolov DV, Nilov DYu, Polyakov MN (2000) C_{60} and C_{78} fullerenes: identification of exohedral substitution isomers. In: Abstracts of IX national conference on crystal growth, Moscow, 16–20 Oct 2000, pp. 650 (in Russian)
7. Yu ND, Sokolov DV, Smolyakov VM, Nikolenko AYu (2007) Role of triangular numbers in the construction of calculation schemes of heterosystems properties. Vestn Tver State Univ Ser Chem 2(30):87–94 (in Russian)
8. Smolyakov VM, Sokolov DV, Nilov DYu (2000) C_{60} and C_{70} fullerenes: formulae of symmetry for generation of exohedral substitution isomers. In: Proceedings of the IV international conference on mathematical modelling, vol 2, MSTU, Stankin, Moscow, pp. 238–242 (in Russian)

Chapter 31
Carboranes and Boranes: Enumeration of Isomer Substitutes and Property Calculation Schemes on the Basis of Pascal Triangle

V.M. Smolyakov, D.V. Sokolov, D.Yu. Nilov, V.V. Grebeshkov, and D.M. Fedin

Abstract A method of determination of the number and type of substituted carboranes, boranes and carbometal borane derivatives (X-, XY-, ... – certain substitutes) is considered. Symmetry formulas Z and chiral and achiral stereo-isomer substitution number generating functions are established. The isomer distribution over families depending on the type and number of the substitutions and the number *m* of the substitution sites are found.

An 11-constant calculation scheme of the properties of $(\Delta_f H^0, S_f^0, \ldots)$ X-substituted (X = CH_3, F,...) para-$C_2H_{2\text{-}k}X_kB_{10}H_{10\text{-}l}X_l$ in consideration of the pair nonvalent interactions over two atoms and three and four atoms in the polyhedron skeletons is obtained on the base of Pascal triangle (K_3) number splitting.

Keywords Carboranes · Boranes · Isomers substitutions · Additive scheme

Nomenclature

X-XY-, XYZ-...	molecular substitutions
Z	symmetry formula
F	generating function
AB, ...	generating function coefficients - isomer substitution number of certain type
$\rho(m)$	number of isomer families
ΔH_f^0	enthalpy of substance formation

V.M. Smolyakov (✉), D.V. Sokolov, D.Yu. Nilov, V.V. Grebeshkov, and D.M. Fedin
Department of Physical Chemistry, Tver State University, Sadovy pereulok, 35, Tver 170002, Russia
e-mail: smolyakov@inbox.ru

S.Yu. Zaginaichenko et al. (eds.), *Carbon Nanomaterials in Clean Energy Hydrogen Systems - II*, NATO Science for Peace and Security Series C: Environmental Security 2, DOI 10.1007/978-94-007-0899-0_31,

K_3	triangular numbers
S_f^0	entropy of substance formations
G	totality of symmetry elements
$f_l^{\alpha} f_m^{\beta}$,	– cyclic indices of symmetry operations, where α, β, ... are the numbers of cycles, *l*, *m*, – cycle lengths
$\|G\|$	is the order of the group the number of the polyhedron symmetry operations
$h^{\alpha}x^{\beta}y^{\gamma}$...	are the families each containing some certain atoms
E,	σ_h, σ_v, σ_d, *i*, S_n, C_n are the point group symmetry operations
Z_1	is the group cyclic index neglecting enantiomery
Z_2	group cyclic index considering enantiomery
$Z^{Achir.}$	is the cyclic index of the symmetry group for achiral isomer substitutions
$Z^{Chir.par}$	is the number of chiral pairs
$\boldsymbol{C}_{2v}$ $\boldsymbol{D}_{5d}$, $\boldsymbol{D}_{3h}$, $\boldsymbol{D}_{4d}$, $\boldsymbol{D}_{5h}$, $\boldsymbol{D}_{2d}$, $\boldsymbol{C}_{5v}$, $\boldsymbol{C}_s$, $\boldsymbol{C}_{1v}$	are the symmetry groups of polyhedrons
P	is the property under study
p_0 p_1, p_2, ...	are the additive scheme parameters
C_n^0, C_n^1, ...	are the coefficients of the additive scheme
k l	are the degrees of substitution
n_X^{B1}, n_X^{C11}, n_{XX2}^{B1-B2}	are the coefficients of the additive scheme.

Subscript

MG	molecular graphs
MLS	the method of least squares.

31.1 Introduction

Rigorous information on the number and shape of different theoretically possible isomers is a prerequisite necessary for the determination of physico-chemical characteristics of substances typical of inorganic, organic and physical chemistry. For complex molecules it is a difficult task to find the shape of all structural and stereoisomers without having special algorithms.

The forthcoming of D. Poya's theorem on the calculation (enumeration) proved to be an important step not only for the graph theory but also for the mathematics in large [1–3]. The theory of enumeration is successfully used as a part of combinatorial analysis in statistical physics, structural chemistry, molecular biology, etc.

The theorem enables to calculate arbitrary constructions including chemical isomers. D. Poya successfully combined the classical method of generating function with the main results of group permutation theory [4, 5]. In the method of generating function the sequence a_0, a_1, a_2, ... of the values determining the number of the objects of different types is replaced by a function $F(\mathrm{x}) = a_0 + a_1\mathrm{x} + a_2\mathrm{x}_2 + \ldots$ obeying the rules of ordinary polynomials.

The aim of the present paper is to perform a combinatory study of isomer substitutions of polyhedral clozo-carboranes and boranes making use of the D. Poya's theory.

Problems:

1. on the example of ortho-carborane $C_2B_{10}H_{12}$ present method for determining the type and number of isomers of molecular replacement polyhedra;
2. to obtain formulas of symmetry groups and generating functions for determining the number of chiral and achiral isomers of substitution;
3. calculate the number of isomers in families, distribute them to the types of substitution $\rho(m)$ and depending on the number m of possible sites of substitution;
4. to form a number of homologues of X-substituted methyl-carborane CH_3-$C_2H_{2\text{-}k}X_kB_{10}H_{10\text{-}l}X_l$ and build a additive scheme to assess their $\Delta_f H^0_{298\,\mathrm{K,gas}}$;
5. to obtain calculation scheme for properties ($\Delta_f H^0, S^0_f, \ldots$) of X-substituted para-$C_2H_{2\text{-}k}X_kB_{10}H_{10\text{-}l}X_l$ on the basis of partitioning triangular (K_3) Pascal triangle numbers taking into account valence interactions, pair nonvalent interactions over two atoms, over three and four atoms in the skeleton of the polyhedron.

31.1.1 General Scheme of Derivation of Chemical Isomers of Substitution in Imitation of Carborane orto-$C_2B_{10}H_{12}$

Polyhedral closo-carborane is a special class of boron compounds. Stereochemical feature of the polyhedral closo-carboranes $C_2B_nH_{n+2}$ is that they contain carbon atoms with coordination numbers 4, 5 and 6 [6, 7]. Derivation of substitution isomers at the vertices of a polyhedron (or molecules) can be drawn on the basis of the theory of Polya enumeration [3, 4].

In this theory, the symmetry group G of the polyhedron is assumed to be known. The basis for the enumeration of chemical substitution isomers of is the notion of symmetry groups of models of the parent molecule (or molecular graph of a polyhedron). At sequential implementation of all possible operations exchange of places on the tops of the possible replacement of the polyhedron takes place.

These exchanges form cycles (in the mathematical sense) that is denoted as symbols of the form $f_l^{\alpha}, f_m^{\beta}, \ldots$, where α, β, ... – a number of cycles resulting from the implementation of this symmetry operation; l, m, ... – the order cycle, i.e. a number of vertices of a polyhedron-atoms involved in the cyclic exchange.

Postulate 1. If K_i identical (nonidentical) symmetry operations i correspond a geometric model, then the cycle index of the group (or a formula for symmetry) of all symmetry operations of the polyhedron can be represented as [3]

$$Z_1(G) = \sum_i \frac{1}{K_i} \times \sum_i K_g f_l^{\alpha}(g) f_m^{\beta}(g) \ldots, \tag{31.1}$$

where K_i - order of the group, K_g - the number of g similar operations. With the help of the cyclic index of corresponding group of permutations one can proceed from a combinatorial counting of a number of all configurations to the calculation of their equivalence classes. With permutations of the type

$$f_l^{\alpha} = (h^l + x^l + y^l + \ldots)^{\alpha}, f_m^{\beta} = (h^m + x^m + y^m + \ldots)^{\beta}, \tag{31.2}$$

formula (31.1) is transformed into the generating function

$$h^{\nu} + \mathrm{A}h^{\nu-1}x + \mathrm{B}h^{\nu-2}x^2 + \ \ldots, \tag{31.3}$$

where coefficients (1, A, B, ...) are equal to the number of substitution isomers of the type $\mathrm{h}^{\alpha}\mathrm{x}^{\beta}\mathrm{y}^{\gamma}$..., and can be calculated from the ratio for the polynomial coefficients.

$$(h + x + y \ldots)^n = \sum_{\alpha,\beta,\gamma} n!/(\alpha!\beta!\chi) \cdot h^{\alpha}x^{\beta}y^{\gamma} \ldots \tag{31.4}$$

Postulate 2. If for the symmetry group of the original polyhedron Z_2 one takes the rotation group (a subgroup of its point group), the cyclic code (31.1) and the generating function (31.3) will include mirror isomers [4, 24].

Postulate 3. If from a group set (31.1) one selects operations of the second kind only (operations reflect σ_h, σ_v, σ_d, i and S_n) and takes them as a source of the symmetry group of the polyhedron (as a subgroup of its point group), the cyclic code (31.1) and generating function (31.3) will include only $Z^{\mathrm{Achir.}}$ achiral isomers [5].

Postulate 4. A number of chiral pairs $Z^{\mathrm{Chir.par}}$ (only left or right spatial isomer with opposite) of polyhedron or molecule is computed as half the difference between operations of the symmetry group of rotations and operations of reflections of polyhedron or molecule.

Chirality is a feature of an object of not being identical to its mirror image. Chirality is a necessary and sufficient condition for enantiomers. Stereoisomers, which are not mirror images of each other are called diastereoisomers. Molecules containing the plane, the center of symmetry or mirror-rotation axis of symmetry are achiral. Enantiomers are stereoisomers that can not be combined with the imposition, but can be associated with both the object and its mirror image.

Substituted polyhedra with m places of possible replacement decay ρ (m) families (h^{m}, $\mathrm{h}^{\mathrm{m}-1}\mathrm{x}$, $\mathrm{h}^{\mathrm{m}-2}\mathrm{x}^2$,...), corresponding to the decomposition of m into whole positive parts: $\rho(1) = 1$, $\rho(2) = 2$, $\rho(3) = 3$, $\rho(4) = 5$, $\rho(5) = 7$, $\rho(6) = 11$, $\rho(7) = 15$, $\rho(8) = 22$, $\rho(9) = 30$, $\rho(10) = 42$, $\rho(11) = 56$, $\rho(12) = 77$, $\rho(13) = 101$, $\rho(14) = 135$, $\rho(15) = 176$, $\rho(16) = 231$, $\rho(17) = 297$, ..., $\rho(19) = 490$, $\rho(20) = 627$.

31.1.2 Derivation of Substitution Isomers

Let's demonstrate procedure of finding of substitution isomers of Polya's theorem on the example of a molecule of ortho-carborane $C_2B_{10}H_{12}$ (Fig. 31.1). The point group of this molecule has C_{2v} symmetry elements E, σ_v, C_2, associated with the following symmetry operations E, ${}^1\sigma_v$, ${}^2\sigma_v$, C_2 [8, 9].

Operations of symmetry C_{2v} induce on the set of places of substitution of H atoms in molecule of ortho-$C_2B_{10}H_{12} = \{1, 2, \ldots 12\}$ substitution f_l^α (see above).

The identity operation E leaves unchanged all places of substitution of ortho-$C_2B_{10}H_{12}$ and gives 12 cycles of the first order:

$$E \Rightarrow (1)(2)(3)\ldots\ldots(11)(12) \Rightarrow f_1^{12}.$$

Operation C_2 rearranges six times by two substitution sites, forming six cycles of second order:

$$C_2 \Rightarrow (17)(26)(3\ 11)(4\ 12)\ (58)(9\ 10) \Rightarrow f_2^6.$$

In operations ${}^1\sigma_v$ and ${}^2\sigma_v$ four substitution seats remain unchanged, while the remaining pairs share with each other:

$${}^1\sigma_v \Rightarrow (1)(26)(35)(4)(7)(8\ 11)(9\ 10)(12) \Rightarrow f_2^4 f_1^4;$$

$${}^2\sigma_v \Rightarrow (17)(2)(38)(4\ 12)(5\ 11)(6)\ (9)(10) \Rightarrow f_2^4 f_1^4;$$

The sum of all f_l^α, divided by the number of symmetry operations gives the cycle index of C_{2v} to determine the isomers excluding enantiomers

$$Z_1(C_{2v}) = 1/4\left(f_1^{12} + f_2^6 + 2f_2^4 f_1^4\right). \tag{31.5}$$

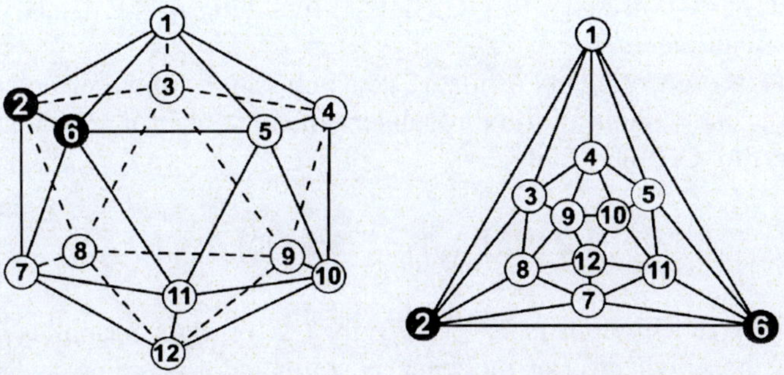

Fig. 31.1 Molecule and the graph of carborane ortho-$C_2B_{10}H_{12}$ C_{2v}

Using replacement of the form (31.2) in (31.5), we obtain the generating function of the numbers of isomers of carborane substitution of ortho-$C_2B_{10}H_{12}$:

$$\Phi(C_{2v}) = (1/4)\{(h + x + \ldots)^{12} + (h^2 + x^2 + \ldots)^6 + 2(h^2 + x^2 + \ldots)^4(h + x + \ldots)^4\}. \quad (31.6)$$

The coefficient of $h^k x^l y^m \ldots$ (31.6) (after reduction of similar) is equal to the number of isomers of substituted form of ortho- $C_{2\text{-}k}X_kB_{10\text{-}l\text{-}m}X_lY_m$. . ..For example, the number of isomers of substitution ortho- $C_2H_2B_{10}H_8X$ (coefficient $h^{10}x^1$) will be:

$$(1/4)(12!/11!/1!) + (1/2)(4!/4!/0!^*4!/3!/1!) = 5,$$

namely - 1-X-$C_2B_{10}H_{11}$, 2-X-$C_2B_{10}H_{11}$, 3-X-$C_2B_{10}H_{11}$, 4-X-$C_2B_{10}H_{11}$, 9-X-$C_2B_{10}H_{11}$.

In calculating isomers according to formula (31.5) one does not include optical isomers. In determining a number of substitution isomers of ortho-$C_2B_{10}H_{12}$ subject to enantiomers operations of reflection are excluded (σ_h, σ_v, σ_d, i, S_4). Cycle index of rotations C_2 of parent polyhedron has the form

$$Z_2(C_2) = 1/2(f_1^{12} + f_2^6). \quad (31.7)$$

With substitutions (31.2) in this formula one obtains generating function of the number of isomers of substituted ortho-$C_2B_{10}H_{12}$ subject to features of chirality

$$\Phi(C_2) = (1/2)\{(h + x + \ldots)^{12} + (h^2 + x^2 + \ldots)^6\}. \quad (31.8)$$

This function includes mirror isomers. According to (31.8) the number of substitution isomers of ortho-$C_2H_2B_{10}H_8X$ (coefficient $h^{10}x^1$) will be: (1/2)(12!/11!/1!) = 6. These isomers are: 1-X-$C_2B_{10}H_{11}$, 2-X-$C_2B_{10}H_{11}$, 3-X-$C_2B_{10}H_{11}$, 4-X-$C_2B_{10}H_{11}$,5-X-$C_2B_{10}H_{11}$, 9-X-$C_2B_{10}H_{11}$, including 4-X-$C_2B_{10}H_{11}$ and 5-X-$C_2B_{10}H_{11}$ - enantiomers.

If one selects (according to Postulate 3) only operations of the second kind of the group set $\boldsymbol{C}_{2v}$, one obtains a formula to determine a number of achiral substitution isomers of ortho–$C_2B_{10}H_{12}$ only

$$Z^{A\,chir.}(C_{2v}) = 1/2(2f_2^4f_1^4). \quad (31.9)$$

The number of chiral pairs $Z^{Chir.par}$ (only left or right spatial isomers with opposite) of the polyhedron or molecule is computed (according to Postulate 4) as half the difference between operations of the symmetry group of rotations and operations of reflections of polyhedron or molecule.

$$Z^{\text{Chir.par}}(C_{2v}) = 1/4\left(f_1^{12} + f_2^{6} - 2\,f_2^{4}f_1^{4}\right). \tag{31.10}$$

31.1.3 Distribution of Isomers into Families

Substituted polyhedrons ortho-$C_2B_{10}H_{12}$ fall (according to number of partitions of number 12 into positive integers) into to 77 families: $h^{12}x$, $h^{11}x$, $h^{10}x^2$, $h^{10}xy$, h^9x^3, h^9x^2y, h^9xyz, h^8x^4, ... $h^2xyzuvwfqrt$, hxyzuvwfqrts (denoting possible types of substitutions). With the help of formulas (31.5)– (31.10) it is easy to determine the number of substitution isomers of polyhedron ortho- $C_2B_{10}H_{12}$ of any type, that for a number of families is shown in Table 31.1.

Total number of m substituted ortho- $C_2B_{10}H_{12}$ is calculated by the formula

$$N_m = \sum_{i=0}^{k} N_m^i \ (\text{m} = 1, 2, \ldots) \tag{31.11}$$

Table 31.2 shows distribution of substitution isomers of ortho- $C_2B_{10}H_{12}$ $\boldsymbol{C_{2v}}$ by a number of possible places of substitution m, calculated from (31.5), (31.7), (31.9), (31.11).

31.2 Para-carborane para–$C_2B_{10}H_{12}$ $\boldsymbol{D_{5d}}$

The symmetry group of molecule para–$C_2B_{10}H_{12}$ is $\boldsymbol{D_{5d}}$. Transactions of the molecular symmetry of para-carborane (Fig. 31.2) correspond to cyclic codes: $E \Rightarrow f_1^{12}$; $2C_5 \Rightarrow f_5^2f_1^2$; $2C_5^3 \Rightarrow f_5^2f_1^2$; $5C_2 \Rightarrow f_2^6$; $i \Rightarrow S_2 \Rightarrow f_2^6$; $2S_{10} \Rightarrow f_{10}^1f_2^1$; $2S_{10}^3 \Rightarrow f_{10}^1f_2^1$; $5\sigma_d \Rightarrow 5f_2^4f_1^4$; and formula for determining the symmetry of substitution isomers of para-carborane $\boldsymbol{D_{5d}}$ regardless chirality features will be as follows [10, 11].

$$Z_1(D_{5d}) = 1/20\left\{f_1^{12} + 4f_5^2f_1^2 + 6f_2^6 + 4f_{10}^1f_2^1 + 5f_2^4f_1^4\right\}. \tag{31.12}$$

With symmetry (31.12) and substituting (31.2) in formula, one obtains generating function of type (31.3), coefficients of which are equal to the number of substitution isomers.

For the rotation group $\boldsymbol{D_{5d}}$ (while operations of mirroring of σ_d, и S_n are excluded) cycle index of calculation of the para-carborane $C_2B_{10}H_{12}$ taking into the account mirror isomers will be as follows

$$Z_2(D_5) = 1/10\left(f_1^{12} + 4f_5^2f_1^2 + 5f_2^6\right). \tag{31.13}$$

Table 31.1 A numbers of substitution isomers of ortho–$C_2B_{10}H_{12}$ C_{2v} in families, calculated from (31.5), (31.7), (31.9), (31.10)

Family[a]	Number of isomers: Excluding enantiomers	Subject to enantiomers	Achiral	Chiral pairs
$h^{12}x^0$	1	1	1	0
$h^{11}x$	5	6	4	1
$h^{10}x^2$	23	36	10	13
$h^{10}xy$	39	66	12	27
h^9x^3	65	110	20	45
h^9x^2y	179	330	28	151
h^9xyz	342	660	24	318
h^8x^4	143	255	31	112
h^8x^3y	521	990	52	469
$h^8x^2y^2$	783	1500	66	717
h^8x^2yz	1515	2970	60	1455
h^8xyzu	2982	5940	24	2958
h^7x^5	218	396	40	178
h^7x^4y	1026	1980	72	954
$h^7x^3y^2$	2036	3960	112	1924
h^7x^3yz	4008	7920	96	3912
$h^7x^2y^2z$	6012	11880	144	5868
h^7x^2yzu	11928	23760	96	11832
h^7xyzuv	23760	47520	0	23760
h^6x^6	258	472	44	214
h^6x^5y	1430	2772	88	1342
$h^6x^4y^2$	3554	6960	148	3406
...	...	...	...	...
h^5x^2yzuvw	498960	997920	0	498960
$h^5xyzuvwf$	997920	1995840	0	997920
$h^4x^4y^4$	8802	17370	234	8568
$h^4x^4y^3z$	34806	69300	312	34494
$h^4x^4y^2z^2$	52266	104040	492	51774
...	...	...	...	...
$h^4x^2y^2z^2u^2$	312456	623880	1032	311424
$h^4x^2y^2z^2uv$	624060	1247400	720	623340
$h^4x^2y^2zuvw$	1247544	2494800	288	1247256
$h^4x^2yzuvwf$	2494800	4989600	0	2494800
$h^4xyzuvwfq$	4989600	9979200	0	4989600
$h^3x^3y^3z^3$	92688	184800	576	92112
$h^3x^3y^3z^2u$	277488	554400	576	276912
$h^3x^3y^3zuv$	554400	1108800	0	554400
...	...	...	...	...
$h^2xyzuvwfqrt$	59875200	119750400	0	59875200
hxyzuvwfqrts	119750400	239500800	0	119750400

[a] Each family contains in its structure atoms of one type (h), two (h and x), three (h, x and y), ... etc. 12 different types

Table 31.2 Distributions of achiral isomers and isomers with antipodes of ortho– $C_2B_{10}H_{12}$ $\boldsymbol{C_{2v}}$ by the numbers *m* of possible places of substitution, calculated from (31.5), (31.7), (31.9), (31.11)

The symmetry formulas	Types of subst.	The number *m* of the substitution places								
		m = 1	m = 2	m = 3	m = 4	m = 5	m = 6	…	m = 12	Sum
$Z_1(\boldsymbol{C_{2v}})$	X	5	23	65	143	218	258	…	1	1167
$Z_2(\boldsymbol{C_2})$	X	6	36	110	255	396	472	…	1	2079
$Z^{Achir.}(\boldsymbol{C_{2v}})$	X	4	10	20	31	40	44	…	1	255
$Z_1(\boldsymbol{C_{2v}})$	XY	10	85	488	2111	6560	15192	…	1168	136322
$Z_2(\boldsymbol{C_2})$	XY	12	138	880	3,990	12672	29648	…	2080	266084
$Z^{Achir.}(\boldsymbol{C_{2v}})$	XY	8	32	96	232	448	736	…	256	6560
$Z_1(\boldsymbol{C_{2v}})$	XYZ	15	186	1611	10449	49086	170532	…	136323	4228095
$Z_2(\boldsymbol{C_2})$	XYZ	18	306	2970	20115	96228	337068	…	266085	8390655
$Z^{Achir.}(\boldsymbol{C_{2v}})$	XYZ	12	66	252	783	1944	3996	…	6561	65535

Similar arguments can be made for all the objects described below

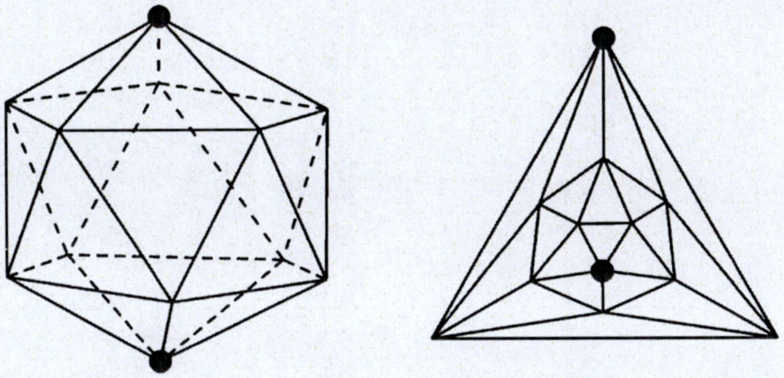

Fig. 31.2 Molecule and the graph of para-carborane $C_2B_{10}H_{12}$ $\boldsymbol{D_{5d}}$

Extracting from the group set $\boldsymbol{D_{5d}}$ second kind operations only, one obtains the formula for the determination of the achiral para-$C_2B_{10}H_{12}$ substitution isomers

$$Z^{A\,chir.}(D_{5d}) = 1/10\left(f_2^6 + 4f_{10}^1 f_2^1 + 5f_2^4 f_1^4\right). \qquad (31.14)$$

The numbers of isomers of substitution para-$C_2B_{10}H_{12}$ $\boldsymbol{D_{5d}}$ in families, calculated from (31.12)–(31.14), is shown in Table 31.3, and distributions of isomers replacement para-$C_2B_{10}H_{12}$ by numbers *m* of possible places of substitution - in Table 31.4.

Table 31.3 The numbers of substitution isomers of para-$C_2B_{10}H_{12}$ $\boldsymbol{D}_{5h}$ in families, calculated from (31.12)–(31.14)

Family	Number of isomers			
	Excluding enantiomers	Subject to enantiomers	Chiral pairs	Achiral
$h^{12}x^0$	1	1	0	1
$h^{11}x$	2	2	0	2
$h^{10}x^2$	10	8	2	6
$h^{10}xy$	14	10	4	6
h^9x^3	22	16	6	10
h^9x^2y	66	40	26	14
h^9xyz	132	72	60	12
h^8x^4	57	37	20	17
h^8x^3y	198	112	86	26
$h^8x^2y^2$	312	174	138	36
h^8x^2yz	594	312	282	30
h^8xyzu	1188	600	588	12
h^7x^5	80	50	30	20
h^7x^4y	396	216	180	36
$h^7x^3y^2$	792	424	368	56
h^7x^3yz	1584	816	768	48
$h^7x^2y^2z$	2376	1224	1152	72
h^7x^2yzu	4752	2400	2352	48
h^7xyzuv	9504	4752	4752	0
h^6x^6	104	64	40	24
h^6x^5y	556	300	256	44
$h^6x^4y^2$	1416	748	668	80
...	...	...	...	...
h^5x^2yzuvw	199584	99792	99792	0
$h^5xyzuvwf$	399168	199584	199584	0
$h^4x^4y^4$	3510	1818	1692	126
$h^4x^4y^3z$	13860	7008	6852	156
$h^4x^4y^2z^2$	20880	10572	10308	264
...	...	...	...	...
$h^4x^2y^2z^2u^2$	124920	62736	62184	552
$h^4x^2y^2z^2uv$	249480	124920	124560	360
$h^4x^2y^2zuvw$	498960	249552	249408	144
$h^4x^2yzuvwf$	997920	498960	498960	0
$h^4xyzuvwfq$	1995840	997920	997920	0
$h^3x^3y^3z^3$	36960	18624	18336	288
$h^3x^3y^3z^2u$	110880	55584	55296	288
$h^3x^3y^3zuv$	221760	110880	110880	0
...	...	...	...	...
$h^2xyzuvwfqrt$	23950080	11975040	11975040	0
$hxyzuvwfqrts$	47900160	23950080	23950080	0

Table 31.4 Distributions of achiral isomers and isomers with antipodes of para- $C_2B_{10}H_{12}$ $\boldsymbol{D}_{5h}$ by the numbers *m* of possible places of substitution, calculated from (31.12)–(31.14)

Symmetry formula	Types of subst.	The number *m* of the substitution places								
		m = 1	m = 2	m = 3	m = 4	m = 5	m = 6	...	m = 12	Sum
$Z_1(\boldsymbol{D}_{5h})$	X	2	8	16	37	50	64	...	1	291
$Z_2(\boldsymbol{D}_5)$	X	1	2	10	22	57	80	...	1	447
$Z^{Achir.}(\boldsymbol{D}_{5h})$	X	2	6	10	17	20	24	...	1	135
$Z_1(\boldsymbol{D}_{5h})$	XY	4	26	112	472	1380	3192	...	292	28448
$Z_2(\boldsymbol{D}_5)$	XY	1	4	34	176	822	2536	...	448	53540
$Z^{Achir.}(\boldsymbol{D}_{5h})$	XY	4	18	48	122	224	384	...	136	3356
$Z_1(\boldsymbol{D}_{5h})$	XYZ	6	54	360	2241	10110	34848	...	28449	856527
$Z_2(\boldsymbol{D}_5)$	XYZ	1	6	72	594	4077	19248	...	53541	1679871
$Z^{Achir.}(D_{5h})$	XYZ	6	36	126	405	972	2052	...	3357	33183

31.3 Metha–Carborane Metha–2,4-$C_2B_{10}H_{12}$ $\boldsymbol{C}_{1v}$

The symmetry group of molecule metha–$C_2B_{10}H_{12}$ is $\boldsymbol{C}_{1v}$. Transactions of the molecular symmetry of metha-carborane (Fig. 31.3) correspond to cyclic codes: $E \Rightarrow f_1^{12}$; $\sigma_v \Rightarrow f_2^4 f_1^4$; and formula for determining the symmetry of substitution isomers of metha-carborane $\boldsymbol{C}_{1v}$ regardless chirality features will be as follows [10–12]

$$Z_1(C_{1v}) = 1/2\{f_1^{12} + f_2^4 f_1^4\}. \tag{31.15}$$

With symmetry (31.15) and substituting (31.2) in formula, one obtains generating function of type (31.3),

$$F(C_{1v}) = (1/2)\{(h + x + \ldots)^{12} + (h^2 + x^2 + \ldots)^4 (h + x + \ldots)^4\}, \tag{31.16}$$

coefficients of which are equal to the number of substitution isomers metha–$C_2B_{10}H_{12}$ regardless chirality features. The formula for the determination of the achiral metha-$C_2B_{10}H_{12}$ substitution isomers

$$Z^{A\,chir.}(C_{1v}) = f_2^4 f_1^4. \tag{31.17}$$

For the rotation group $\boldsymbol{C}_{1v}$ cycle index of calculation of the metha-carborane $C_2B_{10}H_{12}$ taking into the account mirror isomers will be as follows

$$Z_2(C_1) = f_1^{16}. \tag{31.18}$$

Number of $Z^{Chir.par}$ isomers in families $(\pi\text{-}C_5H_5)Fe(\pi\text{-}B_{11}H_{11})$ can be calculated from the ratio

$$Z^{Chir.par}(C_{1v}) = 1/2\left[Z_2(C_1) - Z^{A\,chir.}(C_{1v})\right]. \tag{31.19}$$

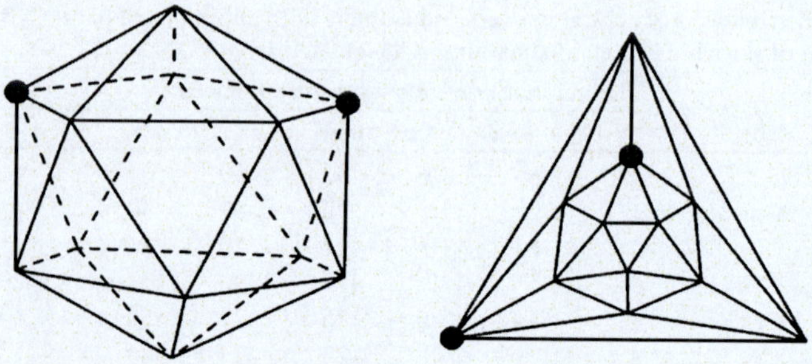

Fig. 31.3 Molecule and the graph of metha-carborane $C_2B_{10}H_{12}$ $\boldsymbol{C}_{1v}$

Table 31.5 The numbers of substitution isomers of metha-$C_2B_{10}H_{12}$ $\boldsymbol{C}_{1v}$ in families, calculated from (31.16), (31.19), (31.17)

	Number of isomers				Number of isomers		
Family	Excluding enantiomers	Chiral pairs	Achir.	Family	Excluding enantiomers	Chiral pairs	Achir.
h^{12}	1	0	1	h^6x^5y	2816	2728	88
$h^{11}x$	8	4*	4	$h^6x^4y^2$	7004	6856	148
$h^{10}x^2$	38	28	10	...	...	...	...
$h^{10}xy$	72	60	12	h^5x^2yzuvw	997920	997920	0
h^9x^3	120	100	20	$h^5xyzuvwf$	1995840	1995840	0
h^9x^2y	344	316	28	$h^4x^4y^4$	17442	17208	234
h^9xyz	672	648	24	$h^4x^4y^3z$	69456	69144	312
h^8x^4	263	232	31	$h^4x^4y^2z^2$	104196	103704	492
h^8x^3y	1016	964	52	...	...	...	...
$h^8x^2y^2$	1518	1452	66	$h^4x^2y^2z^2u^2$	624216	623184	1032
h^8x^2yz	3000	2940	60	$h^4x^2y^2z^2uv$	1247760	1247040	720
h^8xyzu	5952	5928	24	$h^4x^2y^2zuvw$	2494944	2494656	288
h^7x^5	416	376	40	$h^4x^2yzuvwf$	4989600	4989600	0
h^7x^4y	2016	1944	72	$h^4xyzuvwfq$	9979200	9979200	0
$h^7x^3y^2$	4016	3904	112	$h^3x^3y^3z^3$	185088	184512	576
h^7x^3yz	7968	7872	96	$h^3x^3y^3z^2u$	554688	554112	576
$h^7x^2y^2z$	11952	11808	144	$h^3x^3y^3zuv$	1108800	1108800	0
h^7x^2yzu	23808	23712	96	...	...	...	...
h^7xyzuv	47520	47520	0	$h^2xyzuvwfqrt$	119750400	119750400	0
h^6x^6	484	440	44	$hxyzuvwfqrts$	239500800	239500800	0

* Four of the eight X-substituted metha-carborane have enantiomers (4-X-$C_2B_{10}H_9$, 6-X-$C_2B_{10}H_9$, 9-X-$C_2B_{10}H_9$, 10-X-$C_2B_{10}H_9$)

The numbers of isomers of substitution metha-$C_2B_{10}H_{12}$ $\boldsymbol{C}_{1v}$ in families, calculated from (31.16), (31.19), (31.17), is shown in Table 31.5, and distributions of isomers replacement metha-$C_2B_{10}H_{12}$ by numbers m of possible places of substitution - in Table 31.6.

Table 31.6 Distributions of isomers of metha-$C_2B_{10}H_{12}$ $\boldsymbol{C}_{1v}$ by the numbers *m* of possible places of substitution, calculated from (31.15), (31.18), (31.19)

Symmetry formula	Types of subst.	The number *m* of the substitution places m = 1	m = 2	m = 3	m = 4	m = 5	m = 6	...	m = 12	Sum
$Z_1(\boldsymbol{C}_{1v})$	X	8	38	120	263	416	484	...	1	2175
$Z_2(\boldsymbol{C}_1)$	X	12	66	220	495	792	924	...	1	4095
$Z^{Achir.}(\boldsymbol{C}_{1v})$	X	4	10	20	31	40	44	...	1	255
$Z_1(\boldsymbol{C}_{1v})$	XY	16	148	928	4076	12896	29936	...	2176	269000
$Z_2(\boldsymbol{C}_1)$	XY	24	264	1760	7920	25344	59136	...	4096	531440
$Z^{Achir.}(\boldsymbol{C}_{1v})$	XY	8	32	96	232	448	736	...	256	6560
$Z_1(\boldsymbol{C}_{1v})$	XYZ	24	330	3096	20439	97200	338796	...	269001	8421375
$Z_2(\boldsymbol{C}_1)$	XYZ	36	594	5940	40095	192456	673596	...	531441	16777215
$Z^{Achir.}(\boldsymbol{C}_{1v})$	XYZ	12	66	252	783	1944	3996	...	6561	65535

31.4 Para-carborane 1,8-B_6Me_2 Based on the Ditrigonal Dipyramid $\boldsymbol{D}_{3h}$

The symmetry group of molecules of para-carborane 1,8-$B_6Me_2H_8$ (Fig. 31.4) based on ditrigonalnoy dipiramidy $\boldsymbol{D}_{3h}$ (at the base – ditrigon).

12-hexagon ditrigonalnoy dipiramidy is meeting in the hexagonal system. In cross section of it there is ditrigon, crystal symmetry class $\bar{6}m2 = \boldsymbol{D}_{3h}$. Transactions symmetry group $\boldsymbol{D}_{3h}$ molecule of para-carborane 1,8-B_6Me_2 correspond cyclic indices: $E \Rightarrow f_1^8$; $2C_3 \Rightarrow f_3^2f_1^2$; $3C_2 \Rightarrow f_2^3f_1^2$; $\sigma_h \Rightarrow f_2^1f_1^6$; $3\sigma_v \Rightarrow 3f_2^2f_1^4$, $2S_3 \Rightarrow f_3^2f_2^1$. Symmetry formula for determining a number of substitution isomers of para-carborane $\boldsymbol{D}_{3h}$ regardless chirality features will be

$$Z_1(D_{3h}) = 1/12\left(f_1^8 + 2f_3^2f_1^2 + 3f_2^3f_1^2 + f_2^1f_1^6 + 3f_2^2f_1^4 + 2f_3^2f_2^1\right). \quad (31.20)$$

For the rotation group $\boldsymbol{D}_3$ cyclic index for calculation of isomers of the para-carborane 1,8-B_6Me_2 will be written as

$$Z_2(D_3) = 1/6\left(f_1^8 + 2f_3^2f_1^2 + 3f_2^3f_1^2\right). \quad (31.21)$$

Extracting from the group set $\boldsymbol{D}_{3h}$ second kind operations only, one obtains the formula for the determination of the achiral para-carborane 1,8-B_6Me_2 substitution isomers

$$Z^{A\,chir.}(D_{3h}) = 1/6\left(f_2^1f_1^6 + 3f_2^2f_1^4 + 2f_3^2f_2^1\right). \quad (31.22)$$

The numbers of substitution isomers of para-carborane 1,8-B_6Me_2 $\boldsymbol{D}_{3h}$ in families, calculated from (31.20)–(31.22), is shawn in Table 31.7, and the distributions of isomers replacement 1,8-B_6Me_2 by the numbers *m* of possible places of substitution - in Table 31.8.

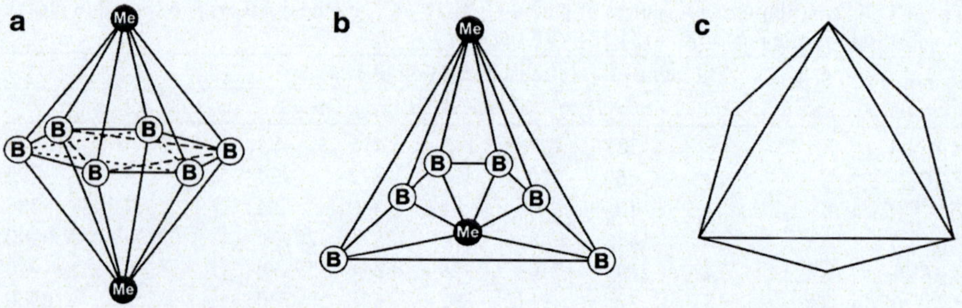

Fig. 31.4 Molecule 1,8-B_6Me_2 (**a**), the graph (**b**), ditrigon (**c**) $\boldsymbol{D_{3h}}$

Table 31.7 Numbers of substitution isomers of para-carborane 1,8-B_6Me_2 $\boldsymbol{D_{3h}}$ in families, calculated from (31.20)–(31.22)

Family	Number of isomers: Excluding enantiomers	Subject to enantiomers	Achiral	Family	Number of isomers: Excluding enantiomers	Subject to enantiomers	Achiral
h^7x	3	3	3	$h^3x^3y^2$	72	100	44
h^6x^2	7	7	7	$h^3x^3y\ z$	126	188	64
h^6xy	11	11	11	$h^3x^2y^2z$	182	286	78
h^5x^3	12	13	11	h^3x^2yzu	332	560	104
h^5x^2y	26	31	21	h^3xyzuv	620	1120	120
h^5xyz	44	56	32	$h^2x^2y^2z^2$	264	432	96
h^4x^4	14	16	12	$h^2x^2y^2zu$	486	846	126
h^4x^3y	40	51	29	h^2x^2yzuv	912	1680	144
$h^4x^2y^2$	56	76	36	$h^2xyzuvw$	1740	3360	120
h^4x^2yz	98	143	53	hxyzuvwf	3360	6720	0
h^4xyzu	176	280	72				

Table 31.8 Distributions of achiral isomers and isomers with antipodes of para-carborane 1,8-B_6Me_2 $\boldsymbol{D_{3h}}$ by the numbers *m* of possible places of substitution, calculated from (31.20)–(31.22)

Symmetry formula	Types of subst.	The number *m* of the substitution places: m = 1	m = 2	m = 3	m = 4	m = 5	m = 6	...	m = 8	Sum
$Z_1(\boldsymbol{D_{3h}})$	X	3	7	12	14	12	7	...	1	59
$Z_2(\boldsymbol{D_3})$	X	3	7	13	16	13	7	...	1	63
$Z^{Achir.}(\boldsymbol{D_{3h}})$	X	3	7	11	12	11	7	...	1	55
$Z_1(\boldsymbol{D_{3h}})$	XY	6	25	76	164	248	250	...	60	989
$Z_2(\boldsymbol{D_3})$	XY	6	25	88	210	328	328	...	64	1241
$Z^{Achir.}(\boldsymbol{D_{3h}})$	XY	6	25	64	118	168	172	...	56	737
$Z_1(\boldsymbol{D_{3h}})$	XYZ	9	54	236	744	1632	2379	...	990	8159
$Z_2(\boldsymbol{D_3})$	XYZ	9	54	281	1011	2367	3540	...	1242	11519
$Z^{Achir.}(\boldsymbol{D_{3h}})$	XYZ	9	54	191	477	897	1218	...	738	4799

31.5 Dekagidroklovodekaborat – ion $B_{10}H_{10}^{2-}$ Based on Bimodal Square Antiprism D_{4d}

Dekagidroklovodekaborat – ion $B_{10}H_{10}^{2-}$ is a polyhedron of bimodal square antiprism $\boldsymbol{D}_{4d}$ (Fig. 31.5) in combination with 2 square pyramids $\rho(10) = 42$ [13].

The symmetry group of polyhedron of doubly charged anion $B_{10}H_{10}^{2-}$ is $\boldsymbol{D}_{4d}$. The cyclic indices corresponding to the symmetry operations are: $E \Rightarrow f_1^{10}$, $C_2 \Rightarrow f_1^2 f_2^4, 2C_4 \Rightarrow 2f_1^2 f_4^2$, $4C_2 \Rightarrow 4f_2^5$, $4\sigma_d \Rightarrow 4f_1^4 f_2^3$, $2S_8 \Rightarrow 2f_2^1 f_8^1$, $4\sigma_d \Rightarrow 4f_1^4 f_2^3$ and the symmetry formula without taking into account chirality features is

$$Z_1(D_{4d}) = 1/16\left(f_1^{10} + 4f_1^4 f_2^3 + 2f_1^2 f_4^2 + f_1^2 f_2^4 + 4f_2^1 f_8^1 + 4f_2^5\right), \tag{31.23}$$

For the rotation group $\boldsymbol{D}_4$ the symmetry formula taking into the account mirror isomers of the anion $B_{10}H_{10}^{2-}$ will be written as

$$Z_2(D_4) = 1/8\left(f_1^{10} + f_1^2 f_2^4 + 2f_1^2 f_4^2 + 4f_2^5\right), \tag{31.24}$$

Extracting from the group set $\boldsymbol{D}_{4d}$ second kind operations only, one obtains formula for determination of the achiral charged anion $B_{10}H_{10}^{2-}$ substitution isomers

$$Z^{\text{A chir.}}(D_{4d}) = 1/8\left(4f_1^4 f_2^3 + 4f_2^1 f_8^1\right), \tag{31.25}$$

The numbers of substitution isomers of anion $B_{10}H_{10}^{2-}$ D_{4d} in some families, calculated from (31.23)–(31.25), given in Table 31.9, and the distributions of isomers replacement anion $B_{10}H_{10}^{2-}$ by numbers m of possible places of substitution – in Table 31.10.

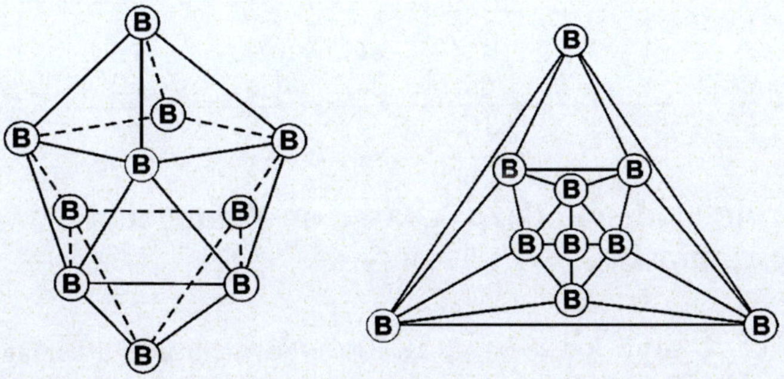

Fig. 31.5 Molecule and the graph of doubly charged anion $B_{10}H_{10}^{2-}$ D_{4d}

Table 31.9 Numbers of substitution isomers of anion $B_{10}H_{10}^{2-}$ $\boldsymbol{D}_{4d}$ in families, calculated from (31.23)–(31.25)

Family	Number of isomers: Excluding enantiomers	Subject to enantiomers	Achiral	Family	Number of isomers: Excluding enantiomers	Subject to enantiomers	Achiral
h^{10}	1	1	1	$h^4x^3y^3$	288	528	288
h^9x	2	2	2	$h^4x^3y^2z$	819	1578	819
h^8x^2	7	9	7	h^4x^3yzv	1593	3150	1593
h^8xy	9	12	9	$h^4x^2y^2z^2$	1242	2400	1242
h^7x^3	12	16	12	$h^4x^2y^2zv$	2400	4728	2400
h^7x^2y	29	46	29	h^4x^2yzvw	4743	9450	4743
h^7xyz	51	90	51	$h^4xyzvwf$	9450	18900	9450
h^6x^4	22	33	22	$h^3x^3y^3z$	1086	2100	1086
h^6x^3y	63	106	63	$h^3x^3y^2z^2$	1626	3156	1626
$h^6x^2y^2$	97	170	97	$h^3x^3y^2zv$	3186	6300	3186
h^6x^2yz	170	316	170	h^3x^3yzvw	6300	12600	6300
h^6xyzv	321	630	321	$h^3x^2y^2z^2v$	4788	9456	4788
h^5x^5	23	34	23	$h^3x^2y^2zvw$	9486	18900	9486
h^5x^4y	92	160	92	h^3x^2yzvwf	18900	37800	18900
$h^5x^3y^2$	177	318	177	$h^3xyzvwfq$	37800	75600	37800
h^5x^3yz	333	630	333	$h^2x^2y^2z^2v^2$	7215	14250	7215
$h^5x^2y^2z$	498	948	498	$h^2x^2y^2z^2vw$	14250	28356	14250
h^5x^2yzv	963	1890	963	$h^2x^2y^2zvwf$	28386	56700	28386
h^5xyzvw	1890	3780	1890	$h^2x^2yzvwfq$	56700	113400	56700
$h^4x^4y^2$	226	413	226	$h^2xyzvwfqr$	113400	226800	113400
h^4x^4yz	413	790	413	hxyzvwfqrs	226800	453600	226800

Table 31.10 Distributions of achiral isomers and isomers with antipodes anion $B_{10}H_{10}^{2-}$ D_{4d} by numbers *m* of possible places of substitution, calculated from (31.23) and (31.24)

Symmetry formulas	Types of subst.	The number *m* of the substitution places: m = 1	m = 2	m = 3	m=4	m = 5	m = 6	...	m = 10	Sum
$Z_1(\boldsymbol{D}_{4d})$	X	2	7	12	22	23	22	...	1	110
$Z_2(\boldsymbol{D}_{4})$	X	2	9	16	33	34	33	...	1	155
$Z_1(\boldsymbol{D}_{4d})$	XY	4	23	82	267	584	968	...	111	4355
$Z_2(\boldsymbol{D}_{4})$	XY	4	30	124	448	1024	1740	...	156	7613
$Z_1(\boldsymbol{D}_{4d})$	XYZ	6	48	261	1245	4176	10233	...	4356	70179

31.6 Borohydride ion $B_7H_7^{2-}$ Based on Pentagonal Dipiramid $\boldsymbol{D}_{5h}$

Structure of borohydride ion $B_7H_7^{2-}$ (Fig. 31.6). Polyhedron is a pentagonal symmetry dipiramida $\boldsymbol{D}_{5h}$ [13, 14]

The symmetry group of borohydride ion $B_7H_7^{2-}$ is $\boldsymbol{D}_{5h}$. The cyclic indices corresponding to the symmetry operations are: $E \Rightarrow f_1^7$, $2C_5 \Rightarrow 2f_5^1f_1^2$,

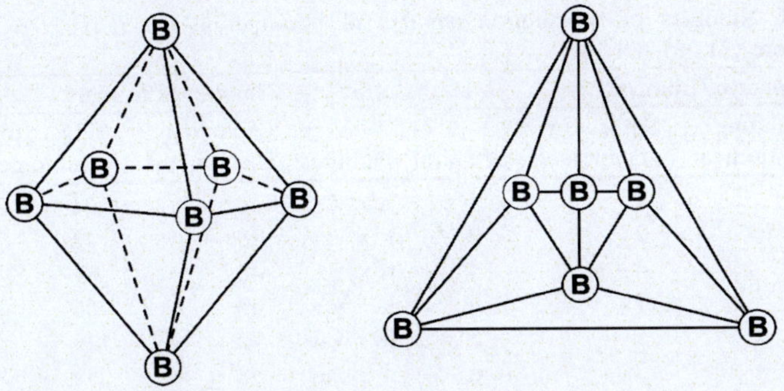

Fig. 31.6 Molecule and the graph of borohydride ion $B_7H_7^{2-}$ D_{5h}

$2C_5^2 \Rightarrow 2f_5^1f_1^2,\ 5C_2 \Rightarrow 5f_2^3f_1^1,\ \sigma_h \Rightarrow f_2^1f_1^5,\ 2S_5 \Rightarrow 2f_5^1f_2^1,\ 2S_5^3 \Rightarrow 2f_5^1f_2^1,\ 5\sigma_v \Rightarrow 5f_2^2f_1^3$ and the symmetry formula regardless chirality features is as follows

$$Z_1(D_{5h}) = 1/20\left(f_1^7 + 4f_5^1f_1^2 + 5f_2^3f_1^1 + f_2^1f_1^5 + 4f_5^1f_2^1 + 5f_2^2f_1^3\right); \qquad (31.26)$$

For the rotation group D_5 the symmetry formula taking into the account mirror isomers of the ion $B_7H_7^{2-}$ will be as follows

$$Z_2(D_5) = 1/10\left(f_1^7 + 4f_5^1f_1^2 + 4f_2^3f_1^1\right); \qquad (31.27)$$

Extracting from the group set D_{5h} second kind operations only, one obtains formula for determination of the achiral ion $B_7H_7^{2-}$ substitution isomers

$$Z^{\text{A chir.}}(D_{5h}) = 1/10\left(f_2^1f_1^5 + 4f_5^1f_2^1 + 5f_2^2f_1^3\right); \qquad (31.28)$$

The numbers of substitution isomers of ion $B_7H_7^{2-}$ D_{5h} in some families, calculated from (31.26)–(31.28), are shown in Table 31.11, and the distributions of isomers replacement ion $B_7H_7^{2-}$ by numbers $m = 7$ of possible places of substitution are shown in Table 31.12.

31.7 Disubstituted Carborane $C_2B_8H_{10}$ Based on Bipyramidal Triangular Dodecahedron C_{2v}

If the opposite faces of a triangular dodecahedron D_{2d} form three-pitched roofs, one obtains a triangular bipyramidal dodecahedron C_{2v} (Fig. 31.7) [15, 16].

The symmetry group of disubstituted carborane $C_2B_8H_{10}$ is C_{2V}. The cyclic indices corresponding to the symmetry operations are: $E \Rightarrow f_1^{10}$, $C_2 \Rightarrow f_2^5$, $\sigma_v \Rightarrow f_2^3f_1^4$, $\sigma_v \Rightarrow f_2^2f_1^6$ and the symmetry formula regardless chirality features is

$$Z_1(C_{2v}) = 1/4\left(f_1^{10} + f_2^5 + f_2^3f_1^4 + f_2^2f_1^6\right). \qquad (31.29)$$

Table 31.11 Numbers of substitution isomers of borohydride ion $B_7H_7^{2-}$ D_{5h} in families, calculated from (31.26)–(31.28)

Family	Number of isomers Excluding enantiomers	Subject to enantiomers	Achiral	Families	Number of isomers Excluding enantiomers	Subject to enantiomers	Achiral
h^7	1	1	1	$h^3x^2y^2$	18	24	12
h^6x	2	2	2	h^3x^2yz	28	42	14
h^5x^2	4	4	4	h^3xyzu	48	84	12
h5xy	5	5	5	$h2x^2y^2z$	42	66	18
h^4x^3	5	5	5	$h2x^2yzu$	72	126	18
h^4x^2y	10	12	8	h^2xyzuv	132	252	12
h^4xy	15	21	9	hxyzuvw	252	504	0
h^3x^3y	12	14	10				

Table 31.12 Distribution of substitution isomers of borohydride ion $B_7H_7^{2-}$ D_{5h} by numbers m of possible places of substitution, calculated from (31.26) and (31.27)

Symmetry formulas	Types of subst.	The number m of the substitution places m = 1	m = 2	m = 3	m = 4	m = 5	m = 6	m = 7	Sum
$Z_1(\boldsymbol{D}_{5h})$	X	2	4	5	5	4	2	1	23
$Z_2(\boldsymbol{D}_5)$	X	2	4	5	5	4	2	1	23
$Z_1(\boldsymbol{D}_{5h})$	XY	4	13	30	52	64	46	24	234
$Z_2(\boldsymbol{D}_5)$	XY	4	13	34	62	80	52	24	270
$Z_1(\boldsymbol{D}_{5h})$	XYZ	6	27	90	225	390	387	234	1360
$Z_2(\boldsymbol{D}_5)$	XYZ	6	27	108	297	552	531	270	1792

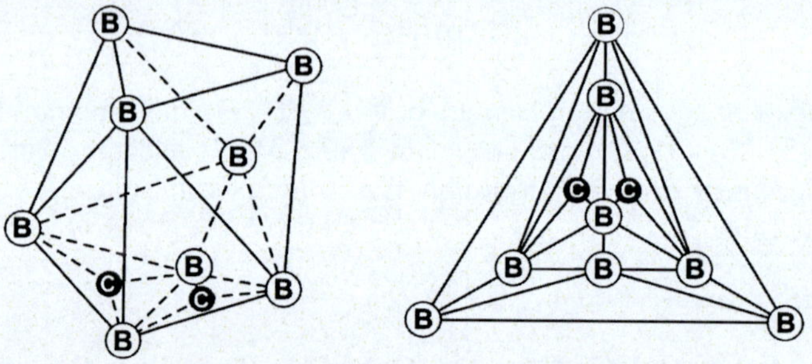

Fig. 31.7 Molecule and the graph of disubstituted carborane $C_2B_8H_{10}$ $\boldsymbol{C}_{2v}$

For the rotation group $\boldsymbol{C}_2$ the symmetry formula taking into the account mirror isomers of the disubstituted carborane $C_2B_8H_{10}$ will be as follows

$$Z_2(C_2) = 1/2\left(f_1^{10} + f_2^5\right). \tag{31.30}$$

Extracting from the group set $\boldsymbol{C}_{2v}$ second kind operations only, one obtains formula for determination of the achiral $C_2B_8H_{10}$ substitution isomers

$$Z^{\text{A chir.}}(C_{2v}) = 1/2\left(f_2^3 f_1^4 + f_2^2 f_1^6\right). \tag{31.31}$$

Substituted carborane (Fig. 31.7) fall into $\rho(m) = 42$ families by number of partitions of the number 10 into positive parts: $h^{10}, h^9x, h^8x^2, h^8xy, h^7x^3, h^7x^2y, h^7xyz, h^6x^4, h^6x^3y, h^6x^2y^2$, h6$x^2yz$, $h^6xyzu, h^5x^5, h^5x^4y, h^5x^3y^2, h^5x^3yz, h^5x^2y^2z, h^5x^2yzu, h^5xyzuv, h^4x^4y^2, h^4x^4yz, h^4x^3y^3, h^4x^3y^2z, h^4x^3yzu, h^4x^2y^2z^2, h^4x^2y^2zu, h^4x^2yzuv, h^4xyzuv, h^3x^3y^3z, h^3x^3y^2z^2, h^3x^3y^2zu, h^3x^3yzuv, h^3x^2y^2z^2u, h^3x^2y^2zuv, h^3x^2yzuvw, h^3xyzuvwf, h^2x^2y^2z^2u^2, h^2x^2y^2z^2uv, h^2x^2y^2zuvw, h^2x^2yzuvwf, h^2xyzuvwfq, hxyzuvwfqr$, containing from (31.24), respectively 1, 5, 19, 33, 42, 114, 216, 72, 250, 370, 702, 1,356, 82, 366, 706,1,368, 2,022, 3,948, 7,740, 882, 1,698, 1,154, 3,330, 6,528, 4,962, 9,726, 19,188, 37,980, 4,416, 6,568, 12,936, 25,560, 19,320, 38,256, 75,960, 151,200, 28,920, 57,312, 113,976, 227,160, 453,600, 907,200 isomers, among which, respectively 1, 5, 13, 21, 24, 48, 72, 34, 80, 100, 144, 192, 38, 102, 152, 216, 264, 336, 360, 174, 246, 208, 360, 456, 444, 552, 576, 360, 432, 536, 672, 720, 840, 912, 720, 0, 1,080, 1,224, 1,152, 720, 0, there are 0 achiral isomers.

31.8 Borohydride ion $B_9H_9^{2-}$ Based on Trimorph Trigonal Prism D_{3h}

The structure of borohydride ion $B_9H_9^{2-}$ (Fig. 31.8) is a trimorph trigonal prism with symmetry $\boldsymbol{D}_{3h}$.

The symmetry group of borohydride ion $B_9H_9^{2-}$ is $\boldsymbol{D}_{3h}$. The cyclic indices corresponding to the symmetry operations are: $E \Rightarrow f_1^9$, $2C_3 \Rightarrow 2f_3^3$, $3C_2 \Rightarrow 3f_2^4f_1^1$, $\sigma_h \Rightarrow f_2^3f_1^3$, $2S_3 \Rightarrow 2f_3^1f_6^1$, $3\sigma_v \Rightarrow 3f_2^3f_1^3$ and the symmetry formula regardless chirality features is

$$Z_1(D_{3h}) = 1/12\left(f_1^9 + 2f_3^3 + 3f_1^1f_2^4 + 4f_1^3f_2^3 + 2f_3^1f_6^1\right). \qquad (31.32)$$

For the rotation group $\boldsymbol{D}_3$ the symmetry formula taking into the account mirror isomers of the ion $B_9H_9^{2-}$ will be as follows

$$Z_2(D_3) = 1/6\left(f_1^9 + 2f_3^3 + 3f_2^4f_1^1\right); \qquad (31.33)$$

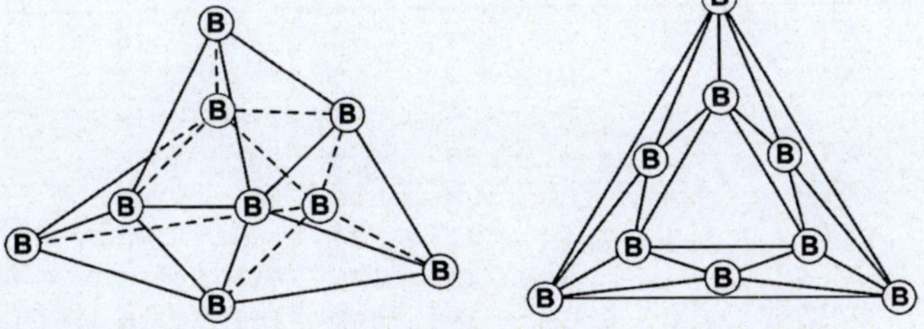

Fig. 31.8 Molecule and the graph of borohydride ion $B_9H_9^{2-}$ D_{3h}

Extracting from the group set $\boldsymbol{D}_{3h}$ second kind operations only, one obtains formula for determination of the achiral ion $B_9H_9^{2-}$ substitution isomers

$$Z^{\text{A chir.}}(D_{3h}) = 1/6\left(4f_2^3f_1^3 + 2f_3^1f_6^1\right). \quad (31.34)$$

Number of $Z^{\text{Chir.par}}$ isomers $B_9H_9^{2-}$ in families can be calculated from the ratio

$$Z^{\text{Chir.par}}(D_{3h}) = 1/2\left[Z(\mathbf{D}_3) - Z^{\text{A chir.}}(D_{3h})\right]. \quad (31.35)$$

The numbers of substitution isomers of ion $B_9H_9^{2-}$ D_{3h} in some families, calculated from (31.32), (31.35), (31.34) is shown in Table 31.13, and distributions of substitution ion $B_9H_9^{2-}$ by numbers *m* of possible places of substitution is shown in Table 31.14.

Table 31.13 Numbers of substitution isomers of ion $B_9H_9^{2-}$ (trimorph trigonal prism $\boldsymbol{D}_{3h}$) in some families, calculated from (31.32), (31.35), (31.34)

	Number of isomers				Number of isomers		
Family	Excluding enantiomers	Chiral pairs	Achiral	Family	Excluding enantiomers	Chiral pairs	Achiral
h^9	1	0	1	$h^4x^2y^2z$	330	306	24
h^8x	2	0	2	h^4x^2yzu	636	624	12
h^7x^2	6	2	4	h^4xyzuv	1260	1260	0
h^7xy	8	4	4	$h^3x^3y^3$	153	129	24
h^6x^3	12	5	7	$h^3x^3y^2z$	432	408	24
h^6x^2y	26	18	8	h^3x^3yzu	840	840	0
h^5xyz	44	40	4	$h^3x^2y^2z^2$	656	616	40
h^5x^4	16	8	8	$h^3x^2y^2zu$	1272	1248	24
h^5x^3y	48	36	12	h^3x^2yzuv	2520	2520	0
$h^5x^2y^2$	74	58	16	$h^3xyzuvw$	5040	5040	0
h^5x^2yz	132	120	12	$h^2x^2y^2z^2v$	1920	1872	48
h^5xyzu	252	252	0	$h^2x^2y^2zvw$	3792	3768	24
h^4x^4y	60	48	12	h^2x^2yzvwf	7560	7560	0
$h^4x^3y^2$	118	98	20	$h^2xyzvwfq$	15120	15120	0
h^4x^3yz	216	204	12	hxyzvwfqr	30240	30240	0

Table 31.14 Distributions of substitution isomers of ion $B_9H_9^{2-}$ D_{3h} by numbers *m* of possible places of substitution, calculated from (31.32), (31.35), (31.34)

Symmetry formulas	Types of subst.	The number *m* of the substitution places								
		m = 1	m = 2	m = 3	m = 4	m = 5	m = 6	...	m = 9	Sum
$Z_1(\boldsymbol{D}_{3h})$	X	1	2	6	12	16	16	...	1	73
$Z^{\text{Chir.par}}(\boldsymbol{D}_{3h})$	X	0	0	2	5	8	8	...	0	29
$Z^{\text{Achir.}}(\boldsymbol{D}_{3h})$	X	1	2	4	7	8	8	...	1	43
$Z_1(\boldsymbol{D}_{3h})$	XY	1	4	20	76	202	388	...	74	1949
$Z^{\text{Chir.par}}(\boldsymbol{D}_{3h})$	XY	0	0	8	46	146	308	...	30	1460
$Z^{\text{Achir.}}(\boldsymbol{D}_{3h})$	XY	1	4	12	30	56	80	...	44	488
$Z_1(\boldsymbol{D}_{3h})$	XYZ	1	6	42	236	954	2754	...	1950	23479
$Z^{\text{Chir.par}}(\boldsymbol{D}_{3h})$	XYZ	0	0	18	163	774	2430	...	1461	20743
$Z^{\text{Achir.}}(\boldsymbol{D}_{3h})$	XYZ	1	6	24	73	180	324	...	489	2735

31.9 Hypothetical Model of Carbometallic Derivative of Boranes

(π-C_5H_5)Fe(π-$B_{11}H_{11}$) C_{5v} For (π-C_5H_5)Fe(π-$B_{11}H_{11}$) (Fig. 31.9) the symmetry group is C_{5v}: E, $4C_5$, $5\sigma_V$. The cyclic indices corresponding to the symmetry operations are: $E \Rightarrow f_1^{16}$, $4C_5 \Rightarrow f_1^1 f_5^3$, $5\sigma_v \Rightarrow f_1^4 f_2^6$.

Cyclic indices of the symmetry group C_{5v} to determine the number of isomers (π-C_5H_5)Fe(π-$B_{11}H_{11}$) regardless chirality feature has the form

$$Z_1(C_{5v}) = 1/10\left(f_1^{16} + 5f_1^4 f_2^6 + 4f_1^1 f_5^3\right). \tag{31.36}$$

For the rotation group C_5 the symmetry formula taking into the account mirror isomers of the (π-C_5H_5)Fe(π-$B_{11}H_{11}$) will be as follows

$$Z_2(C_5) = 1/5\left(f_1^{16} + 4f_1^1 f_5^3\right). \tag{31.37}$$

To determine the number of achiral isomers (only the operation of the second kind), formula of symmetry has the form:

$$Z^{\text{A chir.}}(C_{5v}) = f_1^4 f_2^6. \tag{31.38}$$

Number of $Z^{\text{Chir.par}}$ isomers in families (π-C_5H_5)Fe(π-$B_{11}H_{11}$) can be calculated from the ratio

$$Z^{\text{Chir.par}}(C_{5v}) = 1/2\left[Z(C_{5v}) - Z^{\text{A chir.}}(C_{5v})\right]. \tag{31.39}$$

Substituted (π-C_5H_5)Fe(π-$B_{11}H_{11}$) fall into $\rho(16) = 231$ family. The results of the calculation of X-, XY-, ... substituted (π-C_5H_5)Fe(π-$B_{11}H_{11}$) in the families listed in Table 31.15, and the distribution of substitution isomers (π-C_5H_5)Fe(π-$B_{11}H_{11}$) C_{5v} by number m of possible sites of substitution is shown in Table 31.16.

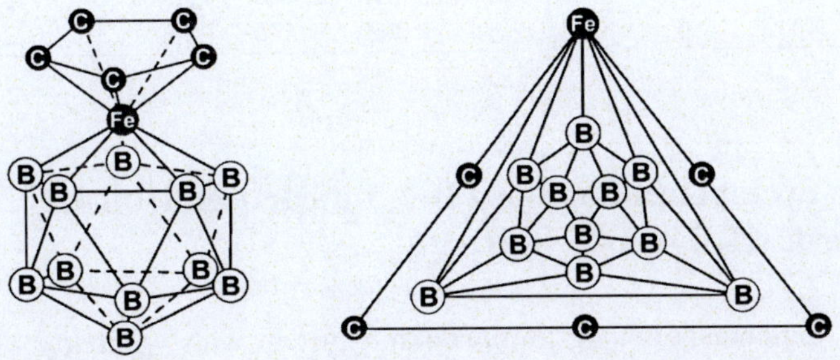

Fig. 31.9 Molecule and the graph of carbometallic derivative of boranes (π-C_5H_5)Fe(π-$B_{11}H_{11}$) C_{5v}

Table 31.15 Numbers of substitution isomers of (π-C_5H_5)Fe(π-$B_{11}H_{11}$) C_{5v} in some families, calculated from (31.36), (31.38), (31.39)

Families	$Z(C_{5v})$	$Z^{Chir.par}(C_{5v})$	$Z^{Achir.}$	Families	$Z(C_{5v})$	$Z^{Chir.par}(C_{5v})$	$Z^{Achir.}$
h^{16}	1	0	1	$h^{10}x^4y^2$	12,198	11,826	372
$h^{15}x$	4	0	4	$h^{10}x^4yz$	24,150	23,898	252
$h^{14}x^2$	18	6	12	$h^{10}x^3y^3$	16,220	15,812	408
$h^{14}xy$	30	18	12	$h^{10}x^3y^2z$	48,276	47,820	456
$h^{13}x^3$	70	42	28	$h^{10}x^3yzu$	96,168	96,024	144
$h^{13}x^2y$	186	150	36	$h^{10}x^2y^2z^2$	72,456	71,688	768
$h^{13}xyz$	348	324	24	$h^{10}x^2y^2zu$	144,396	143,892	504
$h^{12}x^4$	208	156	52	$h^{10}x^2yzuv$	288,360	288,216	144
$h^{12}x^3y$	766	690	76	$h^{10}xyzuvw$	576,576	576,576	0
$h^{12}x^2y^2$	1,146	1,038	108	h^9x^7	1,214	1,074	140
$h^{12}x^2yz$	2,226	2,142	84	h^9x^6y	8,138	7,878	260
$h^{12}xyzu$	4,380	4,356	24	$h^9x^5y^2$	24,294	23,754	540
$h^{11}x^5$	480	396	84	h^9x^5yz	48,228	47,868	360
$h^{11}x^4y$	2,250	2,118	132	$h^9x^4y^3$	40,370	39,710	660
$h^{11}x^3y^2$	4,476	4,260	216	$h^9x^4y^2z$	120,510	119,730	780
$h^{11}x^3yz$	8,808	8,664	144	h^9x^4yzu	240,420	240,060	360
$h^{11}x^2y^2z$	13,236	12,972	264	$h^9x^3y^3z$	160,520	159,800	720
$h^{11}x^2yzu$	26,280	26,136	144	$h^9x^3y^2z^2$	240,900	239,580	1,320
$h^{11}xyzuv$	52,416	52,416	0	$h^9x^3y^2zu$	480,840	480,120	720
$h^{10}x^6$	860	744	116	h^9x^3yzuv	960,960	960,960	0
$h^{10}x^5y$	4,908	4,704	204				

Table 31.16 Distribution of substitution isomers (π-C_5H_5)Fe(π-$C_2B_9H_{11}$) C_{5v} by numbers m of possible places of substitution, calculated from (31.37), (31.36), (31.38)

Symmetry formulas	Types of subst.	The number m of the substitution places								
		m = 1	m = 2	m = 3	m = 4	m = 5	m = 6	...	m = 16	Sum
$Z_1(C_{5v})$	X	4	18	70	208	480	860	...	1	7,071
$Z_2(C_5)$	X	4	24	112	364	876	1604	...	1	13,118
$Z^{Achir.}(C_{5v})$	X	4	12	28	52	84	116	...	1	1,023
$Z_1(C_{5v})$	XY	0	30	372	2,678	13,452	50,432	...	7,070	4,320,085
$Z_2(C_5)$	XY	0	48	672	5,096	26,208	99,304	...	13,118	8,583,168
$Z^{Achir.}(C_{5v})$	XY	0	12	72	260	696	1,560	...	1,022	57,000

31.10 π-cyclopentadienil-π-(1)-2,3-dicarbolliliron-III (π-C_5H_5)Fe(π-$C_2B_9H_{11}$) C_s

Crystals of compounds belong to monoclinic system with symmetry $P2_1/c$, with four molecules per unit cell. The structure of π-cyclopentadienil-π-(1)-2,3-dicarbolliliron-III (Fig. 31.10) was studied in [17] using X-ray diffraction.

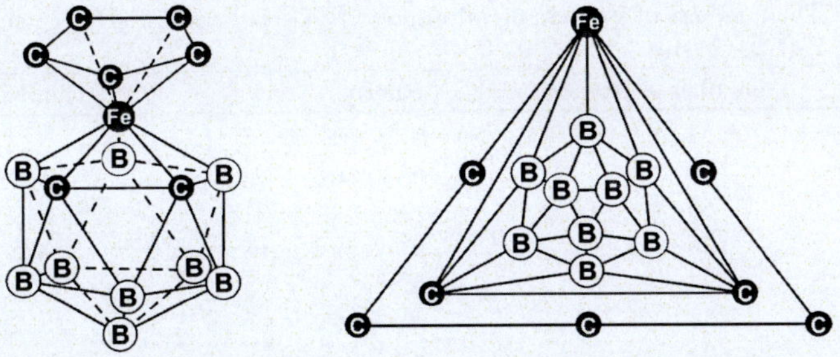

Fig. 31.10 Molecule and the graph of π-cyclopentadienil-π-(1)-2,3-dicarbolliliron-III (π-C5H5) Fe(π-C2B9H11) Cs

The symmetry operations of the group $\boldsymbol{C_s}$ (E, C_s) induce in the places of possible substitutions (π-C_5H_5)Fe(π-$C_2B_9H_{11}$) following permutation: $E \Rightarrow f_1^{16}$, $C_s \Rightarrow f_2^6 f_1^4$. Symmetry formulas $\boldsymbol{C_s}$ for determination of number of substitution isomers(π-C_5H_5)Fe(π-$C_2B_9H_{11}$), excluding chirality features will be as follows:

$$Z_1(C_s) = 1/2\left(f_1^{16} + f_2^6 f_1^4\right). \tag{31.40}$$

Using substitutions (31.2) in formula $Z_2(C_s)$ one obtains the generating function

$$F(C_s) = 1/2\{(h + x + \ldots)^{16} + (h^2 + x^2 + \ldots)^6 (h + x\ldots)^4\}, \tag{31.41}$$

in which coefficients (1, A, B, ...) (after activation of similar) are equal to the number of achiral substitution isomers of type $h^{\alpha}x^{\beta}y^{\gamma}$... .

Substituted (π-C_5H_5)Fe(π-$B_9C_2H_{11}$) also fall into $\rho(16) = 231$ family.

The results of the calculation of X-, XY-, ... substituted (π-C_5H_5)Fe(π-$C_2B_9H_{11}$) $\boldsymbol{C_s}$ in the families listed in Table 31.17, and distribution of isomers of $\boldsymbol{C_s}$ by number m of the substitution places is shown in Table 31.18.

31.11 Additive Scheme for Calculation of the Properties of para-$C_2H_{2\text{-}k}X_kB_{10}H_{10\text{-}l}X_l$ on the Basis of Partitioning Numbers of Pascal Triangle

Calculation scheme for properties ($\Delta_f H^0, S_f^0, \ldots$) of X-substituted para-$C_2H_{2\text{-}k}X_kB_{10}H_{10\text{-}l}X_l$ can obtained on the basis of partitioning triangular (K_3) Pascal triangle numbers taking into account valence interactions, pair nonvalent interactions over two atoms, over three and four atoms in the skeleton of the polyhedron

Table 31.17 The numbers of isomers of substitution (π-C_5H_5)Fe(π-$C_2B_9H_{11}$) in some families, calculated by $Z_1(\boldsymbol{C}_s)$, (31.40)

Family	Number of isomers	Family	Number of isomers
h^{16}	1	$h^{11}x^2yzu$	131112
$h^{15}x$	10	$h^{11}xyzuv$	262080
$h^{14}x^2$	66	…	…
$h^{14}xy$	126	$h^4xyzuvwfgqrts$	435,891,456,000
$h^{13}x^3$	294	$h^3x^3y^3z^3u^3v$	1,345,344,000
$h^{13}x^2y$	858	$h^3x^3y^3z^3u^2v^2$	2,018,016,000
$h^{13}xyz$	1,692	$h^3x^3y^3z^3u^2vw$	4,036,032,000
$h^{12}x^4$	936	$h^3x^3y^3z^3uvwf$	8,072,064,000
$h^{12}x^3y$	3,678	$h^3x^3y^3z^2u^2v^2w$	6,054,048,000
$h^{12}x^2y^2$	5,514	$h^3x^3y^3z^2u^2vwf$	12,108,096,000
$h^{12}x^2yz$	10,962	$h^3x^3y^3z^2uvwfg$	2,421,6192,000
$h^{12}xyzu$	21,852	$h^3x^3y^3zuvwfgq$	48,432,384,000
$h^{11}x^5$	2,226	$h^3x^3y^2z^2u^2v^2w^{\ 2}$	9,081,079,200
$h^{11}x^4y$	10,986	…	…
$h^{11}x^3y^2$	21,948	$h^2x^2yzuvwfgqrtsk$	2,615,348,736,000
$h^{11}x^3y\ z$	43,752	$h^2xyzuvwfgqrtskm$	5,230,697,472,000
$h^{11}x^2y^{\ 2}z$	65,652	hxyzuvwfgqrtskmn	10,461,394,944,000

Table 31.18 Distribution of achiral substitution isomers (π-C_5H_5)Fe(π-$C_2B_9H_{11}$) $\boldsymbol{C}_s$ by numbers *m* of possible places of substitution, calculated from (31.40)

	The number *m* of the substitution places					
Types of subst.	m = 1	m = 2	m = 3	m = 4	…	Sum
X	16	120	560	1,820	…	65,535
XY	32	480	4480	29,120	…	43,046,720
XYZ	48	1,080	15,120	147,420	…	4,294,967,295
XYZU	64	1,920	35,840	465,920	…	152,587,890,624
all subst. different	256	30,720	2,293,760	119,275,520	…	48,661,191,875,666,899,999

[10, 13, 18]. The 1,12-$C_2B_{10}H_{12}$ carborane has the form of icosahedron $\boldsymbol{D}_{5d}$. Pascal triangle row elements (C_n^m, $m \leq n$) were used [18, 19] to describe the structural elements of each molecule and building of the additive scheme matrix for a number of substitution isomers of the basic compound with specified symmetry ($\boldsymbol{D}_{5d}$). Each coefficient of the scheme presents the number of superpositions of the given subgraph onto the molecular graph [19, 20]. The triangle column elements are structure invariants. If the property (***P***) of X-substituted is the sum of contributions of the structural elements, then [21, 22]

$$P = C_n^0 p_0 + C_n^1 p_1 + \ldots + C_n^{n-1} p_{n-1} + C_n^n p_n, \tag{31.42}$$

where $p_0, p_1, p_2, \ldots$ are the parameters, and $C_n^0, C_n^1, \ldots$ are the coefficicients, where $C_n^2, C_n^3, C_n^4, \ldots$ are the numbers K_3, etc. When splitting C_n^1, C_n^2, C_n^3 in (31.42) one obtains [13] scheme (in the pair approximation) for the calculation of the properties of ***P*** 292 of X-Substituted para-carborane.

In the graph interpretation of the method of construction of additive schemes is based on decomposition of polygonal numbers of Pascal's triangle [19, 20]. In particular, the property *P* for each molecule of the homologous series can be represented as a linear function of number of structural elements (nodes, paths of length one, two, three, etc.), the sum of which is equal to the number of triangular $K_3 = n\ (n+1)//2$, where $n = 1,2,3$ .... In the pair approximation for all heteroatomic molecular graphs of para-carborane all the (heterogeneous) subgraphs of length two, the length of three, ... etc. are written down. Molecular graphs themselves are also considered subgraphs, because, for example, first 11 counts of molecules X-substituted para-1,12-$C_2B_{10}H_{12}$ (X = CH_3, F, ...) are subgraphs of the rest 239 (see Table 31.19). Thus, for example, counts 1,2-X_2-$C_2B_{10}H_{10}$ and 2,3-X_2-$C_2B_{10}H_{10}$ are subgraphs of the molecule 1,2,3-X_3-$C_2B_{10}H_9$ etc. Each factor of the scheme (in other words, the number of ways of imposing a subgraph of a certain length (type) i_1, i_2, ... in the MG) is the result of decomposition of triangular numbers [19, 20].

Table 31.19 Coefficients of the additive scheme (31.45) for calculation of features of X-substituted para-carborane 1,12-$C_2B_{10}H_{12}$ excluding chirality features (X = CH_3, C_2H_5, F, ...)

	Coefficients of the additive scheme (31.40)											
Substituted of para-carborane	1	2	3	4	5	6	7	8	9	10	*11*	K_3
$C_2B_{10}H_{12}$	1	0	0	0	0	0	0	0	0	0	0	0
1-X-$C_2B_{10}H_{11}$	1	1	0	0	0	0	0	0	0	0	0	0
2-X-$C_2B_{10}H_{11}$	1	0	1	0	0	0	0	0	0	0	0	0
1,2-X_2-$C_2B_{10}H_{10}$	1	1	1	1	0	0	0	0	0	0	0	1
1,7-X_2-$C_2B_{10}H_{10}$	1	1	1	0	1	0	0	0	0	0	0	1
1,12-X_2-$C_2B_{10}H_{10}$*	1	2	0	0	0	1	0	0	0	0	0	1
2,3-X_2-$C_2B_{10}H_{10}$	1	0	2	0	0	0	1	0	0	0	0	1
2,4-X_2-$C_2B_{10}H_{10}$*	1	0	2	0	0	0	0	1	0	0	0	1
2,7-X_2-$C_2B_{10}H_{10}$	1	0	2	0	0	0	0	0	1	0	0	1
2,8-X_2-$C_2B_{10}H_{10}$	1	0	2	0	0	0	0	0	0	1	0	1
2,9-X_2-$C_2B_{10}H_{10}$	1	0	2	0	0	0	0	0	0	0	1	1
1,2,3-X_3-$C_2B_{10}H_9$	1	1	2	2	0	0	1	0	0	0	0	3
1,2,4-X_3-$C_2B_{10}H_9$	1	1	2	2	0	0	0	1	0	0	0	3
1,2,7-X_3-$C_2B_{10}H_9$	1	1	2	1	1	0	0	0	1	0	0	3
1,2,8-X_3-$C_2B_{10}H_9$	1	1	2	1	1	0	0	0	0	1	0	3
1,2,9-X_3-$C_2B_{10}H_9$*	1	1	2	1	1	0	0	0	0	0	1	3
1,2,12-X_3-$C_2B_{10}H_9$*	1	2	1	1	1	1	0	0	0	0	0	3
1,7,8-X_3-$C_2B_{10}H_9$	1	1	2	0	2	0	1	0	0	0	0	3
1,7,9-X_3-$C_2B_{10}H_9$	1	1	2	0	2	0	0	1	0	0	0	3
2,3,4-X_3-$C_2B_{10}H_9$	1	0	3	0	0	0	2	1	0	0	0	3
2,3,5-X_3-$C_2B_{10}H_9$	1	0	3	0	0	0	1	2	0	0	0	3
2,3,7-X_3-$C_2B_{10}H_9$	1	0	3	0	0	0	1	0	2	0	0	3
2,3,8-X_3-$C_2B_{10}H_9$	1	0	3	0	0	0	1	0	1	1	0	3
2,3,9-X_3-$C_2B_{10}H_9$*	1	0	3	0	0	0	1	0	0	1	1	3
2,4,7-X_3-$C_2B_{10}H_9$*	1	0	3	0	0	0	0	1	1	1	0	3

(continued)

Table 31.19 (continued)

Substituted of para-carborane	Coefficients of the additive scheme (31.40) 1	2	3	4	5	6	7	8	9	10	*11*	K_3
2,4,9-X_3-$C_2B_{10}H_9$*	1	0	3	0	0	0	0	1	1	0	1	3
2,4,10-X_3-$C_2B_{10}H_8$*	1	0	3	0	0	0	0	1	0	2	0	3
1,2,3,4-X_4-$C_2B_{10}H_8$	1	1	3	3	0	0	2	1	0	0	0	6
1,2,3,5-X_4-$C_2B_{10}H_8$	1	1	3	3	0	0	1	2	0	0	0	6
1,2,3,7-X_4-$C_2B_{10}H_8$	1	1	3	2	1	0	1	0	2	0	0	6
1,2,3,8-X_4-$C_2B_{10}H_8$	1	1	3	2	1	0	1	0	1	1	0	6
…	1	0	4	0	0	0	0	2	2	1	1	6
2,4,7,9-X_4-$C_2B_{10}H_8$	…	…	…	…	…	…	…	…	…	…	…	…
2,4,7,10-X_4-$C_2B_{10}H_8$	1	0	4	0	0	0	0	2	1	3	0	6
2,4,9,11-X_4-$C_2B_{10}H_8$	1	0	4	0	0	0	0	2	2	0	2	6
…	…	…	…	…	…	…	…	…	…	…	…	…
1,2,3,4,5,6,7,8.9,10-X_{10}- $C_2B_{10}H_2$	1	2	9	9	9	1	8	8	8	8	4	55
1,2,3,4,5,6,7,8.9,10,11,12-X_{12}- C_2B_{10}	1	2	10	10	10	1	10	10	10	10	5	66

*Two of the eight di-substituted para-carborane have enantiomers (2,10-X_2-$C_2B_{10}H_{10}$; 2,11-X_2-$C_2B_{10}H_{10}$), and of all the three-substituted para-carborane - six enantiomers (1,2,10-X3-$C_2B_{10}H_9$; 1,2,11-X3-$C_2B_{10}H_9$; 2,3.10-X3-$C_2B_{10}H_9$; 2,3,11-X3-$C_2B_{10}H_9$; 2,4,8-X3-$C_2B_{10}H_9$; 3,4,8-X3-$C_2B_{10}H_9$)

If a property *P* para-carborane is the amount of contributions made by elements of structure, represented as a sum of subgraphs (vertices and paths) of different lengths in the MG, then for the properties (*P*) para-carborane one obtains the scheme

$$P(napa - 1, 12 - C_2B_{10}H_{12}X) = n_O p_O + K_3 p_l. \qquad (31.43)$$

Here, p_O, and p_l - empirical parameters; $n_O = 1$, $K_3 = 1/2[n(n+1)]$ - triangular number (the total number of bonds C-X, C-B, and all pairs of X ... X directly unconnected atoms in the para-carborane (number of ways of applying subgraphs of length 1, 2, 3, 4 and 5 on the investigated graph). Scheme includes two parameters.

If in (31.43) one isolates C-X and C-B from K_3 it results in a simple chart with four parameters

$$P(1, 12 - C_2B_{10}H_{12}X) = n_O p_O + n_{C-X} p_{C-X} + n_{B-X} p_{B-X} + K_3' p'_l. \qquad (31.44)$$

Here p_O, $p_{C\text{-}X}$, $p_{B\text{-}X}$ and p'_1 - empirical parameters; $n = 1$, $n_{C\text{-}X}$, $n_{C\text{-}B}$ - the number of bonds C-X and B-X in the molecule of para-carborane, and $K_3' = 1/2[n(n-1)]$- triangular number (total number of all pairs of X ... X directly unconnected atoms in a molecule of para-carborane (i.e. the number of ways of applying subgraphs of length 3, 4 and 5 on the investigated graph). Scheme includes four parameters.

However, schemes (31.43) and (31.44) do not distinguish isomers in the series (di-substituted, three-substituted, etc.). If we expand (31.44) by number K'_3, one obtains the scheme...

$$\begin{aligned}\boldsymbol{P}(\boldsymbol{D}_{5\mathrm{d}})(\textit{para}\text{-1,12-}C_2H_{2-k}X_kB_{10}H_{10-l}X_l) = {} & p_{\text{II-1,12-C2B10}} + n^{C}_{1\text{-X1}}p^{C}_{1\text{-X1}} + n^{B}_{11\text{-X1}}p^{B}_{11\text{-X1}} \\ & + n^{C\text{-}B}_{1,2\text{-X2}}p^{C\text{-}B}_{1,2\text{-X2}} + n^{C\text{-}B\text{-}B}_{1,7\text{-X2}}p^{C\text{-}B\text{-}B}_{1,7\text{-X2}} + n^{C\text{-}B\text{-}B\text{-}C}_{1,12\text{-X2}}p^{C\text{-}B\text{-}B\text{-}C}_{1,12\text{-X2}} \\ & + n^{B\text{-}B}_{2,3\text{-X2}}p^{B\text{-}B}_{2,3\text{-X2}} + n^{B\text{-}C\text{-}B}_{2,4\text{-X2}}p^{B\text{-}C\text{-}B}_{2,4\text{-X2}} + n^{B\text{-}B}_{2,7\text{-X2}}p^{B\text{-}B}_{2,7\text{-X2}} \\ & + n^{B\text{-}B}_{2,8\text{-X2}}p^{B\text{-}B}_{2,8\text{-X2}} + n^{B\text{-}B\text{-}B}_{2,9\text{-X2}}p^{B\text{-}B\text{-}B}_{2,9\text{-X2}}.\end{aligned} \tag{31.45}$$

Here $n^{C}_{1-X1}, n^{B}_{11-X1}, n^{C-B}_{1,2-X2}, n^{C-B-B}_{1,7-X2}, \ldots$ are the coefficients expressed through unbonded atoms pair interactions in the molecule; $p^{B}_{11-X1}, p^{C-B}_{1,2-X2}, n^{C-B-B}_{1,7-X2}, \ldots$ are the empirical parameters, determined by the method of least squares (MLS) from experimental data [12] for the physical-chemical properties of the P X-substituted carborane (X = CH_3, C_2H_5, F. Cl,...) 1,12-$C_2B_{10}H_{10}$. Scheme (31.45), contains 11 parameters (see Table 31.20) and distinguishes these isomers of X-substituted para-1,12-$C_2B_{10}H_{10}$. Formula (31.45), taking into account all pairwise non-valent interactions of atoms (H ... H, X ... H and X ... X) and can be used as "working".

Table 31.20 Numbers and parameters of the equation (31.45) for calculation of physicochemical properties of X-substituted para-1,12-$C_2B_{10}H_{12}$

1	2	3	
$p_{\text{II-1,12-C2B10}}$	n^{C}_{1-X1} p^{C}_{1-X1}	n^{B}_{11-X1} p^{B}_{11-X1}	
4	5	6	7
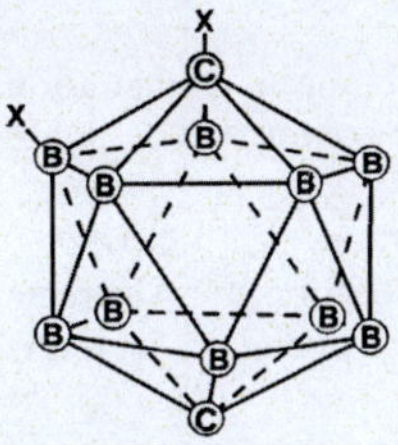	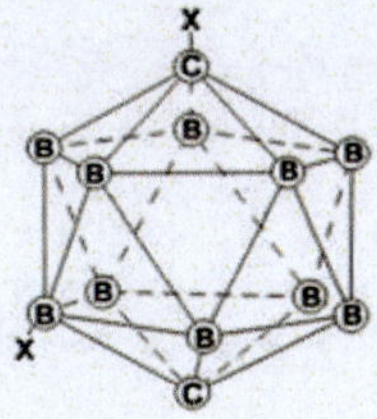	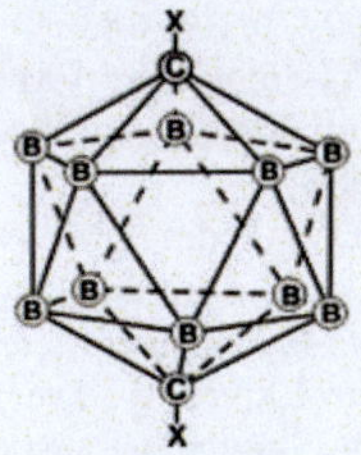	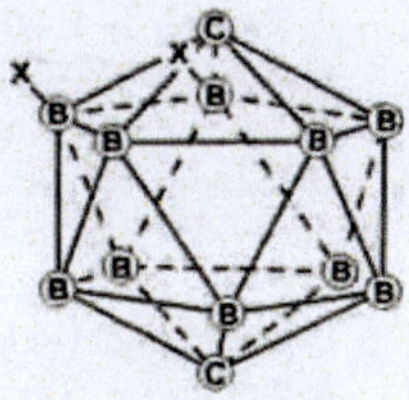
1,2-X2-$C_2B_{10}H_{10}$ $n^{C-B}_{1,2-X2}$ $p^{C-B}_{1,2-X2}$	1,7-X2-$C_2B_{10}H_{10}$ $n^{C-B-B}_{1,7-X2}$ $p^{C-B-B}_{1,7-X2}$	1,12-X2-$C_2B_{10}H_{10}$ $n^{C-B-B-C}_{1,12-X2}$ $p^{C-B-B-C}_{1,12-X2}$	2,3-X2-$C_2B_{10}H_{10}$ $n^{B-B}_{2,3-X2}$ $p^{B-B}_{2,3-X2}$
8	9	10	11
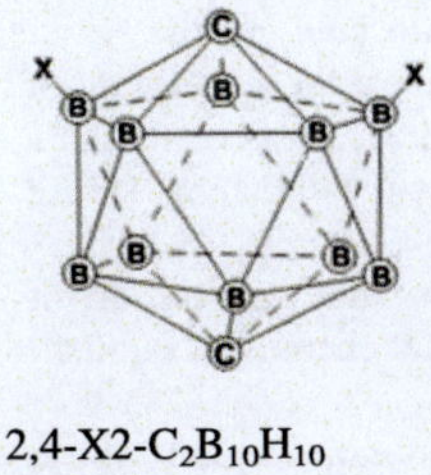	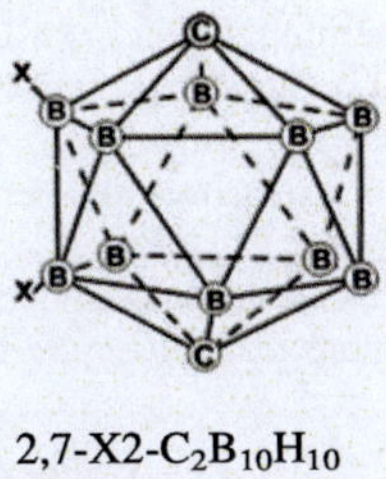	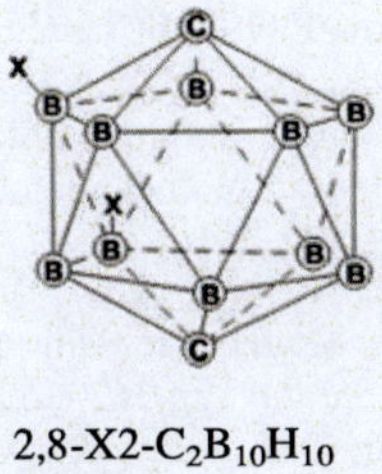	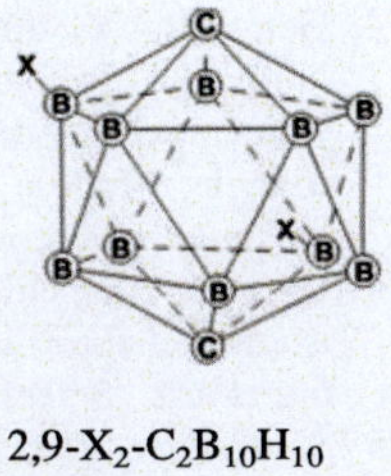
2,4-X2-$C_2B_{10}H_{10}$ $n^{B-C-B}_{2,4-X2}$ $p^{B-C-B}_{2,4-X2}$	2,7-X2-$C_2B_{10}H_{10}$ $n^{B-B}_{2,7-X2}$ $p^{B-B}_{2,7-X2}$	2,8-X2-$C_2B_{10}H_{10}$ $n^{B-B}_{2,8-X2}$ $p^{B-B}_{2,8-X2}$	2,9-X_2-$C_2B_{10}H_{10}$ $n^{B-B-B}_{2,9-X2}$ $p^{B-B}_{2,9-X2}$

31.12 Conclusions

The results of X-, XY-,. . .enumeration of (X,Y = CH_3, F,. . .) carboranes, boranes and carbometallic derivative of boranes, and their distribution over families and depending on the number *m* of the substitution places are important for the formation of the files of homological series for new molecular [23] and crystal structures and development of mathematical models for the prognosis of their physico-chemical properties.

References

1. Bekkenbah E (1968) Applied combinatorial mathematics. Mir, Moscow, p 363
2. Gavrilova G, Gavrilova G (1979) Enumeration problems of combinatorial analysis, Sattransfers. Mir, Moscow, p 365
3. Harrari F, Palmer E (1977) Enumeration of graphs. Mir, Moscow, p 324
4. Polya G (1937) Acta Math B 68:145–254
5. Polya G (1936) Kristallogr B A93:415–443
6. Smolyakov VM, Sokolov DV, Nilov DY (2007) In: Abstracts of the XVIII Mendeleev conference on general and applied chemistry, vol. 2, Moscow, 23–28 Sept 2007, p 524
7. Sokolov DV, Nilov DY, Smolyakov VM (2007) Enumeration and identification of basic structure isomer substitutes of known symmetry group considering different isomer types, In: Abstracts XVI international conference on chemical thermodynamics (RCCT 2007), vol. 1, Suzdal, 1–6 July, 2007, p. 2/S–183
8. Sokolov DV, Nilov DY, Smolyakov VM (2007) Ortho– and metha–carboranes: enumeration of isomer substitutes. In: Abstracts of the IV All-Russian youth science conference on under the sign of Σ, Omsk, p. 1
9. Nilov DY, Sokolov DV, Fedin DM, Nikolenko AY (2007) Carboranes 1,2,3-$C_9B_9H_{12}$, 1,2,3,9,11,12- $C_6B_6H_{12}$, 1,2,3-$CB_9H_{11}Cl$: generation of isomers of substitution. In: Abstracts of the regional scientific-technical conference of young scientists physics, chemistry and new technologies, Tver State University, Tver, p 43
10. Nilov DY, Sokolov DV, Smolyakov VM (2007) Para-carborane: enumeration of isomers and the assessment framework of properties based on the Pascal's triangle. In: Abstracts of the IV All-Russian youth science conference on under the sign of Σ, Omsk, p 2
11. Sokolov DV, Nilov DY, Nikolenko AY, (2007) Carboranes 1,12-$CSiB_{10}H_{12}$ and 1,2-$FB_{10}H_{10}Cl$: generation of isomers of substitution. In: Abstracts of the regional scientific-technical conference of young scientists physics, chemistry and new technologies, Tver State University, Tver, p 62
12. Gerasimov YI, Akishina P (1984) Chemical thermodynamics. MSU, Moscow, p 320
13. Nilov DY, Sokolov DV, Smolyakov VM (2007) Carboranes and boranes: the generation of isomers and the substitution graph model calculation of properties on the basis of Pascal's triangle, Modern problems of theoretical and experimental chemistry. Publishing House "Science", Saratov, pp 360–361
14. Sokolov DV, Nilov DY, Smolyakov VM (2007) Boranes and carboranes: generation of isomers of substitution, Modern problems of theoretical and experimental chemistry. Publishing House "Science", Saratov, pp 367–369
15. Smolyakov VM, Fedin DM, Nilov DY, Grebeshkov VV (2009) Carboranes based on the threepyramidal dodecahedron: the generation of isomers of substitution. Collected papers at the international conference on actual problems of chemical science, practice and education, Kursk, pp 252–255

16. Smolyakov VM, Nilov DY, Grebeshkov VV (2009) Carboranes on the basis of bypyramidal trigonal dodecahedron: enumeration of isomers of substitutions, Abstracts. In: Proceedings of the XVII international conference on chemical thermodynamics in Russia, vol. 2, Kazan, 2009, p 466
17. Zalkin A, Templeton DH, Hopkins TEJ (1965) Am Chem Soc 87:3988–3990
18. Nilov DY, Sokolov DV, Smolyakov VM (2007) Boranes and carboranes: enumeration of isomers of substitution and the construction of additive schemes on the basis of Pascal's triangle. In: Proceedings of the XVI international conference on the chemical thermodynamics in Russia, Suzdal, 2007, № 1. p 159
19. Nilov DYu, Sokolov DV, Smolyakov VM, Nikolenko AYu (2007) Vestn Tver State Univ Ser Chem 2(30):87–94
20. Smolyakov VM, Sokolov DV, Nilov DY (2000) In: Proceedings of the IV international conference on mathematical modelling, vol 2, Moscow, MSTU, Stankin, p 238–242
21. Smolyakov VM, Sokolov DV, Nilov DY, Polyakov MN (2009) Metod of additive scheme construction on the base of Paskal triangle multiangular number splitting. In: Academician Bannykh OA (ed) Proceedings of the III international conference on deformation and fracture of materials and nanomaterials, vol II, Moscow, 2009, pp 388–389
22. Smolyakov VM, Sokolov DV, Nilov DY, Polyakov MN (2009) Method of additive scheme construction on the base of Pascal triangle multiangular number splitting, Abstracts. In: Proceedings of the XVII international conference on chemical thermodynamics in Russia, vol. 2, Kazan, 2009, p 464
23. Smolyakov VM, Sokolov DV, Nilov DYu (2009) Recurrence formulas for the transfer of hetero-isomers of molecules and their radicals. Vestn Tver State Univ Tver Tver State Univ 39:65–78
24. Galchenko GL, Tamm NB, Pavlovich VK, Olshevskaya VA, Sacharin LI (1990) 9-Methyl-o-carborane-12. Synthesis and determination of thermochemical properties, Organometallic Chemistry, M.: Nauka, USSR, 1990. 3(2): 414–418

Chapter 32
Thermodynamic Properties and Phase Equilibriums in Ternary Alloys of the Al-C-3D-Metal Systems

N.E. Vovkotrub, V.S. Sudavtsova, M.A. Shevchenko, Yu.V. Lagodyuk, and V.G. Kudin

Abstract The thermodynamic properties of liquid alloys of the 3d-Me-C systems are calculated using the described algorithm and some experimental data. The analysis of own and literary data has allowed to establish the most reliable mixing enthalpies of binary boundary systems Al-3d-metal. This allowed to estimate the thermodynamic properties of ternary alloys of the Al-C-Sc(Ti, V, Cr, Mn, Fe, Co, Ni, Cu) systems by Kohler equation and to find a regularity of its changing.

Keywords Al-C-3d metal systems · Mixing enthalpy · Heat effect · Electronegativity · Interaction strength

32.1 Introduction

Aluminum and its alloys are widely used in electrical engineering, mechanical engineering, aviation and other fields of national economy, because they have low density, high electrical and thermal conductivity, resistant to oxidation. Carbon is often found in such alloys as an impurity. Therefore, to improve methods of obtaining and operating relevant to know the thermodynamic properties of binary and, in particular, ternary aluminum containing alloys. Studying of these alloys is a difficult experimental task, we estimated the enthalpy of formation of melts of binary systems using a similar parameters of intermetallic compounds.

N.E. Vovkotrub and V.S. Sudavtsova (✉)
Institute for Problems of Materials Science of NAS of Ukraine,
Krzhyzhanovsky str., 3, Kiev 03142, Ukraine
e-mail: vsudavtsova@univ.kiev.ua

M.A. Shevchenko, Yu.V. Lagodyuk, and V.G. Kudin
Department of Chemistry, National Taras Shevchenko University,
64, Vladimirska str., Kiev 01033, Ukraine

S.Yu. Zaginaichenko et al. (eds.), *Carbon Nanomaterials in Clean Energy Hydrogen Systems - II*, NATO Science for Peace and Security Series C: Environmental Security 2,
DOI 10.1007/978-94-007-0899-0_32, © Springer Science+Business Media B.V. 2011

32.2 Prediction Methodic

The standard thermodynamic functions of forming of many binary intermetallics are determined and placed into reference books. We tried to use standard thermodynamic functions of compounds for prediction of molar enthalpies and Gibbs energies of liquid alloys. For this purpose we considered the forming process of 1 mol of liquid alloy A_xB_{1-x} at temperature T from the pure compounds at 298 K:

$$\begin{gathered} xA_{298} + (1-x)B_{298} = A_xB_{1-x} \quad \Delta H^0_{298} \\ xA_{298} \rightarrow xA_T \quad \Delta H_1 = x\Delta H^T_{298,A} \\ (1-x)B_{298} \rightarrow (1-x)B_T \quad \Delta H_2 = (1-x)\Delta H^T_{298,B} \\ (A_xB_{1-x})_{289} \rightarrow (A_xB_{1-x})_T \quad \Delta H^T_{289,A_xB_{1-x}} \\ xA_T + (1-x)B_T = A_xB_{1-x} \quad \Delta H \end{gathered} \tag{32.1}$$

According to the Gess law, the heat effect of the reaction (32.1) can be written so:

$$\Delta H = \Delta_f H^0_{298} + \Delta H^T_{298,A_xB_{1-x}} - x\Delta H^T_{298,A} - (1-x)\Delta H^T_{298,B} \tag{32.2}$$

It is clear from the Eq. 32.2 that for the definition of mixing enthalpy of liquid alloy we ought to know the standard forming enthalpy of this alloy at 298 K and also the heating enthalpies of it and of pure components. When the heating enthalpies of intermetallics and pure components are known, the standard thermodynamic functions of compounds may be used for calculation of mixing enthalpies of liquid alloys with the same composition. Then, it worth saying that the algebraic sum of the last three items of the Eq. 32.2 is often close to zero. So, neglecting them, we can write $\Delta H \approx \Delta H^0_{298}$. This shows us a way to estimate the mixing enthalpies of liquid alloys with the composition which is the same with intermetallics. Owing to the fact that mixing ΔH of alloys varies smoothly with the composition, we can approximate it with the polynomials or other analytical functions.

When the temperature dependence of heat capacity of compound and its heat of melting are identified, so, due to Kirchhof's rule, the standard forming enthalpy can be recalculated into the higher temperature:

$$\Delta H^0_T = \Delta H^0_{298} + \int_{298}^{T} \Delta C_P T + \Delta H_{\text{пл.}}, \tag{32.3}$$

where ΔC_P is an excess heat capacity of intermetallic.

Using the standard forming enthalpies at 298 K of the compounds series and also other thermodynamic characteristics, we calculated mixing enthalpies of liquid alloys of either investigated or unstudied systems. On the Fig. 32.1 calculated and experimental forming enthalpies of binary alloys Fe-S(Si) are compared. It is obvious that they are in a rather good agreement. On the Fig. 32.2 the predicted values of ΔH are

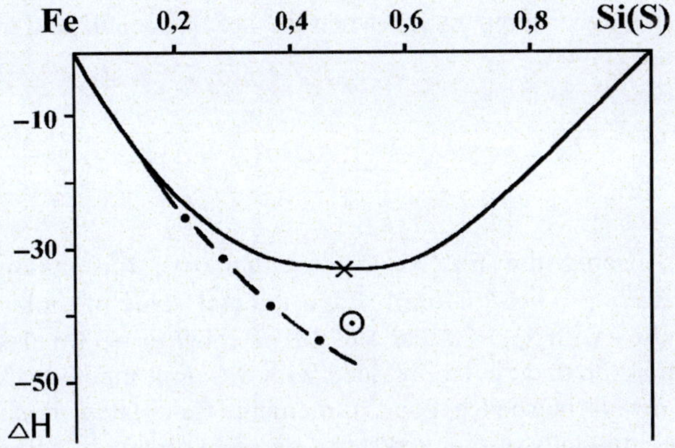

Fig. 32.1 Mixing enthalpies of liquid alloys of the systems Fe-Si(S): full line - calculated from $\Delta_f H^0_{FeS}$, x - experiment [1]; dot - and - dash line - calculated from $\Delta_f H^0_{FeS}$, ⊖ - experiment [2]

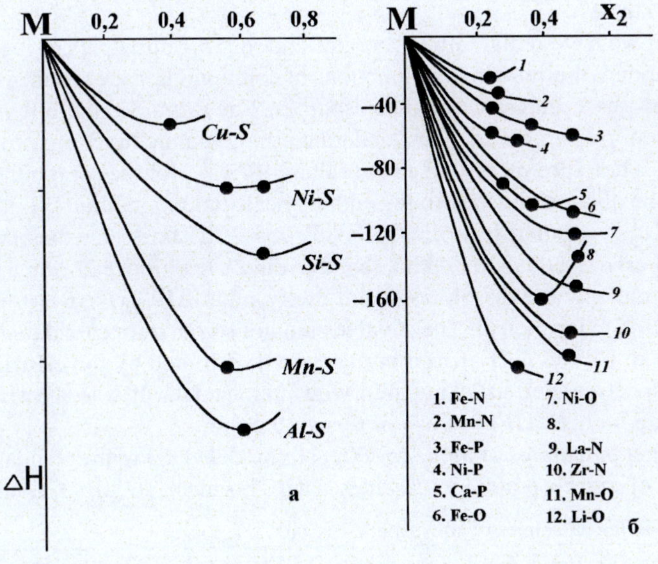

Fig. 32.2 Mixing enthalpies of liquid Me-E alloys, calculated from the standard forming enthalpies of compounds

presented for some unstudied systems, which are important for metallurgy, welding and other industrial fields. By the value of calculated enthalpies it is possible to make conclusions about the energy and interaction type in these alloys, as well as to watch their change during the substitution of one component with another, and to forecast their desulphurating, dephosphorating and desoxidating properties.

We can do analogously in the cases when it is necessary to forecast molar Gibbs energies in liquid alloys. By the Gibbs-Helmholtz equation $\Delta G^0_{298} = \Delta H^0_{298} - T\Delta S^0_{298}$ we can calculate the standard Gibbs energy of compound, and also to recalculate it into the higher temperature, when the heat capacities and melting heats of intermetallic are known:

$$\Delta G_T^0 = \Delta H_{298}^0 + \int_{298}^{T} \Delta C_P dT - T\Delta S_T^0 - T \int_{298}^{T} \frac{\Delta C_P}{T} dT - T\Delta S_{n\pi.}. \qquad (32.4)$$

Using the Eq. 32.4, we calculated, for example, for iron monosilicide $\Delta G_{1873}^0 = -35$ kJ/mol. Experimental value of molar Gibbs energy of the Fe-Si alloy with $x_{Si} = 0.5$ at 1,873 K is equal to -33 ± 1 kJ/mol [1]. It is remarkably that the data, even obtained with different methods at high temperatures, are not always in such a good agreement. Calculated values of $\Delta_m G$ from $\Delta_f G_{298}^0$ of compounds of many systems agree in common well with the experimental data. Due to this fact, it is expedient to use this methodic for the calculation of molar Gibbs energies, and from those – the components activities in alloys of difficultly studying systems.

So, when we have molar Gibbs energies and mixing enthalpies of melts, we can estimate standard thermodynamic functions of compounds, the existence of which is known from phase diagrams or other sources. For example, the alloys Fe-B are inclined to amorphization, so their standard thermodynamic functions are difficult to be obtained. Therefore we estimated the values of ΔH_{298}^0 for the compounds FeB and Fe_2B from the obtained by the mixing enthalpy calorimetry method [3]. It was found that their ΔH_{298}^0 are equal accordingly to -37 and -30 kJ/mol. On the other hand, in [4] ΔH_{298}^0(FeB) is equal to -32.3 (by the direct reaction calorimetry method) or -293 kJ/mol (by the dissolution calorimetry); and $\Delta H_{298}^0(Fe_2B)$=-226 kJ/mol (direct reaction calorimetry). These values are not so far from calculated by us.

Recently in [5] the $\Delta_f H$ of iron borides are determined by the effusion method with mass-spectrometer and compared with literary data. It was shown that these $\Delta_f H$ correlate with $\Delta_{mix} H$ of alloys of this system, too.

The mixing enthalpies of melts M-C(B) are modeled using the standard forming enthalpies of corresponding carbides and borides [6–8]. Calculated and

Table 32.1 Mixing enthalpies of alloys metal-carbon

Система	– ΔH	x_C	Система	– ΔH	x_C
Be-C	80.0	0.66	Al-C	30.0	0.42
Mg-C	29.3	0.66	Sc-C	25	0.66
Ca-C	19.8	0.66	Y-C	37.7	0.66
Sr-C	28.2	0.66	La-C	23.7	0.66
Ba-C	25.1	0.66	Mn-C	4.1	0.15
Si-C	36.0	0.5	Mn-C*	23.5	0.3
Si-C*	13.2	0.1	Cr-C	19.6	0.4
Ti-C	115.8	0.5	Cr-C*	47.0	0.35
Zr-C	100.0	0.5	Mo-C	5.0	0.33
Hf-C	113.0	0.5	W-C	7.6	0.5
V-C	59.2	0.5	Fe-C	−7.0	0.33
Nb-C	70.0	0.5	Fe-C*	10.0	0.2
Ta-C	72.4	0.5	Co-C*	4.2	0.16
B-C	14.3	0.2	Ni-C*	3	0.09

Table 32.2 Mixing enthalpies of alloys of lanthanide-carbon systems

Me	Ce	Ce	Pr	Nd	Sm	Eu	Gd	Dy	Ho	Ho	Er	Tm	Yb	Lu
x_C	0.6	0.66	0.66	0.66	0.6	0.66	0.66	0.66	0.6	0.66	0.66	0.66	0.66	0.66
ΔH	19.4	20.9	20.3	17.4	23.7	20.9	41.8	16.4	46.9	36.3	25.8	27.9	25.1	25.1

Table 32.3 Mixing enthalpies of alloys of actinide-carbon systems

Me	Th	U	U	U	Pu	Pu	Pu
x_B	0.5	0.5	0.6	0.66	0.5	0.6	0.66
– ΔH	61.9	48.5	41.0	30.5	25.1	40.8	11.2

Table 32.4 Mixing enthalpies of binary melts metal-boron

system	Be-B		C-B		
–ΔH	24 (0.4)		14 (0.8)		
system	Mg-B	Al-B	Si-B	Mn-B	Fe-B
–ΔH	18.9 (0.6)	22.4 (0.66) 16.5 (0.4)*	20.0 (0.5) 15.0 (0.4)*	26.5 (0.66) 44.0 (0.4)*	19.2 (0.5) 47.0 (0.4)*
system	Ca-B	Sc-B	Ti-B	V-B	Cr-B
–ΔH	17.1 (0.84)	88.3 (0.66)	97.6 (0.66)	86.5 (0.66)	41.8 (0.66)47.0 (0.25)*
system	Sr-B	Y-B	Zr-B	Nb-B	Mo-B
–ΔH	30.1 (0.84)	14.3 (0.84)	98.6 (0.66)	82.3 (0.66)	35.6 (0.33)
system	Ba-B	La-B	Hf-B	Ta-B	W-B
–ΔH	47.6 (0.84)	46.1 (0.84)	115.4 (0.66)	64.2 (0.66)	35.6 (0.5)

* The molar part of B is in brackets

experimentally determined (*) mixing enthalpies of melts of the binary M-C(B) systems in the extremum points (kJ/mol) and the molar parts of carbon or boron in it are presented in the Table 32.1–32.4 (experimental values are marked off by *). We can see from the Table 32.1 that the mixing enthalpies of alloys of binary systems Mg(Ca, Sr, Ba)-C are near to one another but not to the Be-C. Systems of carbon and metals of the third group are characterized by resembling interaction energies. This may be explained by the near atomic sizes.

In the binary carbon-containing melts with Cr and Mn experimental values of ΔH are more exothermic than calculated. For the carbides of Fe, Co and Ni the standard forming enthalpies are positive [4], but mixing enthalpies of such alloys Fe (Co, Ni)-C are negative. This indicates a different behaviour of these alloys in liquid and solid states.

The strongest interaction of components is characteristic for the alloys of binary systems Ti(Zr, Hf)-C, and this fact can be explained by their electron structure.

The interaction of carbon with V, Nb and Ta is also strong but it is less that for Ti(Zr, Hf).

Mixing enthalpies of the systems lanthanide(actinide)-carbon are not experimentally studied. So as the information about it is important, we calculated mixing enthalpies of these systems. The results are presented in Tables 32.2 and 32.3.

In the actinide-carbon melts the heat effects should be greater than in the latter. Then, the interaction energy increases in the series Pu-C U-C Th-C for equiatomic compositions.

32.3 Modeling Results

The first molar enthalpy of carbon in aluminum (−55 kJ/mol) and enthalpy of formation Al_4C_3 (−18 kJ/mol). Using these data and the type of melting of aluminum carbide, we calculated that the minimum of the enthalpy of mixing

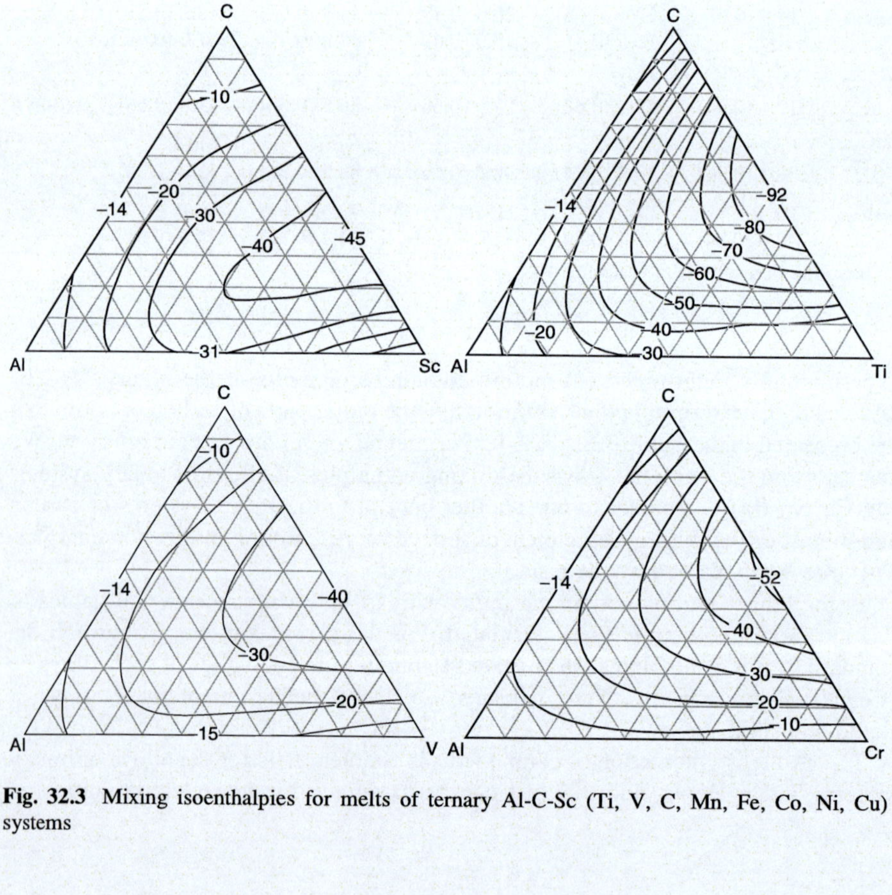

Fig. 32.3 Mixing isoenthalpies for melts of ternary Al-C-Sc (Ti, V, C, Mn, Fe, Co, Ni, Cu) systems

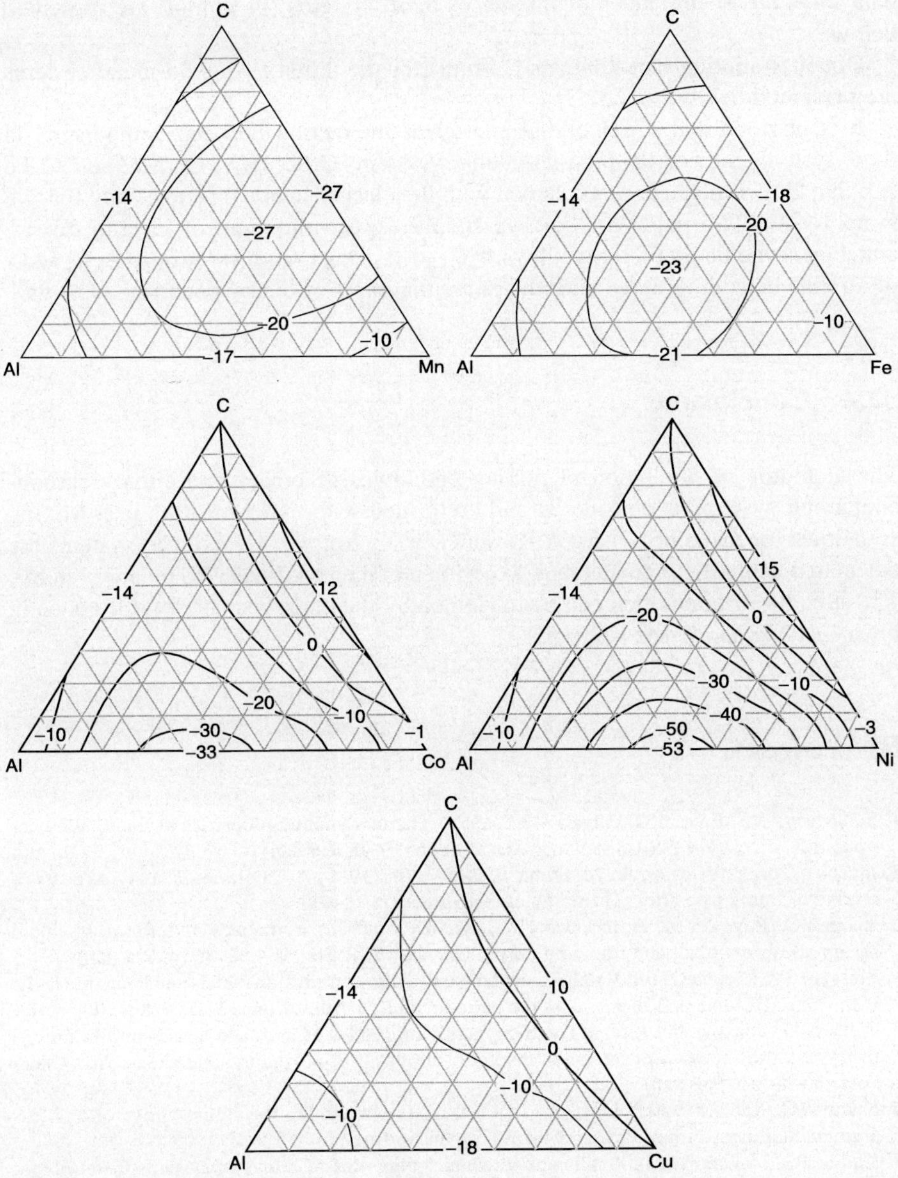

Fig. 32.3 (continued)

of aluminum-carbon melts is −14 kJ/mol. For other binary carbon-containing systems in the same manner were predicted their thermochemical properties. Then, using Kohler, Bonnier-Cabo and other equations, we calculated thermochemical properties of liquid alloys of ternary systems Al-C-3d-metal from these data. Integral isoenthalpies of mixing of these systems (in kJ/mol) are presented below.

Obtained mixing isoenthalpies (kJ/mol) for the ternary Al-C-3d-metal systems are presented below (Fig. 32.3).

It is obvious that the maximal energy of interaction between components in these systems falls on the binary boundary systems C-Sc(Ti, V, Cr, Mn) and Al-Fe (Co, Ni, Cu), which can be explained with the electrochemical factor. Only for two systems (Al-C-Mn and Al-C-Fe) we should expect minimum of integral mixing enthalpy in the field of ternary alloys: $Al_{0,05}C_{0,47}Mn_{0,48}$ and $Al_{0,36}C_{0,26}Fe_{0,38}$. Most of all, this behaviour agree with the phase diagrams of binary boundary systems.

32.4 Conclusion

The methodic of prediction of mixing enthalpies of binary and ternary carbon-containing systems is considered and confirmed with experimental facts. Mixing enthalpies for melts of ternary Al-C-Sc(Ti, V, C, Mn, Fe, Co, Ni, Cu) systems are calculated using the predicted and experimental data for binary boundary systems. The strength of interaction can be explained by the electronegativity difference in binary Al-Me and C-Me systems.

References

1. Sudavtsova VS, Batalin GI, Ulyanov VI (1975) Thermodynamic properties of liquid alloys of iron with silicon. Izv AS USSR Inorg Mat 2(1):66–70 (in Russian)
2. Iguchi I, Tozaki Y, Kakizaki M, Fuma T, Ban-ya Sh (1981) A Calorimetric study of heats of mixing of liquid iron alloys. J Iron Steel Inst Jpn 7(7):925–932
3. Yesin YO, Bayev VM, Petrushevsky MS, Geld PV (1975) Enthalpies of forming of liquid binary alloys of cobalt and iron with boron. Izv AS USSR Metals 4:82–86 (in Russian)
4. Meschel SV, Kleppa OJ (2001) Thermochemistry of alloys transition metals and lantanide melts with some IIIB and IVB elements in the periodic table. J Alloy Comp 321(1):183–200
5. Zaitsev AI, Zaitseva NE (2002) Thermodynamic properties of melt and phase equilibriums in the system iron-boron. Conversion of the liquid alloys Fe-B into amorphous state. J Phys Chem 76(1):33–44, (in Russian)
6. Kudin VG, Makara VA, Sudavtsova VS (2001) Interaction in the liquid alloys of systems aluminum(silicon)-boron. Powder Met 1/2:79–84 (in Russian)
7. Kudin VG, Makara VA (2002) Thermodynamic properties of alloys of systems metal-boron. Inorg Mater 38(3):280–284 (in Russian)
8. Kudin VG, Makara VA (2004) Interaction in alloys of binary systems metal-carbon. Powder Met 1/2:78–82 (in Russian)

Chapter 33
The Agreement Phenomenon of the Component Analysis with Dimensions of the Graphenic-Like Carbon and Boron Nitride Nano Sized Particles

V.V. Garbuz, V.A. Petrova, and A.V. Yakovlev

Abstract The model of the plane graphenic-like sp^2-hybridization monoatomic layers of two stereo isomers such as disordered (d) carbon (d-C) and boron nitride (d-BN) in powders has been presented. The dependences of the component content, dimension and surface characteristics were calculated and determined. All data for d–BN are experimental tested by means of the 15 d–BN various powder examples with disordered graphite-like structure.

Keywords Theory · Graphene-like carbon · Experiments · Nanoparticle · Oxidation · Carbon and boron nitride · Powder

33.1 Introduction

The key information about exclusive chemical properties of the d-C presented in the review by Boehm H.P. [1] was used.

1. Non reversible adsorption of oxygen to carbon-graphenic layers occur in origin above the −40°C.
2. Oxygen extraction out of d-C powders by heating takes place in view of the CO/CO_2 only.
3. Surface of the d-C nano sized particles is chemically inert; it was determinated [1].
4. The oxygen (and other elements) is reacting with the perimetrical atoms of the d-C nano dimension particles only.

The d-BN and d-C are the stereo isomers (space group – p6_3/mmc [2]). All calculations and experimental data of these isomers will be closely connected. It would be interesting to investigate interactions of the chemical elements amount and dimensional characteristics.

V.V. Garbuz, V.A. Petrova (✉), and A.V. Yakovlev
Department for Methods of Analyses of the Inorganic Materials, Institute for Problems of Materials Science of NAS of Ukraine, Krzhizhanovsky str. 3, Kiev 03142, Ukraine
e-mail: wpetrowa@ukr.com

S.Yu. Zaginaichenko et al. (eds.), *Carbon Nanomaterials in Clean Energy Hydrogen Systems - II*, NATO Science for Peace and Security Series C: Environmental Security 2,
DOI 10.1007/978-94-007-0899-0_33,

Experimental data of X-rays diffraction, Infra Red spectrometry, surface properties and quantitative analyses (B, N, C, O, and H) characteristics for nanodimension powders of the disordered (turbostratic) graphite-like boron nitride d-BN has been examined when phase transformations into dense modifications by high temperature shock compression was investigated [3].

The next step of these exams was modeling of all own and literature experimental data [1–3] into suitable form as schematic sketches, mathematician equations and tables about existence levels of plane graphenic sp^2-hybridization layers of the d-C and d-BN.

33.2 Calculations and Experiments

Unlike conventional designs, this model is under perimetrical control of the main informed structural volume element as axes of the 6_3-degree. The chemical bonding with using of geometrical and force characteristics of the inert surface and reactionary active exterior perimeter of carbon [1], boron plus nitrogen as constructing particles, hydrogen and oxygen as donors and acceptors atoms has been analyzed according to the Slater's atom-ionic radii and Allred-Rokhogh's electro-negativities quantities [4–6], shown in Table 33.1.

Physical sense of electro-negativities by Allred-Rokhov [4] is – $\chi_A = Z_{eff.} \cdot e^2/r^2{}_{cov}$, where Z_{eff} – effective charge of atom core; e – electron charge; r – covalent radius.

$$(x_B + x_N)/2 = 2,54, \quad (r_{ai}B + r_{ai}N)/2 = 0,075$$

$(\chi_N - \chi_B) = 1.06$, on the round-side particles would be – B (δ+); N(δ−);C(δ−).

The perimetrical carbon atoms of the d-C and nitrogen(3−) of the d-BN with presence of no paired electrons gain electro negative charge. Analogous atoms of boron(3+) in d-BN particles has positive charge, has shown on the chart (Fig. 33.1).

The Plane Perimetrical Model (PPM) consists of any coexisting levels, such as: the burning of plane perimetrical particles as result of the centre symmetrical interfaced poly-macro-cyclization process. Digital values of the Degree of Macro Cyclization (DMC) equal of the whole numbers natural series 0; 1; 2 . . . n.

These processes (Fig. 33.1) are as result of chain reactions of the polymerization in plane of the sp^2-hybridization for the carbon atoms [7] and as reactions of the

Table 33.1 Geometrical and forces characteristics for the atoms which consisted of graphenic-like mono atom layers of d-C and d-BN

No	Chemical element	The degree of oxidation	Electro-negativities χ_a by Allred-Rokhov [4]	Atom-ionic radii r_{ai} (nm) by Slater [5]
1	C	IV±	2.50	0.070
2	N	III−	3.07	0.065
3	B	III+	2.01	0.085
4	O	II−	3.50	0.060
5	H	I+	2.20	0.025

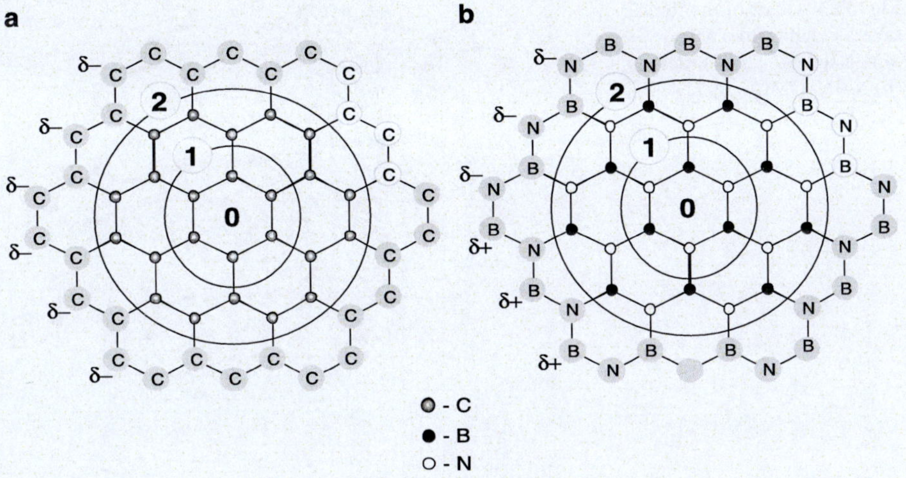

Fig. 33.1 Chart of the d-C (**a**) and d-BN (**b**) particles. DMC – 0; 1; 2

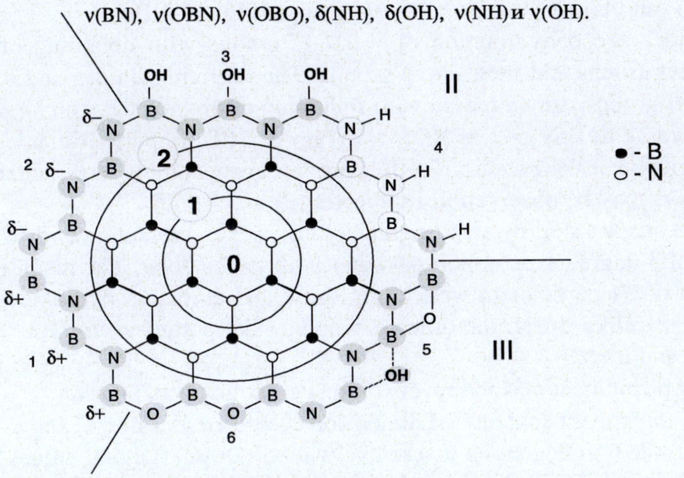

Fig. 33.2 Levels of coexisting of the perimetrical boron and nitrogen atoms in (PPM) d-BN. DMC – 0; 1; 2. I – 1, 2 – starting conditions; II – 3, 4 – perimeter neutralization of charges by hydroxo- and hydrogenation in order to boron(3+) and nitrogen(3-) atoms; III – 5,6 – boron(3+) dehydrotation, thermal dehydrogenation, hydrolysis of the nitrogen(3−), addition of the bridge-like oxy and hydroxyl groups to perimeter boron(3+) atoms

polycondensation with water vapors and CO/CO_2 extraction for the burning of the d-BN in the same plane [8]. In case of this extraction is not fully off, the water vapors in during to high temperature may be to react with its products of d-BN at some degradation levels (Fig. 33.2).

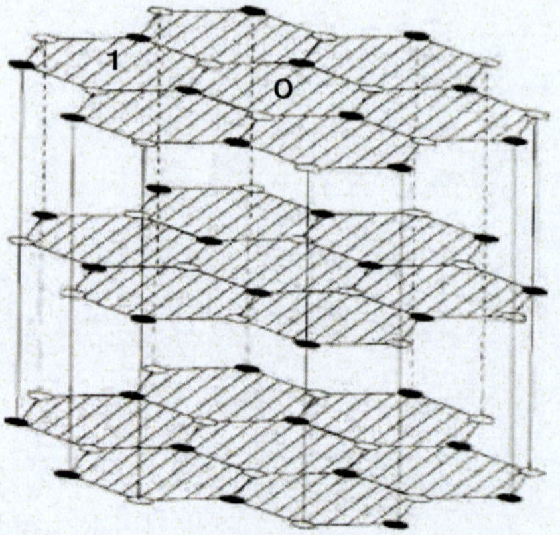

Fig. 33.3 Chart of the three layers graphite-like nano dimension packet of ordered (o) o-BN. DMC – 0; 1

The first level is origin point (Figs. 33.1b and 33.2I). Some d-BN particles in these suitable conditions may be to arrange in the ordered graphite-like nano packet with AA'AA'-type of structure [2] (Fig. 33.3).

The second is as result of the perimetrical hydrolysis with burning of the imides (=N–H) and borohydroxy (=B–OH) groups (Fig. 33.2II).

Third – the dehydrotation of =B–OH groups with obtaining of the bridges hydroxyl groups and then – oxy groups. The changing imides and oxy groups or =B–OH groups and ammonia with following destroy of the stoichiometric course in according to $B/N > 1$ or $B/N < 1$ (Fig. 33.2III). Impurities elements for disordered graphenic-lake status of d-BN particles are becoming more suitable. All these behaviors may be observed in IR spectra [9].

Next one is a step-by-step original hydrolytic fragmentation of d-BN particles in plane of 3-degree axes in boiling water with back cooler. The result is full dissolving of d-BN particles to weak borate acid and ammonia hydroxide in the water solution. Unlike d-BN, the ordered graphite-like particles (Fig. 33.3) stay hydrophobic and inert.

This phenomenon as the reverse fractals burning was named.

The component content and dimension characteristics of d-C and d-BN particles are depended on degree its macro cyclization (DMC). Digital values of the DMC equal to the whole numbers natural series 0; 1; 2 . . . n. The atom quantities: N_C; N_B; N_N; N_O; particular molecules mass $M_{\text{d-C}}$; $M_{\text{d-BN}}$ and amount of elements (% mass. part); X_C; X_B; X_N; X_O operates of some equations.

For d-C:

$$N_{(C)} = 6\sum(2n+1); \tag{33.1}$$

$$N_{(O)} = 6n \tag{33.2}$$

$$M_{d-C} = N_{(C)}A_{(C)} + N_{(O)}A_{(O)} \tag{33.3}$$

$$X_C = N_{(C)}A_{(C)}100\%/M_{d-C} \tag{33.4}$$

$$X_O = N_{(O)}A_{(O)}100\%/M_{d-C} \tag{33.5}$$

For d-BN:

$$N_{(B=N)} = 3 \cdot \sum (2n + 1); \tag{33.6}$$

$$N_O = 3n \tag{33.7}$$

$$M_{d-BN} = N_{(B=N)}(A_B + A_N) + N_O A_O \tag{33.8}$$

$$X_B = N_{(B=N)}A_B 100\%/M_{d-BN} \tag{33.9}$$

$$X_N = N_{(B=N)}A_N 100\%/M_{d-BN} \tag{33.10}$$

$$X_O = N_O A_O 100\%/M_{d-BN} \tag{33.11}$$

where: A_C A_B, A_N, A_O atomic masses.

Average diameter of particles in dependence of DMC with using of inter layer parameter « a – 0.25040 nm [2] » was calculated with using of the circuit chart (Fig. 33.4). It has slightly under values for suitable calculations of the surfaces and densities, but in comparison of X-rays diffraction data its accuracy are better than one degree (Table 33.2 and so on).

$$d_{aav} = a \cdot \cos \cdot \beta \cdot (2n + 1) \tag{33.12}$$

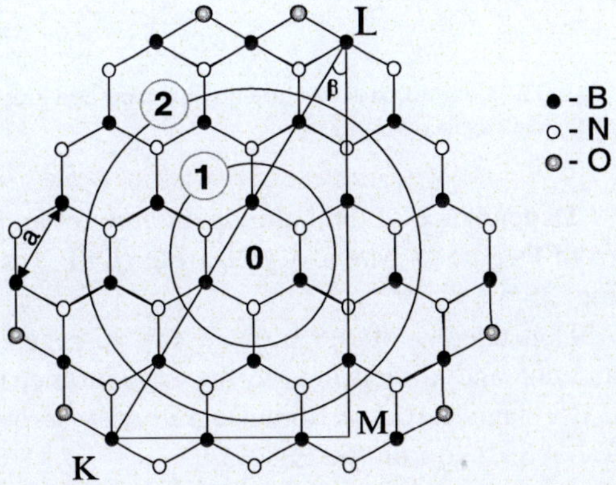

Fig. 33.4 Circuit chart of determination of the average diameter $d_{a\,av.}$ for the d-C and d-BN

Table 33.2 Dependence of the dimension ($d_{a.av}$) and oxygen amount (X_O) as function of DMC for the d-C nanoparticles

No	DMC	$X_O \pm 0.1\%$ (mass. part)	$d_{a.av} \pm 0.4$ (nm)
1	1	25.0	0.7
2	2	19.5	1.5
3	3	15.4	2.4
4	4	12.6	3.3
5	5	10.6	4.2
6	6	9.2	5.0
7	7	8.1	5.9
8	8	7.2	6.8
9	9	6.5	7.7
10	10	5.9	8.6
11	11	5.5	9.4
12	13	4.7	11.2
13	15	4.1	12.9
14	17	3.7	14.7
15	19	3.3	16.4
16	21	3.0	18.2

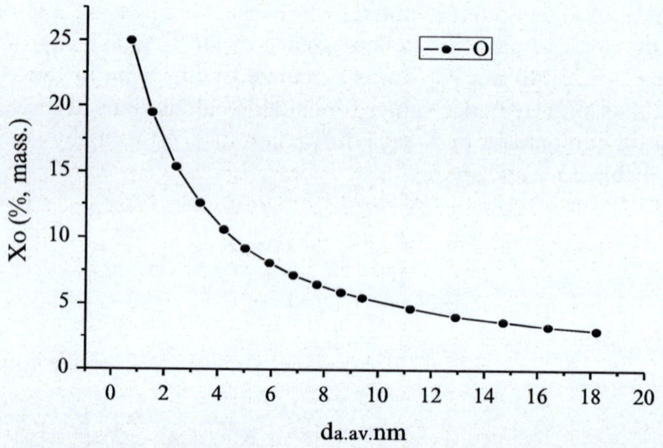

Fig. 33.5 Chart of oxygen amounts in the carbon graphenic layers as function of dimension ($d_{a.av}$) of the d-C nano particles

Dependence of the component amount and dimension of the d-C nano particles from Degree of Macro Cyclization (DMC) is shown in the Table 33.2 and on the Fig. 33.5

Dependence of the boron (X_B), nitrogen (X_N) and oxygen (X_O) component amount and dimension of the stoichiometric course in according to B/N = 1 d-BN nano particles from Degree of Macro Cyclization (DMC) is shown in the Table 33.3 and on the Fig. 33.6.

Table 33.3 Calculated data of component amount and dimension d-BN particles as function of DMC

		Component amount, $X_e \pm 0.02$ (%, mass. part)			
No	DMC	X_B	X_N	X_O	$d_{a.av} \pm 0.2$ nm
1	0	43.55	56.45	0.00	0.2
2	1	37.51	48.61	13.88	0.6
3	2	38.10	49.37	12.53	1.1
4	3	38.86	50.36	10.78	1.5
5	4	39.48	51.17	9.35	1.9
6	5	39.97	51.81	8.22	2.4
7	6	40.37	52.32	7.32	2.8
8	7	40.68	52.73	6.59	3.2
9	8	40.95	53.07	5.99	3.7
10	9	41.17	53.35	5.48	4.1
11	10	41.35	53.59	5.06	4.5
12	11	41.51	53.80	4.69	5.0
13	12	41.65	53.98	4.38	5.4
14	13	41.77	54.13	4.10	5.8
15	14	41.87	54.27	3.86	6.3

In further in case of similar experimental data for creation comparative conditions with calculated ones, all analyses of d-BN samples with non stoichiometric course were reduced in according to B/N = 1.

The experimental and calculated results of full amount and dimensions of particles for 15 powder examples of d-BN are presented in Table 33.4. All reduced experimental data (according to B/N = 1) have shown close agreement with calculated characteristics by represented model.

33.3 Conclusions

At first was mathematically founded the Phenomenon of the compound content dimensional factor $B_xN_yO_z$ and $C_{x'}O_{z'}$– f ($d_{a\ average}$) and average diameter of d-BN and d-C nanoparticles. Experimental data for d-C would be shown in future article after validation of the oxidation presage for the d-C and tasted of the oxygen amounts.

The stability of the d-BN and d-C particles is supported of docking oxygen (and other) atoms to its chemically active perimeters.

This addition causes to be in properties of bertollides for simple substances such as carbon and binary deltoids as BN.

This model is reflected of characteristic aspects of chemical amounts, IR, X-rays and physic-chemical properties of nanosized particles these compounds.

The close agreement of experimental and calculated data characteristics is successful attestation of represented model.

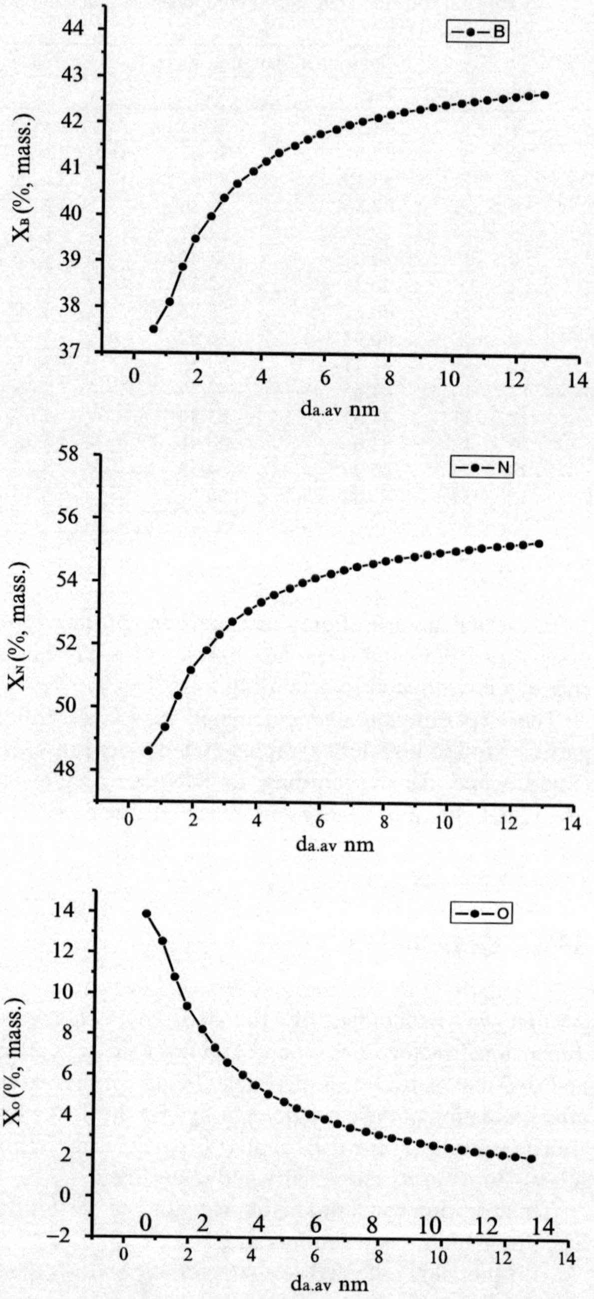

Fig. 33.6 The dimensional dependence of elements: boron, nitrogen and oxygen amount in particles of the d-BN

Table 33.4 Attestation results of calculated Plane Perimetrical Model (PPM) for the d-BN nano particles

	$X_B \pm 0.1\%$(mass. part)			$X_N \pm 0.1\%$(mass. part)			$X_O \pm 0.1\%$(mass. part)			$d_{a.\ av} \pm 0.2$ (nm)	
No	Exp. 1	Red. 2	Calc. 3	Exp. 1	Red. 2	Calc. 3	Exp. 1	Red. 2	Calc. 3	X-rays 4	Calc. 3
1	41.8	42.1	42.1	51.8	54.7	54.6	5.9	3.3	3.3	5–10	7.6
2	40.9	41.1	41.2	51.1	53.4	53.4	7.5	5.5	5.5	4–5	4.1
3	41.2	41.0	41.0	50.1	53.2	53.1	7.1	5.8	6.0	4–5	3.7
4	41.0	41.8	41.9	50.7	54.2	54.3	7.5	3.9	3.9	4–5	6.3
5	41.5	41.5	41.5	53.7	53.7	53.8	3.4	4.6	4.7	4–5	5.0
6	38.6	39.5	39.5	46.5	51.3	51.2	13.1	9.1	9.4	1–3	1.9
7	35.3	39.7	40.0	36.4	51.5	51.8	17.0	8.9	8.2	1–3	2.4
8	31.6	38.0	38.1	30.2	49.2	49.4	18.7	12.8	12.5	1–3	1.1
9	42.8	42.1	42.5	56.0	55.1	55.1	1.2	2.4	2.3	5–10	11.0
10	37.6	43.1	43.2	32.6	55.9	55.9	29.8	0.9	0.9	30–50	29.2
11	43.1	42.5	42.5	56.3	55.1	55.1	0.6	2.4	2.4	5–10	5.0
12	41.4	41.4	41.5	52.5	55.1	55.1	6.1	4.8	4.7	5–10	4.8
13[a]	37.7	37.8	37.8	47.0	49.1	49.0	15.0	13.1	13.2	–	0.9
14	42.6	42.6	42.7	55.3	55.3	55.3	2.0	2.0	2.0	10–15	12.8
15[b]	42.7	42.8	42.8	53.8	55.5	55.5	3.5	1.8	1.8	–	14.7
1	*2.0*	*3.0*	*4.0*	*5.0*	*6.0*	*7.0*	*8.0*	*9.0*	*10.0*	*11*	*12.0*

1. Experimental chemical analysis ($B/N \neq 1$)
2. Reduced about to $B/N = 1$
3. Calculated about to $B/N = 1$

[a] Example by the plasma-chemical was synthesized in the Institute for Problems of Chemical Physics (Russian AS)

[b] Experimental data of the M. Hubaček et al. [9]

References

1. Boehm HP (1966) Chemical identification of the surface groups. Adv Catal 16:179–274
2. Kurdumov AV, Britun VF, Borimchuk NI, Yarosh VV (2005) Martencite and diffuse transformations in carbon and boron nitride at shock compression. – Kiev: Kupriyanova Publishing House. pp.10–68(in Russian)
3. Kurdymov AV, Britun VF, Garbuz VV et al (2009) About impurities in powders of graphite-like boron nitride and its role in phase transformations into dense modifications by high temperature shock compression. Mater Sci Nanostruct 2:25–32 (in Russian)
4. Day K, Selbin J (1976) 2nd Ed. Theoretical inorganic chemistry. New York: Van Nostrand Reinhold, pp 133–142
5. Weinschtein BK, Fridkin VM, Indebom VL (1979) Modern crystallography. Structure of the crystals. Science 2:82–88
6. Barnard AK (1968) Theoretical basis of inorganic chemistry. Moscow: Mir, pp 64–67 (in Russian)
7. Garbuz VV (2008) Methods of gas analysis. In: Ckorohod VV, Gnecin GG (eds). Inorganic material science. Encyclopaedic edition, Kiev: Naukova Dumka, pp 858–875 (in Russian)
8. Kosolapova YaT et al (1985) Non-metal high melting compoundsMoscow: Metallurgy, pp 85–117 (in Russian)
9. Hubaček M, Sato T, Ishii T (1994) A coexistence of boron nitride and boric oxide. J Solid State Chem 109:384–390

Chapter 34
Utilizing the Waste Heat of SOFC by Newly Developed Cogeneration System

Beycan İbrahimoğlu and Sevgi Fettah

Abstract It is known that generated by solid oxide fuel cell (SOFC) high temperature exhaust gas streams leaving the stack has a temperature around 700°C. Starting from this idea, the high temperature gas streams can be used in various applications but in our design it can be used for as an extra power generation in addition to SOFC system. Because of thermoelectric (TE) modules offer many advantages over other technologies, it is selected for power generation in our design.

A TE module can be used for heat recovery as well as converting the heat to electricity directly by principle of Seebeck effect. They can be used to generate additional power with the SOFC exhaust gas. In this study, SOFC-TE modules cogeneration system designed by us, can produce electricity benefiting from the waste heat of SOFC.

In our SOFC design, the exhaust heat is recovered in a heat exchanger to preheat the fuel and air entering to the stack. Other amount of the exhaust heat is utilized in heat recovery unit for recovering the waste heat from there with using TE modules as a heat recovery unit. To cool surface of TE modules, water is chosen as a circulation fluid which comes from water tank. When a temperature differential is established between the hot and cold ends of TE modules, a voltage is generated.

Produced electricity can be used in various applications such as electronic controlling system of SOFC system and operating electrolysers for H_2 and O_2 production. H_2 produced from the electrolyser is sent to the reformer mixing here with H_2 produced from the reformer then are given to SOFC. O_2 produced from the electrolyser is given to the air pump, as a result air is being oxygen-rich air, and then is sent to SOFC. So, it is clearly seen that, O_2 enriched air will provide more efficient operation to SOFC and other aspect is; H_2 and O_2 produced by utilizing the waste heat generated from SOFC, shows the economic relevance of SOFC.

B. İbrahimoğlu (✉)
Department of Mechanical Engineering, Faculty of Engineering
Gazi University, Maltepe, Ankara, Turkey
e-mail: beycanibrahimoglu@yahoo.com

S. Fettah
Clean Energy Laboratory, Vestel Defence Industry, Golbasi, Ankara, Turkey

S.Yu. Zaginaichenko et al. (eds.), *Carbon Nanomaterials in Clean Energy Hydrogen Systems - II*, NATO Science for Peace and Security Series C: Environmental Security 2,
DOI 10.1007/978-94-007-0899-0_34, © Springer Science+Business Media B.V. 2011

Keywords SOFC fuel cell · Cogeneration · Thermoelectric module · Waste heat · Power generation

34.1 Introduction

Thermoelectricity (TE) is known as conversion between electrical energy and heat energy to each other. TE processes are explained by: Joule's law, Peltier effect, Seebeck effect and Thomson effect. TE modules consisting of dozens thermoelements of N-and P-type semiconductors. Thermoelements are connected electrically in series, thermally in parallel for different purposes of usage at different capacities to obtain the thermoelectric modules.

TE modules is used for cooling and heating applications and also can be used in power generation [1] TE modules offer many advantages over other technologies [2]:

- TE devices have no moving parts and, therefore, need substantially less maintenance.
- Life testing has shown the capability of thermoelectric devices to exceed 100,000 h of steady state operation.
- Thermoelectric devices contain no chlorofluorocarbons or other materials that may require periodic replenishment.
- The direction of heat-pumping in a thermoelectric system is fully reversible. Changing the polarity of the DC power supply causes heat to be pumped in the opposite direction – a cooler can then become a heater.
- Precise temperature control to within $\pm 0.1^\circ$C can be maintained using thermoelectric devices and the appropriate support circuitry.
- TE devices can function in environments that are too severe, too sensitive, or too small for conventional refrigeration.
- TE devices are not position-dependent.

Thermoelectric cooler, working with DC voltage, changing the flow direction can easily switch to the cooling or heating regime. Cooling of thermoelectric module is takes place by transporting of heat one surface to the other surface of TE module [3–6]. TE module cooler is used where vapour compression refrigeration systems can not be used; was began to use in small-volume but complex devices applications (cooling of integrated circuits, stabilization of solid-state laser temperature, cooling of chargers, etc...) [7, 8].

34.2 Methods

Solid oxide fuel cell (SOFC) systems high temperature exhaust gas streams leaving the stack have a temperature around 700°C. So, the high temperature gas streams can be used for additional power generation for ex. for usage of electrolyser. Because of thermoelectric (TE) modules offer many advantages over other technologies, it is selected for power generation in our design.

Thermoelectric modules can be used to generate additional power with the SOFC exhaust gas. The aim is benefiting from thermoelectric modules in order for using the waste heat in various applications.

Electrical energy generation is as follows: due to the temperature in the exhaust tank, heat is increasing on the one surface of TE module, the other surface's heat decreasing is required. In our SOFC design, the exhaust heat is recovered in a heat exchanger to preheat the fuel and air entering to the stack. Due to the temperature difference (ΔT) between surfaces of TE module, electrical energy is produced by the Seebeck effect. If cooling of TE modules surface could not achieved in a short time, thermoelectric modules are disturbed. Liquid cooling system is chosen for cooling the TE modules. Water is used as a coolant. Liquid cooling system composed of reservoir to store water and water pipes. Water entering the system with water pipes and left water the chamber after chamber wander.

In our SOFC design, the exhaust heat is recovered in a heat exchanger to preheat the fuel and air entering the stack. Operating principle of system and parts of the system are given below (Fig. 34.1):

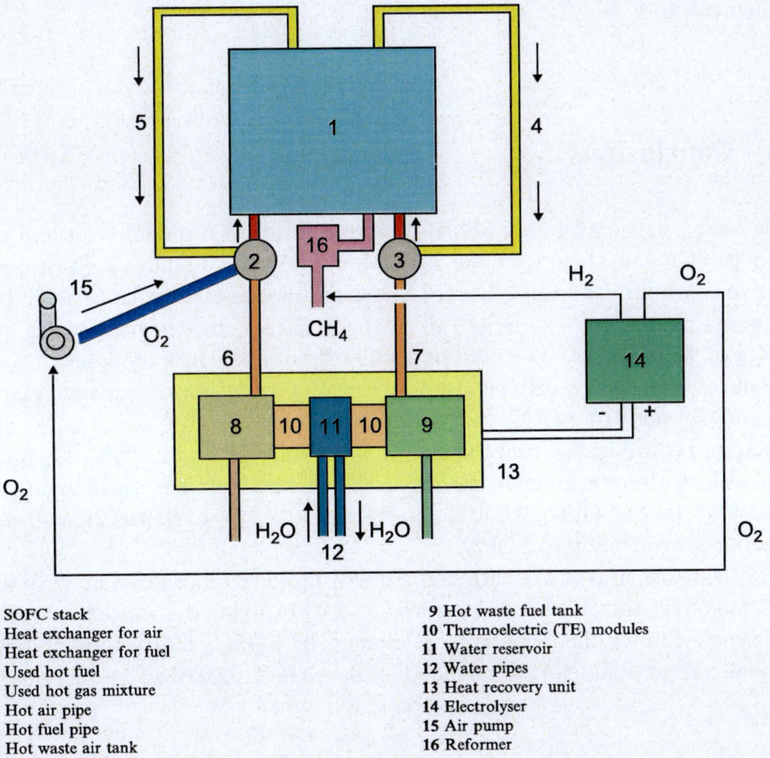

Fig. 34.1 SOFC design. *1* SOFC stack, *2* Heat exchanger for air, *3* Heat exchanger for fuel, *4* Used hot fuel, *5* Used hot gas mixture, *6* Hot air pipe, *7* Hot fuel pipe, *8* Hot waste air tank, *9* Hot waste fuel tank, *10* Thermoelectric (TE) modules, *11* Water reservoir, *12* Water pipes, *13* Heat recovery unit, *14* Electrolyser, *15* Air pump, *16* Reformer

CH4 gas as a fuel, enters to the reformer **(16)** in the reformer, 95–97% is of fuel is converted to hydrogen with a catalyst at 600°C, is fed directly into SOFC **(1)** system. In SOFC **(1)**, reached approximately temperature of 750°C gas mixture is used in the system then, is passing through the fuel pipe **(4)** and sent to the fuel exchanger **(3)**. To benefit from the heat of the gas mixture again, after heat exchanger **(3)** is sent to special copper tank **(9)**. Likewise, air; is sent to the air pump **(15)** after heating by heat exchangers **(2)** is given to SOFC **(1)**. Reacted and reached to high temperature air, is sent to heat exchanger **(2)** again and then a special copper air tank **(8)** is thrown out from there. Exhaust air and gas mixture are heated the copper tanks (**8** and **9**) to 350–400°C. TE modules **(10)** are mounted on the surface of copper tank **(8)**. The one surface of TE modules is heating via heat of copper tank (8). To obtain temperature difference (ΔT) between surfaces of TE modules **(10)** cold water **(12)** is given to other surface of TE module. Due to temperature variation of TE modules **(10)** surfaces, DC power obtained by the Seebeck effect, generated voltage is given directly to the electrolyser **(14)** without any need to rheostat. Obtained hydrogen by electrolyser **(14)**, is given directly to the reformer **(16)**, and obtained oxygen by electrolyser is sent to the air pump **(15)**. O_2 produced from the electrolyser is given to the air pump, as a result air is being oxygen-rich air, and then is sent to SOFC.

34.3 Conclusions

In this study, designed by us SOFC-thermoelectric (TE) modules system to take advantage of exhaust heat from the use of thermoelectric modules is envisaged. Our objective is utilizing the waste heat of SOFC to produce electrical energy to be used for various purposes by principle of Seebeck Effect. In designed by us system, utilizing of waste heat is executed by using of thermoelectric modules.

Produced electricity can be used in various applications such as electronic controlling system of SOFC system and operating electrolysers for H_2 and O_2 production. H_2 produced from the reformer then are given to SOFC. O_2 produced from the electrolyser is given to the air pump, as a result air is being oxygen-rich air, and then is sent to SOFC. So, it is clearly seen that, O_2 enriched air will provide more efficient operation to SOFC.

It is clearly seen that O_2 enriched air will provide more efficient operation to SOFC and other aspect is H_2 and O_2 produced by utilizing the waste heat generated from SOFC, shows the economic relevance of SOFC. Utilization of both the electrical and thermal energy produced by the SOFC, describes co-generation.

References

1. Dişlitaş S, Ahıska R (2006) Microcontroller based thermoelectric generator application. J Sci 19 (2):135–141
2. Riffat SB, Ma X (2003) Thermoelectrics: a review of present and potential applications. Appl Therm Eng 23(8):913–935
3. Güler NF, Ahıska R (2002) Design and testing of a microprocessor-controlled portable thermoelectric medical cooling kit. Appl Therm Eng 22(11):1271–1276
4. Stevens JW (2001) Optimal design of small ΔT thermoelectric generation system. Energ Convers Manage 42(6):709–720
5. Ashika R, Fidan U, Dislitas S (2008) The influence of the different temperature control systems on the thermoelectric medicine KIT's performance. J Fac Eng Arch Gazi Univ 23(2):441–447
6. Ahıska R, Güler İ, Savas Y (1999) Termoelektrik sogutucunun özelliklerinin arastırılması. Politeknik Derg 2(3):89–97
7. Min G, Rowe DM (1995) Peltier devices as generators, CRC handbook of thermoelectrics. CRC Press, London, Chap. 38
8. Rowe DM, Bhandari CM (1983) Modern thermoelectrics, Holt, Rinehart and Winston. Prentice Hall, New York

Chapter 35
NMR Investigations of Hydrogen Intercalates in GaSe Layered Crystals

Yu.I. Zhirko, Z.D. Kovalyuk, V.V. Trachevsky, and A.K. Mel'nik

Abstract The performed investigations of NMR spectra inherent to intercalates in H_XGaSe crystals at various temperatures, hydrogen concentrations and geometry of external magnetic field B enable us to refine the model of hydrogen introduction to layered crystals and offer a scheme of splitting characterizing the level of H_2 molecules in these crystals under different conditions.

It is shown, that in absence of an external magnetic field B molecular orthohydrogen as the diamagnetic gas (owing to symmetry of a crystal and nearest nuclear environment) forms in GaSe crystal three spin sub-lattices in which molecular spins located mutually-perpendicularly to each other and form right-hand rectangular system of coordinates. At $B = 0$ the total projection of two spins inside layer is located against an optical axis C and spin between layers – along axis C. It is offered the scheme of 1D and 3D QW's that is capable to explain localization of hydrogen molecules in layered crystals as well as mechanism of their deintercalation from these crystals with participation of totally symmetrical half-layer lattice $A_1{}^{/1}$-vibration taking part in scattering of H_2 molecules and in formation of oscillating parameters of potential barriers.

Keywords Layered crystal · Hydrogen · NMR · QW

Yu.I. Zhirko (✉)
Institute of Physics, NAS of Ukraine, 46 prosp. Nauky, Kyiv 03028, Ukraine
e-mail: zhirko@nas.gov.ua

Z.D. Kovalyuk
Chernivtsi Department of I.M. Frantsevich, Institute for Materials Science,
NAS of Ukraine, 5 str. Iriny Vil'de, Chernivtsi 58001, Ukraine

V.V. Trachevsky and A.K. Mel'nik
G.V. Kurdyumov Institute for Physics of Metals, NAS of Ukraine,
36 blvrd. Akademika Vernadskogo, Kyiv 03680, Ukraine

S.Yu. Zaginaichenko et al. (eds.), *Carbon Nanomaterials in Clean Energy Hydrogen Systems - II*, NATO Science for Peace and Security Series C: Environmental Security 2,
DOI 10.1007/978-94-007-0899-0_35, © Springer Science+Business Media B.V. 2011

35.1 Introduction

Layered GaSe crystals are semiconductor belonging to A^3B^6 binary chemical compounds and attract investigators interest because of heterostructures based on layered crystals possess good photosensitivity and find their application in solar cells [1–3]. Also they are perspective materials for electric charge capacitors [4].

Due to layered nature of these crystals (weak molecular van-der-Waals bonds between crystal layers and strong ion-covalent ones inside layers), the great attention of researchers is paid to intercalation, that is to insertion of atoms and molecules into interlayer space of layered crystals (the so-called "van-der-Waals gap").

It is noteworthy that the van-der-Waals gap reaches up to 40%...45% of the total crystal bulk, and its surface area inherent in 1 cm^2 measures (2...2.5) 10^3 m^2. Thereof, as it was shown in [5], these layered crystals InSe and GaSe can be used for hydrogen accumulation, and the concentration of it can reach values close to $\boldsymbol{x} = 4.0$, where $\boldsymbol{x}$ is the amount of hydrogen atoms per formula unit of the crystal.

The offered in [5] model of hydrogen introduction into layered crystals was further refined in [6]. In particular, NMR spectra of $H_{1.0}GaSe$ powders [6] under room temperature contain three bands (Fig. 35.1, curve 1) associated with molecular hydrogen. In accord with the model offered in [5] and with account of the band energy shift caused by the influence of matrix crystalline field on H_2 molecules, these bands were identified with H_2 (Fig. 35.1, insert) present in: (i) crystal layers, where strong ion-covalent bonds are active (***L***-band); (ii) interlayer space with weak van-der-Waals bonds (***I***-band); and (iii) the regions between crystalline grains of the powder (***G***-band).

In this case, as shown in [5, 6], at low concentrations ($\boldsymbol{x} \leq 2.0$) molecular hydrogen is predominantly located in interlayer space, while with growing $\boldsymbol{x}$, when this space is filled, it comes into intra-layer space. Indeed, as seen from NMR spectra of powders the integrated intensities of ***G, I, L*** bands for $\boldsymbol{x} = 1.0$ are in proportion 1:9:2.5 between each other. It has been also shown in [6] that

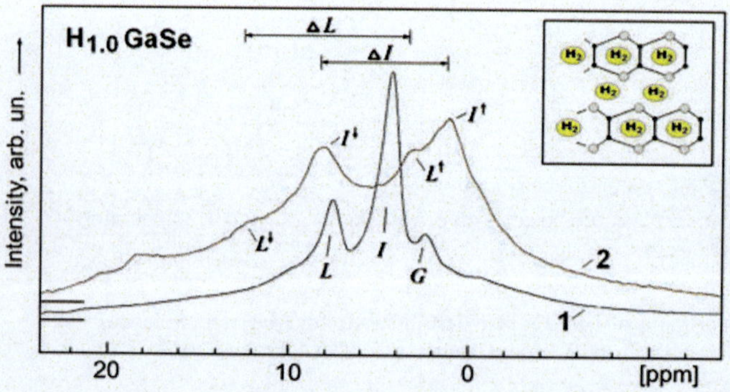

Fig. 35.1 NMR spectra of intercalates in $H_{1.0}GaSe$ single crystals (*curve 2*) and their powders (*curve 1*) at room temperature. Insert. 2D sketch of molecular hydrogen location in $H_{1.0}GaSe$ powders [7]

growth of the hydrogen concentration results in increasing $\boldsymbol{a_0}$ and $\boldsymbol{C_0}$ lattice parameters of the crystal.

This work is aimed at further refining the mechanism and dynamics of hydrogen intercalation and deintercalation from layered crystals, which is based on additional temperature ($T = 295\ldots380\ K$) radiospectroscopy investigations of GaSe single crystals intercalated with hydrogen in concentrations $\boldsymbol{x} = 0\ldots4.0$ and at different geometry of external magnetic field $\boldsymbol{B}$.

35.2 Experimental

Bulk non-doped GaSe single crystals (in ε-modification, see Fig. 35.2a) were grown using the Bridgman method. For further investigations, we prepared the samples of 10–40 μm thickness and with dimensions 5×5 mm^2. Optical and electron microscopy images for single GaSe crystal samples (see Fig. 35.2b, c) demonstrate small spherical inclusion with 100–500 nm diameter attributed with residual Se (see EDS spectra on Fig. 35.2d) located in interlayer space of a crystal. Inclusion of Se is the amorphous phase of red monoclinic β-Se. They are easy eliminated by 2–3 h annealing of a crystal at $T = 400°C$.

The following process for hydrogen intercalation to GaSe samples was realized in electrolytic solution by using the method of "sweeping field" in the galvanostatic mode. Choosing the optimum voltage and current density, we prepared the samples of H_XGaSe intercalates homogeneous in composition over the whole range of concentrations $0 < \boldsymbol{x} \leq 4.0$ where the hydrogen concentration was determined via the amount of electrical charge passing through the sample placed into an electrochemical cell. This method is described in detail in [5].

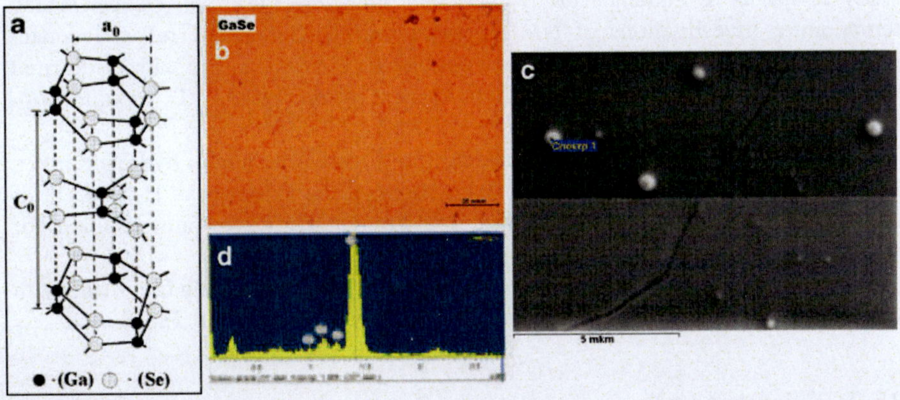

Fig. 35.2 (**a**) Crystal structure for ε-GaSe (spatial group $D_{6h}{}^4$) $C_0 = 15.95$ Å, $\mathbf{a_0} = 3.755$ Å [7]. (**b**) Optical image of single GaSe crystal sample of a 10 mkm thickness obtained at ×1,000 times magnification on optical transmission microscope Primo Star 5 (Carl Zeiss). (**c**) Image of $H_{4.0}GaSe$ crystal sample at ×10,000 magnification, and (**d**) EDS spectra in the region of Se spherical inclusion obtained on scanned electron microscope Zeiss EVO 55 XVP (Carl Zeiss) with INCA ENERGY 450 detector

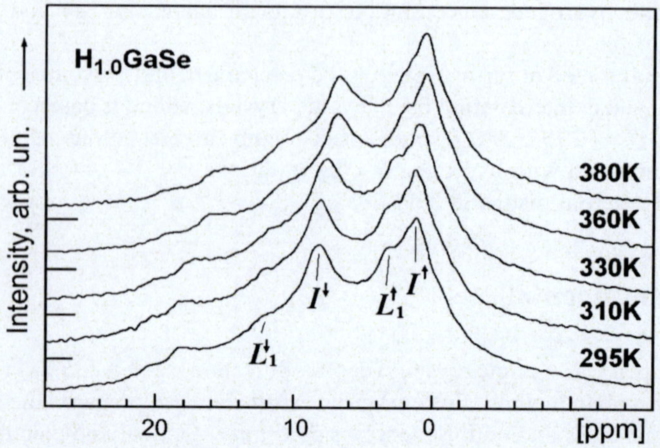

Fig. 35.3 Temperature dependences of the shift and splitting of molecular hydrogen *I*- and L_1-bands for intercalates in $H_{1.0}GaSe$ crystals

Radiospectroscopic investigation of H_XGaSe intercalates were performed using the NMR spectrometer Bruker AvanceTM^{400}. As seen from Fig. 35.1, the spectra of $H_{1.0}GaSe$ single crystals (curve 2) contain only clearly pronounced ***L***- and ***I***-bands inherent to molecular hydrogen located in layer and interlayer spaces, and in external magnetic field these bands are split by doublets $\boldsymbol{L}^{\uparrow}$, $\boldsymbol{L}^{\downarrow}$ and $\boldsymbol{I}^{\uparrow}$, $\boldsymbol{I}^{\downarrow}$. The fact that the split value $\boldsymbol{\Delta L} > \boldsymbol{\Delta I}$ also confirms conclusion [6] that H_2 molecules in H_XGaSe single crystals are located in two different crystal fields – layer and interlayer.

As it was reported in [5], deintercalation of molecular hydrogen from layered crystals takes place at the temperature 110°C and permanent pumping down. To study temperature influence on H_2 behavior in layered crystals, we performed temperature investigations of NMR spectra for $H_{1.0}GaSe$. The respective data have been depicted in Fig. 35.3, where doublet $\boldsymbol{I}^{\uparrow}$, $\boldsymbol{I}^{\downarrow}$ is associated with H_2 located in van-der-Waals gap and (renamed here for next discussion) $\boldsymbol{L_1}^{\uparrow}$, $\boldsymbol{L_1}^{\downarrow}$ doublet with H_2 located in layer cell.

To further refine the considered model [5, 6] for introduction of hydrogen into a layered crystal, we carried out the investigations of the influence of H_2 concentrations on NMR spectra. Depicted in Fig. 35.4 are the NMR spectra of H_XGaSe for the concentrations $\boldsymbol{x} = 1.0$ and $\boldsymbol{x} = 4.0$ ($T = 295$ *K*).

Obtained experimental data maximum of $\boldsymbol{I}^{\uparrow}$, $\boldsymbol{I}^{\downarrow}$ and $\boldsymbol{L_1}^{\uparrow}$, $\boldsymbol{L_1}^{\downarrow}$ bands shifting with temperature (T) and hydrogen concentrations ($\boldsymbol{x}$) was collected in Table 35.1.

35.3 Discussion

It is known that molecular hydrogen at $T \geq 295$ K consist of para-hydrogen (25%) and ortho-hydrogen (75%) molecules. At $T = 0$ all molecules are in para-hydrogen state. Para-hydrogen possesses the total molecular

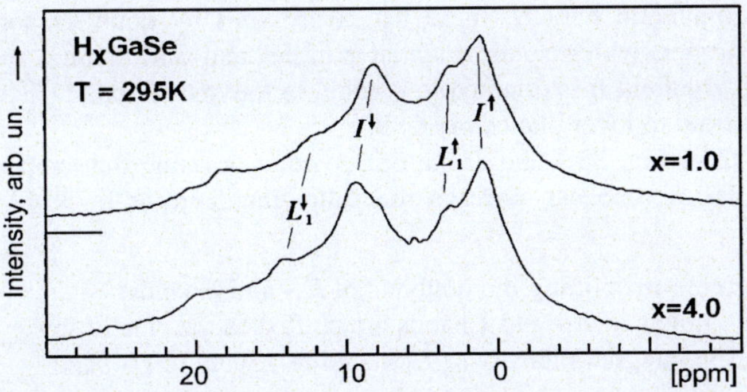

Fig. 35.4 NMR spectra for $x = 1.0$ and $x = 4.0$ hydrogen concentrations in H_XGaSe intercalates at room temperatures

Table 35.1 Data for temperature and concentration dependences of the energy shift and splitting of L_1-, L_2- and I-bands observed for H_XGaSe in external magnetic field of the NMR spectrometer Bruker Avance*TM* 400

		(ppm)								
x	T (K)	$I^{\uparrow}$	$I^{\downarrow}$	ΔI	$L_1^{\uparrow}$	$L_1^{\downarrow}$	ΔL_1	$L_2^{\uparrow}$	$L_2^{\downarrow}$	ΔL_2
4.0	295	1.0	8.4	7.4	3.3	14.05	10.75	8.0	19.72	11.72
1.0	295	1.19	8.19	7.0	3.19	12.42	9.23	7.4	18.33	10.93
	310	1.22	8.10	6.88	3.0	12.12	9.02		18.0	
	330	0.95	7.57	6.62	2.73	12.0	9.27		17.29	
	360	0.48	6.96	6.48	2.0	11.05	9.05		16.15	
	380	0.50	6.74	6.24	2.0	10.84	8.84		15.38	

spin $S_{p\text{-}H2} = S_H(+1/2) + S_H(-1/2) = \uparrow + \downarrow = 0$, while ortho-hydrogen – $S_{o\text{-}H2} = S_H(+1/2) + S_H(+1/2) = \uparrow + \uparrow = 1$. In external magnetic field ***B*** the energy level of a free H_2 molecule is split by three states with spins projection $S_{H2} = (+1, -1, 0)$.

Any hydrogen molecule in crystal experiences action of a crystalline field that causes limitation of its free precession relatively to external magnetic field ***B***, which eventually results in lowering its energy state. When measuring powder NMR spectra of H_XGaSe (analogue to the so-called "magic angle"), optical axes of crystalline grains (of 1-μm diameter, in our case) are chaotically average oriented relatively to external magnetic field ***B***, and in this case the total spin of all ortho-hydrogen dipole molecules in powder was equal to zero that corresponds with a para-hydrogen molecules. The bands ***G, L*** and ***I*** observed in NMR spectra of H_XGaSe powder (Fig. 35.1, curve 1) are related with resonant absorption of hydrogen molecules with total spin equal zero which are located in three different crystalline surroundings, namely: between grains, between layers and inside layers.

In the case of H_XGaSe single crystals, pronounced in the NMR spectra are two bands of molecular hydrogen located in interlayer and intra-layer spaces.

In external magnetic field ***B***, these bands are split by doublets corresponding to ortho-hydrogen which spins oriented parallel and anti-parallel to ***B***. In our experiments, the field ***B*** is oriented in parallel to the optical axis ***C*** that is directed along the normal to layer planes in GaSe.

As seen from Fig. 35.3 and Table 35.1, with increasing the temperature from 295 up to 380 *K*, one can observe the following features in NMR spectra of $H_{1.0}GaSe$:

- some decrease in splitting the doublets of $\boldsymbol{L_1}$- and ***I***-bands;
- the very doublets of $\boldsymbol{L_1}$- and ***I***-bands are shifted to the side of lower energy;
- lowering the integral intensity of $\boldsymbol{L_1}^{\uparrow}$-band relatively to $\boldsymbol{I}^{\uparrow}$- band.

It is also worth to note that the shift of these bands and their splitting value are in linear proportion to the inverse temperature [$\approx$ ***const***$\cdot$(1/*T*)] and the constant of the temperature shift for $\boldsymbol{L_1}^{\uparrow}$ and $\boldsymbol{I}^{\uparrow}$ bands of ortho-hydrogen oriented along the field ***B*** is somewhat lower than that one for opposite orientation.

This behavior of NMR spectra is indicative of the increasing mobility of hydrogen molecules in intra-layer and interlayer spaces with increasing temperature, which results in a reduced influence of matrix crystalline field on hydrogen. As shown by calculations within the framework of the ideal gas model [5], in GaSe crystals (for $\boldsymbol{x} = 1.0$ and $T = 300$ *K*) molecular hydrogen creates the pressure in the interlayer space equal to $P = 17.7$ MPa, which results in growth of the lattice parameter $\boldsymbol{C_0}$. This fact is also confirmed by direct X-ray investigations [6].

When $\boldsymbol{x} = 1.0$ and $T = 380$ *K*, the H_2 pressure value in the interlayer space is 22.4 MPa. In accord with [5], at $T = 384$ *K* (110°C) and weak pumping down hydrogen leaves crystal: for the low concentration $\boldsymbol{x} = 0.1$, its yield is approximately 60%, and for $\boldsymbol{x} = 4.0$, when molecular hydrogen occupies both interlayer and intra-layer spaces, its yield reaches 85–90%.

It seems obvious to expect that growth of the hydrogen concentration resulting in the pressure increase in interlayer space and, respectively, in growing the lattice parameters $\boldsymbol{a_0}$ and $\boldsymbol{C_0}$ [6], should result in the increasing mobility of hydrogen in intra-layer and interlayer spaces, i.e., to drop in splitting of $\boldsymbol{L_1}$- and ***I***-bands as well as to their shift into the range of lower [***ppm***] values. However, as it can be seen in Figs. 35.4 and 35.5, with growing the hydrogen concentration ***x*** the splitting of doublets in $\boldsymbol{L_1}$- and ***I***- bands is increased, and $\boldsymbol{L_1}^{\uparrow}$- , $\boldsymbol{L_1}^{\downarrow}$- and $\boldsymbol{I}^{\downarrow}$-bands (except $\boldsymbol{I}^{\uparrow}$-band) are shifted to the range of higher [***ppm***] values.

It is indicative of the fact that, with increasing the hydrogen concentration, the influence of matrix crystalline field on the hydrogen molecule grows, and fixed orientation of the molecule relatively external magnetic field ***B*** is kept. It should be also noted in accord with [5] that for $\boldsymbol{x} \geq 2.0$ hydrogen, having filled all the translationally ordered states in interlayer space of layered crystal, goes into intra-layer space by virtue of quantum-dimensional effects. It also leads to a growing pressure inside crystalline layers. In this case, as seen in Fig. 35.4 the splitting value for $\boldsymbol{L_1}$-band is larger than that for ***I***-band.

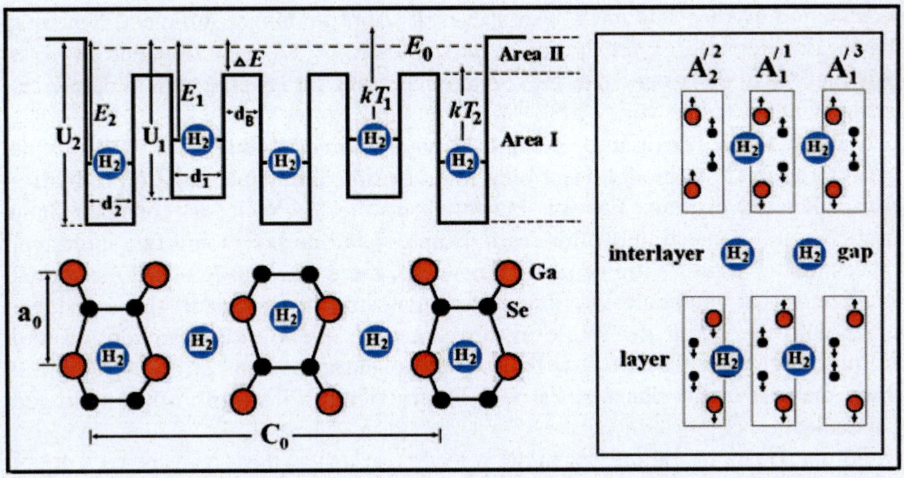

Fig. 35.5 2D sketch and scheme of H_2 molecule 1D-localization in QW's of interlayer and intra-layer spaces of GaSe crystal at $T = 0$. Insert. Motion of unit cell atoms for full symmetric vibrations in ε-GaSe crystal

35.3.1 Discussion of the Model

Note that even at high concentrations ($\boldsymbol{x} = 4.0$) molecular hydrogen in the interlayer space at room temperature can be considered as condensed gas that, with the concentration growth, comes more and more into crystalline layers, which due to increased pressure enhances lattice parameters.

Performed in [5] an estimation was made of the level energy value for hydrogen molecule localization inside a one-dimensional well inherent to the interlayer space. In accord with the known expression

$$E_0 = \frac{\pi^2 \hbar^2}{2M_{H2} d_z^2} \tag{35.1}$$

as well as values for hydrogen molecule mass $\boldsymbol{M}_{\mathrm{H2}} = 3.673$ a.u. and width of the interlayer space $\boldsymbol{d}_{\boldsymbol{Z}} = 0.308$ nm for InSe crystal [8], obtained was the energy of the localization level $\boldsymbol{E}_0 = 1.1$ meV. Note that the considered one-dimensional case describes appearance of the level in the interlayer space in a right manner, since motion of the molecule in the interlayer plane has a quasi-continuous spectrum.

However, the estimation of $\boldsymbol{E}_0$ value performed in [5] explains localization of H_2 in interlayer space only qualitatively. It is quite sufficient to consider processes of two-dimensional exciton movement in the plane of crystalline layers when hydrogen fills the interlayer space. But to explain behavior of NMR spectra for molecular hydrogen with increasing the temperature and concentration as well as its localization in inter- and intra-layer spaces, it seems no longer sufficient. Note that the localization energy for hydrogen molecules $\boldsymbol{E}_0 = 1.1$ meV obtained in [5] is much less than the kinetic energies which necessary for H_2 molecule to deintercalating from crystal($\boldsymbol{kT}$ for $T = 384\ K$ is equal to 33.0 meV).

With this aim, taking into account the data obtained in this work, let us refine the offered in [5, 6] model describing intercalation and deintercalation of hydrogen in layered crystals. For this purpose (see Fig. 35.5), let us consider GaSe single crystal consisting of three cells from crystalline layers and two interlayer spaces that contain H_2 molecules. In this case, there takes place one-dimensional localization of molecules in the direction normal to layers in the interlayer space with the width $\boldsymbol{d}_1$. While in the planes of layers, molecular motion will be quasi-free, i.e., free for low hydrogen concentrations and strongly localized when the rest of the phase space will be practically filled with other hydrogen molecules.

For the H_2 molecule present in the cell of layer space, there takes place a three-dimensional potential well with the energy of a ground localization level (111) in the case of cubic geometry (linear sizes of the well in three directions ***x, y, z*** are the same, i.e. $\boldsymbol{d}_2{}^{X} = \boldsymbol{d}_2{}^{Y} = \boldsymbol{d}_2{}^{Z} \equiv \boldsymbol{d}_2$) that is equal

$$E_0 = \frac{3\pi^2\hbar^2}{2M_{H2}d_2^2}. \tag{35.1a}$$

Thus, we have a set of two type quantum wells, namely: one- and three-dimensional ones, where the hydrogen molecule can be present, being limited with potential barriers of finite width and height. To shorten the description of H_2 molecule behavior, let us simplify the model by changing the 3D well with 1D one. As seen from Eqs. 35.1 and 35.1a, at equal energies of the ground state the only well width should be changed.

With this simplification, let us construct a set of five 1D wells (Fig. 35.5) separated with potential barriers. Since during intercalation the molecular hydrogen penetration into the intra-layer space is more difficult than into the interlayer one, three deeper wells correspond to localization of H_2 inside crystalline layers and two more shallow – to localization of H_2 in interlayer space. This area of quantum wells with localized there H_2 molecules is marked in the figure as Area I. Depicted on a Fig. 35.5 case corresponds to the concentration $\boldsymbol{x} = 4.0$, when $T = 0\ K$; $\boldsymbol{B} = 0$. The rest over-barrier phase space of the layered crystal is designated as Area II and was considered latter.

Before proceeding to calculation of well parameters in Area I, note some additional requirements:

1. When considering quantum well geometrical parameters $\boldsymbol{d}_1$ and $\boldsymbol{d}_2$, it is necessary to take into account the fact that the diameter of a hydrogen molecule $\mathbf{D}_{H2} = 0.148$ nm is comparable with the width of interlayer space $\boldsymbol{d}_1$. Moreover, covalent radii of selenium and gallium atoms $\mathbf{R}_{Se} = 0.116$ nm and $\mathbf{R}_{Ga} = 0.126$ nm in the layered crystal GaSe possess sizes comparable with $\mathbf{D}_{H2}$, too. The

above particle sizes should be taken into account when determining the efficient geometrical sizes of QW's

$$\delta_1 = d_1 - D_{H2} \quad \text{and} \quad \delta_2 = d_2 - D_{H2} \tag{35.2}$$

as well as d_B – the width of the potential barrier between them

$$d_B = [(C_0/2) - 2D_{H2} - \delta_1 - \delta_2]/2, \tag{35.2a}$$

where $C_0 = 1.595$ nm [7] is the lattice parameter of the crystal.

2. Since the molecule spin is equal to integer (0, 1), molecular hydrogen obeys Bose statistics. Then, levels in each of the QW sets are degenerated and possess the ground level E_1 for interlayer space and E_2 for the intra-layer one. Assume also that in every QW the hydrogen molecule possesses only localization level and $E_2 > E_1$. The depths of QW's have finite values, and $U_2 > U_1$. In this case, the localization levels of H_2 molecules can be found as solutions of the quadratic equation

$$E_{(1,2)} \cong U_{(1,2)} - \frac{M_{H2}\delta^2_{(1,2)}}{2\pi^2\hbar^2}U^2_{(1,2)}. \tag{35.3}$$

3. Since at $T = 384\ K$ molecular hydrogen comes out of the interlayer space, let the level energy will be as $E_1 = kT_1 = 33.0$ meV.
4. At the same time, at $T = 384\ K$ and $x = 4.0$ with a weak pumping down approximately 85% of hydrogen comes out of the crystal. Therefore, transfer of hydrogen from the intra-layer space to the interlayer one should take place at $T \leq 384\ K$, i.e. $kT_2 \leq kT_1$.
5. We assume that at temperatures above 330 K molecular H_2 is in a combined state (Area II), i.e., with increasing the temperature up to 380 K, one can observe the only interlayer hydrogen present in a large common well. Thus, the difference between levels is $\Delta E = E_2 - E_1 = 3.0$ meV, where $E_2 = 36.0$ meV.

Indeed, as seen from the NMR spectra (Fig. 35.3), at temperatures $T > 330\ K\ L_1{}^{\uparrow}$-band becomes to vanish, and at $T = 380\ K$ one can observe the only $I^{\uparrow}$-band that (see Table 35.1) practically stops its shifting to the range of lower [*ppm*] values. While $L_1{}^{\downarrow}$ - and $I^{\downarrow}$- bands caused by resonance absorption related with molecular ortho-hydrogen oriented anti-parallel to the external magnetic field ***B*** and being in more low-energy states prolong to shift with increasing temperature.

Thereof, in accord with the set H_2 localization levels E_1 and E_2 and using the expression (35.3), let us determine the parameters of quantum wells in absence of external magnetic field ***B***:

- the interlayer QW depth is $U_1 = 64$ meV, efficient width $\delta_1 = 0.0557$ nm, $d_1 = 0.2037$ nm;
- the intra-layer QW depth is $U_2 = 69$ meV, $\delta_2 = 0.0533$ nm, $d_2 = 0.2013$ nm;
- and in accord with (35.2a), the barrier width between them is $d_B = 0.1473$ nm.

Note that parameters $\boldsymbol{\delta}_1$ and $\boldsymbol{\delta}_2$ obtained without any account of the experimental fact that, with increasing the temperature and hydrogen concentration in the layered crystal, $\boldsymbol{a}_0$ и $\boldsymbol{C}_0$ grow, too, which enhances the well width $\boldsymbol{d}_1$, $\boldsymbol{d}_2$ and changes the barrier thickness $\boldsymbol{d}_{\mathrm{B}}$ in places where hydrogen is present. By other words $\boldsymbol{d}_1$, $\boldsymbol{d}_2$, $\boldsymbol{d}_{\mathrm{B}} = \boldsymbol{f}(\boldsymbol{T}, \boldsymbol{x})$.

35.3.2 The Case $B = 0, T > 0, x > 0$

As it was noted above, growth of the temperature forces the molecular hydrogen to come out of QW's to Area II and fill the phase space of the layered crystal, and for sufficient concentrations when $\boldsymbol{x} \to 4.0$ creates in it some level more $\boldsymbol{E}_0$ with the energy ≤ 1.0 meV. For $\boldsymbol{x} \to 0$, the level energy $\boldsymbol{E}_0 \to 0$: H_2 behavior is close to the quasi-free one limited by the crystal bulk and atoms of its composition. By this cause, at low concentrations the hydrogen amount leaving the crystal does not exceed 60%...70%, while at high concentrations it reaches 80%...90%.

With increasing the temperature, the molecule kinetic energy in the QW can be enhanced due to scattering at potential well walls with absorption of vibration energy quantum from crystalline lattice. When $T = 0$, the potential barrier walls in the first approximation are rigid and fixed (we do not take into account zero vibrations of atoms). But increase in temperature means that phonons of the crystalline lattice become active, the density of population for i-phonon branch Ω_i grows with temperature in accord with the law

$$n_i = \frac{1}{e^{\hbar\Omega_i/kT} - 1}. \tag{35.4}$$

Periodical shifts of atoms in lattice that are caused by presence of phonons, results in changing the potential barrier width $\boldsymbol{d}_{\mathrm{B}}$, which one influence the energy of hydrogen molecule localization: for inelastic H_2 reflection from the barrier wall (with absorption of a phonon) the molecule can pass to Area II.

The process of H_2 molecule scattering by lattice vibrations will be the most efficient when: (i) the molecule kinetic energy in the well is close to the phonon energy, and (ii) the momentum of the molecule reflected from the barrier wall coincides (is summed up) with the barrier wall momentum (phonon momentum), which is possible when lattice atoms in the course of vibrations change the width of the QW and barrier.

In the case of GaSe crystal, the unit cell of which consists of eight atoms located in two crystal layers, there exist 24 normal vibrations of the lattice [9].

Three acoustic vibrations of them for $\boldsymbol{k} = 0$, where $\boldsymbol{k}$ is the wave vector of the crystal reciprocal lattice, possess the energy equal to zero and do not directly contribute to the molecule kinetic energy. All the doubly-degenerated $\boldsymbol{E}$-vibrations also do not essentially contribute to the molecule kinetic energy, as they do not take part in significant changing $\boldsymbol{d}_1$, $\boldsymbol{d}_2$ and $\boldsymbol{d}_{\mathrm{B}}$ width. Only three transverse (totally symmetrical) $\boldsymbol{A}$-vibrations can provide direct contribution to changing the molecule kinetic energy because they can change QW width. Motion of unite cell atoms during this vibration is shown on Insert of Fig. 35.5.

The first totally symmetrical $\boldsymbol{A}_2{}^{/2}$-vibration (interlayer) is not valid here, because: (i) it possesses the low energy (40 cm^{-1}) to scatter H_2 molecule from QW; (ii) interlayer space is changed, but the thickness of the crystalline layer (barrier width $\boldsymbol{d}_B$ and $\boldsymbol{d}_2$) remains constant.

The second $\boldsymbol{A}_1{}^{/1}$-vibration (half-layer with energy 133 cm^{-1}) can change width $\boldsymbol{d}_1$, $\boldsymbol{d}_2$ of QW's but remain unchanged barrier width $\boldsymbol{d}_B$ between them. It was an active at $T = 300\ K$ to take part in molecule scattering from both type QW's.

The third $\boldsymbol{A}_1{}^{/3}$-vibration (intra-layer with energy 307 cm^{-1}) are realized with changing both the volume of the crystalline layer cell and the thickness of the interlayer space, but $\boldsymbol{A}_1{}^{/3}$-vibration is still insufficiently active even at temperatures close to 380 K.

Therefore, we assume that the totally symmetrical half-layer $\boldsymbol{A}_1{}^{/1}$-vibration takes main part in the processes of hydrogen molecule scattering in the wells of two kinds. Also, this vibration takes very active part in the processes of exciton decay [10].

35.3.3 *The Case* $B \neq 0, T > 0, x > 0$

As seen from Figs. 35.3 and 35.4, the application of external magnetic field $\boldsymbol{B}$ results in vanishing degeneration and splitting of the ortho-hydrogen molecule level in every well. There arise doublets of $\boldsymbol{L}_1$ and $\boldsymbol{I}$-bands for ortho-hydrogen molecules in NMR spectra of H_XGaSe crystals, these molecules being in two different layer and interlayer crystalline fields. Energy levels of para-hydrogen molecules with zero spin (not observed in NMR spectra of single crystal) remain the same.

At the same time as molecular hydrogen in a gaseous-like state possess diamagnetic properties the amount of ortho-hydrogen molecules oriented in parallel and anti-parallel direction to applied external magnetic field $\boldsymbol{B}$ must be equal one to another. Really on Figs. 35.3 and 35.4 one can see that summarized integral intensities of $\boldsymbol{I}^{\uparrow}$ and $\boldsymbol{L}^{\uparrow}$ bands for ortho-hydrogen oriented along $\boldsymbol{B}$ are in well coincidence with summarized integral intensities of $\boldsymbol{I}^{\downarrow}$ and $\boldsymbol{L}^{\downarrow}$ bands for ortho-hydrogen oriented opposite $\boldsymbol{B}$.

Finally the NMR investigations of $H_{1.0}$GaSe single crystals at $T = 295\ K$ with different orientation of external magnetic field $\boldsymbol{B} \perp \boldsymbol{C}$ and $\boldsymbol{B} \| \boldsymbol{C}$ (where $\boldsymbol{C}$ is a crystal axis oriented normally to crystal payer plane) were conducted. Experimental data (see Fig. 35.6) cannot find energetic shift of H_2 absorption bands maximums with changing $\boldsymbol{B}$ orientation. But at the same time slight redistribution of integral intensities for $\boldsymbol{I}$ and $\boldsymbol{L}_1$ bands occurred. Thus in $\boldsymbol{B} \perp \boldsymbol{C}$ geometry $\boldsymbol{I}^{\downarrow}$-band increased and are compatible (in intensities) with $\boldsymbol{I}^{\uparrow}$-band but $\boldsymbol{L}_1{}^{\downarrow}$-band decreased. For $\boldsymbol{B} \| \boldsymbol{C}$ geometry their mutual redistribution are reciprocal.

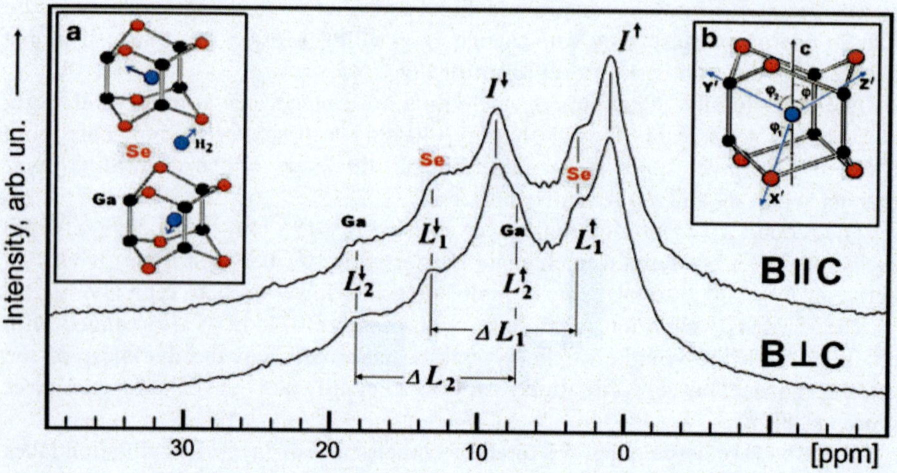

Fig. 35.6 NMR spectra of $H_{1.0}GaSe$ single crystal at $T = 295\ K$ for longitudinal and transverse orientation of applied external magnetic field $\boldsymbol{B}$ according to crystal axis $\boldsymbol{C}$. Insert. (**a**) 2D sketch presenting spin orientation of ortho-hydrogen molecule in a layer and intra-layer crystal space. (**b**) Projection of ortho-hydrogen molecule spins in interlayer and intra-layer space of GaSe crystal on right-hand orthogonal coordinate system $X'Y'Z'$

Also in geometry $\boldsymbol{B}\perp\boldsymbol{C}$ we can clearly observe additional one doublet of $L_2^{\uparrow}$ and $L_2^{\downarrow}$ bands. Note that these additional $L_2^{\uparrow}$and $L_2^{\downarrow}$ bands also observed in Figs. 35.3 and 35.4 but they are not so pronounced because of some intermediate $\boldsymbol{B}\angle\boldsymbol{C}$ geometry. Temperature and concentration dependencies for $L_2^{\uparrow}$and $L_2^{\downarrow}$ bands are also collected in Table 35.1. It is important that L_2 doublet has greater shift than L_1 doublet and $\Delta L_2 > \Delta L_1$.

Two L_1 and L_2 absorption doublets for hydrogen molecule in a crystal layer evidenced about presence of two different position of H_2 inside a crystal cell. Really in layer space (see insert of a Fig. 35.6) for H_2 molecule located in center of a cell there are two environments consisting from atoms Ga and Se. The nearest sub-lattice consisted from Ga atoms give the greatest contribution of a crystal field into chemical shift and is identified by us to a doublet L_2. The Se atoms are removed further from the centre of a cell: therefore contribution of Se atoms in chemical shift is less. We identify the contribution of Se atoms sub-lattice to a doublet L_1. For everyone sub-lattice there are three equivalent dislocations of a ortho-hydrogen molecule along an $\boldsymbol{C}$ axis and three in opposite.

With nearest-neighbour distances $\boldsymbol{d}$ obtained in [8] for InSe: $d_{\text{In-In}} = 2.79$ Å; $d_{\text{In-Se}} = 2.65$ Å; $\angle\varphi = 119°30'$, layer thickness $d_l = 5.36$ Å, the distance between layers $d_i = 3.08$ Å and interlayer distance $d_{\text{Se-Se}} = 3.80$ Å., one can find an angle $\angle$ between optical axis $\boldsymbol{C}$ and nearest Se ($\angle\varphi_1$) and Ga ($\angle\varphi_2$) atoms by values $\angle\varphi_1 = \pm 40°45'$ and $\angle\varphi_2 = \pm 59°$ respectively.

To define orientation of hydrogen molecule in interlayer space it is essential that each layers are shift one to another and an angle of Ga-Ga-Se bound in layer is about 120°. In this circumstance angles in Ga and Se directions in interlayer space are degenerated, equal to $\angle\varphi = \pm 54^\circ 10'$ and have an intermediate value between $\angle\varphi_1$ and $\angle\varphi_2$ angles.

As it is seen on Fig. 35.6 intra-layer doublets $\boldsymbol{L}_1$ and $\boldsymbol{L}_2$ also differently react to a direction of an external magnetic field $\boldsymbol{B}$ concerning axis $\boldsymbol{C}$. So in geometry $\boldsymbol{B}\perp\boldsymbol{C}$ doublet $\boldsymbol{L}_2$ are more pronounced, and for $\boldsymbol{B}||\boldsymbol{C}$ geometry $\boldsymbol{L}_1$ bands are more appreciable. It allows us to make a conclusion that in external magnetic field $\boldsymbol{B}$ spin orientation of H_2 molecule inside a layer in Ga and Se sub-lattices try to compensate each other. It is possible only in that case, when H_2 spins in Ga and Se sub-lattices have opposite directions concerning axis $\boldsymbol{C}$. Thus in absence of an external magnetic field $\boldsymbol{B}$ a corner of a spin inclination concerning an axis $\boldsymbol{C}$ in direction to Se atom is negative $\angle\varphi_1 = -40^\circ 45'$ and in direction to Ga atom is positive $\angle\varphi_2 = +59^\circ$ respectively. In this case as it is seen on an insert of a Fig. 35.6 corner between H_2 spins in Ga and Se sub-lattices with good accuracy are equal to 90° and their summarized projection on axis $\boldsymbol{C}$ are negative.

In interlayer space at $\boldsymbol{B} = 0$ an angle between spin of ortho-hydrogen molecule and axis $\boldsymbol{C}$ was positive: $\angle\varphi = +54^\circ 10'$. It is not easy to show that in the whole three various directions of H_2 spins orientation in GaSe made a right-hand rectangular system of coordinates $X'Y'Z'$, which allow compensate H_2 spins in absence of external magnetic field $\boldsymbol{B}$. Presented on inserts (a) and (b) of Fig. 35.6 schema of H_2 molecule spin projection clearly shown the reason of $\boldsymbol{I}$, $\boldsymbol{L}_1$ and $\boldsymbol{L}_2$ – band doublets redistribution with changing $\boldsymbol{B}$ orientation according to axis $\boldsymbol{C}$.

Finally, using the data obtained as a result of these investigations, in Fig. 35.7 we have shown the scheme of splitting the level of a hydrogen molecule in H_XGaSe single crystals and powders in external magnetic field $\boldsymbol{B}$ of NMR spectrometer Bruker AvanceTM^{400} with changing temperature and hydrogen concentration. These data was collected in Table 35.1.

In insert the scheme of hydrogen molecule localization in layer and interlayer QW's with account of two Ga and Se sub-lattices of a layer cell are proposed. The given scheme consists of 1D interlayer QW, projection of a 3D crystal cell QW on a figure plane and potential barrier between them, which becomes transparent enough at high kT.

The scheme allows to:

(i) show splitting of H_2 ground state in 3D QW at $\boldsymbol{B} = 0$ and $\boldsymbol{B} > 0$ due to different crystal field of a nearest Ga and Se atomic sub-lattices constituting crystal cell;
(ii) differentiate phase spaces of hydrogen molecule located in 1D and 3D QW's and
(iii) demonstrate process of hydrogen exit from 3D to 1D QW with growing kT. In this case when H_2 molecule exit from 3D to 1D QW it loses energy and localized in interlayer 1D QW. Accordingly the process of hydrogen entry in crystal cells occurs at high enough temperatures and concentration that allows to overcome a potential barrier created by atoms of a crystal cell. Nevertheless some part of hydrogen molecules at deintercalation remains localized in crystal cells.

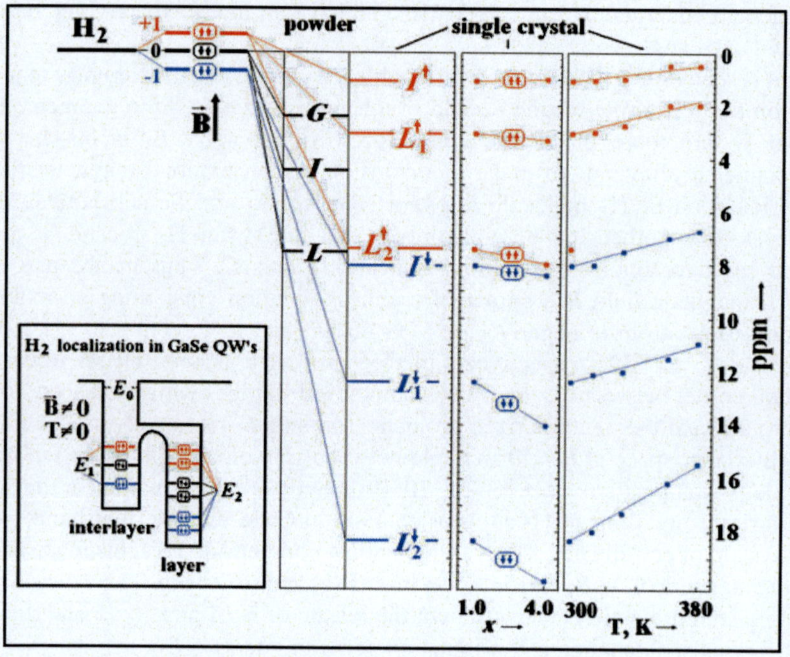

Fig. 35.7 Scheme of splitting the level of a hydrogen molecule in H_XGaSe single crystal and powder in external magnetic field (*B*) with changing the hydrogen concentration (*x*) and temperature (*T*)

Also at large hydrogen concentration and high temperatures, when hydrogen molecules are not located in QW's and posses high kinetic energy in volume of interlayer space appearance of an additional quazi-local level with energy $\boldsymbol{E}_0$ is possible.

35.4 Conclusions

The performed investigations of NMR spectra inherent to intercalates in H_XGaSe crystals at various temperatures, hydrogen concentrations and geometry of external magnetic field $\boldsymbol{B}$ enable us to refine the model of hydrogen introduction to layered crystals and offer a scheme of splitting characterizing the level of H_2 molecules in these crystals under different conditions.

It has been ascertained that in the course of intercalation atomic hydrogen recombine to the molecular state, occupies interlayer space of a crystal, and due to quantum-dimensional effects comes into the interstitial space, occupies centers of crystalline layer cells where its recombination to molecular state takes place. It results in localization of hydrogen molecules, which, in its turn, leads to growth of lattice parameters $\boldsymbol{a}_0$ and $\boldsymbol{C}_0$.

It has been shown, that in absence of an external magnetic field $\boldsymbol{B}$ molecular hydrogen, as the diamagnetic gas (owing to symmetry of a crystal and nearest nuclear environment) forms in GaSe three spin sub-lattices in which spins located mutually-perpendicularly to each other and form right-hand rectangular system of coordinates. At $\boldsymbol{B} = 0$ the total projection of a two spins inside a layer as a whole is located against an optical axis $\boldsymbol{C}$, and between layers along an axis $\boldsymbol{C}$.

It has been offered the scheme of 1D and 3D quantum wells that is capable to explain localization of hydrogen molecules in layered crystals in the course of intercalation as well as mechanism of their deintercalation from the crystals with participation of totally symmetrical lattice $\boldsymbol{A}_1{}^{/1}$-phonons taking part in scattering of molecules and in formation of oscillating parameters of potential barriers and quantum wells. In addition, described is the mechanism of H_2 scattering by $\boldsymbol{A}_1{}^{/1}$-phonons in quantum wells.

35.5 Afterwords

At the same time, conducted in [11] optical (exciton absorption of H_XGaSe at $T \geq 4.5$ *K*) and electrophysical (Hall-effect of H_XInSe at $T \geq 80$ *K*) investigations as well as (to be published) EPR investigations of InSe and GaSe crystals intercalated with hydrogen show that very small fraction of hydrogen ($\approx$0.001%) in H_XInSe and H_XGaSe intercalates, being in atomic states, passivate point lattice defects, too. In InSe crystals of $\boldsymbol{n}$-type, passivation of deep donor levels by hydrogen results in an essential increase of the free carrier concentration (from 10^{15} cm^{-3} up to 10^{16} cm^{-3}) for $T \geq 80$ *K*. When $T = 120$ *K*, the EPR line half-width and $\boldsymbol{g}$-factor of the hydrogen unpaired electron for H_XGaSe are higher than those for H_XInSe.

References

1. Lebedev AA, Rud' V Yu, Rud' Yu V (1998) Photosensitivity of geterostructures porous solicon-layered $A^{III}B^{VI}$ semiconductors. Fiz Tekch Polupr 32:353–355 (in Russian)
2. Martines-Pastor J, Segura A, Valdes JL (1987) Electrical and photovoltaic properties of indium-tin-oxide/p-InSe/Au solar cells. J Appl Phys 62:1477–1483
3. Shigetomi S, Ikari T (2000) Electrical and photovoltaic properties of Cu-doped *p*-GaSe/*n*-InSe heterojunction. J Appl Phys 88:1520–1524
4. Grigorchak II, Zaslonkin AV, Kovalyuk ZD, Mintjanskii IV, Savitskii PI (2002) Galvanic element. Patent of Ukraine No46137, Bulletin № 5
5. Zhirko YuI, Kovalyuk ZD, Pyrlja MM, Boledzyuk VB (2007) Application of layered InSe and GaSe crystals and powders for solid state hydrogen storage. In: Vezirogly TN et al (eds) Hydrogen materials science and chemistry of carbon nano-materials. Springer, Berlin, pp 325–340
6. Zhirko YuI, Kovalyuk ZD, Klad'ko VP, Trachevsky VV, Shapovalova IP, Vorsovsky AL (2007) Investigation of hydrogen intercalation in layerted crystals InSe and GaSe. In:

Proceedings of 15th international conference on hydrogen economy and hydrogen treatment of materials (HTM-2007), Vol 2, Donetsk, 21–25 May 2007, pp 606–610
7. Kuhn A, Chevy A, Chevalier R (1975) Crystal structure and interatomic distance in GaSe. Phys Status Solidi B 31:469–473
8. Olguin D, Cantarero A, Ulrich C, Suassen K (2003) Effect of pressure on structural properties and energy band gaps of γ-InSe. Phys Status Solidi B 235:456–463
9. Belenkii GL, Stopachinskii VB (1983) Electronic and vibration spectra of A^3B^6 layered semiconductors. Usp Fiz Nauk 140:233–270 (in Russian)
10. Zhirko YuI (2000) Investigation of the light absorption mechanisms near exciton resonance in layered crystals. Part 2. N = 1 state exciton absorption in GaSe. Phys Status Solidi B 219:47–61
11. Zhirko Yu, Kovalyuk Z, Zaslonkin A, Boledzyuk V (2010) Photo and electric properties of hydrogen intercalated InSe and GaSe layered crystals. In: Proceedings of international conference Nano/Molecular Photochemistry and Nanomaterials for Green Energy Development (Solar'10), Cairo, 15–17 Feb 2010, pp 48–49

Chapter 36
Nanomaterials as a New Eco-Threat: Chemical and Nanotoxicological Peculiarities

A.I. Kharlamov and A.V. Skripnichenko

Abstract Experts-ecologists of Journal of New Scientist consider two main eco-threats of coming twenty-first century: the toxicity of environment and an acidation of the ocean. Both these threats are the consequences of objective development of humanity. Appearance of new eco-threat is connected to recent opening of novel state of matter: nanomaterials, nanoparticles and nanostructures. Unique feature of these nanoobjects is that their behaviour, chemical and toxicological properties essentially differ from nowadays known substances. These new substances as owing to nanosmall dimension and lightness are distributed everywhere: air, water, food, clothes, textile, cosmetic and packing. Such nanoobjects can easily penetrate in organism of the human being by all accessible routes (nose, mouth and skin) and because of extremely high chemical activity it weaken (or injure) work of various organs. Therefore eco-nanothreat is practically uncontrollable nowadays. However nanoparticles of different substances (carbon, gold, silver, and titanium, silicon, iron and zinc oxides) in huge amount are produced and the first results of study of influence of nanoparticles of some substances on living organism are extraordinarily threatening.

Keywords Nanomaterials · Nanoparticles · Eco-nanothreat · Nanotoxicology · Nanodemocratic threat

36.1 Introduction

Nanomaterials are not quite poison: they can destruct and transform all living organisms.

Quite recently (almost 25 years before) were held unusual experiment in field of physics and after in field of chemistry. By means of electronic microscopes of

A.I. Kharlamov (✉) and A.V. Skripnichenko
Frantsevych Institute for Problems of Materials Science of NAS of Ukraine,
Krzhyzhanovsky str. 3, UA- 03142 Kiev, Ukraine
e-mail: dep73@ipms.kiev.ua

S.Yu. Zaginaichenko et al. (eds.), *Carbon Nanomaterials in Clean Energy Hydrogen Systems - II*, NATO Science for Peace and Security Series C: Environmental Security 2,
DOI 10.1007/978-94-007-0899-0_36,

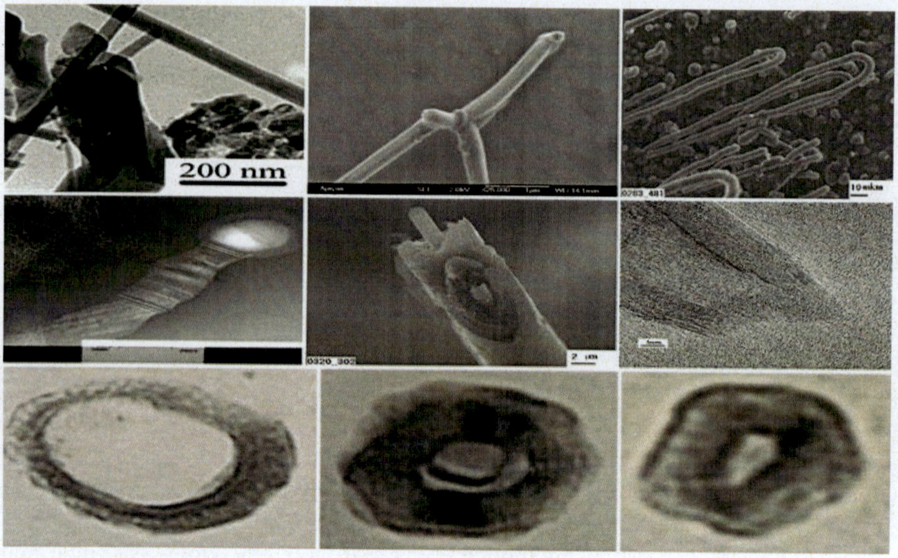

Fig. 36.1 Nanostructures of silicon carbide [1–4], silicon [5] and carbon [6–9] which were synthesized by us

nanodimensional level were revealed nanosized objects such as spheroidical molecules with a hollow core (fullerenes and nanotubes), nanoparticles and nanophases (mainly noble metals) as well as nanostructures of different morphology: multiwalled nanotubes, nanorods, nanowires, nanospheres, onions, toroids (Fig. 36.1). The principle peculiarity of these new substances is that their structure and properties principally differ from the previously known substances.

In addition, their properties depend on size and morphology of nanoobjects. In fact, revealed nanosized objects are the carriers of new unknown yet chemical and physical properties and, consequently, their impact on living organism is uncontrollable nowadays. An epoch of "nano" (nanochemistry, nanotechnology, nanomediciny and, certainly, nanoecology), epoch of forming of nanology [10] as science about the nanoworld (Fig. 36.2) started. Now it is need to create the instrumentations and procedures for correct study of properties and especially toxicological peculiarities of these new substances. However in present time many substances in the shape of nanodimentional particles and structures are produced and used in very large amount as cosmetics, water purification, bulking agent and packing. Therefore today it is important to understand main chemical peculiarities of applied (especially into food and cosmetics) nanoobjects in order to foresee the effects of their harmful impact.

In the present report we will discuss the reasons of appearance of new eco-threat (nanothreat), will present the classification of some known and the most applied and harmful nanomaterials as well as will consider their chemical peculiarities and available nanotoxicological results.

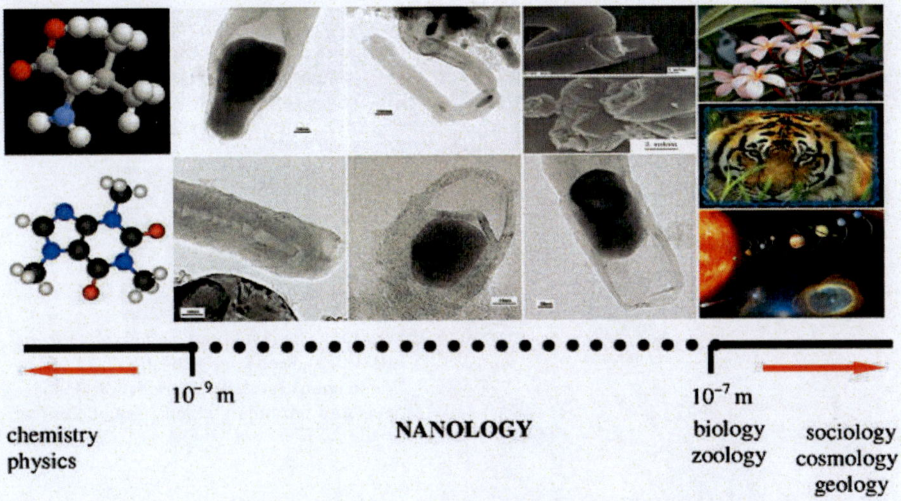

Fig. 36.2 Logical place of nanology as a science about nanoworld among well-known science

36.2 Nanothreat Peculiarity and the Reasons of Its Appearance

Among experts-ecologists the most harmful and probable in twenty-first century are considered eco-threats caused by the contaminations of an environment by toxic substances and the acidation of the ocean. Natural catastrophes caused by global warming and appearance ozone holes at uncontrolled ejection industrial gases are possible also (heat and drought in Russia by summer of 2010 year). These eco-threats are anthropogenic ones and the consequences of their realization may be the poisoning of living organisms and human, warming, heat, drought and famine. In addition the technogenic catastrophes may have a global character also (Chernobul (SSSR 1986 year)), crash of atomic submarine Kursk (Russia 2000 year). We want to tell also about coming demographic threat concerning with mild dying out from the predominance of the rights and the freedoms of human over his religion-ethical regulations. If the level of reproduction on our planet reduces to this day level of Germany the date of the dying out will be 2400 year (John Lesli). In future we will see that it is not the first threat connected with universal development of democracy.

With discovery of nanoworld and following forming of nanochemistry and nanotechnology development it is value to tell about the possibility of integral change of life quality (Fig. 36.3). So, nanomedicine dreams about immortality owing to creation of an opportunity of regular updating of cells can be already soon (approximately in 50 years). Nanodoctors are going to clear internal organs of people as well as find and recover diseased cells. It is proposed that nanorobot enter in human organism with water drink can be introduced. Well-known nanoexpert Kriss Feniks proposed the project of a replacement of human blood by multitude (approximately 500 billion) nanorobots. Weight of such blood composes 2 kg.

Today, with opening of nanoworld, we have a main question:
What carry nanoparticles and nanotechnologies products (ubiquitous nanosystems) for humanity?
HEAVEN OR HELL? **Of course, People majority expect the improvement of life quality**

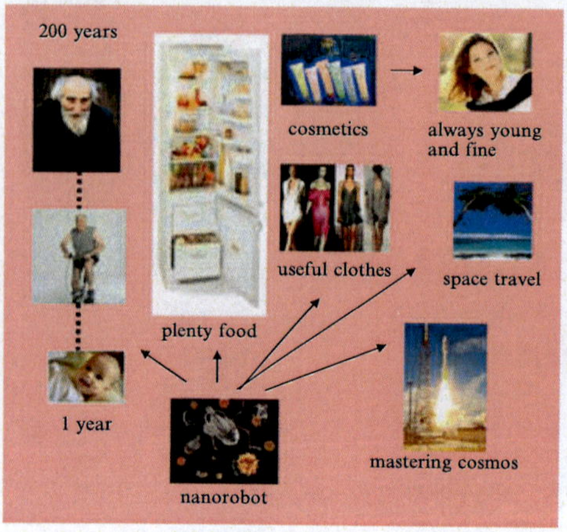

Everybody dreams about abundance (plenty and tasty food), long, healthy and happy life spans (and even immortality), space travels and nanoman which created comfort for Person

Fig. 36.3 Main dreams of people concerning the improving of life quality

Nanorobots will be produced from a sapphire. So blood bacteria, viruses and parasites will not be to contain. Military dreams to made a clever dust (or "flying insects") as nanosystems capable to generate energy, to cooperate among themselves and external world. The flying insects, being precipitation on object transfer the complete information about its state. On a signal this object can be destroyed at any moment. The cost price of such insects will make approximately 10 cents and their nanofactory production will be directly on a field of fight.

However, possible consequences of nanotechnology on depth of influence on progress of humanity may be compared to the opening of a radio-activity or with computer revolution. It is doubtless, that the changes of such scale can carry only positive consequences. These negative consequences may be caused by nanotechnological, nanoecological and nanodemocratic threats. The hazards connected to industrial production of very dangerous nanoobjects and extraordinary smart nanosystems may be unavoidable in the future. Well-known nanologist Drexler thinks that nanorobots as the collectors of new nanosystems can be programmed on self-manufacture. At self-duplication they can consume usual alive essences and to transform all biomass of the Earth in "grey goo".

Nanotechnology will allow essentially to reconstruct human organism and to operate energy balance without the use of products of a food and, consequently, will be created nanohuman. These people of the future (nanopeople) by means of interfaces will cooperate with the computer (nanorobots) directly through nervous system. Since nanoprogress is created predominantly simultaneously with development and deepening of democracy personal nanosensors of a residence, intelligence and feelings will the main threats at nanodemocracy epoch. Therefore, nanodemocracy is nonviolent manipulation by means of ubiquitous nanosystems (pervasive nanosensors of intellect and feeling) consciousness of the overpowering majority of nanopeople. So, created in Australia the microrobot (by length 0.25 mm) has an independent source of energy, camera and system of communication. This microrobot is capable to travel controlled inside man organism and to fulfil specific tasks. So, the robots with the mechanism of genetic prompting can be nanoweapon of ethnic assignment.

Uncontrolled distribution of nanomaterials on environment and their penetration into human organism may lead to not only poisoning but to different pathologies (to Koschei immortal). The reason of such powerful impact of nanomaterials is caused by shapes multiplicity and chemical features of new carriers of properties.

36.3 Nanochemical and Nanotoxicological Peculiarities of Nanomaterials

In short let is note that nanomaterials can differ on nanostructured materials (units of building are nanostructures and nanoparticles), powdery materials (partially contain nanosized objects as ingredients) and isolated nanostructures as the components at building of more complex systems. Therefore, nanomaterials comprise nanoobjects always. Nanoobjects as new state of the matter are classified on four groups concerning their shape, size and structure (Fig. 36.4).

In present time at huge manufacture such size of nanoobjects are distributed everywhere (air, water, food, clothes, cosmetics and packing) and they are capable to penetrate in organism of a man by all possible routes (Fig. 36.5):

- by respiratory organs (adsorbing on a huge surface lungs, it is easy to be soaked up in blood, passing a liver as a purifying barrier); a digestive tract; through a skin.

Uncontrollable penetration of hard revealed new substances present unpredictable threat for human organism as far as they have very unusual chemical and physical peculiarities.

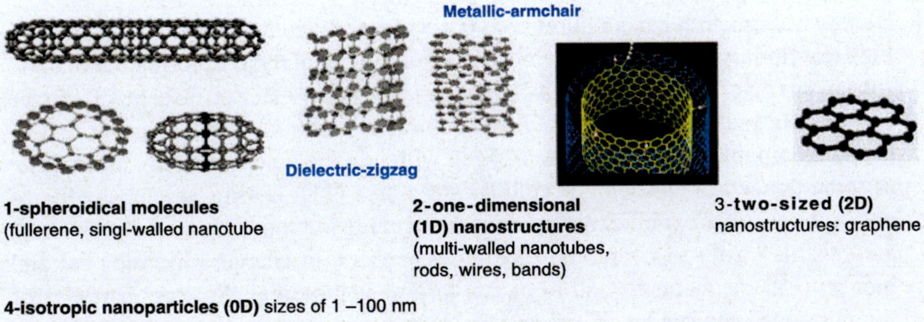

Fig. 36.4 Schema of classification of nanoobjects

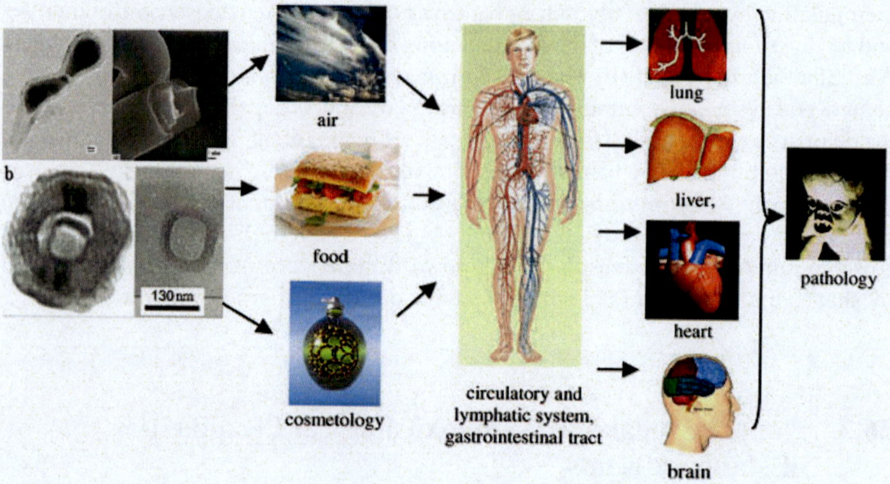

Fig. 36.5 Routes of penetration of nanoparticles and nanostructures into human organism

1. The dimension of some nanoparticles is compared with the size of molecules. (Diameter of fullerene molecule C_{60} is 0,7 nm whereas the size of viruses is 20–500 nm and bacterium is 100–5,000 nm). That is why it is very hard to detect and remove them from an environment by traditional methods of filtration. However such size of nanoparticles of titanium and zinc oxides (TiO_2 and ZnO) are successfully applied in cosmetics, in particular, in anti-sun creams. Nanoparticles from such creams and the clothes also easily (in contrast to bacteria and viruses) penetrate through large (in comparison with the size of a number of nanoparticles) pores of human skin (Fig. 36.6).
2. Suddenness of the change of properties of nanoparticles into living organism as far as the properties of, for example, nanophase depend from the size of this particle. Here we would like to remind the difference between the terms "nano-particle" and "nanophase". A term "nanoparticle" means that it is only nanosized part of macrophase. Nanophase is a nanosized part of macrophase but it has principally another properties or these properties depend on its size.
3. Unusual morphology of nanoobjects (Fig. 36.7);
4. Nanoparticles surface has a number of free valences therefore they have extremely high reactionary ability and in living organism can be catalysts of formation of toxic substances. Besides, carbon molecules as very good electron acceptors can be carcinogenic substances. Probably as acceptors of electrons carbon various nanos-tructures capable to induce active oxygen with unpaired electron (radical) that can at some concentration damage cellular structures [11]. Owing to nanosmall size and high reactionary ability nanoparticles and nanostructures of various substances are capable easily to overcome biological barriers, to change physiological and biochemical mechanisms and to cause various pathologies. So, nanoparticles of titanium and zinc oxides (TiO2 and ZnO) catalyse photooxidation [12] whereas nanoparticles of iron oxides can cause metallic (zinc) fever [13, 14]. Nanoparticles

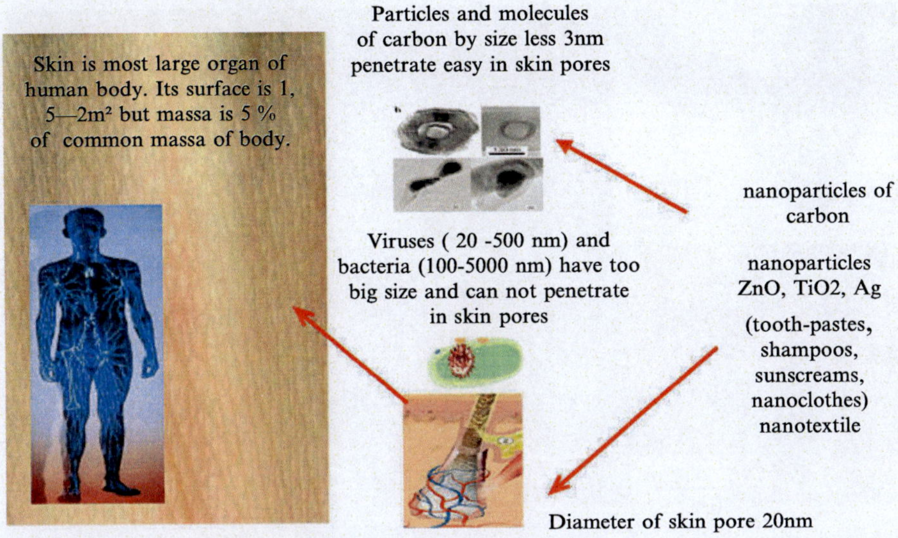

Fig. 36.6 Diagram of penetration of nanopaticles through skin pores

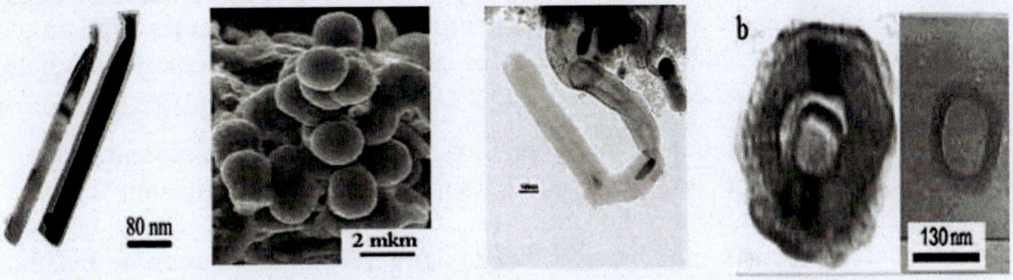

Fig. 36.7 Basic morphologies of nanostructures: rods, onions, tube and toroid

of aluminium promote overgrowth of a organism fabric for the account of formation and duplication of cells [15]. Nanoparticles of iron cause inflammatory processes in a stomach and intestine (mice, birds and fish), and also change during formation, development and maturing of blood cells [16]. It is need to know also that different food packings comprise nanoparticles of selenium, silver, zinc oxide, titanium dioxide and many other inorganic substances, which impact on health of the human being can be very harmful. Nanoparticles of zinc and iron oxide (sizes more than 22 nm) cause an induction of the active forms of oxygen in cells of rats, anemia (decreasing of concentration of haemoglobin in blood) and infringement of system of curtailing of blood [17–19]. So, single-walled carbon nanotubes activate a beginning of neoplasms growth into kidney [11, 20].

5. Carbon nanoparticles (especially partially destroyed) can be as capacious containers for carcinogenic substances and can exert double influence on health of

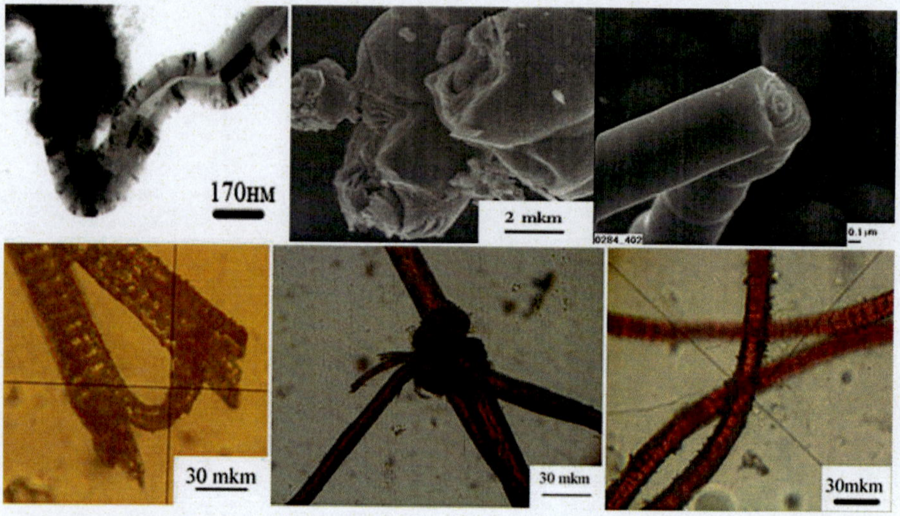

Fig. 36.8 Carbon nanostructures as containers of molecules of thread-like colored crystals

the people. So, by us was shown [21, 22] that on surface of carbon structures (Fig. 36.8) obtained at pyrolysis of hydrocarbons the molecules of various substances are located. So, we established that thread-like colored crystals from prepared benzene (toluene) extracts grow (Fig. 36.8).

The preliminary researches have shown [21, 22] that these transparent macro-threads consist of more thin microthread and contain practically only carbon. Therefore, flying particles and nanostructures of carbon (fume, soot) always contain nanoamount of another substances which toxicity very hard to examine but the result of their impact can be very harmful.

36.4 Conclusions

Appearance of new eco-threat namely nanothreat caused by recent opening of new substances (spatial molecules, nanostructures, nanoparticle, nanophases and nano-materials) which toxicological properties are dangerously and extremely unpredict-able. These nanoobjects are distributed everywhere (air, water, textile, clothes, foodstuff, cosmetics and packing) and very easy penetrate into human organism through skin pores, mouth, nose and eyes. Because of nanosmall dimension and extremely high reactivity they are capable destroy different human organs. In near future at successful nanotechnology progress can appear the new nanothreats (nanothechnological and nanodemocratic threats) which will be peculiar to only for developed as rule democratic countries.

References

1. Kharlamov AI, Kirillova NV, Karachevtseva LA, Kharlamova AA (2003) Low-temperature reactions between vaporizing silicon and carbon. Theor Exp Chem 39(6):374–379
2. Kholmanov I, Kharlamov AI, Milani P et al (2002) A simple method for the synthesis of silicon carbide. nanorods. J Nanosci Nanotechnol 2(5):453–456
3. Kharlamov AI, Kirillova NV (2002) Gas-phase reactions of formation of silicon carbide nanofilaments from silicon and carbon powders. Theor Exp Chem 38(1):59–63
4. Kharlamov AI, Kirillova NV, Loytchenko SV et al (2002) Synthesis of elongated nanostructures of silicon carbide from powdery silicon and carbon. Reports of Academia of Science of Ukraine 2002(10): 98–105 (in Russian)
5. Kharlamov AI, Kirillova NV, Karachevtseva LA, Fomenko VV,Bondarenko ME (2006) Vapor-gaseous process of low-thermal (1200 C) transformation of polycrystalline silicon to its highly orient anisotropic particles. Reports of Academia of Science of Ukraine 2006(12): 48–55 (in Russian)
6. Kharlamov AI, Loythenko S V, Kirillova NV, Kaverina SV Fomenko VV (2004) Toroidal nanostructures of carbon. Single-walled 4 –, 5 – and 6 hedrons and nanorings. Report of Academia of Science of Ukraine (1):95 –100 (in Russian)
7. Kharlamov AI, Kirillova NV, Ushkalov LN (2006) Simultaneous growth of spheroidal and tubular carbon structures during the pyrolysis of benzene. Theor Exp Chem 42(2):90–95
8. Kharlamov AI, Ushkalov LN, Kirillova NV, Fomenko VV, Gubareny NI Skripnichenko AV (2006) Synthesis of onion nanostructures of carbon at pyrolysys of aromatic hydrocarbons. Report of Academia of Science of Ukraine (3):97–103 (in Russian)
9. Kharlamov AI, Kharlamova GA, Kirillova NV, Fomenko VV (2008) Persistent organic pollutants at nanotechnology and their impact on people health. In: Mehmetli E, Koumanova B (eds) The fate of persistent organic pollutants in the environment. Springer, Dordrecht, pp 425–441
10. Kharlamov AI, Kirillova NV (2009) Fullerenes and fullerenes hydrides as products of transformation (polycondensation) of aromatic hydrocarbons. Report of Academia of Science of Ukraine (5):112–120 (in Russian)
11. Donaldson K, Aitken R, Tran L, Stone V, Duffin R, Forrest G, Alexander A (2006) Carbon nanotubes: review of their properties in relation to pulmonary toxicology and workplace safety. Toxicol Sci 92(1):5–22
12. Alt V, Bechert Th, Steinrücke P, Wagener M, Seidel P, Dingeldein E, Domann E, Schnettler R (2004) An in vitro assessment of the antibacterial properties and cytotoxicity of nanoparticulate silver bone cement. Biomaterials 25(18):4383–4391
13. Oberdörster G, Oberdörster E, Oberdörster J (2005) Nanotoxicology: an emerging discipline from studies of ultrafine particles. Environ Health Perspect 113:823–839
14. Hoet P, Bruske-Holfeld I, Salata O (2004) Nanoparticles – known and unknown health risks. J Nanobiotechnol 2:12
15. Chen L (2008) Manufactured aluminum oxide nanoparticles decrease expression of tight junction proteins in brain vasculature. Neuroimmune Pharmacol 3:286–295
16. Kovalenko LV, Folmanis GE (2006) Biologically active nanopowders of iron. Nauka, Moscow, p 124, in Russian
17. Wang B (2006) Acute toxicity of nano- and micro-scale zinc powder in healthy adult mice. Toxicol Lett 161:115–123
18. Zhu M-T, Feng WY, Wang B, Wang T-Ch, Gu Y-Q, Wang M, Wang Y, Ouyang H, Zhao Y-L, Chai Z-F (2008) Comparative study of pulmonary responses to nano- and submicron-sized ferric oxide in rats. Toxicology 247:102–111
19. Heinlaan M, Ivask A, Blinov I, Dubourguier H-Ch, Kahru A (2008) Toxicity of nanosized and bulk ZnO, CuO and TiO2 to bacteria Vibrio fischeri and crustaceans, Daphnia magna and Thamnocephalus platyurus. Chemosphere 71:1308–1316

20. Jiang J, Oberdrster G, Elder A, Gelein R, Mercer P, Biswas P (2008) Does nanoparticle activity depend upon size and crystal phase? Nanotoxicology 2(1):33–42
21. Kharlamova G, Kharlamov A, Kirillova N, Skripnichenko A (2008) Novel transparent molecular crystals of carbon. In: Vaseashta A, Mihailescu I (eds) Functionalized nanoscale materials, devices, and systems. Springer, Dordrecht, pp 373–379
22. Kharlamov AI, Kirillova NV, Zaytseva ZA (2007) Novel state of carbon: transparent thread-like anisotropic crystals. Report of Academia of Science of Ukraine (5):101–106 (in Russian)

Chapter 37
The Structural Properties of the Sp^1-Carbon Based Materials: Linear Carbon Chains, Carbyne Crystals and a New Carbon Material – Two Dimentional Ordered Linear-Chain Carbon

J.G. Korobova, M.B. Guseva, D.I. Bazhanov, and V.V. Khvostov

Abstract In this paper we have presented experimental diagnostic data: electron diffraction, ESCA and Raman spectroscopy for different materials based on sp^1-carbon – carbyne crystals and a new film carbon material – two-dimensional ordered linear-chain carbon. We have described their structural models, based on this data. The stability of atomic structure and electronic properties of those materials have been investigated theoretically by means of first-principals calculations based on density functional theory. Also there were theoretically studied short and infinite linear carbon chains with different geometry (straight, kinked and branched). The obtained results have shown that kinks in single carbon chains could be stabilized by hydrogen impurities. A fundamental role of hydrogen on structure stabilization of two-dimensional ordered linear-chain carbon films has been found also. The influence of nitrogen impurities (found in our experiments) on shape and stability of a crystal structure of carbyne has been investigated.

Keywords Electronic properties · Carbine crystal · Film · Linear – chain · Nitrogen · Stability · Shape · Electron diffraction

37.1 Introduction

Since 1959 when material based on sp^1-carbon was obtained for the first time (it was carbyne), problem of formation and stability of a linear-chain carbon is extensively studying [1, 2]. Then, in 1959, there were obtained small (less 100 nm) carbyne crystals by means of chemical methods with following heating [3]. From that time till now numerous methods for obtaining sp^1-carbon were described but always they got small nanocrystals needed heating during formation process.

J.G. Korobova (✉), M.B. Guseva, D.I. Bazhanov, and V.V. Khvostov
Physical Department, Lomonosov Moscow State University, Leninskie gory, house 1, building 2, 119899 Moscow, Russia
e-mail: KorobovaJG@yandex.ru

S.Yu. Zaginaichenko et al. (eds.), *Carbon Nanomaterials in Clean Energy Hydrogen Systems - II*, NATO Science for Peace and Security Series C: Environmental Security 2, DOI 10.1007/978-94-007-0899-0_37,

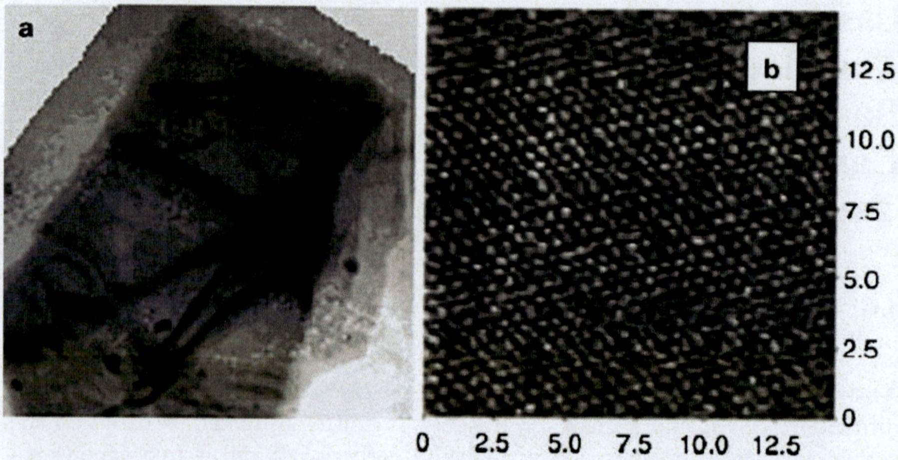

Fig. 37.1 (**a**) Carbyne crystal phase contrast picture; (**b**) AFM picture of TDO LCC film

Basically, the problem of instability of chain-like carbon is related to low stability of the free ensemble of the free linear-chain carbon clusters with sp^1-type electron hybridization. The stability of linear-chain carbon can be improved if the chains grow on the surface of a substrate, as one end of the carbon chain is fixed on the substrate surface. The stability of chainlike carbon can also be enhanced as a result of the parallel growth of carbon chains in the perpendicular direction to the substrate surface [4].

In 1990 s in our laboratory a new form of sp^1-carbon was evaporated by means of ion-assisted condensation of carbon on NaCl surface [4]. It was film structure named two dimensional ordered linear-chain carbon (TDO LCC). Later it was proposed a new pulsed-plasma condensation method [5], which allowed depositing such films on large surfaces and at high growth rate (Fig. 37.1).

In this work we have investigated sp^1-carbon based materials experimentally and theoretically. First of all, we would like to describe experimental data and structural models of carbyne and TDO LCC.

Carbynes are linear allotropic form of carbon [1]. In contrast with graphite and diamond whose carbon atoms are in state of sp^2 and sp^3-hybridization of atomic orbitals the structure of carbyne is characterized by sp^1- hybridization of atomic orbitals. There are some models to describe different diffraction patterns of carbyne obtained in different methods [6, 7] but the most suitable for experimental results were described in our previous experimental work [4]. Carbyne in that work was prepared using two different techniques: (A) low-temperature carbonization of a polyvinylidene fluoride (PVDF) from using a chemical dehydrofluorination reaction [8]; (B) ion sputtering of graphite in combination with a bombardment of growing carbon film by Ar + ions [9]. Both methods gave similar diffraction pattern (see Fig. 37.2a) [4]. The diffraction pattern of the microinclusions of carbyne single crystals demonstrates an excellent sharpness (see Fig. 37.2b).

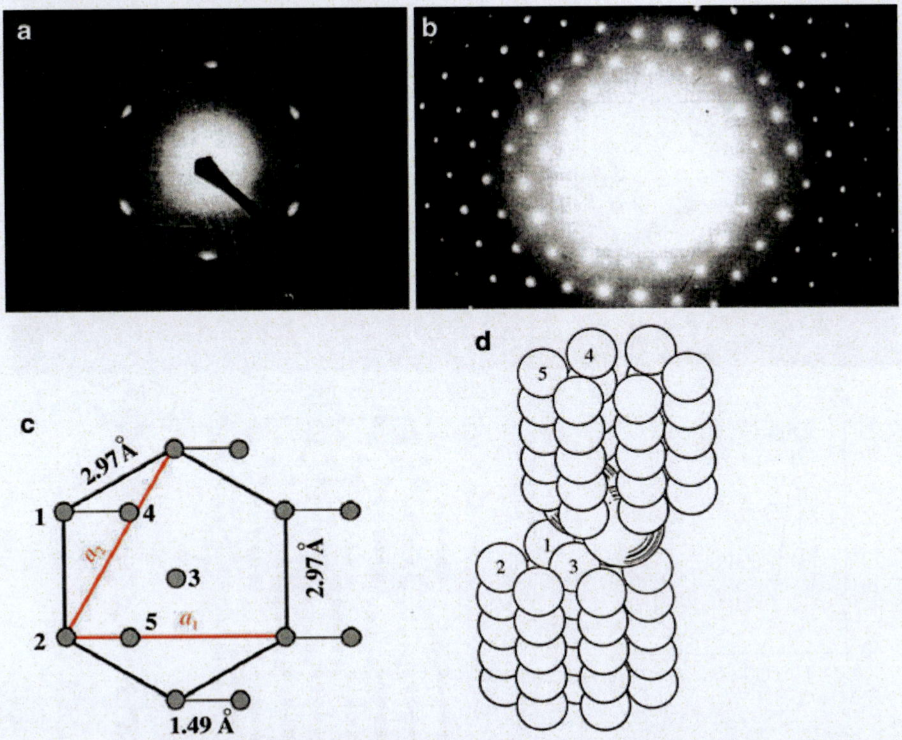

Fig. 37.2 (**a**) Electron diffraction pattern of the carbyne prepared using method A; (**b**) electron diffraction pattern of a single crystal microinclusion in the carbyne prepared using method B; (**c**) model of the position of carbon chains in a unit cell of carbyne crystal in (000*l*) projection; (**d**) spatial model of carbyne structure [4]

Structural model appropriate to observed experimental results and based on the analysis of the Patterson function for X-ray diffraction is shown in the Fig. 37.2c (projection (000 *l*)) and 2d. The chains (1) and (2) are located at the corners of a hexagon; one is located at the center (3), and two others ((4) and (5)) are shifted by 1,49 Å along a translation vector relative to corner chains. The chains are located in adjacent layers (i.e. the chains (1), (2), and (3) are located in the lower layer, while the chains (4) and (5) are located in the upper one), see Fig. 37.2d. So carbyne has a supercell consist of two subcells: (I) – hexagonal packed carbon chains with one chain in the middle of the hexagon and (II) – hexagonal packed carbon chains without one chain in the middle of the hexagon shifted in 1.49 Å as shown in Fig. 37.2c.

Films of TDO LCC are a new carbon form accurately described in our earlier experimental paper [10]. Then we had found that these films of TDO LCC were characterized by unique electron diffraction, as shown in Fig. 37.3.

The electron diffraction pattern consists of only one bright narrow diffraction ring (Fig. 37.3a) or six point reflections equally spaced in this ring (Fig. 37.3 b). The electron diffraction pattern shown in Fig. 37.3b corresponds to the hexagonal

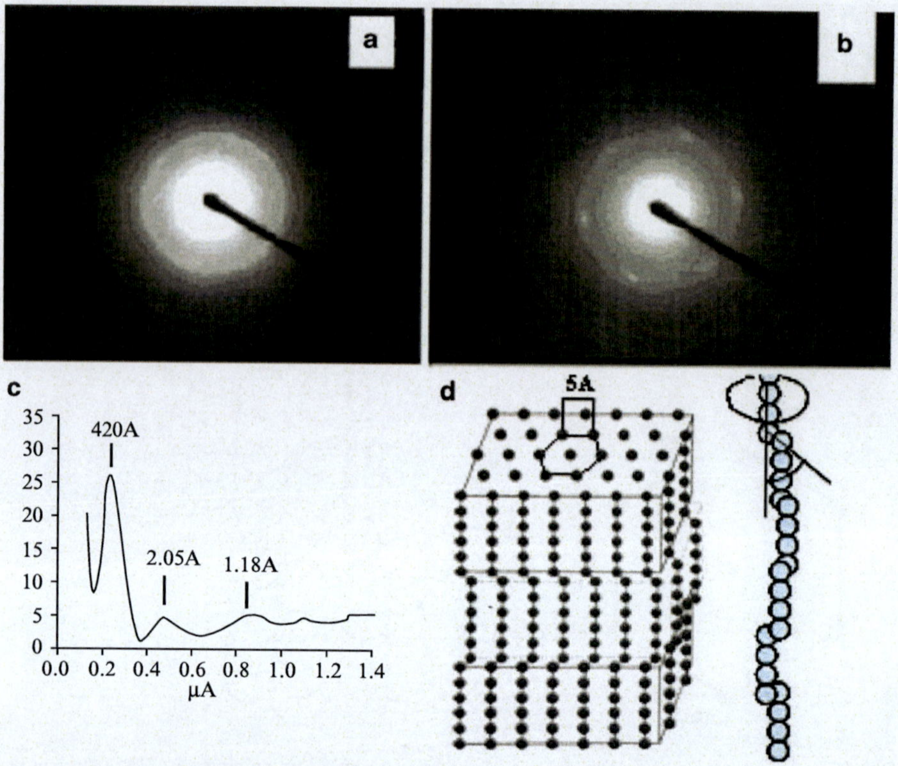

Fig. 37.3 (**a**), (**b**) Electron diffraction patterns of the TDO LCC films; (**c**) the distribution of the electron diffraction intensity; (**d**) model of atomic structure of TDO LCC film [10]

crystal lattice with the lattice parameter varying from 0.490 to 0.507 nm and indicates a high degree of ordering in the film. The distribution of electron diffraction intensity shown in Fig. 37.3c reveals the specific features of the obtained diffraction patterns. There is strong attenuation of higher-order reflections, so that the intensity of the second order reflections is decreased by an order of magnitude compared to the first-order reflections. This can be attributed to the layered structure of the film with small (0.09 nm) random displacements between the layers that the film consists of [4]. It should be noted that obtained inter-plane spacing in the range between 0.412 and 0.456 nm is a characteristic feature of only one known carbon form that is carbyne with the sp^1 -type of electron hybridization of carbon atoms.

Since films TDO LCC are multi-layer structure, each layer consists of linear carbon atomic chains. We proposed that chains in the layer are linked together by weak Van der Waals forces. They are oriented normal to the surface of the layer (substrate). Chains in adjacent layers are shifted relative to each other in the bond length between atoms and rotated around the axis. These chains are densely packed in a hexagonal lattice. In addition, these chains are not straight, but curved and kinks correspond to the layers in the film. The distance between the chains is in the range from 0.490 to 0.503 nm. The layers are randomly shifted relative to each other. In other words, the film structure consists of identical curved chains, and kinks are randomly oriented with respect to the axes.

Earlier, we've found, that Electronic Spectroscopy for Chemical Analysis (ESCA) spectrum of the TDO LCC film having a thickness of 50 nm indicates that the film consists of carbon with purity higher than 99% (Fig. 37.4a) [10].

The Raman spectrum of the sp^1-hybridized carbon–chain film consists of two broad maximums at 1,550 cm^{-1} and 2,100 cm^{-1} [10]. The high-frequency band in the region of 2,100 cm^{-1} is commonly attributed to stretching vibrations frequency of long sp^1-hybridized carbon chains. The other peak at 1,550–1,580 cm^{-1} corresponds to the position of the sp^2-carbon bond vibration frequency and, therefore, it is conventionally considered to be a feature of sp^2-carbon phase. However, this peak can alternatively be explained by the presence of regular kinks in the

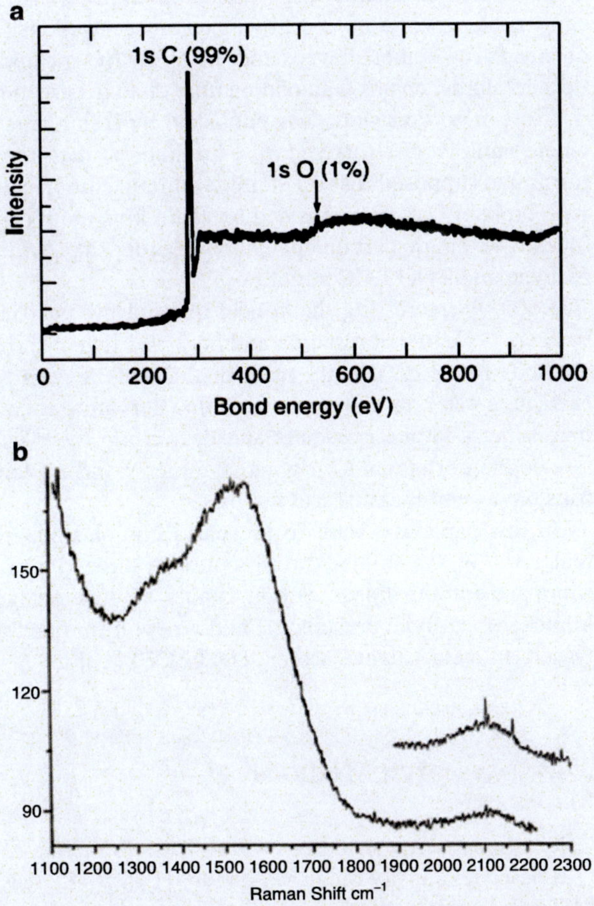

Fig. 37.4 (**a**) ESCA spectrum of the TDO LCC film [6]; (**b**) Raman spectrum of the TDO LCC film

polycumulene carbon chains [2]. Recently we've repeated Raman spectroscopy diagnostic of a film with thickness of 100 nm on Si substrate and again found presence of all feaches, described above (see Fig. 37.4b).

This new carbon films have demonstrated a lot of magnificent properties such as extremely low work function (0.43 eV) and anisotropy of electrical conductivity: in one direction (across film) it is dielectrics while in other one (normal to film surface) quantum (tunnel) conductivity takes a place [10]. In experiments on biomedical examination of TDO LCC, the absence of protein denaturation was found on its surface. It was also characterized by exceptionally low blood coagulation potential, thus having perfect blood compatibility [10]. So, there is a wild field of different applications of films TDO LCC: in micro and nanoelectronics for creation nanotransistors, as perfect cold cathodes and in medicine.

The sp^1-carbon based materials were investigated by numerous experimental methods such as TEM, Auger and Raman spectroscopy, X-ray diffraction, EELS [4, 10], so one can find rich experimental data to describe them. Unfortunately there are not so much theoretical studies of sp^1-carbon based materials. The first investigation of short carbon chains was carried out by Pitzer and co-workers [11] in 1959 after carbyne was obtained. The authors have found (using LCAO method) that for short carbon chains (less 15 atoms) chains with odd number of atoms are more stable then even ones.

Most important work was published by Heinmann and co-workers in 1983 [7], where authors systematized all experimental data for different described carbyne forms and supposed that for stability of long carbon chains they should to be kinked with probably not compensated bonds in kink atom (Fig. 37.5). So we'd decided to investigate single carbon chains too in order to provide better understanding of the carbyne and TDO LCC structure.

Recently interesting theoretical research was carried out by W. Windl and co-workers [12]. In that research authors investigated carbyne structure using *ab initio* methods based on density functional theory. It was found, that carbon chains in carbyne crystal are curved so that this curvature modifies the lattice constants by increasing a-lattice constants and decreasing the c-lattice constants and also can reduce the hexagonal to trigonal symmetry and explain the fact that both symmetries have been measured in the past.

In this paper we want to present a complex theoretical investigation of such materials. We've studied the structures of three kinds of sp^1-type carbon materials: short and infinite linear carbon chains of different geometry (strait, kinked and branched), carbyne crystals [4] and a new film material named Two-Dimensional Ordered Linear Chain Carbon (TDO LCC) [10].

37.2 Research Methods

Calculations were carried out within VASP (Vienna *ab-initio* simulation package) [13] based on density functional theory [14]. VASP has been installed on the SKIF-MSU-Chebyshev supercomputer complex.

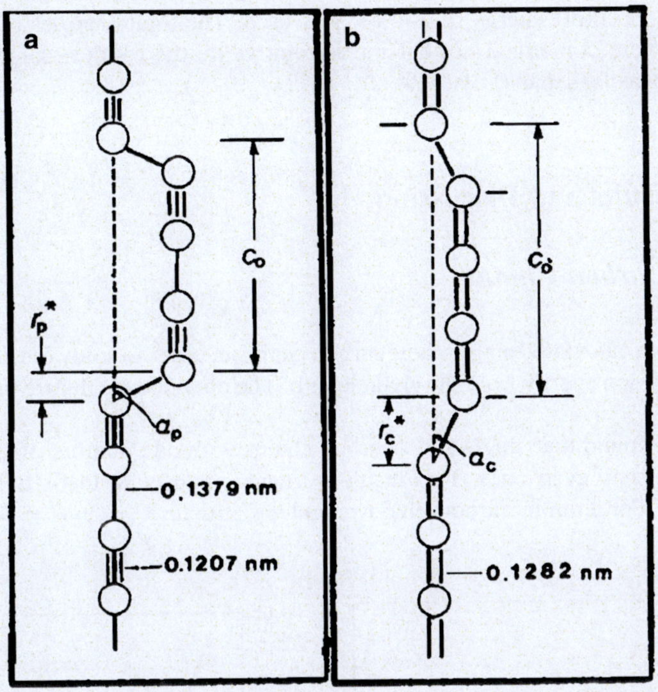

Fig. 37.5 Heinmann's model of kinked carbon chains: (**a**) polyyne type, (**b**) cumulene type [7]

For structural modeling we set the positions, number and type of atoms in initial file. In all calculations initial interatomic distance between carbon atoms (cumulene bond type) was 1.28 Å, between carbon and hydrogen atoms the distance was 1.06 Å, and between carbon and nitrogen atoms it was 1.15 Å. In order to create appropriate models of these structures we used hexagonal supercell for carbyne and TDO LCC while for single chains – cubic one. All atoms were located in such supercells and these cells were infinitely translated in all directions. For the creation of finite structures like short carbon chains and films we used a large supercell where our atomic structures were separated from images by large vacuum spaces in order to avoid interaction between these structures and their images.

Our self-consistent electronic-structure calculations of sp[1]-carbon materials have been performed by means of the projector augmented-wave method [15]. The generalized gradient approximation (GGA) has been applied for exchange-correlation functional using Perdew-Wang'91 treatment. Structural relaxations were performed via a quasi-Newton algorithm using the exact Hellmann-Feynman forces acting on ions. The integration over Brillouin zone (BZ) was performed using the tetrahedron method with Blöchl corrections. BZ sampling was performed using $6 \times 6 \times 6$ k-point mesh in Monkhorst-Pack grid for carbyne crystals, 1x1x10 for infinite carbon chains, $1 \times 1 \times 1$ for short carbon chains and $4 \times 4 \times 1$ for

TDO LCC. A cutoff energy of 250 eV was used. The total energies of the whole supercells were converged up to 1 meV/atom, while the residual force acting on each atom was less than 0.01 eV/Å.

37.3 Results and Discussion

37.3.1 Carbon Chains

At first we've calculated single short carbon chains in order to study the dependence of their cohesion energy from the chain length. The obtained results are presented in Fig. 37.6a.

We have found that single short carbon chains with odd number of atoms were more stable than even ones. This result is in agreement with theoretical work of Pitzer [11]. For infinite carbon chain cumulene structure (=C=C=C=) had the

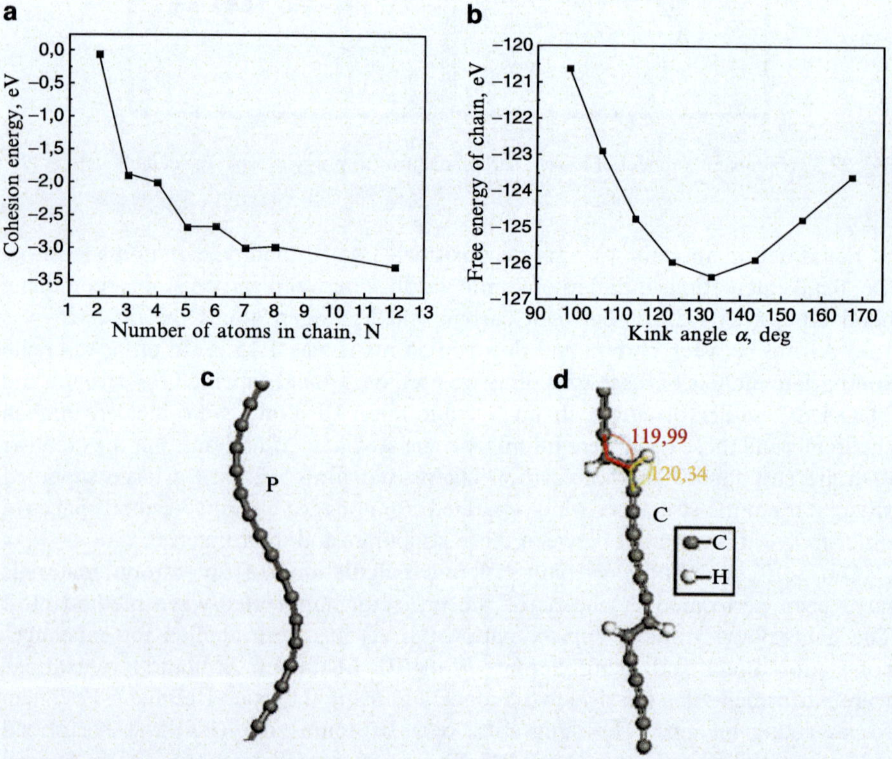

Fig. 37.6 **(a)** Dependence of cohesion energy of carbon chain from number of atoms in it; **(b)** dependence of free energy of carbon chains from kink angle; **(c)** pure kinked carbon chain after ionic relaxation; **(d)** kinked carbon chain with hydrogen after ionic relaxation

lowest energy, while polyyne (−C≡C−C≡) structure was less stable. After that we've studied Heiman-type kinked carbon chains [7]. We studied the dependence of chain stability from the angle of a kink. Results of calculations are presented in Fig. 37.6b. The obtained results showed that the most stable carbon chain had a kink angle 132°. However we found from our calculations that in pure carbon chains Heiman-type kinks were destroyed after ionic relaxation process as shown in Fig. 37.6c. We supposed that according to Heimans idea about not compensated bonds in kink atom there should be admixture atoms, which are bonded with kink atoms. We proposed that such atoms could be hydrogen which can still be presented during formation process of sp^1-carbon material. Since ESCA measurements hadn't showed any impurities in carbon materials we decided that hydrogen was the best candidate on admixture role in such materials because it can't be found by this method. Obtained results (see Fig. 37.6d) demonstrate that hydrogen atoms stabilize the kink formation thus the presence of hydrogen satisfies Heiman-type kinks model.

As the first step in investigation of structural properties of carbyne we have studied the atomic and electronic properties of the branched linear carbon chain structure (see Fig. 37.7).

We proposed that such structure can help us to determine length of strait linear fragment of a carbon chain in carbyne crystal and kink angle. We calculated different types of such chains: a-type – infinite periodic structures with constant number of atoms per each linear fragment ($n = m$, designations see in Fig. 37.7); b-type – infinite periodic structures with constant number of atoms per supercell ($n + m = 24$), but varying n and m; c-type – chains with constant number of atoms per supercell ($n + m = 24$), but varying n and m, and with hydrogen atom per each end of carbon chain. The results of our calculations are presented in Fig. 37.8.

The structure of a-type chains after ionic relaxation is shown in Fig. 37.8b. One can see from Fig. 37.8b that after ionic relaxation of a-type chains we had different type of carbon bonds in different parts of structure: there was polyyne type in

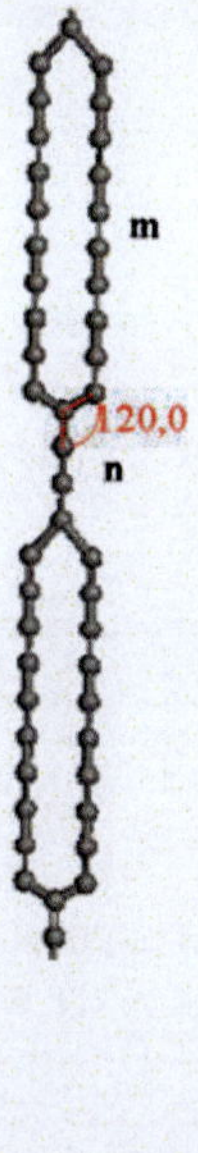

Fig. 37.7 The initial structure of a branched carbon chain (before ionic relaxation)

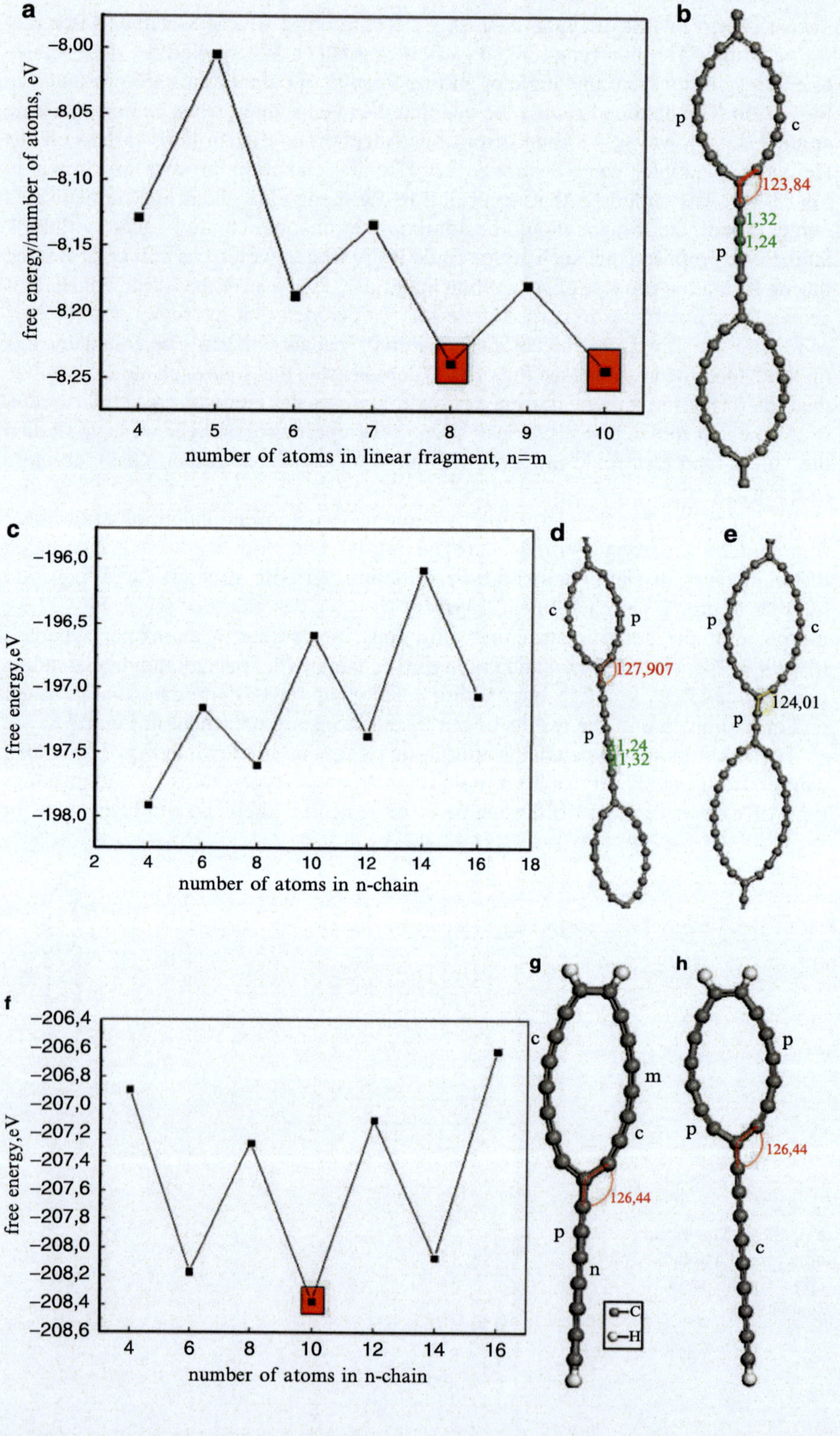
a
free energy/number of atoms, eV
-8,00
-8,05
-8,10
-8,15
-8,20
-8,25
4 5 6 7 8 9 10
number of atoms in linear fragment, n=m
b
p
c
123,84
1,32
1,24
p
c
free energy,eV
-196,0
-196,5
-197,0
-197,5
-198,0
2 4 6 8 10 12 14 16 18
number of atoms in n-chain
d
c
p
127,907
p
1,24
1,32
e
p
c
124,01
p
f
free energy,eV
-206,4
-206,6
-206,8
-207,0
-207,2
-207,4
-207,6
-207,8
-208,0
-208,2
-208,4
-208,6
4 6 8 10 12 14 16
number of atoms in n-chain
g
c
m
c
126,44
p
n
C
H
h
p
p
126,44
c

n-chain and there were polyyne and cumulene types in opposite sides of ellipse. This was obtained for all examined numbers of atoms in *n*-chain (from 4 to 10). Since carbon atom has a valence IV, it ought to has such valence in the branching point of a chain too, but one bond was occupied by polyyne *n*-chain thus for two other branches carbon atom had only three bonds. These three bonds were divided between two branches thus one branch had one bond (single) while the other one had two (double). We suppose that this is the reason why in opposite parts of the ellipse we had cumulene and polyyne bond types. Figure 37.8a shows the dependence of free energy per atom of a system from the number of atoms in linear fragments of the branched chain $n = m$. One can see that the structures which linear fragments had even number of atoms had the lower energy, and the lowest energy we had obtained for 8 and 10-atomic chains fragments ($|E_8\text{-}E_{10}| < 0{,}001$ eV).

The results we got for b-type calculations are presented in Fig. 37.8 (c, d, e). The structure of relaxed chains is similar to a-type calculations. We had a constant number of atoms per supercell (24) thus we could investigate only linear fragments with even number of atoms *n* and *m*. We obtained that fragments with 4, 8, 12, and 16 atoms in *n*-chain were more energetically preferable for infinite branched chain-like structure than 6, 10 and 14 atomic ones. We again received a polyyne structure in the *n*-type chain. Kink angle was in a range 123°÷130°.

The obtained results for finite branched chains with hydrogen (c-type) are presented in Fig. 37.8 (f, g, h). Figure 37.8f demonstrates that according to our calculations 6, 10, and 14 atoms in *n*-chain were more energetically preferable than 4, 8, 12 and 16 ones in the case of single branched chains with hydrogen in opposite to infinite pure ones. In all cases we obtained kink angle about 126°. It is interesting to note, that we got polyyne *n*-chains and cumulene ellipse for 4, 8, 12, and 16 atoms per *n*-chain. While we got cumulene *n*-chains and polyyne ellipse for 6, 10, and 14 atoms per *n*-chain.

37.3.2 Carbyne

In this part we would like to present our results for carbyne crystal structure calculations. The typical result for carbyne structure after ionic relaxation is shown in Fig. 37.9a, b.

As a result of ionic relaxation we obtained that carbon chains in subcell (I) were curved in order to produce bonds between subcells. In subcell (I) we got cumulene type of chains while in subcell (II) we got polyyne one. The obtained distances and angles in carbyne structure are presented in Fig. 37.9a, b. This result is similar to one obtained by

Fig. 37.8 (**a**) Dependence of free energy of infinite branched carbon chain from number of atoms in *n*-chain in case $n = m$; (**b**) typical relaxed structure of such chain; (**c**) dependence of free energy of infinite branched carbon chain from number of atoms in *n*-chain in case $n + m = 24$; (**d**) typical relaxed structure of chains for $n = 6, 10, 14$; (**e**) typical relaxed structure of chains for $n = 4, 8, 12, 16$; (**f**) dependence of free energy of finite branched carbon chain with hydrogen from number of atoms in *n*-chain in case $n + m$=24; (**g**) typical relaxed structure of chains with hydrogen for $n = 4, 8, 12, 16$; (**h**) typical relaxed structure of chains with hydrogen for $n = 6, 10, 14$

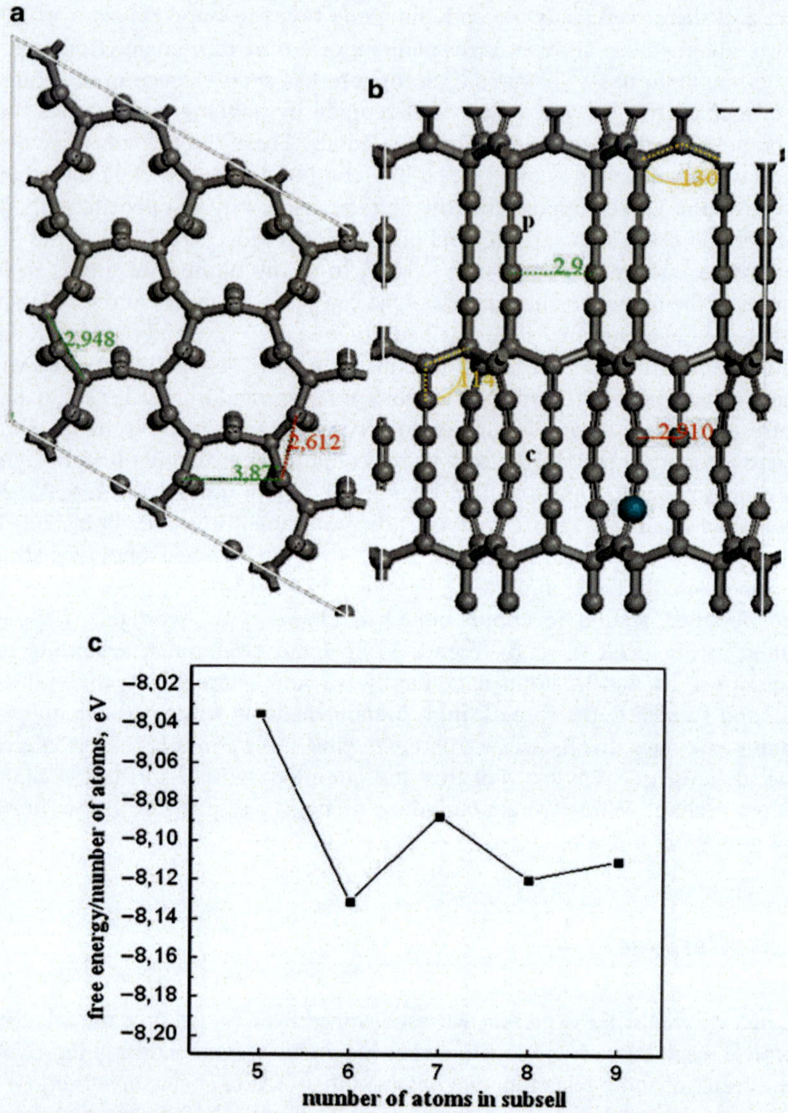

Fig. 37.9 Typical form of relaxed carbyne structure (**a**) above (000 *l* projection); (**b**) side; (**c**) dependence of free energy of carbyne from number of atoms in chains in subcell

W. Windl et al. by the same methods [12]. We studied the stability of such carbyne model in comparison to diamond and found that it was metastable. Also we investigated dependence of stability of carbyne crystal from the number of atoms in chains in each subcell (see Fig. 37.9c). The most stable were carbynes with even number of atoms per chain fragment in subcell. Carbyne with 6-atomic linear chain fragments had the lowest energy. This is in a good agreement with our calculations of branched carbon chains and experimental results [4].

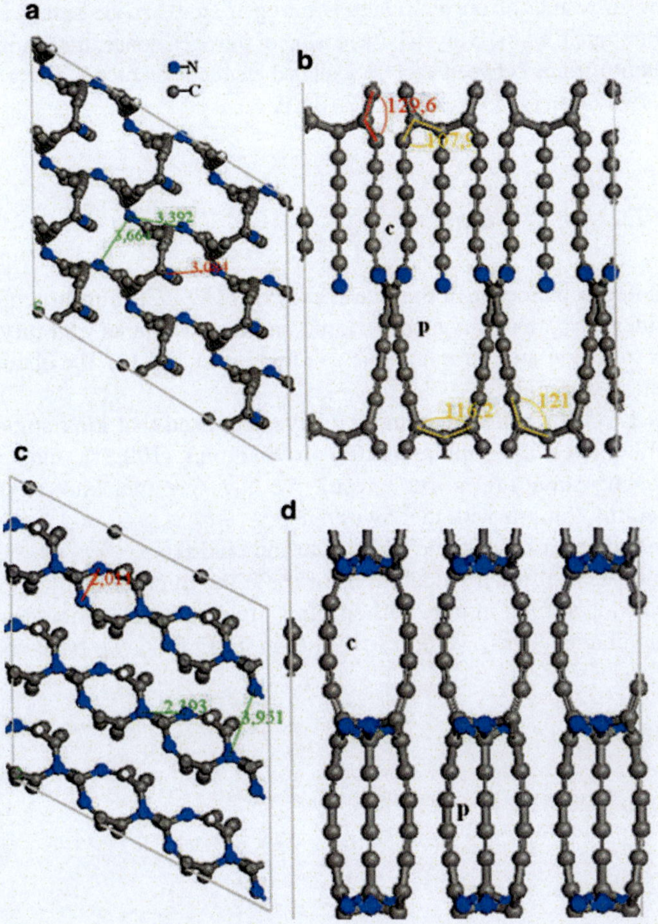

Fig. 37.10 Relaxed carbyne structure with nitrogen: implanted between supercells (**a**) above; (**b**) side; implanted between subcells (**a**) above; (**b**)side

There was found small amount of nitrogen in experimental dataset [16]. We proposed that nitrogen could play significant role in carbyne crystal structure by stabilization of a kink formation. Nitrogen atoms can occupy not compensated bonds of carbon atom in kink analogous to our results for hydrogen in single chains. The obtained results for N-doped carbyne crystal are presented in Fig. 37.10.

We have investigated two types of N-doped carbyne crystal structure: first, when nitrogen atoms were implanted between each carbyne supercell (Fig. 37.10a, b) and second, when nitrogen atoms were implanted between each carbyne subcell (Fig. 37.10c, d). Carbyne structure with nitrogen found to be stable, but metastable in comparison with diamond.

According to results of our calculations nitrogen atoms made strong influence on chain configuration in carbyne crystal. Increasing of nitrogen concentration, i.e. implantation of admixture atoms between each subcell had resulted destruction of carbyne crystal and formation of clusters, shown in Fig 37.10c, d.

37.3.3 TDO LCC Films

We have studied dependence of the free energy of TDO LCC film structure from the length of linear fragment in chains, kink angle and chains chirality. Notation conventions for calculated structure one can find in Fig. 37.11. The obtained results are presented in Table 37.1.

For TDO LCC structure modeling we have proposed that kink angle in chains should be like ones in graphite (120°) or diamond (109.5°), thus we started calculations with such angles. As a result we have got that kink angle is about 130° in the configurations with the lowest energy. That is in a good agreement with our results obtained for the single kinked carbon chains.

We have observed in our calculations of the film structure that kinks in carbon chains were straightening in pure carbon films after ionic relaxation as they did in single carbon chains. Thus, we proposed that in TDO LCC we have analogous to

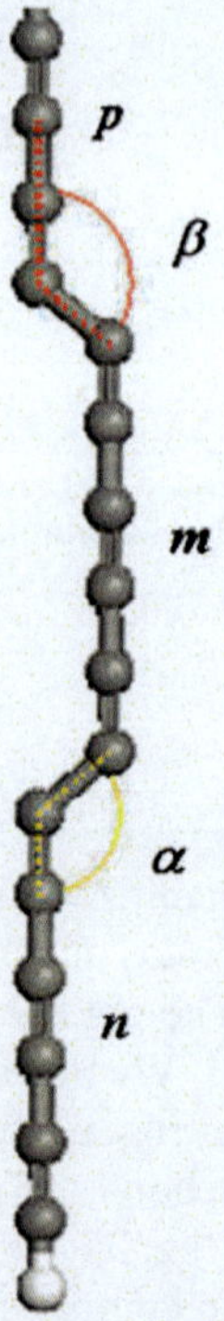

Fig. 37.11 Notation conventions for calculation of TDO LCC structure

Table 37.1 Dependence of TDO LCC free energy from length of linear fragments in chains, kink angle and chains chirality

Number of atoms in linear fragments and kink angles, m_n_p_α_β	Free energy, eV	Kink angles after ionic relaxation, α β °	
6_6_4_109	−385,458	125.8	124.7
6_6_4_109_120	−386,536	125.8	128.9
6_6_4_109_120[a]	−386,478	125.7	128.8
6_6_4_109[a]	−385,418	125.8	124.5
6_6_4_120	−387,562	129.9	128.9
8_6_2_109	−385,394	125.7	124.3
8_6_2_109[a]	−385,706	125.7	123.9
8_6_2_120	−387,525	129.7	128.7
8_6_2_120_109	−386,466	129.8	124.2
8_6_2_120_109[a]	−386,722	129.9	123.9
8_6_2_120[a]	−385,706	125.8	123.9
8_8_0_109	−391,200	125.4	
8_8_0_120	−392,228	129.7	

[a] Chains with chirality

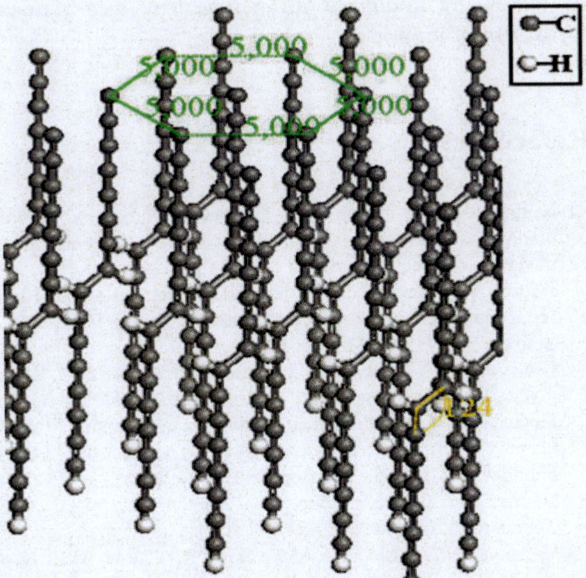

Fig. 37.12 Relaxed structure of TDO LCC film with hydrogen

single chains situation: there should be admixture atoms for stabilization of the kinks. We added to model the hydrogen atoms, bonded with kink carbon atoms (Fig. 37.12). Such kinked structure with hydrogen became stable and kinks satisfied to Heinmann's model. We hadn't gotten experimentally predicted kinks in films without hydrogen in our calculations, so now we are going to make experiments in searching hydrogen in films.

37.4 Conclusions

In this work we have investigated stability of different sp[1]-carbon based materials: chains (strait, kinked and branched), carbyne crystals and TDO LCC films. We found different type of a bonding between carbon atoms (polyyne and cumulene) within one chain or carbyne crystal. Also we obtained that the linear fragments of the chains with 6 and 8 atoms in the length are the most energetically preferable in TDO LCC. It has been found that kink in the atomic structure of linear carbon chain in single chains and in TDO LCC films could be formed by hydrogen impurities. The obtained results have showed that the most preferable kink angle is about 130° for single carbon chains and for in TDO LCC film structures. We have investigated carbyne crystal structure and the influence of nitrogen impurities on it. We have found that in carbyne crystal the carbon chains are curved and probably have the length of a linear fragment of 6 atoms. It has been found that nitrogen impurities changed the shape and the lattice constants of carbyne crystal. Moreover we have revealed that nitrogen impurities didn't stabilized the kinks in the structure of carbyne crystal in opposite to hydrogen. The obtained results have showed that the increasing amount of nitrogen destroyed carbyne crystal structure and leaded to formation of atomic clusters.

References

1. Kudryavtsev Yu, Evsyukov S, Guseva M, Babaev V, Khvostov V (1997) Carbyne – a linear chainlike carbon allotrope. In: Thrower PA (ed) Chemistry and physics of carbon, a series of advances. Marcel Dekker, New York, pp 1–69
2. Babaev V, Guseva M (2000) Ion assisted deposition. In: Heimann RB, Evsyukov SE, Kavan L (eds) Carbyne and carbynoid structures. Kluwer, Dordrecht
3. Sladkov AM (1981) Sov Sci Rev B 3:75–110
4. Kudryavtsev Y, Evsyukov S, Guseva M, Babaev V, Khvostov V(1996) Oriented carbyne layers. Carbon 30: 213–221; Kudryavtsev YP, Heimann RB, Evsyukov SE(1996) Carbynes: advances in the field of linear carbon chain compounds. J Mat Sci 30:5557–5571
5. Guseva M, Babaev V, Novikov N(1997) Tetracarbon. PCT Patent, International Application Number PCT/IB96/01487 from 18 Dec 1996; WO 97/25078, 17 July 1997; Guseva M, Babaev V, Novikov N, Tetracarbon. US Patent 6,355,350 B1; Guseva M, Babaev V, Novikov N, Tetracarbon. US Patent 6,454,797 B2
6. Kasatochkin VI, Sladkov AM et al (1967) Dokl Akad Nauk SSSR 177(2):358–360; Kasatochkin V I, Melnichenko VM, Elizen VM(1975) Electron diffraction by single crystals of carbine. Polym Sci USSR 17(9): 2167–2173
7. Heimann RB, Kleiman J, Salansky NM (1983) Aunified structural approach to liner carbon polytypes. Nature 306:164–167
8. Korshak VV, Kudryavtsev YP Korshak YV, Evsyukov SE et al (1988) Formation of β- carbine by denydrogalogenation. Macromol Chem Rapid Commun 9(3):135–140
9. Guseva MB, Babaev VG et al (1983) In questions of atomic science and technique physics of radiation damage and radiational materials science physico-technical institute. Khar'kov 25 (2):92 (in Russian)
10. V.G. Babaev, M.B. Guseva, N.D. Novikov, V.V. Khvostov et al (2006) p. 219 Franco Cataldo, Polyynes Synthesis, Properties, and Aplications, CRC Press Taylor & Francis Group

11. Pitzer K, Clementy E (1959) J Am Chem Soc 81(17):4477–4485
12. Windl W, Weiqi Luo (2009) First principles study of the structure and stability of carbines. Carbon 47(2):367–383
13. G. Kresse, J. Furthmuller (1996) Phys Rev B 54, 11169
14. Kohn W (1999) Electronic structure of matter-wave functions and density functionals, (Nobel lecture). Rev Mod Phys 71:1253–1266
15. Blochl PE (1994) Projector augmented-wave method. Phys Rev B 50(24):17953–17979
16. Guseva MB, Babaev VG et al (1995) Formational of oriented nitrogen-doped carbon films. JETP Lett 62(9):715–718

Chapter 38
Internal Stresses and Hydrogen Permeability of Hollow Cylinder

N.M. Vlasov

Abstract Hydrogen permeability of hollow cylinder at presence of internal stresses of various physical nature (thermal, concentration and residual) has been investigated. The kinetics of the diffusion processes has been described by a parabolic equation under corresponding initial and boundary conditions. The first invariant of the tensor of internal stresses in the hollow cylinder has a logarithmic dependence on the radial coordinate. Such dependence allows the exact solution of the diffusion kinetics problem to be obtained.

Keywords Internal stresses · Diffusion kinetics · Hydrogen permeability

38.1 Introduction

The strength of structural components depends on the level and character distribution of internal stresses. They occur in the presence of non-uniform deformation. The main types of the internal stresses are thermal, concentration and residual [1]. The stresses change the properties, such as the yield point, as a result of diffusion processes. Various mechanisms are responsible for material property changes: decrease of the energy of surface fracture, corrosion cracking and hydrogen embrittlement. These mechanisms also include diffusion of interstitial impurities (e.g. hydrogen and oxygen) in the internal stress field. Hydrogen is dominant among these interstitial impurities. This dominance is caused by the high diffusion mobility of hydrogen atoms over a wide temperature range. At room temperature, for example, the diffusion coefficient of hydrogen atoms is greater by several orders of magnitude than the diffusion coefficients of other substitutional and/or interstitial

N.M. Vlasov (✉)
Regional Educational Centre of Science, Moscow, State Open University,
K. Gotvalda 2/40, Podolsk, Moscow Region 142114, Russia
e-mail: tsolga@mail.ru

S.Yu. Zaginaichenko et al. (eds.), *Carbon Nanomaterials in Clean Energy Hydrogen Systems - II*, NATO Science for Peace and Security Series C: Environmental Security 2,
DOI 10.1007/978-94-007-0899-0_38,

impurities. The diffusion kinetics is described by a parabolic equation under corresponding initial and boundary conditions. In general, the internal stresses fields have complex spatial dependence. A notable exception is the internal stress field with logarithmic coordinate dependence in a cylindrical coordinate system. Such dependence allows the exact solution of the diffusion kinetics problem to be obtained.

The objective of this paper is to simulate material diffusion permeability in a internal stress field with a logarithmic coordinate dependence. Hydrogen permeability in a hollow cylinder is considered as an example.

38.2 Physical Meaning of Internal Stresses

Internal stresses occur within a material in the presence of non-uniform deformation. Typical examples of such stresses are thermal, concentration and residual stresses. The thermal stress field is caused by non-uniform distribution of temperature. In some cases thermal stresses have a logarithmic spatial variation. Thermal stresses in a hollow cylinder are an example of such dependence. It is known that interaction of an impurity atom with the thermal stress field depends on the first invariant of the stress tensor. For the thermal stresses in the hollow cylinder, this is determined by

$$\sigma_{ll} = \frac{2\alpha\mu(1+\nu)(T_1 - T_2)}{(1-\nu)\ln R/r_0}\left\{1 + 2\ln\frac{r}{R} + \frac{2r_0^2}{R^2 - r_0^2}\ln\frac{r_0}{R}\right\}, \tag{38.1}$$

where α is the coefficient of linear expansion, μ is the shear modulus, ν is the Poisson's ratio, r_0 and R are inner and outer radii of the cylinder, T_1 and T_2 are temperatures on the inside and outside surfaces of the cylinder, respectively ($T_1 > T_2$). Relation (38.1) applicable in the case of plane strain under stress-free boundary conditions. Such coordinate dependence is characteristic of the concentration stresses in the hollow cylinder. The concentration stress field is caused by non-uniform distribution of impurities. The first invariant of the concentration stress tensor for the plane strain condition also has a logarithmic coordinate dependence

$$\sigma_{ll} = \frac{2\beta\mu(1+\nu)(C_1 - C_2)}{(1-\nu)\ln R/r_0}\left\{1 + 2\ln\frac{r}{R} + \frac{2r_0^2}{R^2 - r_0^2}\ln\frac{r_0}{R}\right\}, \tag{38.2}$$

where β is the relative change of parameter of crystal lattice, C_1 and C_2 are concentration of impurities on the inside and outside surfaces of the cylinder ($C_1 + C_2$). The remaining notations correspond to the ones adopted before. Let us consider the following version of the residual stresses formation. The cutting edges of the cylinder are moved apart by angle ω, and the missing material is placed there.

During this operation the area near the outer cylindrical surface is compressed, whereas tension occurs near the inner cylindrical surface. The first invariant of the residual stress tensor for the plane strain condition also has a logarithmic coordinate dependence [2]

$$\sigma_{ll} = \frac{\omega\mu(1+v)}{2\pi(1-v)l}\left\{1 + 2\,ln\frac{r}{R} + \frac{2r_0^2}{R^2 - r_0^2}\,ln\frac{r_0}{R}\right\}, \tag{38.3}$$

where ω is an angle of the cutting edge opening of the cylinder. The remaining notations correspond to the ones adopted before.

38.3 Hydrogen Permeability of Hollow Cylinder

The elastic interaction of the hydrogen atoms with the internal stresses is defined by known relation

$$V = -\frac{\sigma_{ll}}{3}\delta\upsilon, \tag{38.4}$$

where σ_{ll} is the first invariant of tensor stresses, $\delta\upsilon$ is the volume change caused by the introduction of a hydrogen atom. For $\sigma_{ll} > 0$ (tension stresses) and $\delta\upsilon > 0$ (an hydrogen atom increases the crystal lattice parameter) potential V takes a negative value. It corresponds to the attraction of the hydrogen atom to the tension stress area and its displacement from the compression stresses area. Relation (38.3) takes into account only the dimensional effect in the energy of the hydrogen atom connection with the internal stress field. The other types of interactions (module, electrostatic, chemical) can be easily estimated by renormalization of the constants in relation (38.3).

The hydrogen concentration field is determined from the solution of the equation of a parabolic type under the corresponding initial and boundary conditions [3]

$$\frac{1}{D}\frac{\partial C}{\partial t} = \Delta C + \frac{\nabla(C\nabla V)}{kT}, \quad r_0 < r < R, C(r,0) = 0,$$
$$C(r_0, t) = C_o,\ C(R,t) = 0 \tag{38.5}$$

where D is the coefficient of hydrogen atoms diffusion, k is the Boltzmann constant, T is the absolute temperature, C_0 is the concentration of hydrogen atoms on an inside surface of the cylinder. The physical meaning initial and boundary conditions is clear. At start time the hydrogen atoms concentration in the cylinder are absent. Zero concentration of hydrogen atoms remains and on an outer surface. On an inner surface of the cylinder constant concentration of hydrogen atoms is supported. Equation 38.5 connects physics and mechanics when simulating material diffusion permeability in the internal stress field. The complex coordinate dependence of the potential V does not allow us to obtain an analytical solution equation (38.5). This difficulty can be overcome for the logarithmic coordinate dependence of potential V. This is caused by the fact that in this case potential V is a harmonic function and its gradient is inversely proportional to the radius in the polar coordinate system. This simplifies the solution for the diffusion kinetics. From (38.1), (38.2), (38.3) and (38.4) after simple mathematical manipulations we obtain

$$\frac{1}{D}\frac{\partial C}{\partial t}=\frac{\partial^2 C}{\partial r^2}+\frac{1-\beta_1-\beta_2-\beta_3}{r}\frac{\partial C}{\partial r},\quad r_0<r<R$$
$$C(r,0)=0,\ C(r_0,t)=C_o,\ C(R,t)=0 \tag{38.6}$$

$$\beta_1=\frac{4\alpha(T_1-T_2)\mu(1+\nu)\delta\upsilon}{3kT(1-\nu)ln R/r_0},$$

$$\beta_2=\frac{4\beta(C_1-C_2)\mu(1+\nu)\delta\upsilon}{3kT(1-\nu)ln R/r_0},$$

$$\beta_3=\frac{\omega\mu(1+\nu)\delta\upsilon}{3\pi kT(1-\nu)}.$$

The value $\frac{\omega}{2\pi}$ present relative deformation while forming residual stresses in the cylinder (the angle ω is measured in radians).

The dimensionless parameters of the problem are defined a contribution of thermal stress (β_1), concentration stress (β_2) and residual stress (β_3) by changing hydrogen permeability of the cylinder. We will consider two cases of the distribution of the internal stresses. For the first case ($\beta_1+\beta_2+\beta_3=1$) we get

$$\frac{1}{D}\frac{\partial C}{\partial t}=\frac{\partial^2 C}{\partial r^2},\quad r_0<r<R$$

$$C(r,0)=0,\ C(r_0,t)=C_o,\ C(R,t)=0. \tag{38.7}$$

If the internal stresses changes a sign ($\beta_1+\beta_2+\beta_3=-1$) we will receive

$$\frac{1}{D}\frac{\partial C}{\partial t}=\frac{\partial^2 C}{\partial r^2}+\frac{2}{r}\frac{\partial C}{\partial r},\quad r_0<r<R$$

$$C(r,0)=0,\ C(r_0,t)=C_o,\ C(R,t)=0. \tag{38.8}$$

The solution of the problems (38.7) and (38.8) are known. For convenience we will write down in a kind

$$C=C_0 f(r,t)\ (a),\quad C=C_0\frac{r_0}{r}f(r,t)\ (b),$$

where

$$f(r,t) = \frac{R/r_0 - r/r_0}{R/r_0 - 1} - \frac{2}{\pi}\sum_{n=1}^{\infty}\frac{\sin\frac{\pi n(r/r_0-1)}{R/r_0-1}}{n} exp\left(-\frac{\pi^2 n^2 Dt}{r_0^2(R/r_0-1)^2}\right). \qquad (38.9)$$

The diffusion flows of the hydrogen atoms though outer surface in the condition f(R,t) = 0 for relation (9a) and (9b) are the following

$$|\vec{j}_1| = DC_0\frac{\partial f(r,t)}{\partial r}|_{r=R},$$

$$|\vec{j}_2| = DC_0\frac{r_0}{R}\frac{\partial f(r,t)}{\partial r}|_{r=R},$$

where $|\vec{j}|_1$ and $|\vec{j}|_2$ are diffusion flows of the hydrogen atoms through outer surface for two cases respectively. Their relation characterizes of the hydrogen permeability of the cylinder depending on sign of the internal stresses

$$\frac{|\vec{j}_1|}{|\vec{j}_2|} = \frac{R}{r_0}.$$

Within the framework of the accepted model it is shown, that the relation of corresponding flows of hydrogen atoms at change of a sign on the internal stresses dependents only on the relation of the characteristic sizes of the cylinder. Simplicity of the received expression is a consequence of the accepted conditions: the sum of dimensionless parameters is equal to unit ($|\beta_1 + \beta_2 + \beta_3 = 1|$). Other values of these parameters will not change a qualitative picture of the hydrogen permeability of the cylinder, and will be changed only of numerical value of the relation diffusion flows.

38.4 Conclusions

Hydrogen permeability of the hollow cylinder at presence of the internal stresses various physical nature (thermal, concentration and residual) has been investigated. The kinetics of the diffusion processes has been described by a parabolic equation under corresponding initial and boundary condition. The first invariant of the tensor of the internal stresses in the cylinder has a logarithmic dependence on the radial coordinate. Such dependence allows for an analytical solution of the diffusion kinetics problem. This results from the fact that potential V is a harmonic function, and its gradient ∇V is inversely proportional to the radius in the polar coordinate system. The hydrogen permeability of the hollow cylinder is defined by two factors: the concentration of hydrogen atoms on the inner surface and influence of the internal stresses in volume.

References

1. Timoshenko SP, Goodier JN (1970) Theory of elasticity. McGraw-Hill, New York/Toronto/London
2. Teodosiu C (1982) Elastic models of crystal defects. Springer, Heidelberg/Berlin/New York, 336 p
3. Vlasov NM, Fedik II (2009) Simulation of material fracture in the field of thermal stresses. J Therm Stresses 32(8):755–767

Chapter 39
The Platinum Catalyst for Micro Fuel Cells on the Basis of Silicon

V. Starkov and A. Teterskiy

Abstract In the structure of the fuel cells (FC) catalyst is one of the most responsible elements. Especially important catalyst in low-temperature FC, to whom FC with proton exchange membrane fuel cell (PEMFC). In addition to the need for effective functioning at room temperatures, catalysts must ensure the longevity of work in conditions of aggressive impacts electrolyte and reaction products in FC. This communication presents the results of research and development of Pt catalysts on the surface of the silicon electrodes and inner surface since porous Si electrodes of micro FC. Structural characteristics of Pt cluster on the Si surface researched using optical and scanner managed by electron microscopy, and x-ray analysis. It is shown that placing Pt directly on the surface of the Si does not provide the necessary durable electrodes. Suggested before generating catalytic layer Pt handle Si surface primer. New technology has demonstrated high electrocatalytically activity Si electrodes for more than 700 h of observation.

Keywords Fuel cell · Silicon · Platinum · Transmission electron microscopy · Energy-dispersive spectroscopy · Electron diffraction · Surface

39.1 Introduction

Last decade the sustainable direction fundamental and applied researches connected with miniaturization of proton exchange membrane fuel cells (PEMFC) was generated. This direction is the scientific basis for industrial production of power sources generation. Cartridge hydrogen-air/O_2 or liquid fuel cells capable of compared to traditional more effectively provide power portable electronic device with power up to 25 W called micro fuel cells (μ-FC). Progress on the creation of industrial

V. Starkov (✉) and A. Teterskiy
Institute of Problems of Microelectronics Technology and Super Purity Materials RAS (IPMT RAS), Chernogolovka 142432, Russia
e-mail: starka@iptm.ru

S.Yu. Zaginaichenko et al. (eds.), *Carbon Nanomaterials in Clean Energy Hydrogen Systems - II*, NATO Science for Peace and Security Series C: Environmental Security 2,
DOI 10.1007/978-94-007-0899-0_39,

technology μ-FC associated with the development of constructive-technological option μ-FC based on silicon and, in particular the porous silicon and technologies carbon nanotubes, nanocomposite membranes and electrocatalysts. It is necessary to note that up to now it is not proposed to this technology, which would assume controlled shaping of the necessary structure "three-phase boundary" in the solid-polymeric fuel cells generally, and in the structure of μ-FC in particular. The nano-structure electrocatalyst is the key element of three-phase boundary. First of all catalyst must have largest possible area of catalytic surface. Furthermore it must once only ensure electrical contact with the electronic (electrodes) and proton (by electrolyte) conductors, and also with the fuel (on the anodic electrode), or the oxidizer (on the cathodes electrode). As fuel can be used as gaseous sources, for example hydrogen and its mixes, and sources of hydrogen in the form of liquid hydrocarbons. Methanol is the most common liquid source of hydrogen. It is necessary to note that is developed the construction-engineering version of μ-FC on the basis of the technology of porous silicon (firm "NEAH" [1]), in which the problem of three-phase boundary is solved due to the use of selectivity of catalysts and application of a methanol mixture sulfuric or phosphoric acids as the electrolyte. However, use for the wide application of fuel cartridges with a similar composition causes large doubt.

This work presents the result of studies the effective electro catalyst creation on the surface of silicon and an internal surface of porous silicon electrodes. It is possible to consider these researches as a necessary step of controlled technology the three-phase boundaries structure in μ-FC creation.

39.2 Experimental

Porous silicon (PSi). In a number of the numerous applications of porous silicon special position can engage use PSi during the miniaturization PEMFC . Structures PSi can selectively be formed on the silicon wafers in the form of porous layers or porous membranes. For the application in the structure PEMFC as the electrodes, porous structures must ensure the necessary permeability for the fuel, the oxidizer and the vapors of water. Therefore the porosity of such structures must exceed (50%–60%). Electrodes must have the highest possible electrical conductivity. Furthermore, as the matrix for positioning the electro-catalyst, electrodes must possess strongly developed surface and ensure electrochemical stability with the work under the conditions of aggressive medium. If we examine physical properties PSi [2], then it is simple to note that the one-time fulfilling of the requirements enumerated above is contradictory. Thus, the most developed surface (to 1,000 m^2/cm^3) possess the nano-porous structures of silicon, which are formed, as a rule, with the anodic etching of silicon with the high conductivity (ρ_v = of 0.01 Ω cm). Would seem these characteristics they must completely favorably affect such parameters of μ-FC as the decrease of ohmic losses on the electrodes and an increase in the area of the catalytic surface of gas-diffusion layers.

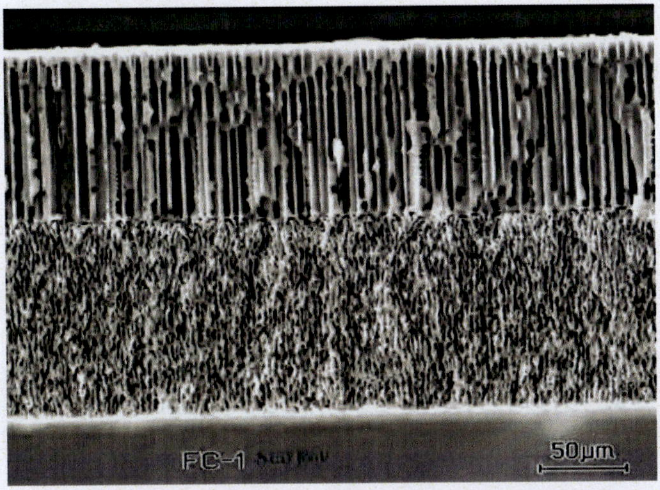

Fig. 39.1 Scanning electron micrograph of the GPSi-2 membrane's cleavage surface. First layer has 130 μm pores thickness and 5 μm average pore diameter. Morphology of first layer pores is direct and perpendicular to the etching plane. Second layer has 150 μm pores thickness and 1.5–2 μm average pore diameter. Morphology of these pores is sponge

However, nano-porous layers with the porosity are more than 50%, formed by the anodic etching of silicon with the low specific resistance are characterized also by very low electrical conductivity (specific resistance of layers it reaches 10^{12} Ω cm). Therefore the application of these porous layers as the matrix for positioning the electro-catalyst of gas-diffusion and catalytic layers into hydrogen-air/O_2 μ-FC is not promising. In our experiments the macroporous structures, formed on the silicon wafers of the p-type conductivity, were investigated. The membranes were gradient-porous two-layered structures (GPSi-2) Fig. 39.1. [3].

The porosity of upper layer composes 70%, and lower – 55% are determined by the regime of etching. The lower nonporous part of the silicon wafer was moved away by mechanical polishing. The thickness of the layer and membrane is determined by the selection of the regimes of anodic etching. For decreasing the resistance entire surface of the membrane was alloyed by boron according to the diffusion method from the gas phase at temperatures in the range (900–1,140°C).

Synthesis of platinum catalyst. The catalytic clusters of platinum can be substituted to the surface of silicon by different methods, among which in the micro-technology of silicon are most common the vacuum sputtering and CVD process, which can be stimulated additionally by other actions. An essential drawback in these methods is the practically insurmountable difficulty with the formation of catalyst on the internal surface of the times, which possess the large aspect ratio (Fig. 39.1). Studies showed that of the nano-clusters platinum on the surface of silicon lose the catalytic activity after the short period of the time (Fig. 39.2, curve 1). In the works [4,5] we have proposed before the synthesis of Pt catalyst to form on the surface of silicon the Au film. Film was formed with chemical deposition and

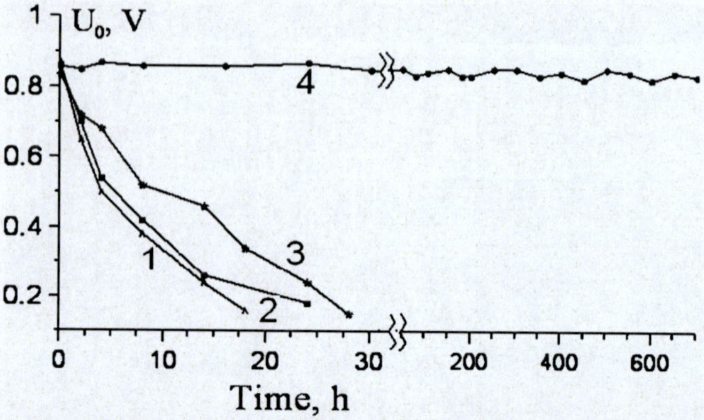

Fig. 39.2 Effectiveness Si electrodes by catalysts Pt (1) and Pt/Au (4) from the time, determined from the value of the U_0 of the "opened electric circuit" half-cell of μ-FC

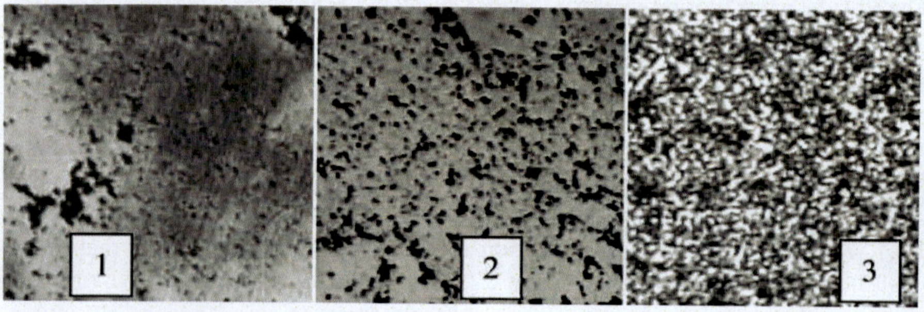

Fig. 39.3 Optical image of the catalytic surface of the silicon electrodes, formed from Pt of the precursors of Table 39.1

after precipitation into the porous base layer, was subjected photon annealing [5]. Film had a structure in the form of islets with a height of 40–60 nm.

Platinum catalyst on the surface of times was formed by restoring the catalyst from the solutions of the Pt-containing (platiniferous) precursors, substituted to the surface of rapids. After drying in air at a temperature of less than 80°C the electrodes were exposed by photon annealing [5].

Figure 39.3 depicts the optical images of the surface of silicon surface with the platinum catalyst, synthesized from the precursors of Table 39.1.

It is evident that during one and the same technological performance, to the uniformity according to the sizes and the structure of the clusters of platinum on the surface of silicon the essential influence renders the composition of the utilized precursor. The best results are achieved with the use of the solution, in composition of which is included the formic acid (model 3, Table 39.1, Fig. 39.3).

Analysis of the results. On the technology proposed the Pt catalyst was synthesized in the porous structure of silicon electrode. Figure 39.4a shows the TEM image of

Table 39.1 Composition of the platiniferous precursors

No	$(NH_4)_2PtCl_6$	H_2O	Ethanol	Izopropanol	Formic acid	Ethilenglycol
1	23 mg	0.5 ml	–	3.5 ml	–	–
2	23 mg	0.5 ml	4 ml	–	–	0.15 ml
3	23 mg	0.5 ml	–	–	4 ml	–

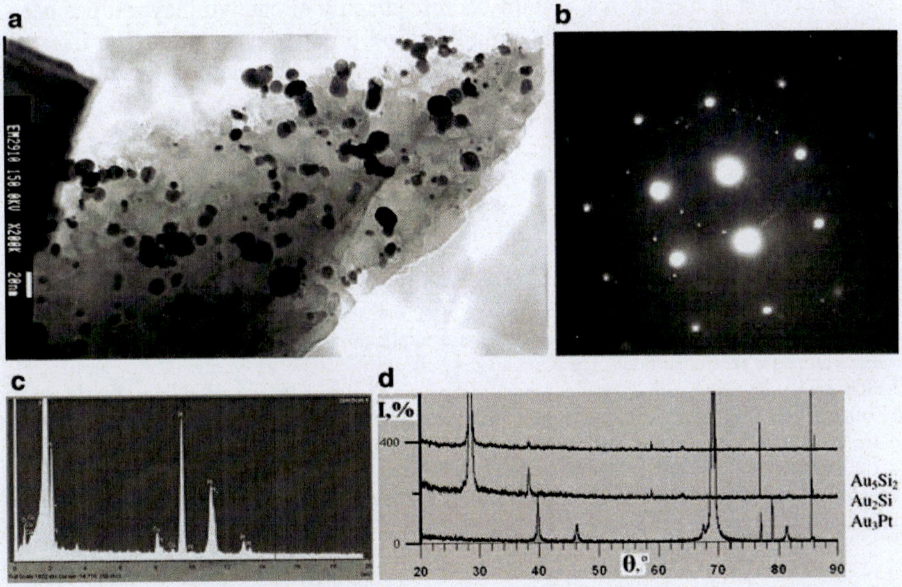

Fig. 39.4 TEM image of Pt catalyst in the times of silicon electrode (**a**); (**b**) the electron diffraction pattern from the Pt clusters; (**c**) EDS spectra of Pt particles; (**d**) the spectrum of X-ray diffraction

platinum catalyst in the times of GPSi-2 structure, depicted to Fig. 39.1. The average Pt particle size (4–7 nm). Energy-dispersive spectroscopy (EDS) (Fig. 39.4c) and data on the electron diffraction (Fig. 39.4b) testify about the polycrystalline structure of the clusters of platinum and the crystal structure of silicon walls rapids. Au concentration is small and below the limit of the detection of Au is located. However, the presence of gold was fixed by the method of X-ray diffraction (Fig. 39.4d). Masses spectral analysis fixed the content in the porous membrane 3.5% masses Pt and 0.004% of masses Au. The catalyst, formed in the porous structure of silicon that preliminarily modified by the gold-bearing primer, demonstrated the best stability in the time with testing of the cathodes electrode of the half-cell FC (Fig. 39.2, curve 4). These results were compared with the analogous data on the temporary stability GPSi-2 of cathodes structures without the application of primers (Fig. 39.2, curve 1) and with the application of primers in the form of coatings In_2O_3 (Fig. 39.2, curve 2), or SnO_2 (Fig. 39.2, curve 3).

39.3 Conclusions

1. It has been shown that Pt catalyst on the surface of porous silicon electrode structures under the conditions of aggressive action, for example 0.5–1 M of solution H_2SO_4 as the electrolyte of the half-cell of μ-FC, rapidly loses electro-catalytic activity.
2. For increasing the catalyst stability before shaping of catalytic layer it has been proposed to conduct the surface modification of silicon, including the internal surface of pores, by the gold-bearing primer and by photon annealing after working.

References

1. Ohlsen L (2004) Silicon-based micro fuel cells for mobile electronic applications. In: Micro nano breakthrough conference, Portland, 28 July 2004. www.neahpower.com
2. Starkov V (2009) Production, properties, and application of porous silicon. All materials. Encyclopedic reference book, No 4, pp 13–21
3. Starkov V, Gavrilin E (2007) Gralient- porous structure of silicon. Phys Stat Sol (c)4(6):2026–2028
4. Teterskiy A, Starkov V (2010) Synthesis of Pt catalysts on surface of porous silicon electrodes. In: Proceeding of meeting: Fundamental problems of solid state ionics, Chernogolovka, p 322 (in Russian)
5. Starkov V, Teterskiy A Claim for obtaining of RF patent on invention. No 2010119817 from 19.05.2010

Chapter 40
Studies of Hydrogen Sorption on Mesoporous Carbon Composite Modified with Adsorbed Palladium

G.M. Telbiz, V.I. Gerda, N.G. Kobylinska, V.M. Zaitsev, and J. Fraissard

Abstract Ordered mesoporous carbon composite based on the matrix synthesis protocol using mesoporous oxides as a template was prepared and step-by-step structure formation was followed by X-ray diffraction, FTIR, N_2 adsorption desorption and TPD mass-spectrometry. It was shown that property of composite is change considerably by Pd incorporation. Transmission electron microscopy and Raman spectroscopy show that the framework of the highly ordered mesoporous carbon/Pd consists of an aligned carbon phase with a graphite mode region. The potential application prepared composite for hydrogen storage is discussed.

Keywords Pd nanoparticles · Carbonization · Hydrogen adsorption · Synthesis · Structure · Ordered carbon composite · Graphite · Temperature · TEM microscopy

G.M. Telbiz
L. V. Pisarzhevsky Institute of Physical Chemistry of NASU,
pr. Nauki 31, Kyiv 03028, Ukraine

V.I. Gerda
Department of Earth Sciences – Geochemistry, Faculty of Geosciences,
Utrecht University, P.O. Box 80021, 3508 TA Utrecht, The Netherlands
and
Department of Chemistry, National Taras Shevchenko University,
64 Vladimirska str, Kiev, Ukraine

N.G. Kobylinska (✉) and V.M. Zaitsev
Department of Chemistry, National Taras Shevchenko University,
64 Vladimirska str, Kiev, Ukraine
e-mail: kobilinskaya@univ.kiev.ua

J. Fraissard
Laboratoire de Physique Quantique, University P. and M. Curie,
ESPCI, 10 Rue Vauquelin, Paris, France

S.Yu. Zaginaichenko et al. (eds.), *Carbon Nanomaterials in Clean Energy Hydrogen Systems - II*, NATO Science for Peace and Security Series C: Environmental Security 2,
DOI 10.1007/978-94-007-0899-0_40,

40.1 Introduction

The problem of storing hydrogen is one of the major technological barriers currently preventing the widespread adoption of hydrogen as an energy carrier. For safe and efficient hydrogen storage, developments of new hydrogen storage materials are currently being researched [1–3].

Practical problems with the storage of hydrogen appear to make solid state storage the most attractive proposed solution, and so the search for potential solid state hydrogen storage media is attracting a significant amount of research interest. Most hydrogen absorbing metals are very reactive, easily forming surface oxides, which then block the uptake of hydrogen. They also require costly time consuming activation. The last possibility, hydrogen adsorption in porous solids such as activated carbons, has been studied since the 1960s [4]. Recent research has focused on finding the ideal adsorbent that can be used at room temperature and allows the storage of desirable amounts of H_2. This has generated interest in the use of carbon materials for hydrogen storage.

Hydrogen storage in a carbon-based material offers significant advantages associated with its low mass density. Early experimental data for hydrogen storage in carbon nano-materials was initially promising. However, the more recent experimental data obtained on various carbon nanostructures are contradictory with large variance in the amount of hydrogen stored [6]. Using mesoporous oxides as a template is particularly attractive due to the possibility of structural order and diversity in achieving novel carbon nanoarchitectures [8]. These materials have aroused considerable interest due to their important fundamental implications and as a potential selective catalysts and "containers" for the storage of hydrogen and methane [5–7]. Controlling preparation and doping carbon nanostructures are key factors in defining mentioned above applications. For example, various transition and noble metals can be supported by carbon nanostructures using the wet impregnation technique [9–15]. Indeed, carbon with a graphitic structure involves a unique metal-support interaction, resulting in quite distinct catalytic behavior. For example, carbon-Pd interactions were discussed in terms of Pd particle size distribution and geometry which, in turn, are related to hydrogenation activity/selectivity [16, 17]. Experimental and theoretical investigations reveal controversies concerning the value of hydrogen storage capacity and the pore characteristics of carbon materials as well as metal loading [18–20]. In the present work ordered mesoporous carbons (like CMK-1) were synthesized using a mesoporous silica as a matrix, according to ref [2, 6, 21]. Mesoporous carbon formation was checked step-by-step. The effects of finely dispersed metallic palladium on the microstructure and sorption properties of carbon composite were also investigated. Potential applications of ordered carbon composite for hydrogen storage are discussed.

40.2 Experimental

40.2.1 *Materials Preparation*

Mesoporous silica (MCM-48) was prepared following a hydrothermal synthesis procedure using sodium silicate as the silica source [20]. The molar composition of the starting mixture synthesis was 1TEOS:0.54 NaOH:0.5 CTABr:110 H_2O. After homogenization of the mixture and hydrothermal treatment in autoclave, the product was filtered, carefully washed in water, dried at ambient temperature and finally calcined in air at 823 K. After template removal, the silica was impregnated with an aqueous solution of sucrose and sulphuric acid [2, 6]. Five milligram of MCM-48 was mixed with an aqueous solution prepared from 6.25 g of sucrose, 0.7 g of H_2SO_4 and 30 ml of H_2O. The resultant mixture was dried for 5 h at 373 K, and then at 443 K. After this partial carbonization, the sample was blended with a solution containing 3.75 g of sucrose, 0.4 g of H_2SO_4 and 30 ml of H_2O. The resultant mixture was dried again for 6 h. The resultant powder was calcined and carbonized under vacuum in a quartz reactor at various temperatures (973, 1,173 and 1,373 K). The blackish composite was obtained in 50% ethanol. The sample was filtered, washed in ethanol and dried at 373 K. For Pd loading, 0.1 g of mesoporous carbon was mixed with $H_2[PdCl_4]$ in acetone solution, ($C_{Pd2+} =$ 1.0009 mg/ml). The volume of solution was varied depending on the degree of Pd loading required (1% and 5%). The mixture was stirred for 3 h and dried until the acetone had evaporated.

40.2.2 *Characterization Techniques*

XRD patterns were obtained with a CuKα source (DRON-3 M diffractometer). The hydrogen adsorption capacity was measured with a volumetric device. 100 mg of carbon sample were put in a cell immersed in a Dewar vessel with liquid nitrogen (77 K), and pure hydrogen was passed at atmospheric pressure. After 45 min the solid-hydrogen equilibrium was reached and the sample cell was isolated. The hydrogen adsorption capacity was computed from the amount of gas desorbed from the material at 313 K. This experimental procedure avoids any error related to temperature gradients between the different parts of the apparatus. The specific surface area and pore size distribution of the mesoporous materials were calculated from nitrogen isotherms at 77 K using DFT model (Micromeritics ASAP 2000) [23]. Before starting this adsorption, the samples were outgassed under vacuum at 423 K during 18 h. FTIR spectra were obtained on a BRUKER IFS 66 spectrometer.

For TEM measurements, powders were deposited on a grid with a holey carbon film and transferred to a JEOL 2000 electron microscope operating at 80 kV. Changes in the amount of gas-phase products during sucrose carbonization were recorded by temperature programmed desorption monitored by a MX-7304 mass-spectrometer.

40.3 Results and Discussion

Figure 40.1 shows the X-ray diffraction patterns of as-synthesized and calcined mesoporous silica and systematic transformation of structure during the removal of silica framework after carbonization of sucrose impregnated in the pores. The sharp diffraction peaks (curve 1,2) show a good correspondence to cubic *Ia3d* symmetry, implying that the matrix is mesoporous. The product maintained the cubic structure without collapse after calcination at 823 K demonstrating a good thermal stability. The intensity change in the XRD pattern (lattice contraction and intensity loss) illustrated that channel systems separated by the silica walls were statistically equally filled with carbon(curve 3,4). Structure formed in these porous systems is obviously disconnected and capable to change their position with respect to one another when the matrix is removed. When the silica walls were destructed, produced CMK-1, a new reflection corresponding of the (110) carbon networks forbidden for cubic *Ia3d* appears. This result shows change of symmetry from cubic *Ia3d* to another cubic structure $I4_132$(curve 5), systematic transformation to new ordered mesoporous structure. The XRD of carbon material showed no patterns in the region 2θ greater than 10°, indicating that the carbon framework was atomically disordered.

FTIR-spectra confirmed XRD data of the systematic structural changes and show the possibility of the checking feature of the forming like CMK-1 structure and illustrate all reactions and processes, which pass in structure of matrix step

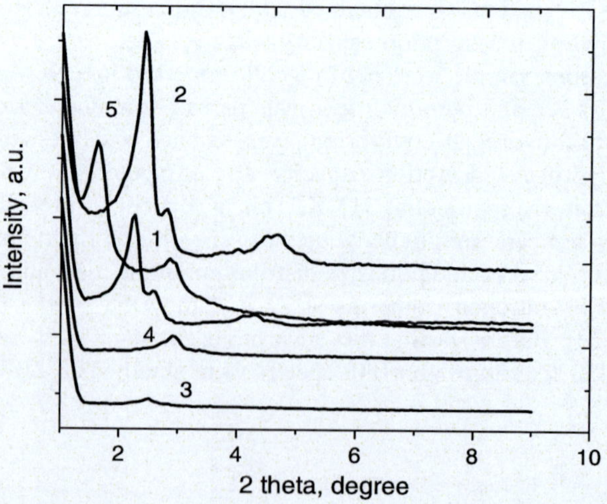

Fig. 40.1 XRD patterns for mesoporous silica(MCM-48) and ordered mesoporous carbon synthesized with sucrose: *1*. As synthesis; *2*. Calcined, 823 K; *3*. MCM-48, $C_{12}H_{22}O_{11} + H_2SO_4$ at 433 K; *4*. After vacuum carbonization at 1,173 K; *5*. Ordered mesoporous carbon

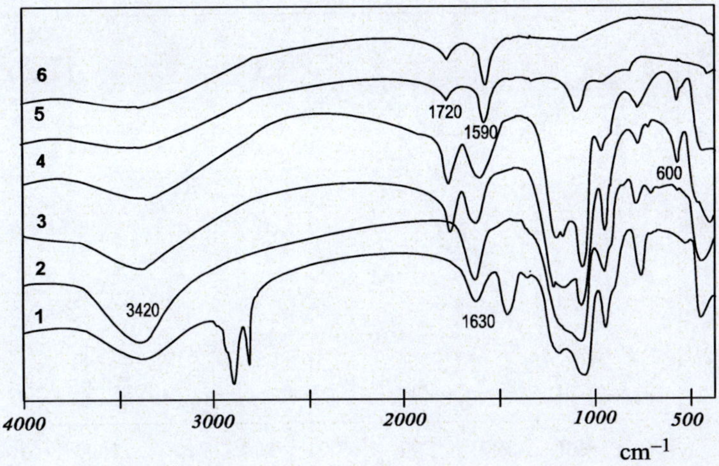

Fig. 40.2 IR-spectra registered after various stages of matrix (MCM-48) treatment and step by step formation: *1*. as synthesized matrix; *2*. calcined at 823 K; *3*. after sucrose + H_2SO_4, 423 K treatment; *4*. reloaded sucrose, 433 K; *5*. carbonization in vacuum, 1,173 K; *6*. fin such as al composite, after dissolved silica matrix

by step (Fig. 40.2). Calcination of mesostructured matrix at 823 K (curve 2) results in complete releasing of porous structure from template, disappearance of the absorption band ($V_{C—H}$), characteristic for the template molecules at 2,800–3,000 cm^{-1}. When the template was removed, ($V_{C—H}$) vibration bands disappeared and absorption bands that corresponding to oscillation of physically adsorbed water bands at 3,420 and 1,630 cm^{-1} was observed.

Carbonization result in the complete destruction of structure of MCM-48 (curve 5) and band at 1,590 cm^{-1} which characterizes the unsaturated carbon structures appeared in the spectra. After dissolution of the silicate matrix (curve 6) only absorption bands with at 1,590 and 1,720 cm^{-1} detected in a spectrum, that can testify to the completeness of carbonization and formation of the carbon silica matrix that resulted in appearance and increased of the band at 1,590 cm^{-1} which characterizes the unsaturated carbon structures. After removed of silica frameworks remain only the absorption bands at 1,590 and 1,720 cm^{-1}, that characterized formation of the mesoporous carbon remain in the spectrum.

After dissolution of the silicate matrix(curve 6) only absorption bands with at 1,590 and 1,720 cm^{-1} detected in a spectrum, that can testify to the completeness of carbonization and formation of carbon. The synthesis information concerning to effect of the sucrose/sulfuric acid, calcination temperature and optimum amounts of sucrose can be used for control to obtain highly ordered mesoporous carbon. With this purpose, we continuously detected r changes in the gas phase by mass spectrometer. Special attention has been paid to the formation of the volatile products which appeared at heating and carbonization sucrose in a vacuum with speed of heating 9.7 K/min in 303–1,023 K.

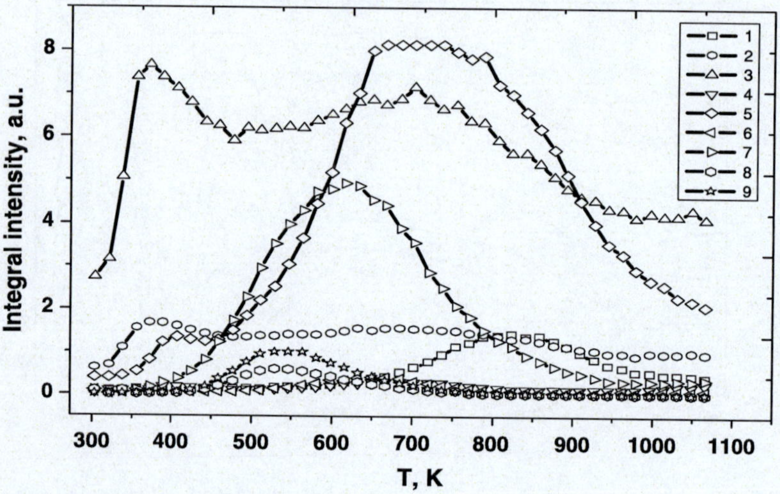

Fig. 40.3 Changes in the amount of gas phase products (calculated from MS data) during the reaction the sucrose carbonization for the ions with *m/z*: *1.* 15 (CH_3^+); *2.* 17 (OH^+); *3.* 18(H_2O^+); *4.* 27 ($C_2H_3^+$); *5.* 28(CO^+); *6.* 41(C_3H5^+); *7.* 44($CO_2{}^+$); *8.* 48(SO^+); *9.* 64($SO_2{}^+$)

As evident from Fig. 40.3, in this temperature range the basic volatile products of thermolysis of sucrose are H_2O, CO and CO_2. Their liberation observed at all time heating, especially it noticeably for H_2O, CO. In case of CO_2, we observed intensive take off at 473–773 K with the maximum about 603 K. The basic process of thermodestruction of sucrose from TPDMS data begins at a temperature 450 K. Thus, at the low temperature main process of thermolysis are dehydratation and insignificant release of the CO. When the temperature increased, we detected complex process, which are linked with splitting of the heterocyclic group and their destruction (CO release). Character of curves from m/z = 48 and 64, corresponding SO and SO_2, at 323–613 K testify to destruction of the sulfonate group.

Interestingly, that according to TPDMS data thermolysis not completed at 1,023 K, such as we detected noticeable release of H_2O and CO. Thus, we can stated, that for successfully completion of a process of sucrose carbonization and formation of the ordered porous carbon structure, additional treatment at a high temperature is necessary. A periodic nature of mesoporous carbon was also confirmed using Raman spectroscopy. In the visible Raman spectrum vibrational "graphite" mode appeared at 1,582 and 1,355 cm^{-1} bands due to highly oriented pyrolytic graphite.

Concerning the porosity, for all materials nitrogen adsorption exhibits type IV isotherms (not shown here), presenting a sharp step at a relative pressure of 0.3–0.4 attributed to capillary condensation in the ordered mesoporous structure. The results of calculations are given in Table 40.1. Pore size analysis shows that both silica matrix and carbon are structurally microporous and mesoporous, but the microporosity is much less important for carbon. The average pore size and the total surface area increase from 2.12 to 3.03 nm and from 1,310 to 1,801 m^2/g for matrix and for carbon, respectively. For the Pd composite, reduction of Pd with

Table 40.1 Structural and sorption properties of pure and metal-doped materials

Samples	S_{tot} (m^2/g)	S_{micro} (m^2/g)	S_{meso} (m^2/g)	S_{macro} (m^2/g)	V_{tot} (cm^3/g)	V_{micro} (cm^3/g)	V_{meso} (cm^3/g)	V_{macro} (cm^3/g)	D (nm)	H_2 capacity, (wt.%)
Pure MCM-48	1,310	400	900	6	1.16	0.10	0.92	0.14	2.12	0.51 ± 0.02
Pure CMK-1	1,800	155	1,580	70	1.15	0.10	1.00	0.05	3.03	1.80 ± 0.04
CMK-1/ Pd1%	1,795	150	1,575	70	1.12	0.10	0.97	0.05	3.11	2.07 ± 0.05*
CMK-1/ Pd5%	1,720	135	1,525	65	1.09	0.09	0.96	0.04	3.11	2.17 ± 0.06*

* related to the same amount of CMK-1

hydrogen does not change the textural/structural parameters significantly as compared with those of like CMK-1 (Table 40.1). The low angle XRD patterns of carbon with or without Pd are identical. The specific surface areas and volumes corresponding to the different type of pores are roughly the same. The average pore diameter has the same value (3.11 nm) but this latter corresponds in fact to a broader distribution of pore size which could result from a modification due to the influence of the solution used for Pd loading.

Analysis of TEM images reveals that the size distribution of palladium particles is quite narrow (average size 2–7 nm), and that these latter are partly dispersed within the matrix. Unfortunately TEM shows also some large particles; especially for CMK-1/Pd 5%. These can be detected also with high-angle XRD, with a reflection characteristic of metallic palladium at 39.7°. Of course, these large particles are located outside the mesopores of carbon network.

40.4 Hydrogen Adsorption Measurement

First we compared the results of the effective hydrogen storage by carbon prepared at different temperatures and sucrose carbonization times. Hydrogen adsorption on these ordered carbon materials increases noticeably with the carbonization temperature up to 1,173 K (Fig. 40.4).

At higher temperature, further increase up to 1,373 K has the reverse effect, leading to a decrease in hydrogen adsorption. These results show that the conditions of ordered carbon synthesis have a great influence on the quantity of hydrogen adsorbed by this material. The experimental data for the hydrogen capacities of matrix and carbon at 77 K and atmospheric pressure are presented in Table 40.1. The lowest value is obtained with silica matrix (0.51 wt.%) while the maximum hydrogen sorption on carbon is 1.80 ± 0.04 wt.% at a carbonization temperature of 1,173 K for 3 h. Similar data were observed Gadiou et al. [18] in the case of ordered nanostructure carbons samples. The amounts of adsorbed hydrogen, related to the

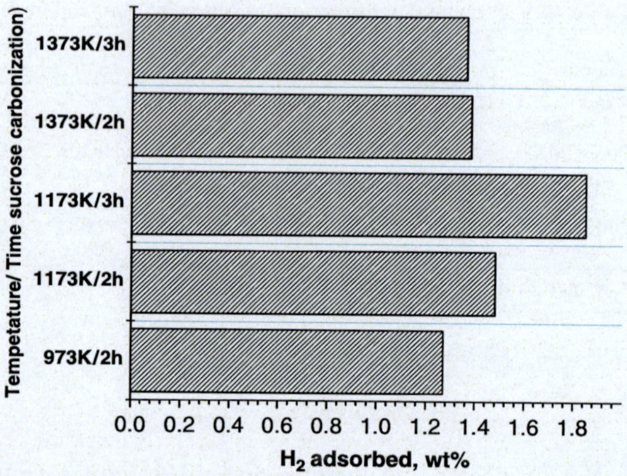

Fig. 40.4 Correlation between temperature/time carbonization of sucrose and storage capacity different CMK-1 at 77 K and ambient pressure

same amount of corresponding CMK-1, are 2.07 and 2.17 wt.% for samples CMK-1 –1% and 5%, respectively. The corresponding increases of hydrogen adsorption due to this doping are approximately 15% and 20% (Table 40.1).

Of course, during hydrogen adsorption there is the formation of β-Pd hydride of approximate formula is Pd-$H_{0.65}$. However, expressed in wt %, this corresponding amount is negligible. This phenomenon can be attributed to the spillover effect, as was proposed in [23]. It was shown in [24] that there is always an equilibrium between H_2 gas, H_2 adsorbed on the Pd surface and H_2 of β-Pd hydride, this latter form being at its maximum value when the surface of the particle is saturated.

$$H_\beta \Leftrightarrow (H)\text{adsorbed} \Leftrightarrow (H_2)\text{gas}$$

Consequently, there is always exchange and desorption of H species to re-form (H_2)gas or to diffuse throughout the carbon structure, as was shown for the case of Pt supported on silica and zeolites [25, 26].

40.5 Conclusions

Matrix synthesis and transformation of structure occurred during the formation of three-dimensional pore structure of ordered mesoporous carbon using mesoporous silica as a template was described. It was shown that for successfully completion of a process of sucrose carbonization the additional ageing at high temperature treatment is the necessary criterion. Modification of the ordered carbon by Pd nanoparticles resulted in increasing of adsorption of hydrogen but more surprisingly that sorptive capacity of H_2 does not depend from Pd concentration (1%–5%). To date we have found no evidence of significant hydrogen adsorption on carbons at ambient temperatures. However, modeling suggests that metal catalyst particles play a crucial role in facilitating adsorption, which may be attributed to the highly crystallized frameworks and the large surface composite areas. Consequently, future work will focus on the experimental determination of the role of the metal catalyst in the interaction of H_2 on ordered carbon composites. The influence of catalyst morphology on the interaction between C and H will be examined such as nature of mesoporous matrix (SiO_2, TiO_2). This approach may be applied to design many other ordered mesoporous crystalline carbon-M*x*O*y* nanocomposites with multi-functional properties.

Acknowledgments This work has been partly supported by the program of the National Academy of Sciences of Ukraine "Fundamental problems of hydrogen energy" and COST program (agreement No D36/003/06).

References

1. Maeland AJ (2001) Hydrides for hydrogen storage. In Recent advances in hydride chemistry (eds M. Peruzzine and M. Poli), Elsevier (Amsterdam - London - New York - Oxford - Paris - Shannon - Tokyo), pp. 531–557
2. Dehouche Z, Djaozandry R, Huot J, Boily S, Goyette J, Bose TK, Schulz R (2000) Influence of cycling on the thermodynamic and structure properties of nanocrystalline magnesium based hydride. J Alloys Compd 305:264–271
3. Fuji H, Orimo S (2003) Hydrogen storage properties in nanostructured magnesium and carbon related materials. Phys B 328:77–80
4. Kidnay A, Hiza M (1967) High pressure adsorption isotherm of neon, hydrogen and helium at 76 K. Adv Cryog Eng 12:730
5. Joo SH, Jun S, Ryoo R (2001) Synthesis of ordered mesoporous carbon molecular sieves CMK-1. Micropor Mesopor Mater 44:153–158
6. Liu R, Ren Y, Shi Y, Zhang F, Zhang L, Tu B, Zhao D (2008) Controlled synthesis of ordered mesoporous C-TiO_2 nanocomposites with crystalline titania frameworks from organic-inorganic-amphiphilic coassembly. Chem Mater 20:1140–1146
7. Schuth F (2001) Non-siliceous mesostructured and mesoporous materials. Chem Mater 13:3184–3195
8. Garsuch A, Klepel O (2005) Synthesis of ordered carbon replicas by using Y-zeolite as template in a batch reactor. Carbon 43:2330–2337
9. Garnsuch A, Böhlmann W, Sattler RR, Fraissard J, Klepel O (2006) ^{129}Xe NMR studies on carbon replicas of Y zeolite. Carbon 44:1173–1179
10. Tiemann M (2008) Repeated templating. Chem Mater 40:961–971
11. Lee D, Kim H, Kang M, Kim Ji M, Lee Ik-Mo (2007) Efficient hydrogenation catalysts of Ni or Pd on nanoporous carbon workable in an acidic condition. Bull Kor Chem Soc 28:2034–2040
12. Saha D, Deng Sh (2009) Hydrogen adsorption in ordered mesoporous carbon doped with Pt, Pd, Ru and Ni. Langmuir 25:12550–12560
13. Zielinski M, Wojcieszak R, Monoeverdi S, Mercy M, Bettahar MM (2007) Hydrogen storage in nickel catalysts supported on activated carbon. Int J Hydrogen Energy 32:1024–1032
14. Huwe H, Froba M (2007) Synthesis and characterization of transition metal and metal oxide nanoparticles inside mesoporous carbon CMK-3. Carbon 45:304–314
15. Park SY, Choi BH, Kang M, Kim JM, Lee IM (2007) Diarylmethanes catalyzed by nickel(II) ion on nanoporous carbon. J Mol Catal A Chem 265:323–329
16. Ros TG, Keller DE, van Dillen AJ, Geus JW, Koningsberger DC (2002) Preparation and activity of small rhodium metal particles on fishbone carbon nanofibres. J Catal 211:85–102

17. Amorin C, Keane MA (2008) Palladium supported on structured and nonstructured carbon: a consideration of Pd particle size and the nature of reactive hydrogen. J Colloid Interface Sci 322:196–208
18. Gadiou R, Saadallah S, Piquero T, David P, Parmentier J, Vix-Guterl C (2005) The influence of textural properties on the adsorption of hydrogen on ordered nanostructured carbons. Micropor Mesopor Mater 79:121–128
19. Orimo S, Meyer G, Fukunaga T, Zuttel A, Schlabch L, Fujii H (1999) Hydrogen in the mechanically prepared nanostructured graphite. Appl Phys Lett 75:3093–3095
20. Panella B, Hirscher M, Roth S (2005) Hydrogen adsorption in different carbon nanostructures. Carbon 43:2209–2214
21. Kresge CT, Leonowicz ME, Roth WJ, Beck JS (1992) Ordered mesoporous molecular sieves synthesized by a liquid-crystal template mechanism. Nature 359:710–714
22. Telbiz GM, Gerda VI, Gun'ko VM (2005) Matrix synthesis and properties of the 3D-nanoordered mesoporous carbon material. Nanosyst Nanomater Nanotechnol 3:661–666 (in Russian)
23. Zacharia R, Rather S, Hwang SW, Nahm KS (2007) Spillover of physisorbed hydrogen from sputter-deposited arrays of platinum nanoparticles to multi-walled carbon nanotubes. Chem Phys Lett 434:286–291
24. Polisset M, Fraissard J (1993) ^{1}H NMR study of hydrogen adsorption on PdNaY: Influence of metal particle size. Colloids Surf A 72:197–206
25. Lenz D, Conner WC, Fraissard J (1989) Hydrogen spillover on silica: III. Detection of spillover by proton NMR. J Catal 117:281–289
26. Fraissard J (1999) NMR studies of supported metal catalysts. Catal Today 51:481–499

Chapter 41
Thermodynamic Properties and Phase Equilibriums in Ternary Alloys of the Al-C-Si(Ge, Sn, Pb) Systems

I.V. Mateyko, V.S. Sudavtsova, M.A. Shevchenko, V.G. Kudin, and N.O. Sharkina

Abstract Some algorithms of thermodynamic modeling by the theory of ideal associated solutions (TIAS) are provided. The thermodynamic properties of liquid alloys of the Al-C system are calculated using the TIAS and some experimental data. The analysis of own and literary data has allowed to establish the most reliable mixing enthalpies of binary boundary systems Al-Si(Ge, Sn, Pb), C-Si. This allowed to estimate the thermodynamic properties of ternary alloys of the Al-C-Si(Ge, Sn, Pb) systems by Kohler equation.

Keywords Theory of ideal associated solutions · Gibbs energy · Binary and ternary systems · Melts · Carbon concentration · Phase diagram

41.1 Introduction

Thermodynamic properties of carbon-containing systems are necessary for forecasting of the direction of different processes, chemical equilibriums deviations etc. Due to the difficulty of their experimental studying, the Al-C and C-Ge(Sn, Pb) melts are not investigated, except some small concentration intervals. These melts can be practically obtained only in the region of small concentration of carbon. So we extrapolated data for the boundary binary systems on the whole concentration interval using also thermochemical properties of intermetallics which form in these systems, and phase diagrams. For this purpose we used thermodynamic modeling by the theory of ideal associated solutions (TIAS).

I.V. Mateyko and V.S. Sudavtsova (✉)
Institute for Problems of Materials Science, NAS of Ukraine,
Krzhyzhanovsky str. 3, Kiev 03142, Ukraine
e-mail: vsudavtsova@univ.kiev.ua

M.A. Shevchenko, V.G. Kudin, and N.O. Sharkina
Department of Chemistry, National Taras Shevchenko University,
64 Vladimirska str., Kiev 01033, Ukraine

S.Yu. Zaginaichenko et al. (eds.), *Carbon Nanomaterials in Clean Energy Hydrogen Systems - II*, NATO Science for Peace and Security Series C: Environmental Security 2,
DOI 10.1007/978-94-007-0899-0_41, © Springer Science+Business Media B.V. 2011

41.2 Bases of TIAS

The TIAS [1, 2] connects negative deviations of thermodynamic properties of melts from the ideal solution with the forming of associates of different kinds of atoms A_iB_j in it. So, the solution A-B is considered as a mixture of monomers A_1,B_1 and associatesA_iB_j, which can form from the latter. The equilibrium composition of that solution is characterized by N reactions

$$i_nA_1 + j_nB_1 \Leftrightarrow A_{i_n}B_{j_n} \quad (n = 1...N)$$

and can be found by solving of the system of N equations, based on the law of mass action:

$$K_n = \frac{x_n}{x_{A_1}^{i_n} x_{B_1}^{j_n}},$$

where K_n is an equilibrium constant of the associate number n forming; x_n is a molar part (which equals to activity) of this associate; x_{A_1},x_{B_1} are molar parts (activities) of monomers in the melt.

The temperature dependence of equilibrium constant of the corresponding associating reaction is bounded up with the thermodynamic properties of associates

$$K_n = e^{\frac{\Delta S_n}{R} - \frac{\Delta H_n}{RT}}$$

where ΔS_n is the entropy of forming of the n-th associate; ΔH_n is the enthalpy of its forming; R is the gas constant; T is the temperature.

In this case the model of thermodynamic properties of the melt is a system of non-linear equations, which parameters are quantity and composition of associates, and their ΔS_n and ΔH_n:

$$x_n = \exp\left(\frac{\Delta S_n}{R} - \frac{\Delta H_n}{RT}\right) a_A^{i_n} a_B^{j_n},$$

$$a_A = x_A\left[1 + \sum_{n=1}^{N}(i_n + j_n - 1)x_n\right] - \sum_{n=1}^{N} i_n x_n,$$

$$a_B = x_B\left[1 + \sum_{n=1}^{N}(i_n + j_n - 1)x_n\right] - \sum_{n=1}^{N} j_n x_n,$$

$$a_A + a_B + \sum_{n=1}^{N} x_n = 1,$$

$$\Delta H = \frac{\sum_{n=1}^{N} \Delta H_n x_n}{1 + \sum_{n=1}^{N} (i_n + j_n - 1) x_n},$$

where $a_A \equiv x_{A_1}$ and $a_B \equiv x_{B_1}$ are activities of melt components, equal to the molar parts of their monomer particles; x_A, x_B are molar parts of components; ΔH is the integral mixing enthalpy.

The physical picture of interaction in the melts gives series of additional conditions to the system of TIAS equations:

$$0 < a_B < 1,$$

$$0 < a_A < 1,$$

$$0 < x_n < 1,$$

$$\Delta H_n < 0,$$

$$\Delta S_n < 0.$$

Also, for the majority of cases it may be taken that the forming enthalpy of each of the associates is more negative than the linear combination of the forming enthalpies of its two neighbor associates. One of the latter has lesser content of the component B, and the second has greater; the monomers A, B are also taken as the associates with $\Delta H_n = 0$. So, if the content of B in the n-th associate is $c_n = \frac{j_n}{i_n + j_n}$, and $c_{n-1} < c_n < c_{n+1}$, then

$$\Delta H_n < \frac{\Delta H_{n-1}(c_{n+1} - c_n) + \Delta H_{n+1}(c_n - c_{n-1})}{c_{n+1} - c_{n-1}}.$$

We have provided the next algorithms to calculate contents of the monomers and associates at fixed temperature and molar parts of components:

1. For each totality of concentrations of associates we can calculate the Gibbs energy of solution: if $\Delta G_n = \Delta H_n - T\Delta S_n$, then

$$\Delta G = RT \frac{x_A \ln x_A + x_B \ln x_B + \sum_{n=1}^{N} x_n \left(\frac{\Delta G_n}{RT} + \ln x_n\right)}{1 + \sum_{n=1}^{N} (i_n + j_n - 1) x_n}.$$

The equilibrium solution has the least value of ΔG among the all possible solutions. So the problem is to find a minimum value of the function of N variables, which is continued, differentiable in the whole definition region except the bounds ($x_n = 0$, $x_A = 0$, or $x_B = 0$), and hasn't any local minimums.

Different numerical methods can be used for this purpose, but they are not of the same effectiveness. For example, the method of steepest descent may give a result quite slowly because of the great difference of partial derivatives of ΔG by the different variables x_n. To obtain a rough approximation, a fast method is a random search (exactly, changing all variables x_n in turn with saving of results which have lower value of ΔG). This method is the most reliable, but it is not advantageous to obtain a precise solution, because it needs nearly the same number of iterations for the each digit in x_n. If we want to obtain a rough approximation (1–2 digits) it doesn't matter, but for 15–16 digits the program runs rather a long time.

2. The Newton method has by far better speed when the values x_n are already close to exact ones. But it cannot be used on the beginning of calculation, except there are only 1–2 associates. This method may give us a point where all the derivatives $\frac{\partial \Delta G}{\partial x_n} = 0$, solving a system of linear equations $\sum_{m=1}^{N} \frac{\partial^2 \Delta G}{\partial x_n \partial x_m} \Delta x_n + \frac{\partial \Delta G}{\partial x_n} = 0$, where $n = 1..N$. However, first, and especially second derivatives of ΔG have too complicated analytical presentation; so, calculation of them as analytically, as numerically, is not rational.

Nevertheless, there is a quite more rational method, which is based on the reduction to zero not $\frac{\partial \Delta G}{\partial x_n}$, but rates of reactions $i_n A_1 + j_n B_1 \Leftrightarrow A_{i_n} B_{j_n}$, which equal to $v_n = k_n (x_{A_1}^{i_n} x_{B_1}^{j_n} e^{-\frac{\Delta G_n}{RT}} - x_n)$. To find a point where all $v_n = 0$, the absolute values of k_n are unnecessary, and we can equate them to 1. So, we need to solve a linear system $\sum_{m=1}^{N} \frac{\partial v_n}{\partial x_m} \Delta x_n + v_n = 0$, where $\frac{\partial v_n}{\partial x_m}$ have rather simple analytical type:

$$\frac{\partial v_n}{\partial x_m} = \begin{cases} x_{A_1}^{i_n - 1} x_{B_1}^{j_n - 1} e^{-\frac{\Delta G_n}{RT}} (i_n p_m x_{B_1} + j_n q_m x_{A_1}), & m \neq n \\ x_{A_1}^{i_n - 1} x_{B_1}^{j_n - 1} e^{-\frac{\Delta G_n}{RT}} (i_n p_m x_{B_1} + j_n q_m x_{A_1}) - 1, & m = n \end{cases},$$

where $p_m = x_A(i_m + j_m - 1) - i_m$, $q_m = x_B(i_m + j_m - 1) - j_m$. We ought to say that the values of p_m, q_m and $e^{-\frac{\Delta G_n}{RT}}$ don't change during all the iterations, so they must be calculated only once at the begin of solving. So, iterations contain simply adding and multiplying, thanks to which the speed of them is high. In the majority of cases we need only 2–4 iterations to obtain a result with a maximal precision (15–16 signs). However, we need to have a rough approximation of x_n at the beginning, found by the other method.

3. A good method to calculate a first approximation is using iterations of the type $\Delta x_n = -\frac{v_n}{\sum_{m=1}^{N} \frac{\partial v_m}{\partial x_m}}$. We also need to verify that ΔG goes down on the each iteration.

So, the optimal strategy of calculation is using of the method (3) to obtain a first approximation, and after that – the method (2). And only in some cases it is necessary to use a slow but the most universal method (1).

When we know concentrations x_n of all associates and monomers, and also Gibbs energies of all solid phases, it is not difficult to prove a possibility of crystallization of them at the given temperature. For this purpose we find the chemical potentials of pure components (monomers) in the solution: $\mu_{A_1} = RT\ln x_{A_1}$, $\mu_{B_1} = RT\ln x_{B_1}$. The crystallization of A will go if the crystallization Gibbs energy is lower than μ_{A_1}: $\Delta G_{cryst}(A) = T\,\Delta S_m(A) - \Delta H_m(A) < \mu_{A_1}$, where melting enthalpy $\Delta H_m(A)$ and entropy $\Delta S_m(A) = \frac{\Delta H_m(A)}{T_m(A)}$ are well-known values. Analogously, crystallization of B needs $\Delta G_{cryst}(B) = T\,\Delta S_m(B) - \Delta H_m(B) < \mu_{B_1}$. Also, crystallization of intermetallic $A_{i_n}B_{j_n}$ needs

$$\Delta_f G(A_{i_n}B_{i_n})^s - \frac{i_n \Delta G_{cryst}(A) + j_n \Delta G_{cryst}(B)}{i_n + j_n} < \frac{\mu_{A_1} i_n + \mu_{B_1} j_n}{i_n + j_n}.$$

In accordance to that we can solve a direct thermodynamic task – to build a phase diagram of A-B system using the properties of associates in this system. To obtain a liquidus line, we should find a temperature at which only a one solid phase can crystallize. It can be done by the binary search, when we take a large temperature interval which contains a liquidus temperature indeed, and halve this interval with the each iteration.

It should be said that the forming enthalpies of solid intermetallics also subordinate to the analogous condition which we have seen for associates:

$$\Delta H_n < \frac{\Delta H_{n-1}(c_{n+1} - c_n) + \Delta H_{n+1}(c_n - c_{n-1})}{c_{n+1} - c_{n-1}}.$$

It is true for all the cases except the cases of intermetallic decay into the others at low temperatures.

If solid phases are not purely stoichiometric but have homogeneity ranges, the calculation will be quite more difficult. Approximately, the forming Gibbs energies for solid solutions based on the A, B, and intermetallic $A_{i_n}B_{j_n}$, which we sign as α, β and γ, can be presented as such functions:

$$\Delta_f G(\alpha) = RT(c\ln c + (1-c)\ln(1-c)) + A_\alpha c + B_\alpha c^2,$$

$$\Delta_f G(\beta) = RT(c\ln c + (1-c)\ln(1-c)) + A_\beta(1-c) + B_\beta(1-c)^2,$$

$$\Delta_f G(\gamma) = \Delta G_{c_n} + A_\gamma(c - c_n) + B_\gamma(c - c_n)^2,$$

where c is a molar part of B in the solid solution α, β or γ; $c_n = \frac{j_n}{i_n + j_n}$.

The coefficients $A_{\alpha,\beta,\gamma}$, $B_{\alpha,\beta,\gamma}$ are fitted in order that the model should agree with the experimental phase diagram. $B_{\alpha,\beta,\gamma}$ are always positive, and when the solubility range become thinner, $B_{\alpha,\beta,\gamma} \to +\infty$. Even this simplified model of solid solutions needs rather a large quantity of calculations, connected with binary search of the composition of solid solution, which is in equilibrium with melt.

So, all methods above illustrate how to find thermodynamic properties of melts of A-B system and the phase diagram of it using properties of associates by TIAS. Actually, we should also solve a reverse task, because properties of associates mostly are inaccessible for direct study. For this purpose we use an optimization, which means diminution of the sum of squares of errors of calculated by TIAS mixing enthalpies and non-variant points on the phase diagram, in compare with experiment. For many systems we also can use experimental values of forming enthalpies and entropies of solid intermetallics. For this optimization, we use the random search and Hook-Jeaves methods. Sometimes, the optimization is the quickest when these two methods are used in turn.

A final result is a set of values of the forming enthalpies and entropies of associates and solid intermetallics, which agrees with all known experimental data for the liquid and solid alloys A-B. However, sometimes the whole agreement cannot be obtained. The reason of it may be incorrect set of associates, or unreliability of experimental data.

The composition of associates is not always the same with the composition of intermetallics which form in the system. Firstly, in the melts A-B in the beginning forms the simplest associate AB, and its concentration is often the greatest among another associates. It isn't strange because the minimum of integral mixing enthalpy is usually not far from the equimolar composition, and for the majority of metal-metal systems – it is in the range $0.33 < x_B < 0.67$. A closeness of minimum of ΔH to $x_B = 0.33$ or 0.67 means heightened stability of associates A_2B and AB_2 accordingly. Finally, in the melts of many systems there are also associates A_3B, AB_3 in the comparably small quantities, and sometimes even A_4B or AB_4. Growing number of atoms in the associate simultaneously causes a lowering of its entropy, though the lowering of enthalpy is the greatest for the simplest associates AB, A_2B and AB_2. So, associates with a large number of atoms are disadvantageous, especially at high temperatures. Due to that, in the metallic systems can exist only very small concentrations of associates A_2B_2, A_2B_3, A_3B_2, A_5B, AB_5. It is characteristic that if the size factor is great (for example, $r_B > r_A$), more stable are associates with many small atoms (A_2B, A_3B, A_4B).

So, in spite of the fact that the TIAS method guarantees an extend adequacy of thermodynamic properties of melts when all the associates are taken into account, for the practical use their number should be not greater than 4–5. If we take greater quantity of associates, the calculation runs much slower; but if we take a smaller quantity – we can lose exactness of correlation with experimental data.

For the systems which are not enough studied experimentally, the agreement after optimization is also imperfect. In this case the search of the most trustworthy data and comparing with analogous metallic systems allows us to find errors and to correct some thermodynamic properties or a phase diagram. Also the TIAS optimization allows to extrapolate thermodynamic properties into the unstudied range of concentrations.

41.3 Modeling results

The Al_4C_3 intermetallic at high temperatures with the transition into liquid state should dissociate into pure components and also smaller associates AlC and Al_2C. Calculated activities of associates and pure components, modeled for the temperature 1,800 K (liquid or undercooled alloys), are presented on the Fig. 41.1.

It is obvious that the Al_4C_3 associate is not very stable in the liquid phase. This fact can be explained with the low probability of congregation of seven atoms in the molecule at high temperatures. Activities of aluminum and carbon show moderate negative deviations from the Raoult's law.

The mixing Gibbs energies, partial and integral enthalpies of the Al-C liquid or undercooled melts at 1,800 K are presented on the Fig. 41.2.

Using Kohler (Fig. 41.3a) and Bonnier-Caboz (Fig. 41.3b) equations, we can estimate thermochemical properties of the ternary systems Al-C-Si(Ge, Sn, Pb) using the experimental and calculated data for the binary boundary systems Al-C (Si, Ge, Sn, Pb), C-Si(Ge, Sn, Pb) [3–6].

Obtained mixing isoenthalpies (kJ/mol) for the ternary Al-C-Si(Ge, Sn, Pb) systems are presented below (Fig. 41.4).

We should say that there are no exact data for the C-Ge(Sn, Pb) systems. It is known that there is no mixibility and no compounds in solid state, and the solubility of carbon in the liquid Ge, Sn, Pb is very low even at the high temperatures. So, the value of $\Delta H_{max} = 10$ kJ/mol were taken for them due to the existence of some systems (for example, Al-Pb), which have also a great range of mixibility absence.

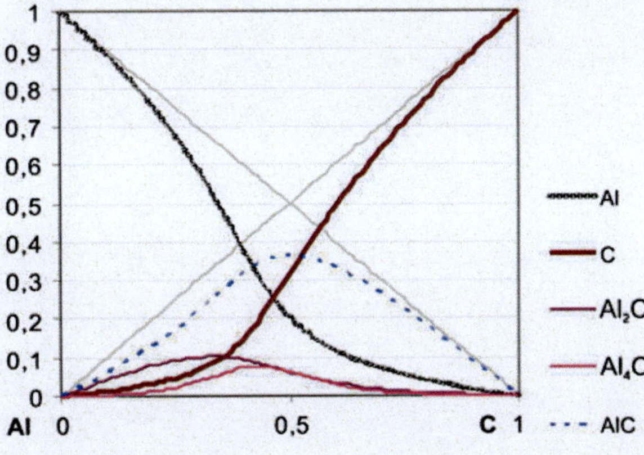

Fig. 41.1 Activities of associates and pure components at 1,800 K in the Al-C melts

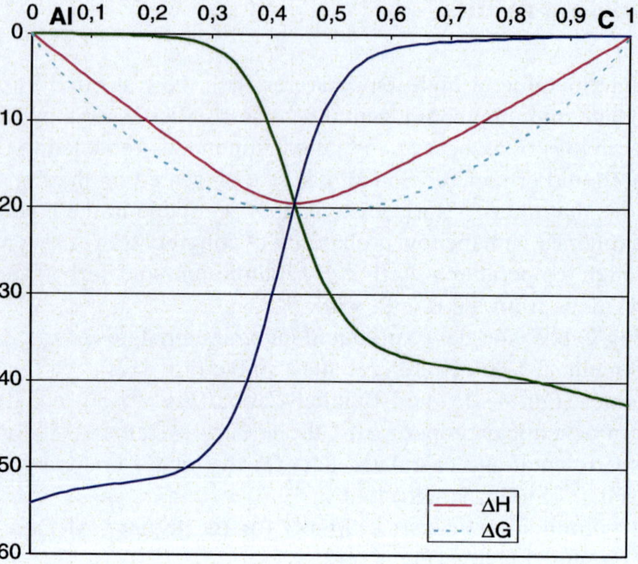

Fig. 41.2 Integral Gibbs energies, partial and integral enthalpies of mixing of the Al-C melts at 1,800 K

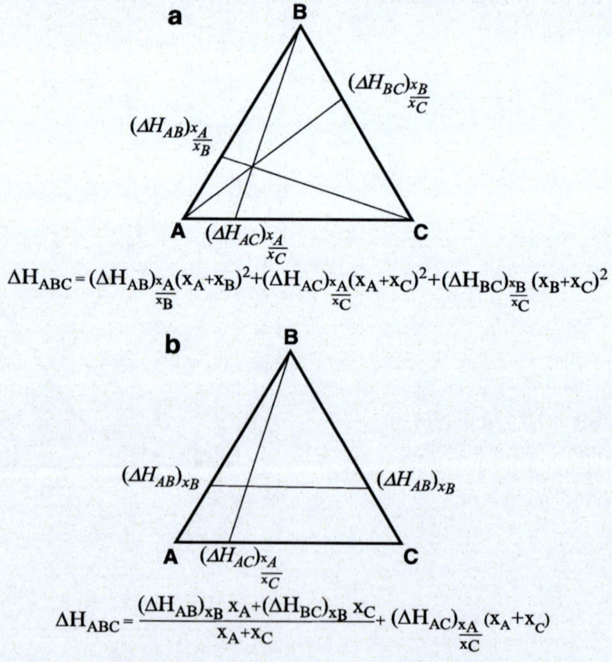

Fig. 41.3 Scheme of excess thermodynamic properties calculation by Kohler and Bonnier-Caboz equations

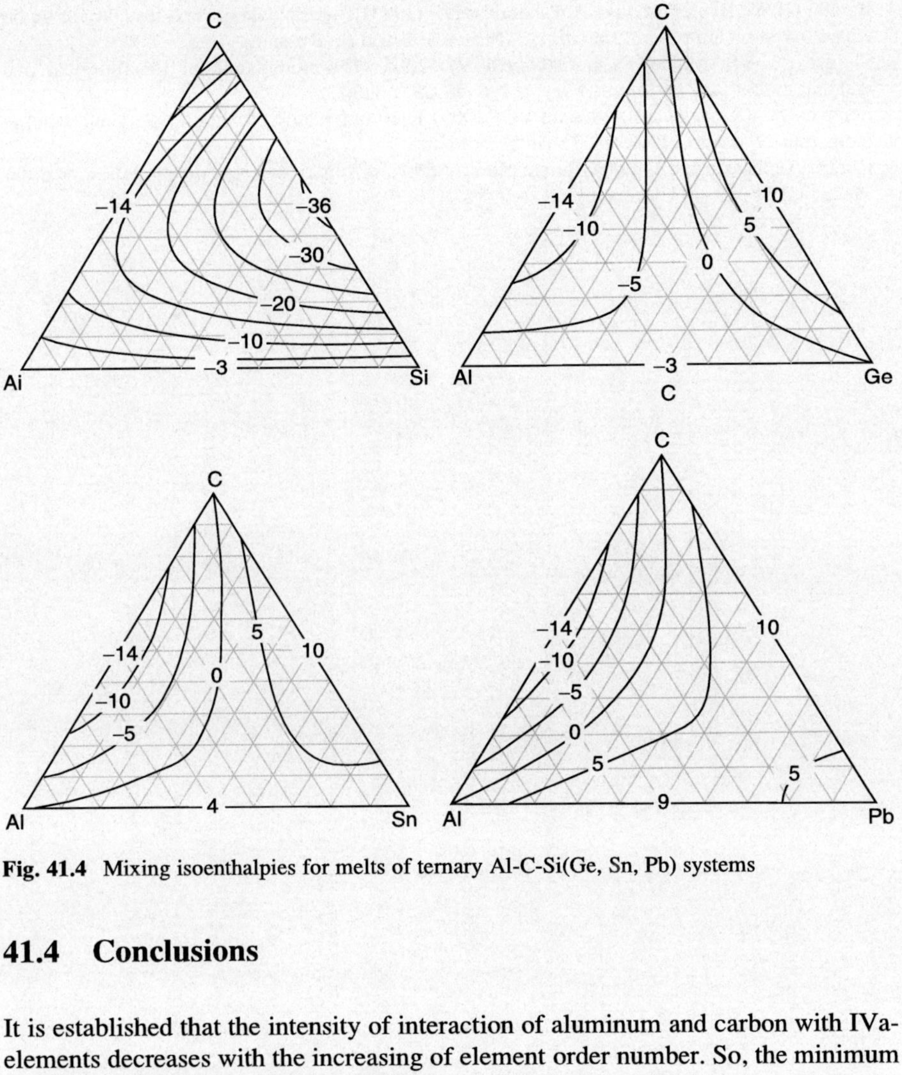

Fig. 41.4 Mixing isoenthalpies for melts of ternary Al-C-Si(Ge, Sn, Pb) systems

41.4 Conclusions

It is established that the intensity of interaction of aluminum and carbon with IVa-elements decreases with the increasing of element order number. So, the minimum of integral mixing enthalpy is situated on the boundary binary system C-Si for the ternary Al-C-Si system. For the other ones mixing enthalpies are alternating and the minimum of them is on the Al-C system.

References

1. Alger MM (1986) Associated solution model for compound – forming liquid metal mixtures. Z Metallkd 77(2):72–79
2. Krull H-G, Singh RN, Sommer F (2000) Generalised association model. Z Metallkd 91 (5):356–365

3. Batalin GI, Ye Byeloborodova A, Kazimirov VP (1983) Thermodynamic and structure of liquid alloys based on aluminum. Metallurgy, Moscow, p 160 (in Russian)
4. Sudavtsova VS, Makara VA, Galynych VI (2005) Thermodynamic of metallurgical and welding melts (part 1). Logos, Kiev, p 190 (in Ukrainian)
5. Sudavtsova VS, Makara VA, Kudin VG (2005) Thermodynamic of metallurgical and welding melts (part 3). Logos, Kiev, p 216 (in Ukrainian)
6. Kudin VG, Makara VA (2004) Interaction in alloys of binary systems metal-carbon. Powder Metall 1/2:78–82 (in Ukrainian)

Chapter 42
Increase of Solubility of Hydrogen in Electrolytic Alloys Ni–B

A.V. Zvyagintseva and Y.N. Shalimov

Abstract The kinetics of process of interaction of hydrogen with electrochemical systems on the basis of nickel is considered. Experimental and physical and mathematical modeling of influence of distribution of a current on an electrode surface on the hydrogen maintenance on surface elements is made. It is positioned that owing to non-uniformity of distribution of a current on a flat electrode structures of the nickeliferous coverings having various stoichiometry of composition and structure are formed.

Influence of an alloying component of boron on process of interaction of hydrogen with nickel and probability of formation of bridging structures Ni-B-H is considered. Results of analytical calculations of definition of temperature drop in an cathode bed of an electrode from technological modes of electrochemical formation of alloy Ni-B are resulted. Possibility without barrier electron transition through double electric bed is shown.

Keywords Solubility · Alloy nickel–boron · Hydrogen accumulation · Transfer of the masses · Current distribution on a surface of an electrode · The histogram · Without barrier conduction of an electron · Temperature at cathodes bed · A relaxation spectrum

42.1 Introduction

Dissolution of hydrogen in electrolytic metals and alloys has specific features owing to simultaneous formation of metal structure and the formation of metal compounds with hydrogen. The majority of researchers connect probability of such formations with presence of structural defects. In particular, the interaction of hydrogen can be carried out on boundary of grain and places dislocations, and

A.V. Zvyagintseva (✉) and Y.N. Shalimov
Voronezh State Technical University, Moscow av. 14, Voronezh 394026, Russia
e-mail: zvygincevaav@mail.ru; shalimov-yn@mail.ru

S.Yu. Zaginaichenko et al. (eds.), *Carbon Nanomaterials in Clean Energy Hydrogen Systems - II*, NATO Science for Peace and Security Series C: Environmental Security 2,
DOI 10.1007/978-94-007-0899-0_42, © Springer Science+Business Media B.V. 2011

other structural defects. Depending on the conditions of electrolysis electrolytic metals have brightly - clear tendency to the formation of structures such as metal-hydrogen in the zone of crystallization with the changing flow of substance. As a rule, in these superficial layers, there are gradients of velocity and temperature of the electrolyte. For this reason there is heterogeneity in the distribution of dissolved hydrogen on the thickness and the metal surface. In the future, dissolved hydrogen can be redistributed over the volume of metal due to temperature, electromagnetic fields or other nature (force, pressure).

However, the connection of the metal-hydride causes a change in the mechanical and physical-chemical properties of the coating. Therefore, we believe that we must consider kinetics of the interaction between hydrogen and electrolytic metals and alloys, as well as the influence of character of the interaction of hydrogen compounds on the structure of the metal. It is interesting to consider the diffusion of hydrogen in the period after crystallization as a further change in the structure of the metal may lead to normalization or to destruction of the formed system .This question is important not only in terms of the theory of hydrogen interactions, but also of great interest to researchers working on creating systems of hydrogen storage. The decision of this problem creates for the development of practical technology in hydrogen power.

42.2 Results and Discussion

Let's consider in general view kinetics of electrochemical reactions. We will define components of the basic components mass-transfer in classical system.

The given system of the equations underlies the decision of problems diffusion kinetics.

$$\frac{\partial C_k}{dt} = -div\vec{j}_k^{(\Sigma)},$$

where the index k - applies to all particles involved in the electrochemical reaction (reactants and products), the value represents the total stream of diffusion, migration and convection of the component. Magnitude:

$$\vec{j}_k^{(\Sigma)} = \vec{j}_k^{(\ddot{A})} + \vec{j}_k^{(\dot{I})} + \vec{j}_k^{(\hat{\Sigma})},$$

represents the total stream of diffusion, migration and convection of the component.

In general, the mass transfer is determined by three mechanisms:

(1) molecular diffusion – the relocation of particles of matter under the influence concentration gradient; (2) migration – the relocation of charged particles under the influence electric field; (3) convection – the transfer of matter along with the flow of moving fluid [1].

In our case, the main discharging particles are ions Ni^{2+}, H_3O^+. Their concentration is very different, so further kinetic calculations must take into account the concentration gradient. Taking into account all types of mass transfer, assuming the independence of the diffusion coefficient of a regulating particle of concentration of a solution, we can write the following expression for concentration distribution of reactive particles k near to an electrode surface:

$$\frac{\partial C_k}{dt} = D_k \nabla^2 C_k + D_k \frac{Z_k \cdot F}{RT} div(C_k \cdot grad\varphi) - \vec{\upsilon} \cdot gradC_k,$$

where $\vec{\upsilon}$ – velocity of fluid, $div\vec{\upsilon} = 0$, velocity of fluid, if the fluid is incompressible; $\frac{\partial C_k}{dt}$ – change of concentration of particles in time; D_k – diffusion coefficient of particles; $\nabla^2 C_k$ – Laplas operator of change of substance concentration, that is the second derivative of change of substance concentration on coordinates in three-dimensional system; Z_k – charge of cations substance; F – Faraday number; R – universal gas constant; T – temperature of the electrolyte solution; $div\ (C_k \cdot grad\ \varphi)$ – the rate of change of concentration of particles from the potential gradient.

In this formula there is not the multiplier defining the factor of viscosity of electrolyte in a frontier layer. Value of temperature in a reaction zone also strongly differs from average value in electrolyte volume. Therefore, the actual change in concentration over time must be different from that calculated by the above formula. In addition, the change in concentration will adjust the Dufour effect, the vector which is directed oppositely vector of displacement due to the forces of an electric field. Besides, during the discharge of particles of different masses at the cathode influences the effect of Sore.

According to effect Sore speed of diffusion of particles of various weight depends on a gradient of temperatures, namely, particles with small molecular weight direct in "hot" areas of space, and with the big weight - in "cold" areas. These phenomena are confirmed by us experimentally, for this purpose we have defined quantity of the dissolved hydrogen in electrolytic metals depending on geometry of the sample.

Figure 42.1 a gives the distribution of current along the surface of the electrode, which implies that the maximum current density corresponds to the angular points of an electrode. In line with this it was determined the concentration of hydrogen at the elementary parts of the cathode, according to which, were constructed histogram distribution of electrode (Fig. 42.1b).

Numbers of elementary sections correspond to the position of the electrode, in which 1 and 5 are at the upper limit of the electrolyte, and 2 and 4 – at the lower boundary of the electrolyte, and 3 – on the center electrode. Such a pattern of distribution, apparently due to the fact that the concentration of discharging ions hydronium H_3O^+ is different from the upper and lower boundaries of the electrode. These results agree well with our experiments conducted to study the distribution of temperature gradient in the electrode layer [2]. In all probability, as expected, maximum concentration of hydrogen will be in areas located near the upper boundary of the electrolyte.

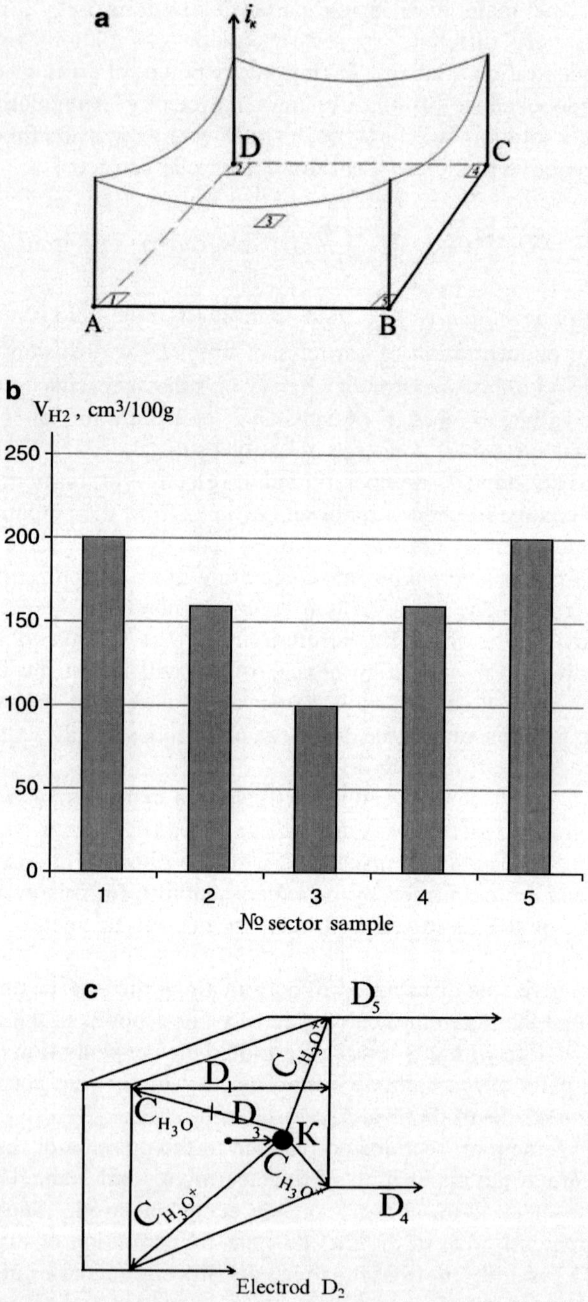

Fig. 42.1 **(a)** Distribution of current on an electrode surface: ABCD – position data of the sample; i_k – density of a cathodic current. **(b)** The diagram of hydrogen abundance on elementary sites of electrode. **(c)** Direction of thermal streams Dufours and movement of particles for the account of effect Sore, where $D_1 = D_5 => D_2 = D_4$, K – point of average temperature of electrolyte

On Fig. 42.1 it is shown, that for the account of effect Sore, ions H_3O^+ direct to these platforms. Thus, the probability of interaction of free atoms of metal with hydrogen ions is maximum on these sites. According to data obtained by us earlier [3] in the process of interaction between the metal (Ni) with hydrogen (Fig. 42.1a, b, c) energy of the resulting compounds are quite different in their values and therefore the interaction of Me-BH may be stoichiometrically inadequate. Apparently, this is the main reason for the difference in energy of bond of the Me-H.

The authors of [4] investigated the influence of concentration of boron compounds on the binding of Ni in obtaining borides of nickel found that increased concentration leads to a more complete binding of nickel. The synthesis of borides of nickel by the same authors found that the reaction product containing occluded hydrogen This circumstance has allowed us to assume, what exactly this hydrogen brings different interpretations in treatment of results stoichiometry of nickel hydrides. In a reducible equation:

$$\begin{aligned}2\mathrm{Ni(CH_3COO)_2} + 6\mathrm{NaBH_4} + 17\mathrm{H_2O} &= \mathrm{Ni_2B} + 5\mathrm{H_3BO_3} + \mathrm{CH_3COONa} \\ &\quad + 2\mathrm{NaOH} + 20,5\mathrm{H_2}.\end{aligned}$$

Most likely hydrogen sorption depends on the number of active sites formation of physical bonds of hydrogen atoms from the alloy components, and eventually may form bridging bonds between individual atoms of hydrogen-type bonds, which are realized in the molecule of aluminum $Al[BH_4]_3$. The block diagram is shown in Fig. 42.2 [4]. Aluminum borohydride has nonsalifiable structure. The molecule is constructed from 3-distorted tetrahedra BH_4 (Fig. 42.2) associated with Al is likely hydrogen bridges.

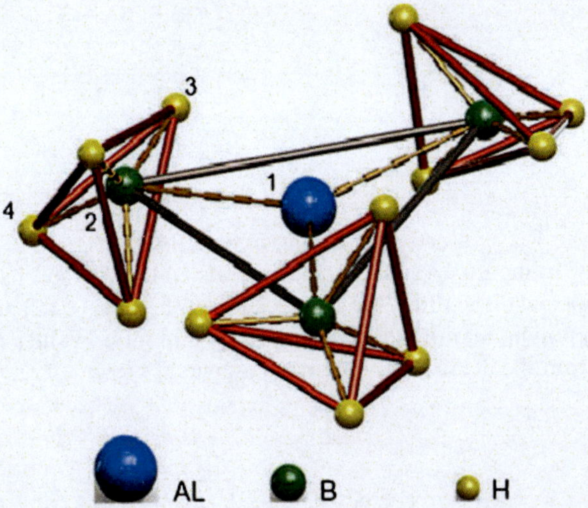

Fig. 42.2 Structure of the molecule $Al(BH_4)_3$ with distances: 1–2 = 2.15 Å; 1–3 = 2.10 Å; 2–3 = 1.28 Å; 2–4 = 1.21 Å; 4 3–4 = 1.93 Å

According to the same author, similar structures are formed by transitive metals of 4–5th period and group of rare-earth elements. Basic distinction between them it is not observed. Some interrelation between propensity to complexation and number of atoms of boron is established.

As in real conditions we deal with final concentration of electrolytes it is possible to assume, that a limiting stage of total electrochemical process is mass-transfer and consequently it is possible to take advantage of following expression:

$$\mu_{Me^{n+}}^{-S} + n \cdot \mu_{\bar{e}}^{-M} = \mu_{Me^0}^{-S},$$

where index S means a finding of components Me^{n+} и Me^0 close to electrode surface (at the boundary of the diffusion layer and the diffuse part of double layer). In this case, you must take into account the order of fairness recovering cations to the interface.

The probability of discharge of the cation depends primarily on the value of the reaction coordinate, i.e. the possibility of barrier-free transfer of an electron through an electric double layer [5]. As the temperature of the barrier-free transition in the range 25–32°C. The actual temperature in the reaction zone at a current density 2–4 A/dm^2 can be determined using the analytical formula:

$$\Delta T_k = \frac{\tau_{el-za}}{a + \frac{1}{i_k} \cdot \tau_{el-za}},$$

where $\tau_{el\text{-}za}$ the duration of electrolysis, min ΔT_k – swing of temperature cathode layer, °C; a – nomogram factor.

Investigated the temperature dependence on the electrode from the process parameters can be expressed in a single graph the three types of curves. Mathematical description of these relationships, according to [6], is defined by the following formulae

$$\Delta T = k \cdot \tau_{el-za}, \tag{42.1}$$

$$\Delta T = \frac{\tau_{el-za}}{a + \frac{1}{i_k} \cdot \tau_{el-za}}, \tag{42.2}$$

$$\Delta T = \frac{\tau_{el-za}}{a + V_{pr} \cdot i_k}, \tag{42.3}$$

где V_{pr} – velocity of moving of electrolyte.

In the Eq. (42.1) factor k depends on density and its decision does not represent special difficulties. For the decision of the Eqs. (42.2) and (42.3) it is possible to use a polynomial in n degree. In this case required values of temperature can be found from the decision of system of a general view:

$$\Delta T = A \cdot i_k + B \cdot i_k^2 + C \cdot i_k^3 + \ldots D \cdot i_k^n \tag{42.4}$$

$$\Delta T = K \cdot \tau_{el-za} + L \cdot \tau_{el-za}^2 + M \cdot \tau_{el-za}^3 + \ldots N \cdot \tau_{el-za}^n \tag{42.5}$$

By specifying in Eq. (42.4) i_k parameter for different values τ_{el} for and in Eq. (42.5) parameter τ_{el} over at different values of current density, we can construct the dependence $\Delta T = f\ (i_k,\ \tau_{el})$ representing the surface temperature field for any randomly selected values of current density and duration of electrolysis. Thus, all technological parameters of process appear are interconnected through factor *a*, which we will define for various values of the specified parameters.

$$a = \frac{\tau_{el-za} - \left(\frac{\tau_{el-za}}{i_k}\right) \cdot \Delta T_k}{\Delta T_k}$$

Let's define value of factor *a* for $i_k = 5\ \text{A/dm}^2$ for different $\tau_{el\text{-}za}$.

According to data presented in the Table 42.1, we can conclude that barrier-free transfer of an electron through an electric double layer in our case is real, because ΔT_k ranges lower than given by the author of [5]. The variation of temperature in the reaction zone is described by the saturation curve, i.e. exponent, a limit which is determined by the size of ~ 12°C, which indicates the possibility of obtaining coatings with a thickness of 30 μm of good quality.

Hydrogen inclusion in deposit Ni-B is possible and at the expense of introduction of atoms in structure of a formed deposit, at presence high-potential defects without formation with components boron and nickel of connections of chemical type. In this case, free hydrogen atoms diffuse into the volume of sediment in the direction of smaller concentrations Thus, arising microdistortions of a lattice of an alloy can prove as the initial centers of corrosion. low-temperature (~ 200°C) long annealing of samples during 30 min, for the purpose of removal of superfluous of hydrogen from metal structure is a necessary processing technological method of increase of corrosion firmness. Authors [6] investigated a relaxation of atoms of hydrogen in the monocrystal nickel sated with hydrogen by electrochemical method. They have established, that the atomic maintenance of hydrogen in volume of the sample made ~p;10^{-4}. Hydride phase $NiH_{0,6}$ was present in surface layer with thickness 10 μm.

Figure 42.3 shows the isochronous spectrum magnetic relaxation of single crystal Ni, doped electrolytic hydrogen. At a temperature of about 205 K (212 K) is a broad relaxation maximum (with semiwidth of peak 50 K), where the position T_{max} shifted markedly change t_2. The relaxation curves are far enough away from one another, both at low temperature and at high-temperature slope of the relaxation peak. Such behavior is characteristic for the diffusion aftereffect. When the

Table 42.1 Value of temperature for various technological modes in cathode layer layer at $i_k = 5$ A/дм^2 ($t_{el\text{-}ta} = 40°C$; $pH_{el\text{-}ta} = 3.5–4.5$)

ΔT_k, °C	Value *a*	$\tau_{el\text{-}za,}$ min	*d*, μm
2.21	0.76	3	3
3.55	0.49	6	6
4.61	0.15	9	9
5.45	–0.20	12	12
6.05	–0.52	15	15
7.85	–2.18	30	30

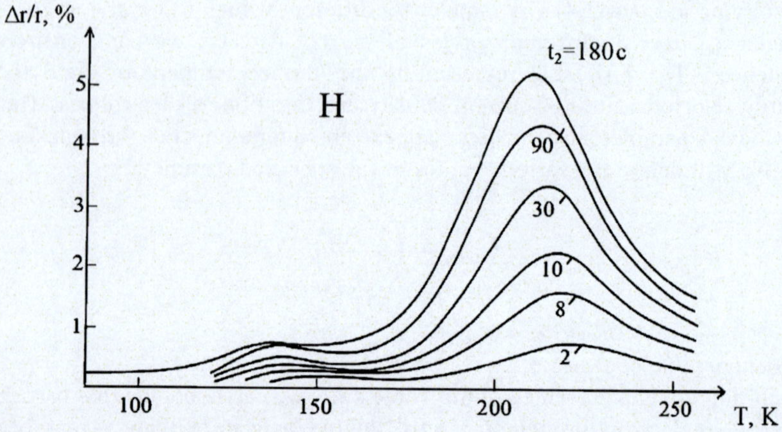

Fig. 42.3 Diffusion consequence of hydrogen in *Ni* ($t_1 = 1$ s), where Δr/r, % – amplitude of peak of an internal friction

orientation aftereffects were observed to peak maxima with half-width 25 K and the coincidence of curves on the high side of the maximum.

Diffusion activation energy (H) was determined by the temperature dependence of isothermal relaxation curves. Values $Q^{(H)} = 0.46 \pm 0.04$ eV have been received. According to other measurements [7, 8], the migration energy for free diffusion of H in Ni is $Q^{(H)} = 0.41$ eV.

In the experimental study of the activation energy of electrolytic chromium, obtained from chromate electrolytes, we have established the importance of the activation energy of diffusion of hydrogen is equal $Q_{Cr}{}^{(H)} = 0.60$ eV [9]. Further investigation showed, that the value of the activation energy determined by technological parameters ramie production of electrolytic chromium.

The divergence between the present results and the values received for free diffusion can be connected with high atomic concentration of hydrogen ($\sim 3{\cdot}10^{-3}$). Under these conditions hydrogen diffusion is influenced by elastic interaction between atoms of hydrogen.

Character of change of dependence of a relative relaxation from temperature of the sample testifies to presence of diffusion of free atoms of hydrogen in structure monocrystal of nickel.

If the sample is annealed at temperatures from 100°C to 200°C, the concentration of hydrogen decreases, and the relaxation peaks shift to lower temperatures (Fig. 42.4). Due to a decrease in the elastic interaction energy of migration in this case also is less than the Q (H) = 0.41 ± 0.04 eV.

Thus, the data and results of our experiment confirmed that the processes of interaction of hydrogen as the single crystals of nickel and electrolytic nickel have a diffusive character. Therefore, for the extraction of hydrogen from the alloy Ni-B should be used on a low-temperature annealing at large time of electrolysis. A special feature of the electrolytic alloy is a big variation in the binding energies for the samples obtained under different conditions of electrolysis.

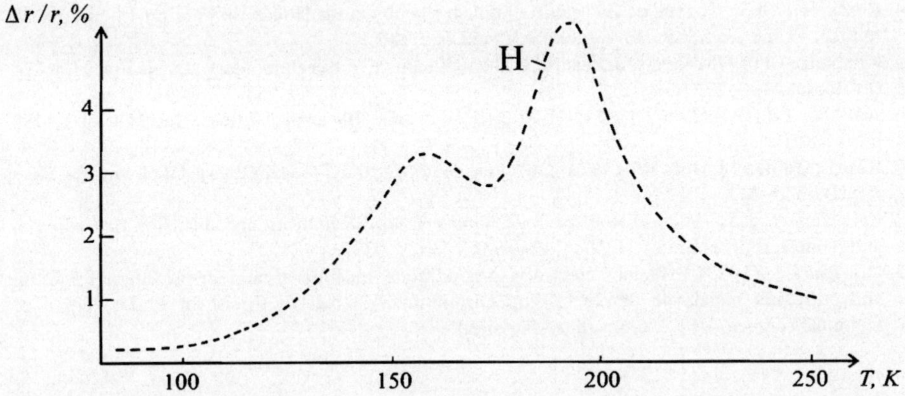

Fig. 42.4 Relaxation spectrum of hydrogen in Ni after annealing at T = 300 K for 200 h. (t_1 = 1 s; t_2 = 180 s)

42.3 Conclusions

1. Boron as an alloying component makes ambiguous influence on process of interaction of hydrogen with metal. It changes its structure and also stoichiometry of electrolytic alloy.
2. Preconditions for formation of sites with various stoichiometric structure of an alloy as a result of non-uniformity of distribution of a current on an electrode surface are created.
3. Depending on the concentration of boron in the local areas of the surface can form the structure, where hydrogen has a very small binding energies to the metal, forming a bridge circuit interaction. In some cases, may be a free allocation of hydrogen even at room temperature.
4. It is shown that in the electrolytes with definite chemical composition is possible barrier-free transition of an electron through an electric double layer, which is confirmed by the results of analytical calculations.
5. The comparative analysis of results on interaction of hydrogen with electrolytic metals and alloys and monocrystal nickel is carried out.

References

1. Damaskin BB, Petriy OA, Tsirlina GA (2008) Electrochemistry. Chemistry, Colossus, Moscow, p 672 (in Russian)
2. Shalimov YN (2007) Influence of thermal and electric fields on electrochemical processes in pulsed electrolysis. Thesis of Dr. Technical science. Ivanovo, p 354 (in Russian)
3. Zvyagintseva AV, Gusev AL, Shalimov YN (2009) Kinetics of the electrochemical hydrogenation of metals in the presence of boron. Altern Energy Ecol 72(4):20–27

4. Remy H (1963) Course of inorganic chemistry. Publishing House of Foreign Literature, M (ed tsp. - Corr ANSSSR Novoselova AV) vol 1, p 920
5. Krishtalik LI (1979) Electrode reaction. The mechanism of the elementary act. Nauka, M, p 224 (in Russian)
6. Amfelda Ed. G, Felkla I (1981) Hydrogen in metals. Basic properties. Mir, M, vol 1, 480 p (in Russian)
7. Katz L, Guinan M, Borg RY (1971) Diffusion of H_2, D_2 and T_2 in single-crystal Ni and Cu. Phys Rev B4:330–341
8. Ebisuzaki Y, Kass WY, O'Keeffee M, Chem Y (1968) Solubility and diffusion of hydrogen and deuterium in platinum. J Chem Phys 49(8):3329–3332
9. Grankin EA (1973) Effect of electrolysis conditions and heat treatment on internal friction and corrosion resistance of electrolytic chromium. Abstract of thesis for Ph.D. degree of technical science. VPI, Voronezh (in Russian)

Chapter 43
Laws of Diffusion of Hydrogen in Electrolytic Alloys on the Basis of Nickel

A.V. Zvyagintseva and Y.N. Shalimov

Abstract The paper discusses the theoretical aspects of diffusion of hydrogen in am electrolytic nickel-based alloys. The analysis of dependence of an overstrain of hydrogen allocation by nature metal is carried out. The problems of heat and mass transfer in the formation of electrochemical systems based on nickel are considered. It is shown that concentration of hydrogen in an electrolytic covering is defined by probability of formation of defective structures in metal and is function of parameters of electrolysis. Experimentally established that the determining factor is the cathode current density and chemical composition of the electrolyte.

Keywords Diffusion · Hydrogen · Nickel · Overstrain · Cathode current · Overstrain temperature coefficient · Heat weight conduction · A concentration gradient

43.1 Introduction

In electrolytic coverings the close attention in works of some authors [1–4] is given to questions of diffusion of hydrogen. Interest to this problem is caused by necessity extraction hydrogen from electrolytic metals for elimination of hydrogen fragility of coverings. Authors underline features of physical and chemical properties of coverings in comparison with the metals received in other ways. One of features of structure of electrolytic coverings is their big heterogeneity on macro- and micro-levels. In this connection heterogeneity play a role of defects of structure on which basically interaction of hydrogen with metal is carried out.

At diffusion of hydrogen under the influence of temperature there is probable a process of decomposition of separate atoms with formation of molecules with their subsequent merge in aggregation (a hydrogen vial). Transport (diffusion) of such molecular structures is interfaced to a number of features, namely separate

A.V. Zvyagintseva (✉) and Y.N. Shalimov
Voronezh state technical university, Moscow av. 14, Voronezh 394026, Russia
e-mail: zvygincevaav@mail.ru; shalimov-yn@mail.ru

S.Yu. Zaginaichenko et al. (eds.), *Carbon Nanomaterials in Clean Energy Hydrogen Systems - II*, NATO Science for Peace and Security Series C: Environmental Security 2,
DOI 10.1007/978-94-007-0899-0_43, © Springer Science+Business Media B.V. 2011

aggregations can move to channels of the closed type with impossibility of their exit for limits of the sample. Apparently, it is a principal cause of discrepancy of results of the experiment, a spent method vacuum extraction and vacuum fusion.

43.2 Results and Discussion

Basic works under the theory of processes of allocation H_2 and O_2 on electrodes from various materials have been spent by A.N. Frumkin with employees [5–7].

Authors notice that mobility and consequently and electric conductivity of ions H_3O^+ and OH^- it is abnormal are great. Because of rather small size of a proton and the big intensity of electric field round it, between a proton and not divided pair electrons oxygen of a molecule H_2O a cart-nikaet covalent ion communication under the scheme [8]:

$$H^+ + H_2O \rightarrow H_3O^+$$

As a result of this transformation the ion of hydrooksonyi is formed. On the basis of the experimental data received with application of the spectral analysis, it is established that formation reaction eczotermical and warmth of reaction has very big size.

$$H^+ + H_2O \rightarrow H_3O^+ + 1110 kDg$$

Usually at calculation of thermal balance of electrode systems (reactors) of warmth of chemical (electrode) reactions are not considered, but their value can exceed energy of all thermal sources which traditionally make it.

The basic influence on gas evolution process on electrodes renders cathodic restoration H_2 and anode oxidation of molecules H_2O and ions OH^-.

The main task is to increase in intensity of gas evolution, it is reduced to reduction of repressure allocation H_2 and O_2 on electrodes (η_{H_2} и η_{O_2}) [5].

Important influence on restoration H_2 and oxidation processes H_2O and OH^- ions is rendered by the material nature electrodes. The size η_{H_2} is connected linear dependence with size of density of a current (i) the known equation of Tafelja [5, 9]

$$\eta_{H_2} = a + b \cdot lg\ i, \tag{43.1}$$

Or

$$i = k \cdot exp(\eta/b), \tag{43.2}$$

where a, b, k – constants, $k = exp(-a/b)$.

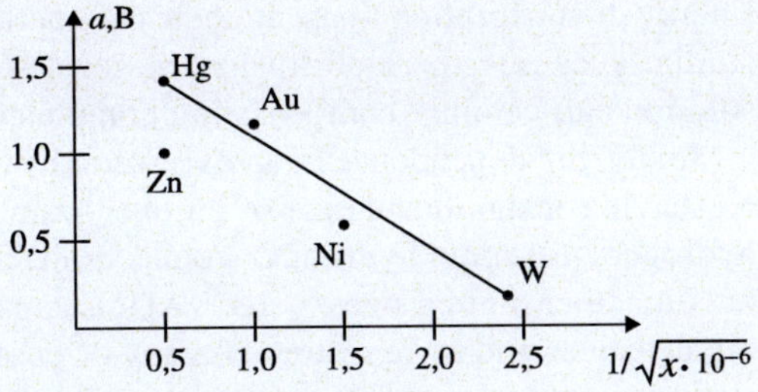

Fig. 43.1 Values of an overstrain segregation of $H_2(\eta_{H_2})$ for various metals at current density j = 1 A/cm^2

The constant a and characterizes degree not-convertibility process on an electrode. In works [5, 6] dependence between a constant and factor of compressibility of metal has been established (χ):

$$a = 2 - \frac{1}{\sqrt{\chi \cdot 10^{-6}}}. \tag{43.3}$$

Graphic representation of the formula (3) is presented in Fig. 43.1.

The analysis of this dependence gives the chance to carry out an optimum choice of a material of an electrode from the point of view of a hydrogen and oxygen overstrain. By comparison of sizes η_{H_2} to the minimum internuclear distance r it is established [5] that the minimum value η_{H_2} corresponds to distance 2.7 Å, at metals with the big or smaller value r conducted-rank η_{H_2} increases. Exact value η depends on a physical condition of a surface (a roughness, presence zapassivirovannyh sites, porosity).

As the electrode temperature considerably differs from electrolyte temperature in volume at the expense of various thermal effects it is necessary to consider influence of temperature on sizes η_{H_2} and η_{O_2}. The quantitative estimation of this influence can be expressed the formula:

$$\left(\frac{\eta}{t}\right)_i = \frac{da}{dt} + \frac{R}{\lambda F} ln\ i, \tag{43.4}$$

where F = 96,500 Kl – number of Faraday; λ – a current share; R = 8.31 Dzh/grad·mol. – universal gas constant.

As $\frac{da}{dt}$ <0, temperature factor on the absolute size it is more at small density of a current and less at the high density.

Proceeding from it, followed expect that at dehydration an ion H_3O^+ absorbed energy will create a negative gradient of temperature in a reaction zone, facilitating transport of ions H_3O^+ in a reaction zone. However, the similar phenomenon is not observed owing to the several reasons: (1) in a zone about an electrode the most probable process is decomposition of atoms of hydrogen with formation of molecules. This reaction is accompanied by heat allocation, therefore it partially compensates thermal energy absorption as a result dehydration an ion hydrooksonyi.

Energy hemosorbtion ions at their interaction with a surface of metals which similarly causes thermodiffusion of ions of hydrogen deep into an electrolyte solution can be other compensating component.

Analyzing dependence of an overstrain by nature metal it is easy to establish that Ni, having insignificant size of an overstrain can actively co-operate with atomic hydrogen, forming hydridical connections. Besides, protonirovannyj hydrogen, having rather limited sizes ($\sim 10^{-5}$ Å) it is capable to get in a crystal lattice on rather big distances and to be restored on any of possible defects with high power level.

The organic law of transfer of heat in not-mobile to environment (molecular spends heat or a conduction) is the law of Fourier according to which the thermal stream is proportional to a temperature gradient:

$$q = -\lambda \cdot grandT \equiv -\lambda \frac{dt}{dy}. \tag{43.5}$$

Here q – a thermal stream, that is quantity heat, transferred through a time unit; grand T – a temperature gradient, that is derivative of temperature on coordinate y, directed on a normal to a surface through which there is heat transfer, and λ – factor of spent heat (a physical constant of substance in which there is heat transfer). The sign a minus is put because heat transfer occurs in a direction in which temperature goes down, that is in a direction of a negative gradient of temperatures.

In actual practice existence of such systems is found out extremely seldom. Basically it is reversible processes with small currents of an exchange. The most simple decision of this problem is realized for electrode system the cylinder-cylinder. Under the law of a curve of saturation the Eq. 43.7 can be copied to the account of motionless electrolyte and dependence of temperature on technological parameters in a kind:

$$\frac{di}{d\tau} = \frac{\lambda}{\rho(T)} \cdot \left(\frac{\partial^2 T(i,\tau)}{dx^2} + \frac{\partial^2 T(i,\tau)}{dy^2} + \frac{\partial^2 T(i,\tau)}{dz^2} \right) - \omega_z \frac{di}{dz} - \frac{1}{\rho(T)} div \sum j_i i_i. \tag{43.6}$$

Classical equation of spent heat [10]:

$$\frac{di}{dt} = \frac{\lambda}{\rho} \nabla^2 t - \left(\omega_x \frac{di}{dx} + \omega_y \frac{di}{dy} + \omega_z \frac{di}{dz} \right) - \frac{1}{\rho} div \sum j_i i_i \tag{43.7}$$

shows that heat-mass carrying over for the account of various making components movement: heat transfers $\frac{\lambda}{\rho} \nabla^2 t$, convective making $(\omega_x \frac{di}{dx} + \omega_y \frac{di}{dy} + \omega_z \frac{di}{dz})$ and chemical component $(\omega_x \frac{di}{dx} + \omega_y \frac{di}{dy} + \omega_z \frac{di}{dz})$ also it will be defined local by temperatures of elementary sites of electrode, and, hence, only ideally the isothermal surface temperature of a field will represent a plane parallel to a surface of electrode. In real electrochemical systems the temperature gradient of the warm field defining character heat-mass carrying over, depends not only on time and co-ordinate of a certain site of an electrode, but, substantially, from technological parameters of process (density of a current, a chemical compound of electrolyte and it heatphysical properties).

Generally the density of a stream of weight for the account molecular and convection of carrying over will be defined by the equation:

$$\vec{j}_i = \vec{j}_{MDi} + \vec{j}_{KDi},$$

where j_i – specific enthalpy i^{go} a component. Generally through the motionless control surface allocated from a mix, it is transferred enthalpies $\Sigma j_i \cdot i_i$.

Passing to concentration parameter it is possible to consider that its influence on transfer of the masses will be defined by Fick's law.

Fick's [9, 10] law according to which diffusion the stream is proportional to gradient of concentration will be the corresponding law for diffusion:

$$j = -DgradC \equiv -D\frac{dc}{dy}, \tag{43.8}$$

where C – concentration of diffusing substances, j – diffusion a stream, that is quantity of substance (in mol's), transferred through a time unit. The factor of proportionality D carries the name of factor of diffusion and represents a physical constant. Laws of Fourier and Fick, in the resulted simple form, are confidants for the description of processes of diffusion of hydrogen and are applicable only at observance of certain conditions. First of all, they concern molecular carrying over to the motionless environment (the length of free run should be small in comparison with all characteristic lengths of a problem).

From results of experiment (Fig. 43.2) follows that with increase i_k from 0.5 to 4.0 A/dm^2 the relative maintenance of hydrogen in covering Ni-B decreases. It is connected with decrease BT_{H2} and pine forest maintenances in covering Ni-B.

Experimental data show that full removal of hydrogen from samples 15 μm [11, 12] in connection with presence of the traps which do not have an open exit to a surface does not occur in the thickness a method of extraction. Therefore the greatest

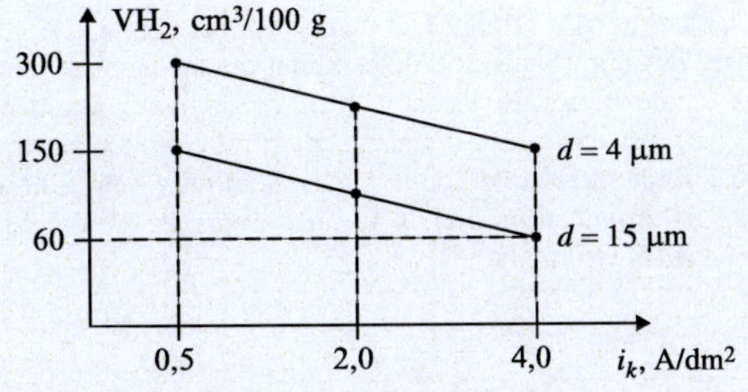

Fig. 43.2 Dependence of hydrogen maintenance in covering Ni-B on i_k

reliability the results received for thin samples (4 μm), for which density of such defects slightly more low because of the limited length of the transport channel possess. It confirms also results of definition extraction hydrogen with a fusion method.

43.3 Conclusions

1. The concentration of hydrogen in an electrolytic deposits is random and is determined by the probability of formation of defect structures in metal.
2. Energy of metal-hydrogen bond depends on the capacity of the results defect and apparently it is spread over a wide range of values. This fact determines the temperature of the extraction of hydrogen from the metal structure.
3. Probability of density distribution of defects in electrolytic deposits is a function of electrolysis parameters. However, the determining factor is the current density and chemical composition of the electrolyte.

References

1. Kuznetsov VV, Haldeev GV, Kichigin VI (1993) Navodorazhivanie metals in elektroli. Mechanical Engineering, Moscow, p 244 (in Russian)
2. Beloglazov SM (1975) Navodorozhivanie of a steel at electrochemical processes. Publishing House Leningrad, Leningrad, pp 1–412 (in Russian)
3. Amfelda G, Fyolklja I (1981) Hydrogen in metals. Applied aspects. Mir, M, vol 2, 430 p
4. Novic A, Berry B (1975) The relaksatsionnye phenomena in crystals of the Lane with English. (ed Nadgorodny EM, Sojfera JM), Atomizdates, M, 472 p (in Russian)
5. Frumkin AN (1987) Electrode processes. The selected works. Science, M, 336 p
6. Levich VG (1952) Physical and chemical hydrodynamics. Publishing House AN of USSR, M, 538 p
7. Polukarov JM, Grishin VV, Antonjan SB (1980) Electrosedimentation of nickel in the conditions of joint action variable and posto-jannogo currents. Electrochemistry 16(3):423–427
8. Harned G, Ouen B (1952) Physical chemistry of solutions of electrolits: the Lane with English. Publishing House of the Foreign Literature, M, 628 p
9. Franc-Kamenetsky DA (1987) Diffusion's and teploperedacha in chemical kinetics. Science, M, p 502 (in Russian)
10. Lykov AV (1967) The theory of heat conductivity. Higher School, M, 600 p
11. Zvyagintseva AV (2008) About the mechanism navodorazhivanija metals in the presence of a pine forest. In the collection of documents and materials "Hydrogen power of the future, нанотехнологии and metals of platinum group in CIS countries". MIREA, M, pp 134–139
12. Zvjagintseva AV, Gusev AL, Shalimov YN (2009) Kinetics of processes of electrochemical hydrogenation metals in the presence of a pine forest. Altern Power Ecol 72(4):20–27 (in Russian)

Author Index

S.Yu. Zaginaichenko et al. (eds.), *Carbon Nanomaterials in Clean Energy Hydrogen Systems - II*, NATO Science for Peace and Security Series C: Environmental Security 2, DOI 10.1007/978-94-007-0899-0, © Springer Science+Business Media B.V. 2011

Subject Index